Temporary Works

Temporary Works
Principles of design and construction

Second edition

Edited by
Peter F Pallett and Ray Filip

Published by ICE Publishing, One Great George Street, Westminster, London SW1P 3AA

Full details of ICE Publishing representatives and distributors can be found at: www.icebookshop.com/bookshop_contact.asp

Other titles by ICE Publishing:

ICE Manual of Health and Safety in Construction, Second edition
C. McAleenan and D. Oloke (eds). ISBN 978-0-7277-6010-4
Civil Excavations and Tunnelling
R. Tatiya. ISBN 978-0-7277-6153-8
Bridge Deck Erection Equipment
Members of IABSE Working Group 6. ISBN 978-0-7277-6193-4

www.icebookshop.com
A catalogue record for this book is available from the British Library

ISBN 978-0-7277-6338-9
© Thomas Telford Limited 2019

ICE Publishing is a division of Thomas Telford Ltd, a wholly-owned subsidiary of the Institution of Civil Engineers (ICE).

All rights, including translation, reserved. Except as permitted by the Copyright, Designs and Patents Act 1988, no part of this publication may be reproduced, stored in a retrieval system or transmitted in any form or by any means, electronic, mechanical, photocopying or otherwise, without the prior written permission of the Publisher, ICE Publishing, One Great George Street, Westminster, London SW1P 3AA.

This book is published on the understanding that the author is solely responsible for the statements made and opinions expressed in it and that its publication does not necessarily imply that such statements and/or opinions are or reflect the views or opinions of the publishers. While every effort has been made to ensure that the statements made and the opinions expressed in this publication provide a safe and accurate guide, no liability or responsibility can be accepted in this respect by the author or publishers.

While every reasonable effort has been undertaken by the author and the publisher to acknowledge copyright on material reproduced, if there has been an oversight please contact the publisher and we will endeavour to correct this upon a reprint.

Cover photo: Deck erection method on the Queensferry Crossing. (Courtesy of Forth Crossing Bridge Constructors)

Commissioning Editor: Michael Fenton
Production Editor: Madhubanti Bhattacharyya
Marketing Specialist: April Asta Brodie

Typeset by Academic + Technical, Bristol
Index created by Pierke Bosschieter
Printed and bound in Great Britain by TJ International, Padstow

Contents

Foreword xv
List of contributors xvii
Introduction xxiii

01 Safety, statutory and contractual obligations 1
Andrew Rattray and Peter F Pallett

1.1. Introduction 1
1.2. Background 3
1.3. Management of temporary works 5
1.4. Construction (Design and Management) Regulations 2015 5
1.5. The Work at Height Regulations 2005 7
1.6. The Health and Safety (Offences) Act 2008 8
1.7. Contractual obligations 9
1.8. Robustness 10
1.9. Public safety 11
1.10. Summary of main points 12
References 12
Further reading 14

02 Management of temporary works 15
Frank Marples

2.1. Introduction 15
2.2. What are (or may be considered as) temporary works? 17
2.3. The parties (who is involved) 20
2.4. The management controls (the who) 23
2.5. Principal activities of the TWC (how temporary works are managed) 26
2.6. Summary – answers to the questions why, what, who and how temporary works are managed 33
References 34

03 Site compounds and set-up 37
Geoff Miles, Edited by: Keith Kirwan

3.1. Introduction 37
3.2. Land and access 44
3.3. Communications, energy, clean water supply and wastewater disposal 47
3.4. Office and welfare accommodation space planning 51
3.5. Materials (distribution, fabrication, handling, storage, testing and unloading) 54
3.6. Hoarding and fencing 55
References 56
Further reading 56

04	**Tower crane bases**	**57**
	Ray Filip and Stuart Marchand	
	4.1. Types of tower crane	58
	4.2. Loading on foundations	59
	4.3. Foundation options	59
	4.4. Foundation design principles	65
	4.5. Foundation construction and inspection	65
	References	66
	Further reading	67
05	**Site roads and working platforms**	**69**
	Christopher Tate	
	5.1. Introduction	69
	5.2. Site roads: principles and design	70
	5.3. Working platforms: principles and design	78
	5.4. Temporary roads for public use	81
	5.5. Construction	82
	References	84
	Further reading	85
06	**Control of groundwater**	**87**
	Neil Smith	
	6.1. Introduction	87
	6.2. Techniques	88
	6.3. Investigation for dewatering	92
	6.4. Analysis and design	93
	References	97
	Further reading	97
07	**Lime and cement stabilisation**	**99**
	Mike Brice and Neil Smith	
	7.1. Introduction	99
	7.2. Materials and their effects on the soil	100
	7.3. The performance of treated soils	103
	7.4. Testing	104
	7.5. Plant	105
	7.6. Health and safety	106
	References	106
	Further reading	106
08	**Jet grouting**	**109**
	John Hislam and Neil Smith	
	8.1. Introduction	109
	8.2. Construction methods	111
	8.3. Design principles	112
	8.4. Monitoring and validation	114
	8.5. Secondary and side effects	114

| | References | 115 |
| | Further reading | 115 |

09   Artificial ground freezing   117
Neil Smith
9.1.	Introduction	117
9.2.	Construction principles	118
9.3.	Design principles	120
9.4.	The effects of freezing and thawing	123
9.5.	Monitoring	125
	Acknowledgements	125
	References	125
	Further reading	125

10   Slope stability in temporary excavations   127
Neil Smith
10.1.	Introduction	127
10.2.	The consequences of failure	129
10.3.	Construction principles	129
10.4.	Design principles – some simple fundamentals	130
10.5.	Monitoring	137
	References	138
	Further reading	139

11   Sheet piling   141
Ray Filip
11.1.	Major alternatives	141
11.2.	Types of steel sheet piling	141
11.3.	Installing sheet piles	142
11.4.	Eurocode 7	148
11.5.	Design	149
11.6.	Inspection and maintenance	154
11.7.	Plastic sheet piles	154
	References	155
	Further reading	156

12   Trenching   159
Ray Filip
12.1.	Introduction	159
12.2.	Techniques	163
12.3.	Design to CIRIA 97 trenching practice	168
12.4.	Controlling water	172
	References	172
	Further reading	173

13   Diaphragm walls   175
Chris Robinson and Andrew Bell
13.1.	Introduction	175

		13.2. Applications	175
		13.3. Construction methods and plant	176
		13.4. Design	186
		References	191
		Further reading	191
14		**Contiguous and secant piled walls**	**193**
		Chris Robinson and Andrew Bell	
		14.1. Introduction	193
		14.2. Applications	195
		14.3. Construction methods and plant	195
		14.4. Design	201
		References	204
		Further reading	205
15		**Caissons and shafts**	**207**
		Andrew Smith	
		15.1. Introduction	207
		15.2. Major alternatives	208
		15.3. Common methods of construction	208
		15.4. Principles of design	216
		References	216
		Further reading	216
16		**Bearing piles**	**217**
		Ray Filip	
		16.1. Introduction	217
		16.2. Types and installation	219
		16.3. Design principles	221
		References	225
		Further reading	226
17		**Jetties and plant platforms**	**229**
		Paul Boddy	
		17.1. Introduction	229
		17.2. Solid structure	229
		17.3. Open jetty structure	231
		17.4. Floating jetties	232
		17.5. Loadings	234
		17.6. Analysis	236
		References	237
		Further reading	237
18		**Floating plant**	**239**
		Paul Boddy	
		18.1. Introduction	239
		18.2. Types and uses	240

	18.3. Design principles	245
	References	249
	Further reading	249

19 Temporary bridging — 251
Bernard Ingham
	19.1. Introduction	251
	19.2. Temporary bridge types	252
	19.3. The design process	256
	19.4. Transportation and construction	259
	19.5. Other applications of temporary bridging parts	262
	References	262
	Further reading	262

20 Heavy moves — 263
Martin Haynes and Andrea Massera,
Edited by: Nick Cook
	20.1. Introduction	263
	20.2. Techniques	264
	20.3. Design	274
	References	279
	Further reading	279

21 Access and proprietary scaffolds — 281
Peter F Pallett and Ian Nicoll
	21.1. Introduction	281
	21.2. Managing scaffolding	282
	21.3. Selection and designation	283
	21.4. Materials and components	286
	21.5. Scaffold design	289
	21.6. Workmanship and inspections	297
	References	297

22 Falsework — 301
Peter F Pallett and Andrew Jones
	22.1. Introduction	301
	22.2. Materials and components	304
	22.3. Loads on falsework	307
	22.4. Falsework design	310
	22.5. Workmanship and inspections	316
	References	317

23 Formwork — 319
Peter F Pallett and Laurie York
	23.1. Introduction	319

	23.2. Vertical formwork	319
	23.3. Economy	320
	23.4. Specifications and finishes	321
	23.5. Tolerances/deviations	322
	23.6. Formwork materials	322
	23.7. Concrete pressure calculation	325
	23.8. Wall formwork design	330
	23.9. Column formwork design	333
	23.10. Striking vertical formwork	334
	23.11. Workmanship/checking	335
	References	335
	Further reading	336

24 Soffit formwork 337
Peter F Pallett

	24.1. Introduction	337
	24.2. Preamble to soffit form design	337
	24.3. Loading on soffit forms	342
	24.4. Design	343
	24.5. Cantilevered soffits	344
	24.6. Striking soffit formwork	345
	24.7. Assessment of concrete strength	351
	24.8. Checking and inspection	352
	References	352

25 Climbing and slip forms 355
Charlie McKillop

	25.1. Introduction	355
	25.2. Climbing and slip-form viability assessment	356
	25.3. Climbing formwork	359
	25.4. Slip forms	361
	25.5. Climbing protection screens	364
	25.6. Checking and inspection	366
	References	366
	Further reading	367

26 Temporary façade retention 369
Ray Filip and Stuart Marchand

	26.1. Introduction	369
	26.2. Philosophy of façade retention	370
	26.3. Types of temporary façade retention schemes	373
	26.4. Loads to be considered	376
	26.5. Design considerations	380
	26.6. Demolition, monitoring and inspection	382
	References	383
	Further reading	383

27	**Bridge installation techniques**	**385**
	Keith Broughton and John Gill	
	27.1. Introduction	385
	27.2. Preparation and selection of installation technique	386
	27.3. Partial deck erection schemes	387
	27.4. Deck erection as a single unit	391
	27.5. Bridge and deck erection by tunnelling and mining	396
	27.6. Segmental bridge construction	399
	References	400
	Further reading	401
28	**Backpropping**	**403**
	Peter F Pallett and David Szembek	
	28.1. Introduction	403
	28.2. The theory	404
	28.3. Loads (actions) to be considered	406
	28.4. Research	407
	28.5. Methodology – multi-storey construction	409
	28.6. Calculation methods – flat slabs	412
	28.7. With one level of backpropping	415
	28.8. With two levels of backpropping	416
	28.9. Worked examples – multi-storey construction	416
	28.10. Methodology – heavy construction	417
	28.11. Conclusion	419
	References	420
	Further reading	420
29	**Pressure testing of pipelines**	**421**
	David Cooper and Stewart Carolan-Evans	
	29.1. Introduction	421
	29.2. Gravity sewer pipelines	422
	29.3. Pressure pipelines	422
	29.4. Design	424
	29.5. Can restraint be provided without temporary works?	425
	29.6. If temporary works are required, what are the options?	426
	29.7. Design of thrust blocks	427
	29.8. Internal (puddle) flanges	428
	29.9. On-site safety considerations	429
	References	431
	Further reading	431
30	**Basement construction**	**433**
	Ray Filip and Chris Robinson	
	30.1. Introduction	433

		30.2.	General planning considerations prior to work commencing	434
		30.3.	Constructing a basement in open cut	435
		30.4.	Constructing a basement in a supported excavation	436
		30.5.	Constructing a basement beneath an existing building or next to adjacent buildings	437
		30.6.	Temporary retaining wall	437
		30.7.	Support scheme to retaining wall	440
		30.8.	Top-down construction	443
		30.9.	Other design considerations	443
		References		444
		Further reading		444
31		**Digital project delivery – visual planning and BIM**		**447**
	Nick Boyle			
		31.1.	Introduction	447
		31.2.	Basics of building information modelling	447
		31.3.	BIM and communication	448
		31.4.	Key issues	449
		31.5.	Methods and techniques	450
		31.6.	Managing and minimising risk	455
		31.7.	Example – temporary substation	456
		31.8.	High-quality animations	458
		31.9.	Operations and future maintenance	461
		31.10.	Communication and engagement	462
		31.11.	Training and support	462
		References		462
		Further reading		462
32		**Rebar stability**		**465**
	Ray Filip and Mark Tyler			
		32.1.	Introduction	465
		32.2.	Potential problems and modes of failure	467
		32.3.	Common solutions	470
		32.4.	Design	471
		32.5.	Design rules	471
		32.6.	Structural behaviour of cages	472
		32.7.	Design solutions – walls	472
		32.8.	Wind loading	474
		32.9.	On-site inspections	475
		32.10.	Ties	476
		References		477
		Further reading		477
33		**Needling and forming openings in walls**		**479**
	Ray Filip			
		33.1.	Introduction	479

	33.2. Assessment of the building	479
	33.3. Is support required?	480
	33.4. Assessing the loads to be supported	481
	33.5. Responsibility for temporary works	482
	33.6. Simple temporary works solutions	483
	33.7. Needling schemes	483
	33.8. Propping to needles	488
	33.9. Sequence of removal of needles	490
	33.10. On-site checklist	490
	References	491
	Further reading	491

34 Temporary works in demolition — **493**

Angus Holdsworth

34.1. Introduction	493
34.2. Understand the structure	493
34.3. Demolition	496
34.4. Temporary works for demolition	498
34.5. Load testing of slabs	503
34.6. Moving plant between floors	504
34.7. Structural stability during demolition	504
34.8. Basement stability and shoring during demolition	505
34.9. Column removal/structural openings	507
34.10. Conclusion	508
References	508
Further reading	509

Index — **511**

Foreword

The first edition of this useful and timely book was published in 2012, and in the six years since then, the work has become a standard reference in the field. During that time, the level of attention given to temporary works in general has increased along with the degree of scrutiny of the competence of those practising in the field. This is to be applauded, although the level of awareness of non- specialist engineers is still too low.

This second edition contains new chapters and extensive rewrites of some of the original chapters. The scope of the work is wide, and the new chapters cover areas where there have recently been problems. It is good to see that these are being addressed.

The management of temporary works is critical to the safe execution of the majority of building and engineering schemes and the widening of the scope of the book can only help in providing guidance for this. What is notable is that the depth of knowledge in a number of fields is still relatively superficial and hopefully some research of these fundamentals can be prompted by a wider exposure to the engineering community.

There is still too much delegation of the responsibility for temporary works to the contractor, with drawings and specifications that do not identify what the thinking or performance requirements for that temporary works is. The designers of permanent works have responsibilities to consider how their creations can be constructed. This may be through early contractor engagement or by clear statement of what the performance requirements are for the temporary condition. There are initiatives to make permanent schemes more material efficient, without considering the implications for the construction and demolition phases.

We are seeing greater levels of awareness and control in temporary works and this is a key part of the management of major works. However, the implementation of the processes required in BS 5975 is still low at the smaller end of the market and we need to be encouraging best practice in this area. Whilst the levels of complexity may be lower and standard solutions may be applicable for some situations, there needs to be appropriate care for any level of temporary works. One of the hurdles for this is the lack of perceived value of engineering from smaller scale clients. It is hard to see how this can be overcome without a more rigorous prosecution of the Construction Design and Management Regulations.

This book is an integral part of the development of the body of knowledge in temporary works. The authors and editors are to be commended on their efforts which have led to an improvement to an already fine volume. It is my hope that through work such as this we can enable all engineers to consider and design for the temporary condition. My thanks go to those who are enabling that process through their efforts.

Tim Lohmann CEng FICE FIStructE
Chairman of the Temporary Works Forum
Director of Wentworth House Partnership

List of contributors

Editors

Eur Ing Peter F Pallett BSc CEng FICE FCS is a civil engineer, educated at Kings College, Taunton and Loughborough University of Technology. He has a total of 48 years' experience and over 30 years' specialising in temporary works worldwide. Self-employed since 1991, he ran temporary works training courses worldwide up to 2017, and formed his own company, Pallett TemporaryWorks Limited, in 2003.

He is a founder member of the BSI committee for falsework (BS 5975), co-author of the National Access and Scaffolding Confederation's TG20:08 on scaffolding and principally known for the Concrete Society publication *Formwork – A Guide to Good Practice*. He is an accredited CUBS 'Expert Witness – Civil Procedures', with expertise in above-ground temporary works.

Ray Filip BEng(Hons) MSc DIC CEng FICE is a civil engineer, educated at the University of Surrey and Imperial College, London. He has over 30 years' construction experience, having worked as a site engineer and Temporary Works Coordinator for principal contractors in the UK and Africa. He also has 10 years' experience in principal contractors' temporary works design offices. In addition, he spent 5 years as the general manager and principal engineer for a specialist subcontractor. In 2007 he formed his own specialist temporary works consultancy, RKF Consult Ltd.

Ray has advised many organisations on temporary works procedures, temporary works design and risk management. He carries out site audits and inspections, and is regularly consulted by the Health and Safety Executive on temporary works issues. He has also provided training courses for many organisations in the UK and Europe.

Ray has been involved in producing a number of technical publications and papers for CIRIA, the Concrete Society, the Contractors Plant Hire Association, the Deep Foundations Institute and *The Structural Engineer* magazine. He is a member of the committee responsible for the British Standard on temporary works (BS 5975).

Contributors

Andrew Bell is a geotechnical engineer and Chartered Civil Engineer with over 20 years experience, gained principally in the design and construction of piled foundations and deep basements. Andrew is currently Chief Engineer at Cementation Skanska.

Paul Boddy MEng CEng MICE is currently the technical director for Interserve Construction Limited, and ultimately responsible for temporary works design and use with the company. Although experienced in all temporary works, his main interest is in floating plant, both big and small. This is borne out of his experiences with contractors and working for British Waterways. He is also an experienced canoeist, which has assisted his expertise in 'going afloat'.

Nick Boyle BA CEng FICE MCIHT is a chartered engineer with over 30 years' experience in civil engineering infrastructure at all stages of a project. He is passionate about integrating design and engineering and using digital technologies to engage the whole project team and achieve the best overall solution for all. He chairs the Balfour Beatty quarterly Safety by Design & Engineering forum, which he has led since 2009. He is an active participant in various Industry and client forums, including the Infrastructure Industry Innovation Platform i3P, the temporary works forum TWf and CIRIA.

Mike Brice BSc MSc CGeol FGS, Associate, Applied Geotechnical Engineering, has been a geotechnical engineer for thirty years and has wide experience of the design and construction of earthworks and soil stabilisation.

Keith Broughton BSc(Hons) CEng MICE retired in 2017, having worked for HOCHTIEF UK Construction Ltd since 1991. He was involved in the engineering and tendering of all of their civil engineering projects, and held the position of technical services director for over 15 years.

Stewart Carolan-Evans BSc(Hons) CEng FICE has been employed for 30 years in heavy civil engineering. He has comprehensive experience of marine, water-related, rail and general civil engineering temporary works design.

Nick Cook has been designing temporary works for 41 years in heavy civil engineering. He has specialised in steel structures and their lifting and moving, including projects such as Sizewell B, Cardiff Millennium Stadium, Second Severn Crossing, Oresund Crossing, Heathrow Terminal 5 roof lifting, the new Heathrow Air Traffic

Control Tower transport and erection and the M56 Thorley Lane movement and installation.

David Cooper BSc(Eng) MICE MCIWEM is the principal civil engineer of Mott MacDonald's design and build arm, Mott MacDonald Bentley Ltd. He has over 30 years' experience working with water industry clients. For the last 18 years he has worked in various design and build teams as a project manager, reinforced concrete structural designer, Temporary Works Designer and in various Construction (Design and Management) Regulations roles, including CDM Coordinator and Principal Designer.

John Gill joined HOCHTIEF UK in 1995, where he is currently the technical services manager, responsible for the design, support and technical delivery of challenging civil projects.

Martin Haynes Sales & Marketing Director, Fagioli Ltd, has covered most aspects of heavy moves in his career since graduating in 1982, with PSC and since their takeover, with Fagioli.

John Hislam BSc MPhil CEng FICE MASCE, Director, Applied Geotechnical Engineering, has over 40 years experience in foundation engineering with both contracting and consulting companies. He has extensive experience in grouting – having worked in both commercial research and the application of grouting in construction and oil exploration – and in piling, inclusive of basement and retaining construction.

Angus Holdsworth BSc(Hons) IEng MICE is the managing director of an engineering consultancy that specialises in demolition, temporary works for demolition and temporary works for construction. He works closely with contractors to develop demolition schemes for technically complex demolition projects, including residential, commercial, industrial, nuclear and rail structures. He has been involved in the engineering required for explosive demolition, pre-weakening, pull downs, high-reach demolition, dismantling and top-down demolition of projects across the UK and as far afield as New Zealand.

Bernard Ingham BTech(Hons) CEng FICE CMgr MCMI after working in permanent bridge design and construction for 10 years with Cleveland Bridge & Engineering and two county councils, worked for the Mabey Group for 27 years, managing bridging and other temporary works projects before setting up his own company.

Andrew Jones BEng CEng FICE has worked for RMD Kwikform for over 20 years and is currently their chief engineer. He is a well-known expert on falsework and formwork, is a member of BSI committee B514/26 and had significant involvement in PAS 8812. He has published several articles and papers, and won a Thomas Telford Premium award for his contribution to the paper 'Re-visiting Bragg to keep UK's temporary works safe under EuroNorms'.

Keith Kirwan has worked in construction for 18 years, most recently with Network Rail working on Crossrail as a construction manager. His background is in heavy civil engineering, with experience in construction, rail, marine, tunnelling, water and energy. He is a project sustainability champion and innovation leader on Crossrail East.

Stuart Marchand MA(Cantab) CEng FICE FIStructE graduated from the university of Cambridge in 1973 and spent most of his first 8 years of experience in heavy civil engineering on road and bridge construction and on the Thames Barrier. He then transferred to the temporary works design sector of the industry, and joined Costain Construction's temporary works department, taking over as the chief engineer in 1991, involved in major developments with deep basements and retained facades. He set up Wentworth House Partnership in 1999, which he has developed as a specialist temporary works and geotechnical consultancy. Stuart has published several papers and contributed to CIRIA and other industry guides and British Standards.

Frank Marples BTech(Hons) CEng CEnv FICE is an engineering consultant with over 40 years' experience in the design and management of systems for temporary works. As the engineering director for Vinci Construction UK Ltd (formerly Norwest Holst), he has been involved in the development of construction methodology and the design of major temporary works, including two cabled-stayed bridges, the movement and incremental launching of bridges, deep excavations including double-walled cofferdams and facade retention schemes. He has had papers published in the *ICE Proceedings* on the management systems for temporary works, and is the chairman of a BSI committee responsible for BS 5975 and its 2019 revision.

Andrea Massera is a Chartered Civil Engineer. He has been responsible for design and fabrication of heavy steel structures and heavy installations internationally since graduating from Pavia (Italy) in 1985. He has been Engineering Director at Fagioli since 2003.

Charlie McKillop BSc CEng MICE is a Chartered Civil Engineer specialising in formwork, falsework and scaffold design. He trained as a draughtsman in the 1970s, detailing pre-cast concrete, and has worked for several international formwork and scaffold suppliers since the early 1980s. He is an active member of the Temporary Works Forum, and sits on the CONSTRUCT council. He currently works for Hünnebeck Forming & Shoring (a BrandSafway company) as their UK & Ireland engineering director.

Geoff Miles MCIOB, was Quality Manager, Costain before his retirement. He previously worked as a Planner, Site Manager and Project Manager and has been responsible for planning and implementing Site Set-Up, i.e. mobilisation and de-mobilisation on more than 15 contracts ranging from £2M to £75M value.

Eur Ing **Ian Nicoll** BSc CEng MICE is a Chartered Civil Engineer with nearly 45 years' experience of heavy industrial construction; much of that time working on projects abroad. He is currently a consultant associated with temporary works, and was formerly the chairman of the NASC Working Group on the development of TG20 for scaffolding.

Andrew Rattray BSc CEng MICE is a Chartered Civil Engineer with over 35 years' experience of the construction industry. He started work as a civil and structural design engineer before joining the Health and Safety Executive (HSE) as a specialist inspector providing professional and technical advice on the health and safety aspects of civil engineering and building-related topics; he retired in 2017. His role at HSE included the investigation of a range of construction accidents and incidents, often involving temporary works. He has been for many years a member of the BSI committee for falsework.

Chris Robinson CEng FICE is a geotechnical engineer and Chartered Civil Engineer with over 20 years' experience, gained principally in the design and construction of piled foundations and deep basements.

Andrew Smith BSc MICE CEng retired as Contracts Director, Joseph Gallagher Ltd. He led tunnelling and shaft projects for more than 40 years with major and specialist contractors.

Neil Smith BSc(Eng) MSc CEng FICE FGS has been a specialist geotechnical engineer from the beginning of his career and has lectured on the geotechnical aspects of temporary works. He is now a senior consultant to Gavin & Doherty Geosolutions. In his 50-year career, Neil has worked for specialist contractors and consultants covering a very wide range of geotechnical work – investigation and specialist processes, interpreting and understanding the ground conditions at sites around the world and applying the knowledge gained to project design. Much of his time now is taken up with the provision of expert witness services on geotechnical aspects of disputes.

David Szembek BSc CEng MICE MHKI is the temporary works manager with Byrne Bros Ltd, a major London-based frame contractor. He has 37 years' experience in the design and management of temporary works for both civil engineering and building projects in the UK and overseas. Recent high-rise construction projects he has been involved with include No. 1 Bank St, Battersea Power Station Phase 1 and The Shard.

Christopher Tate BSc CEng MICE MCIHT is the chief engineer at VolkerFitzpatrick. For more than 40 years he has been involved in the construction of roads and airfield pavements, earthworks and associated temporary works design. Until very recently he ran his own ground investigation and materials testing company.

Mark Tyler BEng(Hons) has 30 years' experience in heavy civil engineering and building construction. He specialises in providing innovative construction methods, temporary works designs and lean construction solutions. His major project experience includes Sizewell 'B', the Second Severn Crossing, the Øresund Tunnel (Denmark) and Heathrow Terminal 5.

Laurie York BSc CEng MICE MCIWEM C.WEM MCIHT after 15 years involved with the supervision of on-site temporary works, spent 20 years designing temporary works, including tower crane foundations, for a major civil engineering contractor, and is now a temporary works consultant.

Introduction

Temporary Works has been produced for practitioners or would-be practitioners to enable them to apply and enhance their practical skills and knowledge, usually developed in the larger area of permanent works, in this specialism. It will also be of interest to all technicians, undergraduates and graduates wishing to broaden their professional development to have a basic understanding of temporary works. It therefore does not go into considerable detail of design, or the minutiae of construction, but instead shows the reader the unique philosophy and applications by reference to current methods that will enable this specialisation to be applied. You will not find complete worked examples, as an assumption has been made that the reader will either be competent in general engineering principles or will wish to develop those skills not yet possessed.

Where appropriate, each chapter has six main headings of introduction, description of topic (how and why it works), any alternatives worth considering, common methods of construction and principles of design, concluding with references.

Our primary target readership could be considered to be engineers working for a contractor or builder, or for a design consultant, or those intending to gain the technical elements of skill needed to perform efficiently in these roles. We would hope that we have also provided something of value to students and their teachers on the more practical engineering courses.

The most economical temporary works can be 'no temporary works'. Often designing out the temporary works is impossible. Where required, it is rarely considered during the planning and preliminary design stages of a project and rarely specified in the pricing for the permanent works – yet the temporary works have a significant impact on the overall cost of a project. This impacts the industry in several ways. Firstly, it has for many years been given minimum coverage in education at colleges and universities, although fortunately that has improved. How many readers had lectures on temporary works while at college? Secondly, it has until recently been considered a contractor's issue and left to the construction phase of a project, with Permanent Works Designers often not even being involved. The result is that the subject of temporary works has expanded into many specialised fields, each having its own experts and, in many cases, being delivered by specialist suppliers.

ICE Publishing realised that to obtain authoritative guidance a single author would not suffice. Murray Grant and Peter Pallett, well-known and respected engineers with detailed knowledge in temporary works were appointed as managing editors for the first edition published in 2012. This second edition is edited by Peter Pallett and Ray Filip, replacing Murray Grant (retired), to manage, coordinate and proof (and, in some cases, write) the detailed chapters. This edition not only brings up to date the topics inherent in temporary works in the original 27 chapters, but introduces a further seven chapters, to expand on the topics covered and introduces several subjects of a technical nature, more often associated with building construction. The result is that several separate authors have contributed, each being arguably the most knowledgeable engineer and practitioner in their field, writing on their own specialised subject. The result is an authoritative work, informed by the latest UK regulations and use of European codes in the UK.

BS 5975:2019 is the source document for recommended procedures for all organisations involved in temporary works. It provides the framework for safe management of the subject, from concept, through design and construction to completion. At the time of editing this book the 2019 edition of BS 5975 was under final preparation. The managing editors have been involved with the significant updates in the 2019 edition and this book has been updated accordingly. A BSI decision to split BS 5975 into a Part 1 for procedures and a Part 2 for falsework design will be undertaken in the future. Readers should be aware, however, that this book refers to BS 5975 as a single document published in 2019.

Scope of temporary works

A precise use of the *Oxford Dictionary* would limit our definition of temporary works to any work that is done to enable the creation of a permanent structure or infrastructure element. However, this would preclude the opportunities that often arise to reduce or avoid the work and costs of creating something that will be removed or redundant once the client's needs have been satisfied. To avoid temporary works is often a broader, and more elegant, approach to construction, with less exposure to risk and savings in overall cost. On several occasions in the preparation of this book, the definition of temporary works has been extended to suggest alternative processes, at all times promoting the safety of operations and introducing correct procedures for temporary works.

Before using the detailed information, readers of this book are advised to consider if there may be an alternative process whereby the cost and effort of providing works that serve no permanent purpose could be reduced or even avoided.

Furthermore, readers are advised not to limit their selection of temporary works to those methods chosen for inclusion in this book (simply because they are those most commonly used), but to explore the wide range of engineering for potential solutions to their specific project.

Design of temporary works

Designers will be aware of the three basic methods of design, namely permissible stress, limit state and 'custom and practice'. Each method has its merits, and no one method should be preferred in temporary works. This book is not a design manual, and identifies the principles of the designs, and not the detailed method of analysis of the members involved.

In Europe few, if any, educational establishments still teach permissible stress. However, at the time of writing, the majority of Temporary Works Designers in the UK are still using permissible stress in temporary works designs. There is limited confidence to date in the UK on the use of limit state for temporary works design, particularly as the parameters and boundary conditions for temporary works are not the same as for permanent designs. For readers unfamiliar with the differences a short resume follows.

Permissible stress design

Where a member has a load at which it fails, and by applying a single factor of safety (usually 2.0 on collapse and 1.2 on overturning), a safe working load is established. The design then involves checking that for all foreseeable conditions the applied service loads and self-weight do not exceed the safe working load. It should be pointed out that nearly all site staff and operatives have an understanding of safe working loads. The member is considered to be elastic and deflections and so on are all checked using the elastic properties of the members.

For example, to lift a pallet of bricks weighing 10 kN (plus lifting attachments) the site staff would refer to the crane capacity table for reach and check that 10 kN is acceptable for that radius. The crane supplier has produced the data sheets in safe working load terms.

Limit state design

This was introduced at the end of the 19th century and considers both the variability of the material from which the member is made and the probability of the loads actually occurring, and applies partial factors to both.

A statistical approach is used to establish the material properties. Several tests to failure are carried out which generate slightly different ultimate values. Statistical analysis is carried out to provide the characteristic strength with an associated confidence limit normally set at 5%. This means that, if tested, 95% of the members will fail above that value.

The design considers the probability that the calculated loads will be exceeded and applies partial safety factors to suit (usually 1.35 on self-weight, and 1.50 on imposed loads) to give the design loads. The process is then repeated for the materials, and the design resistance of the member calculated by dividing the characteristic resistance by a partial factor based on the probability that the material is not as strong as expected.

The design then involves checking that for all foreseeable conditions the design loads are less than the design resistance. This is the check for the ultimate limit state. A further design check is needed on the unfactored loads to check that the serviceability condition is not exceeded.

This means that a like-for-like comparison between permissible stress and limit state is not possible. It goes without saying that designers cannot mix limit state and permissible-stress methods in a design. If designers provide loads to sites they should clearly specify which loads they are providing to avoid confusion.

The European standards adopted in the UK all use limit state design concepts. The scaffolding industry as a case in point, uses the European documents but introduces a conservative 'quasi-permissible stress' approach (see Chapter 21), to provide design data for site use in understandable 'permissible stress' terms.

The crane capacity table discussed earlier will, most likely, have been generated by limit state design, but converted into 'safe working loads' for site use. The same concept occurs in highway bridge design – the signpost for a bridge weight limit does not state the characteristic strength in kilonewtons but the 'safe load' (usually in tonnes) because that is what the users, the public, understand, although European limit state design codes were most probably used in the load verification.

Custom and practice design

There are still examples in temporary works of 'custom and practice' design being the norm. It relies on the experience and knowledge of operatives who have used similar systems before and 'know what works'. In the long distant past someone probably did calculations or tests, but it works. This approach is accepted in the European standard for falsework where support to some in situ concrete building slabs (>300 mm thick and <3.5 m high) are 'custom and practice' as there are no European design rules and 'design' is left to operative experience and knowledge (see Section 22.1.4 in Chapter 22 for class A falsework). Use of column clamps in formwork is another example.

The management systems used (see Chapter 2) should identify and control such systems – 'custom and practice' is probably used more frequently on smaller sites than designers realise – it requires engineering judgement, and organisations should have the relevant procedures to control it.

Peter F Pallett and Ray K Filip

Pallett, Peter F and Filip, Ray
ISBN 978-0-7277-6338-9
https://doi.org/10.1680/twse.63389.001
ICE Publishing: All rights reserved

Chapter 1
Safety, statutory and contractual obligations

Andrew Rattray
HM Principal Specialist Inspector (retired)

Peter F Pallett
Pallett TemporaryWorks Ltd

Safety of temporary works is paramount to protect not only site workers but also the public and others who may be affected by the work. Temporary works should be 'engineered' and given the same degree of care and consideration as the permanent works. It is also important that temporary works are designed to be robust enough to withstand the rigours of site use and detailed to ensure that local or single-component failure does not lead to progressive collapse. Knowledge gained from past collapses or incidents and research has informed current temporary works guidance and standards. BS 5975:2019 (BSI, 2019) is an industry consensus view on good practice, and provides recommendations for procedures for the design, construction, use and dismantling of all types of temporary works. Statutory legislation has a direct effect on the design and operation of temporary works, imposing duties and obligations on many of the parties involved. In addition, all users of temporary works will have contractual obligations under which they work, often imposing specific requirements for the works.

1.1. Introduction

'Temporary works' is a widely used expression in the construction industry and is defined in BS 5975:2019 'Code of practice for temporary works procedures and the permissible stress design of falsework' (BSI, 2019), as 'parts of the works that allow or enable construction of, protect, support or provide access to, the permanent works and which might or might not remain in place at the completion of the works'. The construction of most types of permanent works will require the use of some form of temporary works. Temporary works should be an 'engineered solution' that is used to support or protect an existing structure or the permanent works during construction, or to support an item of plant or equipment or the vertical sides or side slopes of an excavation, or to provide access. It is imperative that the same degree of consideration and care is given to the design and construction of temporary works as to the design and construction of the permanent works. The management of temporary works is discussed in Chapter 2.

Examples of temporary works include (but are not limited to) the following

- *Earthworks*: trenches, excavations, temporary slopes and stockpiles.
- *Structures*: formwork, falsework, propping, façade retention, needling, shoring, edge protection, scaffolding, temporary bridges, site hoarding and signage, site fencing and cofferdams.
- *Equipment/plant*: tower crane foundations, construction hoists, mast-climbing work platforms (MCWPs) and any work to support plant and machinery (e.g. mobile crane outrigger supports, piling and crane platforms, anchorages/ties for MCWPs and hoists).
- *Structures in the temporary condition*: permanent works structures during construction that are unstable or have insufficient strength, and structures during demolition and dismantling (e.g. those subject to pre-weakening).

In order to ensure the strength and stability of any temporary works structure, there are three fundamental aspects that need to be considered, which can be simplified as follows

- *Foundations*: the ability of the ground to carry the loads transmitted from the temporary works structure without failure or excessive deformation or settlement.
- *Structural integrity*: the ability of the temporary works structure itself to carry and transmit loads to the ground via the foundations without failure of the structural elements, including fixings and connections (e.g. by buckling, bending, shear, tension, torsion) and without excessive deflection.
- *Stability*: the ability of the temporary works structure to withstand horizontal or lateral loading without sway, overturning or sliding failure (stability may be inherent in the temporary works structure itself or provided by an existing structure, e.g. the permanent works or a structure being demolished – see Chapter 34 on demolition); there will also be projects where the temporary condition of the permanent works will require temporary works to ensure stability.

Failure to adequately design, construct and maintain temporary works can lead to

- collapse or failure of the temporary works
- structural failures and collapse of the permanent works
- uncontrolled ingress or egress of materials, spoil and water
- collapse of adjacent structures (buildings, transport systems, infrastructure)
- risk of single or multiple fatalities and serious injuries to workers and members of the public
- risk of significant delay and increased costs to construction projects
- significant financial and commercial risks to contractors, subcontractors, designers, suppliers and clients.

The main causes of temporary works failures include

- absence of or an inadequate temporary works procedure

- inadequate site investigation (including geotechnical investigation, identification of underground services, assessment of the structural condition of existing and/or adjacent buildings)
- inadequate or lack of design and/or design brief for the temporary works
- inadequate level of checking of temporary works designs
- lack of awareness on site of temporary works design assumptions
- unavailability of temporary works equipment
- inappropriate use of temporary works equipment
- poorly constructed temporary works and/or absence of checking of adequate erection
- unauthorised changes to an approved temporary works design
- overloading of temporary works (i.e. failure to control loading or lack of awareness of the capacity of the equipment)
- inadequate communication of details of the temporary works design to the erectors
- inadequate foundations for the temporary works
- lack of adequate lateral stability for the temporary works and/or permanent works.

The first code of practice on falsework (BS 5975:2008) to provide recommendations and guidance on the procedural controls for all aspects of temporary works in the construction industry was published in 2008. It also gave specific guidance on the permissible stress design, specification, construction, use and dismantling of falsework. The latest edition of this standard, BS 5975:2019 (BSI, 2019) expands the previous guidance on procedures and includes more detailed guidance on the different organisations involved in temporary works. Correct procedures, and use of relevant technical aspects are the key to the success of falsework and temporary works, all based around good management (see Chapter 2).

1.2. Background

The first report on falsework was by the Joint Committee of the Institution of Structural Engineers and the Concrete Society (CS, 1971) and introduced classes of falsework. Shortly afterwards, the British Standard Code of Practice committee was established and, ironically, a fortnight later a major collapse occurred in the UK by the River Loddon near London. This collapse and other significant falsework collapses in the 1970s, together with an apparent lack of authoritative guidance, led the UK government to set up the Advisory Committee on Falsework, which produced the Bragg Report (Bragg, 1974, 1975), named after the committee chairman. Industry produced the first code of practice (in compliance with one of the recommendations of the Bragg Report) as BS 5975 in 1982, which was informed by the recommendations of the Bragg Report and the earlier Joint Committee report.

The Bragg Report made some very pertinent comments about falsework, which apply equally to all temporary works and which still hold true today, including

> Falsework requires the same skill and attention to detail as the design of permanent structures of like complexity, and indeed falsework should always be

regarded as a structure in its own right, the stability of which at all stages of construction is paramount for safety.

Other key concerns highlighted in the report include

- competency
- design procedures
- design responsibilities
- communication and coordination
- inspection and supervision
- lateral stability.

There have been significant changes to the construction industry since the mid-1970s that have affected how falsework, and temporary works more generally, are dealt with. A paper from the Standing Committee on Structural Safety (SCOSS, 2002), updated in 2010 (SCOSS, 2010), identified the principal changes, which included the following.

- Few contractors now have their own temporary works departments, whereas in the 1970s almost all would design temporary works in-house. The responsibility for temporary works now often falls to a specialist contractor, which can result in a lengthy supply chain. In contrast, suppliers are now providing temporary works design services, based fundamentally on their own equipment.
- In the 1970s, most falsework (and temporary works) was constructed from scaffold tube and fitting components, whereas proprietary systems now dominate the market; the design skills and knowledge of the performance of the systems therefore now tends to lie within the specialist organisations. Furthermore, the increased use of aluminium and certain composites have revolutionised some of the systems now in use.
- There has been a gradual but inexorable loss of traditional skills within the construction industry; in practical terms, this means that the site foreman with a lifetime's experience of 'what works' has largely been lost.
- Procurement routes are now largely chosen to maximise commercial benefit and have little regard to considerations of the flow of information; the difficulties caused by long supply chains are further exacerbated when design and erection responsibility are split and when design/supply briefs do not include site visits or inspections.

Health and Safety Executive (HSE) research (HSE, 2001) into various aspects of falsework produced some worrying findings that were equally applicable to temporary works in general. These included the following

- A lack of understanding, at all levels, of the fundamentals of stability of falsework and the basic principles involved.
- Wind loading was rarely considered.

- There was a lack of clarity in terms of the design brief and coverage of key aspects such as ground conditions.
- The assumptions for the lateral restraint of the falsework made by designers were often ignored or misunderstood by those on site.
- There was a lack of adequate checking and a worrying lack of design expertise.
- Erection accuracy left a lot to be desired.

Based on the research, a number of key concerns were identified

- competency of the falsework/Temporary Works Designer
- sufficiency of information
- adequacy of supervision
- role of the Temporary Works Coordinator
- competency of those erecting falsework/temporary works.

The actions to deal with these concerns are straightforward and require no more than the application of the good practice given in BS 5975:2019 (BSI, 2019). They also fit well with the aspirations of the Construction (Design and Management) Regulations 2015 (HSE, 2015) in respect of their aim of improving the overall coordination and management of health and safety throughout all stages of a construction project. The European limit state design codes, such as BS EN 12811-1:2003 for scaffolding (BSI, 2003) and BS EN 12812:2008 for falsework (BSI, 2008), contain little (if any) reference to procedures or site practice. This means that the procedural items in BS 5975:2019 take on significant importance.

1.3. Management of temporary works

The correct design and execution of temporary works is an essential element of risk prevention and mitigation on a construction site. Section 2 of BS 5975:2019 provides recommendations and guidance for a robust set of procedural controls to be applied to the design, specification, construction, use, maintenance and dismantling of all types of temporary works. Compliance with BS 5975:2019 is not a legal requirement, but the code of practice does provide an industry consensus view on what is considered to be good practice. All temporary works should be an 'engineered solution', and it is imperative that the same degree of consideration and care is given to the design and construction of the temporary works as to the design and construction of the permanent works. See Chapter 2 on the management of temporary works.

1.4. Construction (Design and Management) Regulations 2015

The Construction (Design and Management) Regulations 2015 (CDM 2015) (UK Government, 2015) came into effect on 6 April 2015, replacing the previous 2007 version, which had in turn replaced the original 1994 regulations. The latest revision sought to simplify and rationalise the regulations, primarily based on copying the existing requirements under EC Directive 92/57/EEC on temporary or mobile construction sites (EC, 1992). The HSE guidance on the legal requirements for CDM 2015 is given in publication L153, *Managing Health and Safety in Construction* (HSE, 2015), which describes

- the law that applies to the whole construction process on all construction projects, from concept to completion
- what each duty holder must or should do to comply with the law to ensure projects are carried out in a way that secures health and safety.

The Construction Industry Training Board (CITB) has produced industry guidance, written by representatives of CONIAC (Construction Industry Advisory Committee) with small businesses in mind, for the five duty holders under CDM 2015 and an additional one for workers (CITB, 2015).

In Part 4 of CDM 2015 (HSE, 2015), the regulations are grouped into generally related topics, some of which are particularly relevant to temporary works

- safe places of work
- stability of structures
- demolition or dismantling
- excavations
- cofferdams and caissons.

The available guidance should be read and understood by all those involved in the procurement and use of temporary works. CDM 2015 (HSE, 2015) makes several specific references to the design and management of temporary works; these are considered below.

1.4.1 Designers and principal designers

The definition of 'structure' in CDM 2015 (HSE, 2015) includes 'any formwork, falsework, scaffold or other structure designed or used to provide support or means of access during construction work'. The HSE (2015) and CITB (2015) guidance both refer to Temporary and Permanent Works Designers. It is therefore clear that, under CDM 2015, Temporary Works Designers have exactly the same designer duties as Permanent Works Designers, including the following

- Ensuring that they are competent in their specific field of temporary works design (and able to address the particular relevant health and safety issues).
- Avoiding foreseeable risks, so far as is reasonably practicable, to those involved in the construction, use and dismantling of the temporary works, and providing adequate information on the remaining significant risks, for example, by following the principles of the Eliminate–Reduce–Inform–Control (ERIC) model (CITB, 2015) or any other similar risk-reduction technique as part of the design process.
- Coordinating and cooperating with others, for example, liaising with the Permanent Works Designers to ensure that their designs are compatible and that the permanent works can accommodate any assumed loadings from the temporary works. Examples of loads/forces to be considered would include the lateral restraint of falsework when 'top restrained' and loads on completed slabs by backpropping of multi-storey buildings (see Chapters 22, 24 and 28).

The equal care and consideration required by the designers of temporary works and the permanent works reflects comments made in the Bragg Report (Bragg, 1975) more than 45 years previously.

CDM 2015 (HSE, 2015) introduces the principal designer appointed by the client as the organisation or individual with responsibility for control of the design of the project through the pre-construction phase to delivery of the construction work. The principal designer's duty includes ensuring that all designers, of both the permanent and temporary works, comply with the legal requirements for designers. Although most readers will be aware of who designers are, CDM 2015 includes those who select products for use in construction as having duties as designers, and states that they must take account of the health and safety issues associated with the use of the products specified or selected.

1.4.2 Contractors and principal contractors

All contractors should plan, manage and monitor their construction work so that it is carried out safely and without risks to health. For single contractor projects, the contractor must ensure that a construction-phase plan is drawn up; for projects involving more than one contractor, the principal contractor must ensure that a construction-phase plan is drawn up. CDM 2015 requires the construction-phase plan to include specific measures concerning work involving particular risks, such as

- 'Work which puts workers at risk of burial under earth falls ... or falling from a height, where the risk is particularly aggravated by the nature of the work or processes used or by the environment at the place of work or site.'
- 'Work involving the assembly or dismantling of heavy prefabricated components.'

The majority of temporary works are likely to involve significant risks from working at height, the potential for ground instability and the use of heavy prefabricated components. Consequently, the expectation is that the majority of construction-phase plans will need to contain the management arrangements for dealing with temporary works (i.e. suitable temporary works procedures).

1.5. The Work at Height Regulations 2005

The Work at Height Regulations 2005 (WAHR 2005) (UK Government, 2005, 2007) impose health and safety requirements applicable to all work activity at height, not just in the construction industry. Work at height can take place at any location either above or below ground level, and includes temporary means of access to and egress from such work. The WAHR 2005 implemented the requirements of EC Directive 2001/45/EEC (EC, 2001) and replaced the provisions in the Construction (Health, Safety and Welfare) Regulations 1996 (UK Government, 1996) relating to falls, fragile materials and falling objects.

The WAHR 2005 impose duties on employers and the self-employed to assess the risks from work at height and to organise and plan the work so it is carried out safely. The objective is to make sure that work at height is properly planned, including the selection of relevant equipment, appropriately supervised and carried out in a safe manner. The WAHR 2005 set out a simple hierarchy for managing and selecting equipment for work at height: duty holders must avoid work at height where they can; use work equipment or other measures to prevent falls where they cannot avoid working at height; and, where they cannot eliminate the risk of a fall, use work equipment or other measures to minimise the distance and consequences of a fall should one occur. Priority should be given to collective protection measures over personal protective measures. When working under the control of another person, all employees and the self-employed have a duty to report any activity or equipment which is defective.

In relation to temporary works, the WAHR 2005 place duties on clients and designers to ensure that strength and stability calculations for scaffolding are carried out unless calculations are already available or the scaffold is assembled in conformity with a generally recognised standard configuration (HSE, 2007; schedule 3.2). A recognised standard configuration could be the National Access & Scaffolding Confederation (NASC) Technical Guidance TG20:13 for tube and fitting scaffolds (NASC, 2013) or the manufacturer's guidance for system scaffolds. In such cases calculations have already been prepared and, provided the scaffolding is erected to the stated rules for the solution adopted, further calculations are not required. The WAHR 2005 require that, depending on the complexity of a scaffold, an assembly, use and dismantling plan shall be drawn up by a competent person. (Note that the use in the law of the word 'plan' is a 'plan of work' and not a physical drawing or sketch. The law further uses the phrase 'drawn up', again referring to 'being prepared', i.e. not as in 'a drawing'.) Such a plan should describe the sequence and methods to be adopted when erecting, dismantling and altering the scaffold, if this is not covered by the published guidance referred to above.

The WAHR 2005 state certain requirements for all working platforms. Where there is a risk of falling, there are requirements for guard rails, toe boards, barriers and similar means of protection; the top guard rail should be at least 950 mm high, with intermediate guard rail(s) and toe boards(s) positioned to give a maximum unprotected gap of 470 mm. Where the platform is for a sloping workplace at an angle greater than 10°, the guard rail requirements are more onerous. Although no minimum width of a working platform is stated, the width values given in good practice documents such as BS EN 12811-1:2003 (BSI, 2003) or TG20:13 (NASC, 2013) are used.

1.6. The Health and Safety (Offences) Act 2008

The Health and Safety (Offences) Act 2008 (HSOA 2008) (UK Government, 2008) makes amendments to the Health and Safety at Work etc. Act 1974 (HSWA 1974). HSOA 2008 has resulted in the £20 000 maximum fine that can be given in a magistrates' court applying to most health and safety offences; it does not affect the position in relation to fines for sentences imposed in the Crown Court, where fines are unlimited. HSOA 2008 also makes imprisonment of individuals an option for many health and safety offences, including employees who do not take reasonable care of the health and safety of others, and directors and senior managers. The maximum custodial term that can be given by a magistrates' court is 6 months and by the Crown Court 2 years.

1.7. Contractual obligations
1.7.1 General
All work is carried out by one party instructing a second party to act (i.e. a contract is offered), consideration is shown and, once accepted, forms a contract between the parties. Although a verbal instruction can be considered a contract, the usual form will be a written and legally binding contract between the relevant parties. Employees have contracts of employment, clients have contracts with professional advisors to design structures, and clients also have contracts with organisations to construct the designed permanent works. It should be noted that unless the work is 'design and build' it is unusual for there to be a contract between the professional designer and the organisation carrying out the construction (i.e. the contractor). There may be more extensive contracts for 'design, build and operate', and, in some cases, with finance, often including risk management.

In the UK, there is a separate judicial system for handling the civil procedures involved in disputes both under contract law and the law of tort. Under contract law, both parties accept and agree to liabilities and duties, whereas tort is a violation of a duty established by law. Liability in tort usually arises from a breach of duty established in law, and includes the tort of negligence. However, although the subject of contract and tort law is outside the scope of this book (see ICE, 2011), its implication for temporary works needs to be considered.

Contracts may be based on existing industry formats such as the NEC4 New Engineering Contract (ICE, 2017) which has generally replaced the earlier ICE Conditions of Contract (ICE, 1986–1999). Contracts may also require compliance with detailed specifications. Typical specifications are those from the Highways Agency (2006), the National Building Specification (published annually), the Civil Engineering Specification for the Water Industry (UKWIR, 2011) and the National Structural Concrete Specification for Building Construction (CSG, 2010). The contract will place an obligation on the contractor to comply. This contractual obligation will frequently be passed on to subcontractors and suppliers, therefore obligating their compliance. The contract may also include lists of standards deemed to be included, thus making a particular British Standard a contractual requirement. Hence, although not legally enforceable as a mandatory statutory requirement, use of a particular British Standard would become obligatory under the contract. Failure to adopt the British Standard recommendations would, therefore, risk a court action for damages. This could affect the temporary works design, and such obligations should be included in the temporary works design brief (see Chapter 2 on management).

There will also be contractual obligations when utilising proprietary equipment. A manufacturer, importer or supplier of articles for use at work essentially has a statutory duty under HSWA 1974 to give advice about the equipment's use; and the user, when placing an order (i.e. a contract to purchase or hire such equipment), is obliged contractually to use the equipment as the supplier intended. Often the specific conditions of use assumed by the supplier/importer and incorporated by the Temporary Works Designer are stated on the drawings or in technical datasheets; for example, the use of soffit formwork to provide lateral restraint to falsework when 'top restraint' is assumed (discussed in Chapter 22) places an obligation on the site management and the user to ensure that such considerations exist in practice.

1.7.2   Functions and relationships between parties

The client will have separate contracts with one or more professional advisers and the principal contractor, with implications of privity of contract. The professional adviser may contract with other professional advisers directly. The principal contractor may contract for the supply of labour, materials or both, and also for hire of plant and equipment, with or without labour. Many contractors 'manage' only, and let contracts to subcontractors and other 'contractors'. On design, build and operate projects there will be one main contract, but there may be subsidiary contracts with professional advisers and management organisations.

1.7.3   Responsibilities for the temporary works

Particular contractual issues are the effect of temporary works on the permanent design (discussed in more detail in Chapter 2), and who takes responsibility for the temporary works. Responsibility for the temporary works rests with the principal contractor, and the costs are normally included in the build-up of the tender rates in the contract. Rarely is the temporary works included as a bill item in the schedule of quantities. The law requires Permanent Works Designers to consider loadings from temporary works. Under certain contracts, clauses can require the contractor to provide calculations of the stresses and strains in the permanent works caused by the temporary works; this can be a significant additional cost. For example, on a major bridge it implies that the contractor has to carry out the permanent works design in order to verify the construction sequence and method.

The professional advisor is responsible to the client for ensuring that the contractor's temporary works will produce a finished job which complies with the contract documents, in particular that it is not detrimental to the permanent works. This means that the professional advisor does not have a contractual duty to verify the contractor's temporary works calculations; hence the importance of the independent temporary works design checks recommended in BS 5975 (BSI, 2019) and discussed in Chapter 2. It is always important to establish the responsibilities under the various sub-contracts. Phrases in the contract such as 'in accordance with recognised codes of practice' will make the use of particular codes a contractual requirement.

1.8.   Robustness

Temporary works often comprise a structure with many components, junctions and connections, and, unlike the permanent works, this structure is often reused a number of times and moved from site to site. Temporary works should, therefore, be robust enough to withstand the rigours of site use. Careful attention should be paid to the way in which components, connections and junctions are detailed to reduce the dependence on workmanship. For example, the design of a particular joint or component may be justified by calculations using a minimum thickness of section, whereas engineering judgement would dictate that using a thicker or larger section would reduce the risk of damage during transport and when in use, giving a more robust, and consequently safer, structure. Any critical component or connection should be inherently robust in its own right. Detailing of the temporary works should be such that local or single-component failure does not lead to the progressive collapse of the whole structure. This does not imply that the design of temporary works should be over-conservative but that due consideration should be given to providing alternative load paths so that in the event of the failure of one member or component the load may be redistributed through others. An appreciation of the way the temporary works structure behaves will lead to safe temporary works that do not progressively collapse.

1.9. Public safety

Every year construction work injures and kills people who have no direct connection to it. Some temporary works, such as site hoardings, scaffold fans and public-protection scaffolds, are specifically provided to protect members of the public. However, any construction activity including the erection or dismantling of temporary works has the potential to cause harm, not only to the construction workers but also to members of the public. The law states that business must be conducted without putting members of the public at risk. Providing suitable protection for those actually carrying out construction work will often also provide protection for others who may be affected by the works. An example is climbing protection screens used in multi-storey construction (see Chapter 25). The precautions that need to be taken to adequately protect members of the public may, however, differ from those taken to protect those working on site. Members of the public are likely to be less aware of the dangers involved with construction activities than are those working on site.

In particular, children do not have the ability to perceive danger in the same way as adults do, and may see construction sites as potential playgrounds. While the number of children being killed or injured on construction sites has reduced, there is no room for complacency. Each year, two or three children die after gaining access to construction sites, but many more are injured. Other members of the public have been seriously injured by

- materials or tools falling outside the site boundary
- falling into trenches
- being struck by moving plant and vehicles.

All construction sites require measures to be in place to manage access to the site through well-defined site boundaries and to exclude unauthorised persons such as members of the public. The site boundary should be physically defined, where necessary, by suitable fencing. In populated areas, this will typically mean a 2 m high small-mesh fence or hoarding around the site or work area. Consideration must also be given to the provision of protection from any work activity taking place outside of the site boundary, for example, erection or dismantling of the site fencing or hoarding, utilities excavations, scaffold erection or dismantling and delivery and storage of materials. See Chapter 3 on site compounds and hoardings.

Many hazards have the potential to injure members of the public. In particular, the following need to be considered whenever temporary works are being planned or carried out

- *Falling objects:* ensure that objects cannot fall outside the site boundary.
- *Excavations and openings:* barriers or covers are required.
- *Delivery and site vehicles:* ensure that pedestrians cannot be struck by vehicles entering or leaving site; do not obstruct pavements so that pedestrians are forced into the road.
- *Scaffolding and other access equipment:* prevent people outside the site boundary being struck during the erection, dismantling and use of scaffolding and other access equipment.
- *Slips, trips and falls within pedestrian areas:* inadequate protection of holes, uneven surfaces, poor reinstatement, trailing cables and spillage of materials are common causes.
- *Storing and stacking materials:* keep all materials within the site boundary if possible, or provide protection.

1.10. Summary of main points

- The construction of most types of permanent works will require the use of some form of temporary works.
- All temporary works should be an 'engineered solution'.
- It is imperative that the same degree of consideration and care is given to the design and construction of the temporary works as to the design and construction of the permanent works.
- Key issues for the safety and stability of temporary works are adequate foundations, structural integrity and lateral stability.
- Coordination and cooperation are required between the Temporary and the Permanent Works Designers to ensure that designs are compatible and that the permanent works can accommodate any assumed loadings from the temporary works, for example, the lateral restraint of falsework.
- The majority of construction-phase plans prepared by contractors should contain the management arrangements for controlling the risks associated with temporary works, for example, suitable temporary works procedures.
- The appointment of a competent Temporary Works Coordinator is an essential step for the safe management of substantive temporary works structures.

REFERENCES

Bragg SL (1974) *Interim Report of the Advisory Committee on Falsework*. Department of Employment and Department of the Environment. HMSO, London, UK.

Bragg SL (1975) *Final Report of the Advisory Committee on Falsework*. Department of Employment and Department of the Environment. HMSO, London, UK.

BSI (British Standards Institution) (2003) BS EN 12811-1:2003. Temporary works equipment. Scaffolds. Performance requirements and general design. BSI, London, UK.

BSI (2008) BS EN 12812:2008. Falsework. Performance requirements and general design. BSI, London, UK.

BSI (2011) BS 5975:2008 + A1:2011. Code of practice for temporary works procedures and the permissible stress design of falsework. BSI, London, UK.

BSI (2019) BS 5975:2019. Code of practice for temporary works procedures and the permissible stress design of falsework. BSI, London, UK.

CITB (Construction Industry Training Board) (2015) *The Construction (Design and Management) Regulations 2015. Industry Guidance*. CITB, Bircham Newton, UK.
- CDM15/1: *Industry Guidance for Clients*.
- CDM15/2: *Industry Guidance for Principal Designers*.
- CDM15/3: *Industry Guidance for Contractors*.
- CDM15/4: *Industry Guidance for Designers*.
- CDM15/5: *Industry Guidance for Principal Contractors*.
- CDM15/6: *Industry Guidance for Workers*.

Concrete Society (1971) *Falsework*. Report of the Joint Committee of the Concrete Society and the Institution of Structural Engineers. Concrete Society, Crowthorne, UK, Technical Report TRCS 4.

CSG (Concrete Structures Group) (2010) *National Structural Concrete Specification for Building Construction*, 4th edn. The Concrete Centre, Camberley, UK, Publication CCIP-050.

EC (European Council) (1992) Council Directive 92/57/EEC of 24 June 1992 on the implementation of minimum safety and health requirements at temporary or mobile construction sites (eighth individual Directive within the meaning of Article 16(1) of Directive 89/391/EEC). *Official Journal of the European Communities* **L245**.

EC (European Council) (2001) Directive 2001/45/EC of the European Parliament and of the Council of 27 June 2001 amending Council Directive 89/655/EEC concerning the minimum safety and health requirements for the use of work equipment by workers at work (second individual Directive within the meaning of Article 16(1) of Directive 89/391/EEC) (Text with EEA relevance). *Official Journal of the European Communities* **L195**.

Highways Agency (2006) *Specification for Highway Works. Manual of Contract Documents for Highway Works*. Highways Agency, London, UK.

HSE (Health and Safety Executive) (2001) *Investigation into Aspects of Falsework*. HSE Books, Sudbury, UK, HSE Contract Research Report 394/2001. See www.hse.gov.uk/research/crr_htm/2001/crr01394.htm (accessed 01/08/2018).

HSE (2007) *Managing Health and Safety in Construction. Approved Code of Practice*. HSE Books, Sudbury, UK, Publication L144.

HSE (2015) *Managing Health and Safety in Construction. Construction (Design and Management) Regulations 2015. Guidance on Regulations*. HSE Books, Sudbury, UK, Publication L153.

ICE (Institution of Civil Engineers) (1986–1999) *ICE Conditions of Contract*, 5th, 6th and 7th edns. ICE Publishing, London, UK.

ICE (2011) *ICE Manual of Construction Law*. Thomas Telford, London, UK.

ICE (2017) *NEC4. The Contracts*. ICE Publishing, London, UK.

NASC (National Access & Scaffolding Confederation) (2013) *TG20:13 Operational Guide. A Comprehensive Guide to Good Practice for Tube and Fitting Scaffolding*. NASC, London, UK.

National Building Specification (annually) *Formed Finishes, Section E20*. National Building Specification, Newcastle-upon-Tyne, UK.

SCOSS (Standing Committee on Structural Safety) (2002) *Falsework: Full Circle?* SCOSS, London, UK, SCOSS Topic Paper SC/T/02/01.

SCOSS (Standing Committee on Structural Safety) (2010) *Falsework: Full Circle?* SCOSS, London, UK, SCOSS Topic Paper SC/T/02/01 17.10.02/Rev 20.08.10.

UK Government (1974) Health and Safety at Work etc. Act 1974. The Stationery Office, London, UK.

UK Government (1996) Construction (Health, Safety and Welfare) Regulations 1996. Statutory Instrument 1996/1592. The Stationery Office, London, UK.

UK Government (2005) Work at Height Regulations 2005. Statutory Instrument 2005/735. The Stationery Office, London, UK.

UK Government (2007) Work at Height (Amendment) Regulations 2007. Statutory Instrument 2007/114. The Stationery Office, London, UK.

UK Government (2008) Health and Safety (Offences) Act 2008. The Stationery Office, London, UK.

UK Government (2015) Construction (Design and Management) Regulations 2015. Statutory Instrument 2015/15. The Stationery Office, London, UK.

UKWIR (2011) *Civil Engineering Specification for the Water Industry*, 7th edn. WRc, Swindon, UK.

FURTHER READING

Burrow M, Clark L, Pallett P, Ward R and Thomas D (2005) Falsework verticality: leaning towards danger? *Proceedings of the Institution of Civil Engineers – Civil Engineering* **158(1)**: 41–48.

ICE (Institution of Civil Engineers) (2010) *ICE Manual of Health and Safety in Construction*. Thomas Telford, London, UK.

NASC (National Access & Scaffolding Confederation) (2010) *SG25: Access and Egress from Scaffolds*. NASC, London, UK.

Smith NJ (2006) *Managing Risk in Construction Projects,* 2nd edn. Blackwell, Oxford, UK.

Useful web addresses

Health and Safety Executive (HSE) – books: http://books.hse.gov.uk (accessed 01/08/2018).

Health and Safety Executive (HSE) – Health and safety in the construction industry: http://www.hse.gov.uk/construction/index.htm (accessed 01/08/2018).

Temporary Works Forum: http://www.twforum.org.uk (accessed 01/08/2018).

Temporary Works, Second edition

Pallett, Peter F and Filip, Ray
ISBN 978-0-7277-6338-9
https://doi.org/10.1680/twse.63389.015
ICE Publishing: All rights reserved

Chapter 2
Management of temporary works

Frank Marples
Engineering Consultant, Vinci Construction UK Ltd

This chapter examines why, what, who and how temporary works and the interface between the Permanent Works and Temporary Works Designers are managed and controlled. The importance of the principal contractor (PC), the independence of the PC's Temporary Works Coordinator (TWC) and the support upon which the PC's TWC depends, including the PC's designated individual (DI), the construction team and other organisations involved in the construction project, are discussed. The categories of design check required to ensure that the temporary works can be constructed and can perform their function of ensuring the permanent works are not overstressed (during construction or demolition) and protecting the public and/or the site team (including the operatives) from errors, omissions, misunderstandings and perhaps even the pressures of contract, programme and cost are reviewed.

2.1. Introduction

In the late 1960s and early 1970s there was, unfortunately, a series of significant collapses of falsework and of permanent works in a temporary condition together with associated fatalities. The number and scale of these collapses was sufficient to prompt the UK government to commission an advisory report on the failures. The report, known as the Bragg Report (Bragg, 1975), was an industry milestone and is still held in high regard. It identified a series of causes and made numerous recommendations, the most significant of which were the adoption of procedural controls, coordination between designers of permanent and temporary works, proper management of the process from inception through to loading and unloading, and the appointment of a TWC to manage and be responsible for the process. Initially, BS 5975 used the term 'falsework coordinator', but industry wisely realised that the Bragg Report recommendations for falsework applied equally to all temporary works and the most recent version of this standard, BS 5975:2019 (BSI, 2019), adopts the more general term 'Temporary Works Coordinator'.

In addition to the falsework failures considered in the Bragg Report, around the same time, there were collapses of five major box girder bridges during construction. A Royal Commission (Barber *et al.*, 1971) into the failure of West Gate Bridge concluded that the designers of the bridge failed to consider the construction phases of the box girder. It is also worth noting that the Report highlights that correspondence from site to the design office was not answered but evidently the erection of the box girders still progressed! Proposals by the contractor to add kentledge to the top flange, in order to re-align the plate for welding, were not satisfactorily considered. The conclusions drawn are that: the temporary conditions during erection/construction should be considered as a design case(s); good communication between site and design office (or between design offices) is essential, and any queries should be answered before construction progresses beyond the relevant point (condition); further, that ad-hoc modifications should not be permitted without reference to the designer of the permanent works or temporary works as appropriate. In addition to the former, the commission was of the opinion that when the original contractor was replaced, a more experienced contractor should have been appointed by the client. It was shown that the client had been aware of the new contractor's lack of experience in the type of work to be undertaken.

In 2008, UK recommended procedures for falsework were formally introduced for *all* temporary works in an updated edition of BS 5975 (BS 5975:2008; BSI, 2011); which included a section on procedures and included the permissible stress design of falsework (see Chapter 22 on falsework). Following changes in the Construction (Design and Management) Regulations 2015 (CDM 2015) (HSE, 2015; UK Government, 2015) and defined roles for duty holders including the PC, BS 5975 has been significantly updated (BS 5975:2019; BSI, 2019) and uses the term PC's TWC to denote the individual with overall responsibility for the management and implementation of the temporary works on a project. The BS 5975:2019 recommendations place requirements on the client and designer, and include conditions of the permanent works during construction in the definition of temporary works.

Temporary works and permanent works, and the operatives and public who are entitled to depend on their safe execution and use, are identical insofar as their being dependent on the application of the same engineering principles. It therefore follows that the temporary works and permanent works deserve the same rigour and respect. The guidance to the CDM 2015 (L153; HSE, 2015), is clear, stating that

> Designers should liaise with any other designers, including the principal designer, so that work can be coordinated to establish how different aspects of designs interact and influence health and safety. This includes *Temporary and Permanent Works Designers*. Designers must also cooperate with contractors and principal contractors so that their knowledge and experience about, e.g. the practicalities of building the design is taken into account.

This confirms that both Permanent Works and Temporary Works Designers need to consider the temporary works.

The core Bragg Report recommendations were incorporated in BS 5975 when it was first published in 1982. They remain as valid today as when first written and the recommended minimum lateral stability force was also introduced into the UK National Foreword for the European Falsework Code (BS EN 12812:2008; BSI, 2008). The adoption of the recommended procedural controls for temporary works in general has served industry well, and has been instrumental in preventing further major collapses and failures and in promoting safe practice.

The principal Bragg Report recommendations relate to coordination between the designers of the permanent works and the temporary works (and communication between all parties in general), and anticipated the fundamental principles of CDM 2015 by some 20 years. To ignore the findings of the Bragg Report, therefore, risks designers being at odds with CDM 2015.

Nevertheless, research commissioned by the Health and Safety Executive (Burrow *et al.*, 2005; HSE, 2001) showed that the basic concepts were being forgotten. This was further supported by the Standing Committee for Structural Safety (ICE/SCOSS, 2002) stating that there was 'a lack of understanding of the fundamentals of stability, at all levels of industry' and 'a lack of adequate checking and a worrying lack of design expertise'. All this serves to reinforce the Bragg Report recommendations and to caution against complacency.

BS 5975:2019 has been significantly revised over the previous version with regard to 'procedures in temporary works', in line with the latest revision to CDM 2015 (UK Government, 2015), to ensure its relevance to today's construction industry and that all parties, in addition to the PC, understand they too are involved in temporary works and have a role to play.

Although the use of the procedures in British Standards is not mandatory, every organisation has a legal duty to ensure that it is operating a safe method of work. This is usually embodied in its company procedures, written and specifically adapted for the operations carried out by the particular company or organisation. In many cases, the procedures recommended in BS 5975:2019 are incorporated in quality manuals, best practice guidance and so on, and should be followed. Company procedures are often made binding on subcontractors to ensure that they follow the relevant company procedures and forms.

2.2. What are (or may be considered as) temporary works?

Temporary works are traditionally defined, not only in the Institution of Civil Engineers (ICE) *Conditions of Contract* but also in other forms of contract, in very general terms. The *ICE Conditions of Contract* (ICE, 1986), known as the ICE 5th Edition, defines permanent works, temporary works and the works as follows

> '*Permanent Works*' means the permanent works to be constructed, completed and maintained in accordance with the contract.
> '*Temporary Works*' means all temporary works of every kind required in or about the construction completion and maintenance of the Works.
> '*Works*' means the Permanent Works together with the Temporary Works.

While it is often tempting to provide a list of typical temporary works such as formwork, falsework, excavation support, temporary access, scaffolding, façade retention, temporary slopes, hoardings, cofferdams, trenching, temporary jetties and crane foundations, a list can never be either exhaustive or exclusive and is something that the ICE contractual definition (ICE, 1986) attempts to avoid.

Temporary works could be described as those elements of construction works which are *not* the completed permanent works. The definition of temporary works therefore includes studies of the permanent works in temporary conditions during their phased construction or deconstruction or demolition, whether additional temporary members are used or not. Temporary works provide strength, stability and stiffness as necessary, including construction phases of major bridges to foundations and support of excavations or cut slopes. They are required in civil engineering, building, refurbishment and maintenance (in fact, in all areas of construction), and can include the use of temporary structural members, permanent works in a temporary state and/or geotechnical solutions. Temporary works principles and materials are often applied to similar types of construction, for example, set design for the media (TV, etc.), open-air concerts and temporary grandstands, all of which require temporary structures.

It is evident that the term 'temporary works' is difficult to define as it covers a wide subject matter. The following is put forward to indicate the wide range of temporary works and which may be expected in construction.

The term 'temporary works' includes one or more of the following

- *Incomplete permanent works*, whether additional members are used or not, provide the necessary strength, stiffness and stability in the partially constructed (erected) condition or during deconstruction or demolition. As temporary conditions these are considered to be included in the definition of temporary works.
- *Members*: enable construction of or protect the permanent works or individuals; support the permanent works against loading for which they have not been designed; support or provide access to the permanent works (or individuals) during their construction, and might or might not remain in place at the completion of the works; and support the sides of an excavation, including by cutting to a safe angle.
- *Foundations* (including roadways) for members or machinery used for construction works.

Examples of temporary works include structures (other than the completed permanent works) such as free-standing cores of buildings, construction phases of major bridges, such as cable-stayed or balanced cantilever bridges or bridges installed by launching or sliding, supports, backpropping, tower crane foundations, piling mats (working platforms), earthworks, supports to the vertical sides of excavations and accesses.

Temporary works may be constructed from proprietary equipment or any other materials used in construction works, including steel, concrete, aluminium, timber or plastic. Some of the very first structures ever built used earth ramps for placing of heavy stones – the first temporary works!

Table 2.1 Risk considerations in temporary works

Factor	Effect	Action
Package management and subcontracting	Loss of ownership and therefore of responsibility for the temporary works by the PC. Design interfaces and responsibilities established by a commercial framework, with insufficient attention being paid to the engineering risks.	PC's DI to check the competence of the organisation before appointment.
Supplier design	The advice received will be partial in that the design will inevitably incorporate only the supplier's own equipment and there will be aspects which will either make fundamental assumptions or require separate design. Falsework foundations and top restraint are typical examples.	Introduction of a lead designer to ensure design marked 'by other' is carried out and the interfaces are complete.
Lack of understanding of temporary works and buildability by PWDs of permanent works	Temporary works design will be made more difficult and therefore will carry a higher risk. Interface and interaction between the temporary and the permanent works will not be fully addressed.	A new definition of 'temporary works' includes temporary conditions of the permanent works and the requirement that those who specify methods or construction sequences are responsible for their design check.
Lack of understanding of the importance of procedural control of temporary works by those entering the industry/profession	Temporary works, despite the recommendations of the Bragg Report to the contrary in 1975, is not embedded in civil engineering education.	Organisations should provide their employees with training which covers their role and confirmation that the PC's TWC is responsible for all temporary works on site.
The procedural controls within BS 5975:2019 are not adhered to	The controls, if complied with, reduce the risk of failure to an acceptably low level. Non-compliance, either in part or in the whole, significantly increases the risk of failure.	BS 5975:2019 has attempted to elevate the importance of temporary works to that of health and safety by requiring the DI to be directly responsible to the organisation's supervisory board.

If all the good advice is followed, the implementation of temporary works should be relatively risk free. Regrettably, that is not always the case. The principal factors that continue to put temporary works at risk are listed in Table 2.1.

2.3. The parties (who is involved)
2.3.1 The interface between parties
The principal objective is to provide temporary works that are safe in use, sustainable, do not compromise the integrity of the associated permanent works and, finally, are cost-effective to programme. This requires a managed interface between all the interested parties to take a disciplined and responsible approach. The parties should be informed by the lessons of the past and comply with current regulations and standards of good practice.

This chapter reflects UK good practice, although much of what is written will apply elsewhere. It follows the procedural controls embedded in BS 5975:2019, but makes no attempt either to list or to replicate every aspect.

Traditionally, temporary works were the responsibility of the contractor, with the engineer or the Permanent Works Designer (PWD) only required to provide specific details of the permanent works design to allow the contractor to design the temporary works. Today, CDM 2015 require greater cooperation, liaison and provision of information between all parties, including the client, the principal designer (PD), the PC and any other contractors and designers. For this reason BS 5975:2019 includes clauses on the roles and responsibilities of clients, designers (including the PD), contractors (including the PC, any PC-appointed subcontractors and client-appointed contractors other than the PC) and suppliers. (See Chapter 1 on the contractual and legal aspects.) This requirement for liaison between all parties is reinforced in BS 5975:2019, in which the role of each party is clarified in respect of who does what and how they support the PC's TWC in fulfilling their duties in relation to temporary works. When temporary works may reasonably be expected on a project, clients, and all other organisations appointed, should have a DI responsible for the preparation and implementation of a procedure outlining the organisation's duties in supporting the PC in relation to temporary works. It is also clear from CDM 2015 and BS 5975:2019 that the PC is responsible for the management and implementation of all the temporary works on a site but should accept that other contractors may plan, manage and monitor their own works. These other contractors, possibly specialist contractors, also need procedures to manage their temporary works, but must liaise with the PC's TWC at all times.

The BS 5975:2019 procedures accept that both the client and the PC can contract work to other contractors to manage their organisation's own temporary works, and are required to appoint TWCs themselves. The contractor's TWC will be responsible for coordinating and liaising with the PC's TWC for their organisation's temporary works.

2.3.2 Clients
Clients need to understand the risks associated with temporary works and ensure that the contractual and commercial framework, within which the PC has to work, does not compromise their integrity. Moreover, clients have general responsibilities under

CDM 2015, as explained in the associated HSE guidance (HSE, 2015), and are required to 'make suitable arrangements for managing a project' (so that health, safety and welfare are secured). The arrangements should include the appointment of organisations with the skills, knowledge, experience and capability to manage health and safety risks (HSE, 2015).

2.3.3 Principal designer (PD)

The PD has a duty to plan, manage and monitor the pre-construction phase of a project, to share with the PC information which may affect matters of health and safety during construction, and to ensure that all designers comply with their duties under CDM 2015 (see Regulation 11; UK Government, 2015) in providing information about the design, construction or maintenance of the structure to assist the client, other designers and contractors. This means that the PWD should consider the buildability of any temporary works required and communicate clearly the residual risks, including design assumptions, to the designer of the temporary works. This is clearly indicated as an essential requirement when considering the failures identified in the introduction to this chapter for the West Gate Bridge collapse.

The PD must also liaise with the PC for the duration of the PD's appointment, and share information relevant to the planning, management and monitoring of the construction phase. This means that the PD must ensure that clients and the PWDs share information on the design, including residual risks, to allow the PC, and their Temporary Works Designer (TWD), to carry out their duties accordingly. It is considered that this would include the sharing of computer models of the design and also building information modelling (BIM) (see Chapter 31), where used, to allow the TWD to carry out the necessary checks on the permanent works where the construction method varies from that envisaged by the PWD.

2.3.4 Principal contractor (PC)

During the construction phase the PC is the key organisation in relation to temporary works. The PC should plan, manage and monitor the works and organise cooperation between contractors, including both PC-appointed subcontractors and client-appointed contractors. In relation to temporary works this is taken to mean that the PC's TWC is responsible for managing the PC's temporary works and managing the interfaces between one contractor's (including PC-appointed subcontractors) temporary works and a second organisation's temporary works. CDM 2015 permits competent contractors and subcontractors to plan, manage and monitor (coordinate) their own works, including temporary works.

BS 5975:2019 takes account of CDM 2015 by stating that the PC should check that the subcontractor or contractor is competent, before appointment, and then allow them to coordinate their own temporary works subject to ongoing checks that procedures are being implemented correctly.

As noted above, the PC should also be obtaining relevant information from the PWDs and liaising with the PD to ensure that the temporary works are being managed without detriment to the safety of the workforce or the public.

2.3.5 Temporary works designer (TWD)

Considering the definition of temporary works in Section 2.2 and the issues identified by the investigation into the collapse of the West Gate Bridge (Barber *et al.*, 1971), it is evident that the PWD will, at times, be a designer of temporary works, and should carry out the necessary designs and provide the required output as would be required of a TWD, such as drawings, including sequencing drawings or methodology, and loading information. The PWD is also expected to provide the necessary analytical model to assist the TWD to carry out the necessary checks on the permanent works should the contractor wish to use a different method or sequence of construction than that envisaged by the PWD.

The TWD will most likely be from a contractor, subcontractor or specialist design organisation. The TWD may utilise the services of the proprietary equipment supplier's staff, while retaining overall responsibility for design of the whole temporary works (see 'lead designer' below). Today the design output for the temporary works has moved on from a few hand-drawn sketches on sheet(s) of A4 paper to computer-aided drawings, 3D and 4D modelling, and refined specifications and outline methodology which can often be linked to the output required for the permanent works. Recent advances in hand-held devices allow simple 3D sketches to be prepared, often from the design output, as 'toolbox talk' material for communicating ideas to operatives. This is discussed in more detail in Chapter 31. This improved design output should assist the TWC, and the site team, in implementing the temporary works in a safe manner. It should be noted that the design output does not include calculations.

2.3.6 Lead designer (temporary works)

In temporary works schemes today there are frequently two or more designers, each of whom have responsibility for the design of part of the whole temporary works scheme. Two examples are: (*a*) when a supplier designs falsework which is assumed fixed at the head (known as 'top restrained') but excludes consideration of the additional load case of lateral load on the permanent works from which the falsework derives its lateral stability; and (*b*) where the proprietary supplier has not designed the foundations to the falsework scheme. It is evident that the supplier's designer should provide the direction and magnitude of these two load cases in order that another designer can carry out the necessary designs. The second designer is now known to be the lead designer and has the responsibility not only to design the foundations and check the permanent works can sustain the additional lateral loading, but also to check the derivation of these loads to ensure there is no interface that has not been considered in the design of the temporary works scheme. The supplier's designer should issue a design certificate to confirm that reasonable professional skill and care has been used in the design of the falsework indicated in the design output (drawings, methodology, specification of the falsework, etc.), and the lead designer signs on behalf of the contractor that the whole scheme (falsework, foundations and head restraint) has been designed using reasonable professional skill and care, referencing the supplier's certificate along with any additional design output such as head restraint fixings and foundation details.

The idea of having more than one design certificate for one temporary works scheme was first discussed in the paper 'Improving management controls in the launching of bridges' by Marples and Richings (2014). For the particular scheme discussed there, three certificates were used, one for each section of the design, and these were summarised in one overarching certificate by the PC's TWD.

1. A certificate for the temporary nose, tail and kentledge and the effects on the permanent works – issued by the PC's TWD.
2. A certificate for the hydraulic supports and movement of self-propelled multi-axle transporters based on the output covered in certificate 1, and the design by the specialist subcontractor's TWD.
3. A certificate for the foundations (running surface), based on the output from certificate 2.
4. An overall certificate summarising the three sets of design output certified in certificates 1–3 by the PC's lead designer.

In this approach, which is covered in BS 5975:2019, the final certificate is signed by an individual known as the 'lead designer'.

2.4. The management controls (the who)

The people involved in the procedural controls recommended in BS 5975:2019 are discussed below.

2.4.1 Designated individual (DI)

An organisation which has duties in relation to temporary works, including temporary conditions, should appoint a DI to be responsible for the establishment and maintenance of a procedure for the temporary works that the organisation undertakes.

- *Contracting organisations*: for contractors and suppliers the DI is envisaged as being the company chief engineer or operations director, or someone of similar experience and authority. Where small contractors or builders are involved in temporary works, the DI would be the person responsible for the technical activities of that company.
- *Clients*: the DI role is likely to be filled by a senior individual who has an understanding of the organisation's requirements in relation to the construction projects they undertake, and that person should ensure that the technical obligations and management procedures are fulfilled.
- *Designers*: the DI role is likely to be filled by the senior designer within the organisation.

The BS 5975:2008 + A1:2011 did not restrict the definition of an organisation to contractors, and it was therefore arguable that authorities, government agencies, consulting engineers, clients and suppliers should have their own procedures, reflecting the requirements of the standard and regulations in respect of their activities. As few organisations, other than contractors, had implemented such procedures, BS 5975:2019 has clarified the requirements placed on the other organisations involved in construction projects. It goes even further as it also requires the DI to be a senior member of, or directly responsible to a member of, the organisation's main or supervisory board.

2.4.2 Temporary works coordinator (TWC)

The crucial control is the appointment, for each site, of a TWC. The TWC, who is to be appointed in writing by the DI, is the named person responsible for the safe and timely management of the temporary works on site. It may also be appropriate for the DI to appoint a deputy TWC to control the works when the TWC is absent for any reason. The key point is that this deputy has both the authority and responsibility to act as the TWC when and if the TWC is absent from site for any reason.

BS 5975:2019 makes it clear that, for a given site, the PC should appoint a TWC, known as the PC's TWC. Where the subcontractor or contractor is deemed competent, they may appoint a TWC to coordinate their own temporary works, but the PC's TWC has the responsibility to coordinate the interfaces between contractors and to ensure the contractors are implementing their procedures correctly.

It is important to realise that the TWC manages the process, but should be satisfied that others, on whom they rely, have carried out their duties. This may be by receipt of design and design-check certificates from the TWD or lead TWD. Also, it may be receipt of a design brief from a subcontractor's TWC in order to check the interfaces at the extremities of the subcontractor's work area. The PC's TWC may just accept the contractor or subcontractor TWC's documentation (design brief, design and design-check certificates, and permits to load and unload) if the contractor or subcontractor is working in a self-contained area. It should be noted that the PC's DI is responsible for confirming that the contractor's or subcontractor's procedure is accepted, and the contractor or subcontractor may plan, manage and monitor (i.e. coordinate) their own temporary works within their own sphere of influence. The PC's TWC remains responsible for ensuring they carry out their duties by receiving copies of certificates and carrying out random checks where appropriate. The PC's TWC is always responsible for coordinating temporary works at interfaces between contractors or subcontractors and other contractors, including the PC.

It is only on the TWC's authority that the temporary works are either loaded or unloaded. The most senior site manager, usually a contractor's project manager or site agent on smaller contracts, therefore has to be aware that the TWC's role is critical to the progress of the works and that the TWC is there to protect, in terms of safe working practice, the operations and operatives on site. The project manager must therefore support, and moreover be seen to support, the TWC.

Interestingly, it is the project manager's responsibility to manage the erection of the temporary works in the same way that it is the project manager's responsibility to manage the permanent works. The TWC's role of independence and responsibility for permits to load is thus deliberately separated from the role and pressures of production, therefore minimising the risk of being compromised. In the event that the TWC believes that the support is lacking, that their role is being compromised in some way or that other duties are preventing the proper execution of their role, it is essential that the DI is informed. However, this might not be possible on projects that have few or no engineering staff, where the TWC has responsibility for checking both the temporary works and the progress of the construction works. In such cases particular care is needed to ensure that decisions are not compromised by commercial or other pressures, and that the TWC can seek assistance from the DI. This confirms the earlier point that the DI has to be someone in the organisation with sufficient authority and experience to ensure that the temporary works are not compromised.

The Bragg Report (Bragg, 1975) suggested that the TWC should be a chartered engineer but, recognising that all building projects should have a TWC, BS 5975:2008 + A1:2011 and BS 5975:2019 confirm the TWC role as a management function and that it is not essential for the TWC to have academic or professional qualifications. The competence of the TWC is defined only in general terms in BS 5975:2019, requiring the individual to have the necessary knowledge, skills and experience. Others have tried to expand on the general requirements for a TWC. For example, key attributes of the TWC have been proposed in an ICE discussion paper on the role and competence of the TWC (Marples, 2011). Ideally the TWC should

- have the ability to read and understand the requirements of drawings and specifications
- have the ability to plan and manage people and resources
- have a sound knowledge of legislation, hazards and safe systems of work, and manage health and safety within their area of responsibility
- understand the basic philosophy of procedural control of temporary works as stated in BS 5975:2019
- understand and be able to implement their organisation's temporary works procedures
- use independent judgement and the ability to identify the limits of their personal and the team's knowledge and skills
- have interpersonal skills and the ability to communicate with others at all levels, and the ability to discuss ideas and plans competently and with confidence
- have experience and knowledge relevant to the complexity of the project.

The TWC is a central role in the implementation of temporary works schemes, and it is important that the TWC is a good communicator and conscientiously carries out the duties rather than is capable of designing temporary works.

It is made clear in BS 5975:2019 that the role of TWC does not preclude them from carrying out design or design checking, but this requires a separate specific appointment is made in relation to design by the DI. Obviously any task carried out by the TWC will need to be checked by an independent individual (the TWC cannot design and check their own work). The DI will inevitably apply limits which are defined by the individual's competence in relation to design of temporary works.

As already stated, where contractors other than the PC are contracted to manage temporary works, BS 5975:2019 recommends they appoint a TWC. The training, experience and attributes will be similar to those of a TWC as stated in this chapter.

2.4.3 Temporary works supervisor (TWS)

The concept of a TWS was first introduced in 2011, to recognise that the appointed TWC may not have the capacity to be everywhere – they may be responsible for several smaller sites or for a large complex site that runs over several kilometres. It was never intended that a PC would allow a TWS to be appointed by a subcontractor and be responsible for the coordination of their temporary works.

On both large and small contracts, the PC may find it advantageous for one or more TWSs to be nominated to act as a point of reference and to handle the day-to-day temporary works. This is particularly relevant to organisations operating many small sites, such as utility companies, where small groups of operatives are regularly carrying out routine temporary works, such as trenching, on a daily basis; each group would have a TWS reporting to a central TWC. Any TWS is therefore technically responsible to the TWC on all relevant matters; although on certain sites the TWS may be permitted to sign permits, this should only occur when specifically authorised, and obviously only within the TWS's scope of work and experience.

2.5. Principal activities of the TWC (how temporary works are managed)

The principal activities of the TWC, which are fully defined in BS 5975:2019, are to ensure that

- a register is established and maintained
- a design brief is prepared, in full consultation with all the interested parties
- residual design risks identified by the designer of the permanent works are included in the brief and considered by the designer of the temporary works
- the temporary works are properly designed
- an independent design check is undertaken, working from first principles, not merely an arithmetic check (Note that the designer's calculations are not part of the design output and should, therefore, not be seen by the design checker.)
- the temporary works are in accordance with the checked drawings and sketches and/or design output, and, if so, ensure that a permit to load (bring into use) is issued
- a permit to strike and unload (take out of use) is issued when it can be demonstrated that the permanent structure has gained adequate strength and/or stability.

Each of these key activities is discussed in the following sections.

2.5.1 The temporary works register

The register is the important control document for temporary works on a site. The register should have been started during the tender period, even if the temporary works were included in subcontract packages, and should be updated and maintained throughout a contract. This ensures that the temporary works are properly identified and managed, whatever the method of procurement. The register may include items that, through methods of working, prove not to be required; it is infinitely preferable to have a register that has a number of items which are not required rather than one from which items are missing.

The temporary works register should be a live register that is distributed regularly by the TWC. Moreover, it can be used to demonstrate to interested parties (including Health and Safety Executive inspectors and quality assurance auditors) that the site and the temporary works are being managed safely and properly.

The contractor or subcontractor coordinating their own temporary works should also produce a register. This should be in the same format as the PC's register and should be provided to the PC's TWC at least monthly, but after each update by the relevant TWC.

2.5.2 The design brief

This is probably the most important stage in the whole process; ironically, this is the stage that is probably the most frequently missed, generally by the inclusion of the temporary works packages in subcontracts. 'Why bother with this stage?' or 'We are using a supplier and the representative takes down the details so we don't need one' are common excuses. The first key point is that the designer (and the design checker) are generally remote from site and are therefore at a disadvantage, not being privy to the TWC's detailed knowledge of the requirements and site conditions. The second key point is that having a design brief ensures that the TWC and the site team think the problem through and discuss it with all the interested parties, in particular verifying that the actual site conditions are as assumed. For example, high-level overhead power cables crossing a site are rarely shown on the site plan, yet they affect the type of crane that can be used and will limit the size, handling and location of the temporary works. Checklists are an invaluable starter.

The brief should include the relevant section of the programme, any materials that are available, preferred methods of working and the key information from the designer of the permanent works (e.g. relevant borehole information adjacent to the temporary works, design risks and assumptions, or backpropping requirements or maximum loadings which can be imposed on a suspended slab). It should also define the limits of responsibilities. This is best exemplified by proprietary supplier's designs for falsework, which will, almost without exception, exclude the necessary foundation design and make fundamental assumptions about the ability of the permanent works to provide top restraint. In both cases, it is essential that the TWC provides relevant information to a lead designer and ensures that the brief identifies and resolves these and similar points fully.

It should be noted that TWCs and site personnel who make decisions about methods of working or materials to be used are taking on the role and responsibilities of the designer under CDM 2015 (UK Government, 2015). They may have a greater potential to affect the safety in use of a particular temporary works scheme than the designer who proves, in analytical terms, a concept that has already been decided and included in the brief. This principle is reinforced by a legal maxim in respect of design and build, which predates and even predicted CDM 2015 to the effect that 'he who decides, designs'. This maxim is salutary.

Finally, the brief should set down the required level of information (or output) to be provided by the designer, which should comprise appropriate layouts (such as sketches, 3D graphics and/or clear working drawings) with the particular design risks and design assumptions clearly communicated to those who need to know. This principle applies equally to a standard solution, for which the information source should be available, and the limits of use and the design risks of such a solution should be clearly communicated to the TWC. Interestingly, the design risks for a standard solution will in all probability be more extensive and restrictive than for a specific bespoke design of similar character.

While all this may appear obvious, there are still proprietary suppliers whose design output is limited to a computer printout. It is therefore worth reflecting on the following points.

- Designs of temporary works and permanent works, and the operatives and public who depend on their safe execution, depend on the same engineering principles.
- The project manager and the site team expect to be provided with working drawings and specifications for the permanent works. Is it not reasonable for the same team to be provided with equivalent drawings or sketches and specifications for the temporary works?
- The person undertaking the design check is required to check from first principles, given that the check is not a simple arithmetical check or a check of calculations (ICE/SCOSS, 2002). This therefore presumes that the checker should be working from drawings/sketches.
- The TWC is expected to ensure that the temporary works are inspected before loading. In so doing, it is not unreasonable to assume that the TWC requires the relevant layout and information; this may be from sketches, brochures or a number of drawings (which have been subject to a design check) against which the check prior to loading is carried out.

2.5.3 Design check

The design can be carried out in a number of ways: by individuals on site, by use of a standard solution, by company temporary works offices, by suppliers of proprietary equipment, by specialist consulting engineers or by subcontractors. Each of these will attract different risks, which should be identified in the design brief.

All temporary works designs must be independently checked. The degree of checking should be related to the scale of the temporary works; a simple scheme may be checked by someone in the same office, whereas a more complex scheme might have to be checked by an outside organisation. The level of check should not be confused with, or increased because of, the execution risk (see site controls below, Section 2.5.5). The four recommended categories of design checks for temporary works in BS 5975:2019 are listed in simplified form in Table 2.2.

Table 2.2 Categories of temporary work design check

Design check category	Scope	Independence
0	Restricted to standard solutions	Site issue – by another member of site team or design team
1	Simple design: includes falsework, which does not assume top restraint	By another member of design team
2	More complex design: excavations, structural steelwork, foundations and falsework that assumes top restraint and so on	By individual not involved in design and not consulted by the TWD
3	Complex or innovative design	By another organisation

Data taken from Table 2 BS 5975:2019.
Top-restrained falsework is the method by which the temporary structure is stabilised for lateral movement by connection to external restraints at its head (e.g. to the permanent works of columns or adjacent walls), provided that these elements have been designed to provide the required restraint.

It is important that the principle of a 'higher' level category of check being required when the temporary works are more complex is not misunderstood. Any design check should be carried out, by the checker, to a degree of rigour that the checker considers appropriate to enable the design-check certificate to be signed. The degree of check should not vary depending on whether it is designated category 0, 1, 2 or 3. The checking process may be more straightforward for items that have a lower designated category, but the checker is only ever carrying out a check that is sufficient to enable them to sign, whatever the category. The significance of the various check categories is only that they determine the degree of independence that the checker has in relation to the designer. Some items of temporary works design therefore require greater independence than others for the design check, whether because of the need to involve the thinking or experience of a second organisation or because of contractual requirements or constraints. The risk is that some checkers may consider that a lesser check is acceptable for lower category items – this is incorrect. The checker is always responsible for carrying out the check to the detail necessary to allow them to confirm the adequacy of the temporary works as designed.

When using 'standard solutions' it is *not* a requirement for the originator (a contractor, supplier or trade organisation) to actually issue separate design-check certificates to verify the original calculations for their 'standard solution'. The originator of the solution took on that responsibility when the standard solution was published. There is, however, a requirement to confirm that the user of such a 'standard solution' on the project has it checked independently to confirm they have used the correct data, the right page and that it is actually relevant for the work in question and within the parameters set for that solution (i.e. the conditions are as intended) (see the category 0 check in Table 2.2).

Other forms of contract may have other requirements for the certification of the temporary works, particularly those from Highways England or Network Rail. Temporary works on Highways England contracts which have a public interface will be subject to an Approval in Principle (AIP or equivalent), followed by the associated design and check certificates.

On Network Rail contracts, a Form 3 (F0003) (formerly called Form C) is required for all temporary works that affect the safety of the railways. This is a Network Rail form covering the design, design check, Network Rail approval and, if applicable, issue by the subcontractor with approval of the zone civil engineer.

It should always be remembered that checks that are required by, or undertaken by, other organisations should in no way be considered as an alternative to or as a reason to reduce or omit any of the checking stages that the organisation responsible for the temporary works is required to undertake.

The TWC should provide the checker with the design brief, the relevant layouts and information (drawings and sketches), and the residual design risks identified by the designer of the permanent works; calculations should not be provided. The checker's role is to carry out an independent check of the concept, working from first principles (which is why calculations are not provided). With the increasing use of computers and other design aids, the importance of simple rule-of-thumb checks should not be overlooked. The check is therefore not an arithmetic check of the original designer's calculations, due to the risk that a fundamental error may be repeated by the checker. The checker must issue a check certificate, which lists all the documents and drawings as necessary with their revision status. Indeed, without such a list, the design-check certificate is of questionable value.

Although BS 5975:2019 recommends the four design-check categories shown in Table 2.2, individual organisations may have different views on the scale of works and introduce classifications for the risk level as 'simple' versus 'complex' or 'minor' versus 'major', each of which can have different meanings. This is discussed in Section 2.5.5, and Table 2.3 is relevant. Often the classification of temporary works is defined in other terms, but the essential philosophy remains that all temporary works designs are checked from first principles. It is important not to confuse 'classification of risk' with 'category of design check', and to understand that managing temporary works allows for design-check categories, as in Table 2.2. This is because the same arrangement of temporary works could be located in different parts of a project and be subject to different levels of risk (e.g. in a field compared to by a railway line). Note that use of 'unclassified' can never be used as a scale because, fundamentally, all items of temporary works have to be given a check category (even if it is only 'category 0').

2.5.4 On-site supervision and control

As previously stated, the site project manager is the person responsible for the proper execution of the permanent works on site by the site team. The TWC has to ensure that the project manager and the team are provided with clear drawings and other

Management of temporary works

Table 2.3 Execution risk classes for temporary works

Implementation risk class	Risk	Permits required	Other control measures
Very low	No identified practical mode of failure. No impact if failure occurs.	N/A	Control via RAMS environmental control system. Inspection by site team, not necessarily recorded on the temporary works register and may not require a design brief.
Low	Minor structures with high levels of robustness. Very experienced workforce. Failure is entirely within the site and of low impact. Inconvenient, but personal injury unlikely.	Permits can be signed by a TWC or an authorised TWS.	Follow company procedures, including inspection and test plan.
Medium	Conventional structures. Conventional construction methods. Relatively experienced workforce. Failure would be major, potentially involving injury, fatality or significant economic loss. Would not initiate secondary events.	Permits can be signed by the PC's TWC, or an authorised contractor's TWC.	Follow company procedures, including inspection and test plan.
High	Schemes with dependency on critical structural details, with little or no redundancy, or with stability reliant on critical elements. Inexperienced workforce. Unfamiliar processes or equipment. Failure would be catastrophic in its own right, or if minor might initiate a secondary or chain reaction of major or catastrophic events.	Permits signed by the PC's TWC or an authorised deputy only.	Follow company procedures, including inspection and test plan. PC's DI to ensure the scheme is reviewed. Possible hazard and operability (HAZOP) study or peer review.

Data taken from Table 1 of BS 5975:2019.
Note: The PC's TWC determines the signatory for each permit applicable to the item of temporary works.

information about the temporary works design. This distinction is deliberate. In addition, any TWS (if appointed) will need to be kept informed.

2.5.5 On-site checking

The execution risk, covered in Table 1 of BS 5975:2019 and repeated here as Table 2.3, may depend on the location and the expertise of the team members or the criticality of individual elements of the temporary works. This would require the site controls, as opposed to the design and checking of the design, to be more rigorous and be carried out by more experienced individuals. For example, a major contractor, with many years' experience and competent operatives, would consider a 4 m trench as routine work and relatively low risk in terms of the team's competence and compliance with good practice and procedures. A small house builder, rarely digging deeper than 1 m and whose team has limited experience of significant temporary works, would, however, consider 4 m as high risk. A scheme with critical structural details and little redundancy would be a high risk in relation to execution risk, and would require these members/details to be checked carefully by someone with knowledge and experience of this type of structure. The higher the execution risk then the more experienced the individual who should sign the permits.

In procedural terms, the TWC is responsible for ensuring that an inspection of the completed temporary works is carried out when, in the opinion of the site project manager, they are complete. On large projects the TWC is on site and is part of the site team. It therefore follows that when any substantial temporary works are erected over a period of time the PC's TWC, contractor's or subcontractor's TWC and any relevant TWS should be making regular or informal inspections as the work progresses and should be notifying the site team of any issues observed. As an example, it would be sensible to carry out an inspection of the foundations before falsework is erected. None of this will compromise the TWC's independence or the final inspection. As already discussed, a TWS may be delegated by the TWC to carry out the day-to-day operations on site, which may include the checking of completed temporary works. If the TWS observes differences between the erected temporary works and the checked drawings and sketches, then the TWC should be informed. It may be necessary for the TWC to refer back to the TWD for further guidance.

2.5.6 Permit to load

Finally, and only when the TWC is in all respects satisfied, the TWC (or TWS where authorised) signs and issues to the project manager a permit to load, which will generally be limited in time. If the TWS is authorised to sign the permit, the TWC must also be included in the process as the single point of authority and responsibility. It would be incorrect to issue a permit to load falsework some weeks in advance of a pour date, given the inclination of site teams to use misplaced initiative and 'borrow' key components for another element of temporary works. Falsework permits are therefore normally issued the day before and are valid for the following day only. The actions of anyone on the site team who wilfully loaded temporary works in full knowledge that the permit had been withheld or had not been issued would normally be treated as a disciplinary matter.

The TWC can inevitably come under immense commercial pressure from the site team to sign and avoid delay and cost. It is in these situations that the TWC has to have sufficient strength of character and the confidence and full support of the project manager. Given the consequence of failure, the presumption is that the TWC has to withhold permission if there is sufficient doubt. The TWC and TWS should always be able to rely on support and guidance from their DI in the event that they are encountering difficulties over the temporary works with any member of the site team.

Checklists for some temporary works have been issued; refer to the available guidance on falsework (Concrete Society, 2014a), formwork (Concrete Society, 2014b) and trenching (Irvine and Smith, 2001).

2.5.7 Permit to unload

This is the final stage. While the loading of the temporary works is formally controlled, the unloading or transfer of load to the permanent works must also be assessed and controlled. This stage can generate loadings of the temporary works (and permanent works) that are significantly different from those imposed when they were loaded. A particular case would be the transfer of load through multi-storey structures when backpropping. (See Chapter 28.) It is therefore clear that unloading requires communication and cooperation between the TWC and the designers of the permanent and temporary works, as required by CDM 2015.

For concrete structures, the value and method of assessing the concrete strength at the time of striking should be agreed beforehand with the PWD and carefully controlled on site. Incorrect sequencing and/or order of removal of supports can also seriously compromise the works under construction, so the permit to unload should define any sequence required. Modern methods of strength assessment are now available (see Chapter 24 on soffit formwork).

This section has concentrated on ensuring that the temporary works are safe, prior to unloading or taking out the temporary works, but is the permanent structure itself a stable structure when the temporary works are removed? Once again careful coordination and communication between the PC's TWC and the PWD are essential.

Structures that combine temporary works with prestressing of the permanent works or multi-storey floor slab construction with a requirement for backpropping need careful planning and control. The exact order of removal of supports can affect the load transfer, so the permit to unload should state any required sequence or procedures for striking or unloading the structure.

2.6. Summary – answers to the questions why, what, who and how temporary works are managed

This chapter has provided background information on why the management of temporary works should be controlled, defined temporary works, and outlined how temporary works should be managed and controlled and who by. It has not sought to repeat fully the detailed procedural points in BS 5975:2019. The TWC's role is to protect the team, including the operatives, from errors, omissions, misunderstandings and perhaps even the pressures of the programme and cost. The TWC can be assisted by a TWS, but remains the responsible person.

Failures of temporary works can all too easily cause serious injury or fatalities, with the associated distress to families and, in all probability, legal action. Contracts will incur delay and cost that is difficult to recover. The industry relies on those who work with, within, on or under temporary works, and therefore owes it to them to follow the established procedures. In the event of failure and an enquiry or legal action, individuals must be able to demonstrate that established practice rules have been followed.

Finally, while it can never be an absolute rule, failures are generally caused by a series or combination of factors. It follows that such failures can best be prevented by the consistent application of a series of checks and balances; this is why temporary works are controlled. It is worth repeating the words given in clause 6.1.1.7 of BS 5975:2019

> One of the main aims of any method of work should be to minimise the chance of errors being made and to maximise the chance of errors being discovered if they are made!

REFERENCES

Barber EHE, Bull FB and Shirley-Smith H (1971) *Report of Royal Commission into the Failure of West Gate Bridge*. CH Rixon, Government Printer, Melbourne, Australia.

Bragg SL (1975) *Final Report of the Advisory Committee on Falsework*. Department of Employment and Department of the Environment. HMSO, London, UK.

BSI (British Standards Institution) (2008) BS EN 12812:2008. Falsework – Performance requirements and general design. BSI, London, UK.

BSI (2011) BS 5975:2008 + A1:2011. Code of practice for temporary works procedures and the permissible stress design of falsework. BSI, London, UK.

BSI (2019) BS 5975:2019. Code of practice for temporary works procedures and the permissible stress design of falsework. BSI, London, UK.

Burrow M, Clark L, Pallett P, Ward R and Thomas D (2005) Falsework verticality: leaning towards danger? *Proceedings of the Institution of Civil Engineers – Civil Engineering* **158(1)**: 41–48.

Concrete Society (2014a) *Checklist for Erecting and Dismantling Falsework*. Concrete Society, Crowthorne, UK, PUB CS123.

Concrete Society (2014b) *Checklist for the Assembly, Use and Striking of Formwork*. Concrete Society, Crowthorne, UK, PUB CS144.

HSE (Health and Safety Executive) (2001) *Investigation into Aspects of Falsework*. HSE Books, Sudbury, UK, HSE Contract Research Report 394/2001. See www.hse.gov.uk/research/crr_htm/2001/crr01394.htm (accessed 01/08/2018).

HSE (2015) *Managing Health and Safety in Construction. Construction (Design and Management) Regulations 2015. Guidance on Regulations*. HSE Books, Sudbury, UK, Publication L153.

ICE (Institution of Civil Engineers) (1986) *ICE Conditions of Contract*, 5th edn. Thomas Telford, London, UK.

ICE/SCOSS (Institution of Civil Engineers/Standing Committee on Structural Safety) (2002) *Falsework: Full Circle? Report on Temporary Works.* ICE/SCOSS, London, UK, Topic Paper SC/T/02/01.

Irvine DJ and Smith RJH (2001) *Trenching Practice*, 2nd edn. Construction Industry Research and Information Association (CIRIA), London, UK, Report 97.

Marples F (2011) The role and competence of temporary works coordinators. *Proceedings of the Institution of Civil Engineers – Civil Engineering* **164(2)**: 53.

Marples F and Richings JD (2014) Improving management controls for the launching of bridges. *Proceedings of the Institution of Civil Engineers – Bridge Engineering* **167(2)**: 131–142.

UK Government (2015) Construction (Design and Management) Regulations 2015. Statutory Instrument 2015/15. The Stationery Office, London, UK.

Useful web addresses

CONSTRUCT (Concrete Structures Group): http://www.construct.org.uk (accessed 01/08/2018).

Health and Safety Executive (HSE) – books: http://books.hse.gov.uk (accessed 01/08/2018).

Temporary Works: http://www.temporaryworks.info (accessed 01/08/2018).

Temporary Works Forum: http://www.twforum.org.uk (accessed 01/08/2018).

Temporary Works, Second edition

Pallett, Peter F and Filip, Ray
ISBN 978-0-7277-6338-9
https://doi.org/10.1680/twse.63389.037
ICE Publishing: All rights reserved

Chapter 3
Site compounds and set-up

Geoff Miles
Retired

Edited by: **Keith Kirwan**
Construction Manager, Network Rail

Typically, a construction project manager may be responsible for completing the design and build of a new £40 million facility over a timescale of 2.5 years. In terms of their role and responsibilities, that makes them the equivalent of the CEO of a small to medium-sized 'start-up' enterprise, with revenue of £16 million in the first year. In most other industries, manufacturers already have their permanent, sophisticated, production facilities fully operational before they take on new orders from clients. However, upon award of a new contract, and within just a few weeks, builders and civil engineers have to first create their 'temporary factory and welfare facilities' on a distant site before they can start their client's work. They often establish (or set up) these concurrently with the commencement of delivery of the permanent works. Set-up requires management of a very large number of trades. This chapter provides advice on the range of items that may need to be designed, costed and included in the project preliminaries cost plan.

3.1. Introduction

Apart from the obvious operational needs to provide production facilities, the following legislation imposes obligations on all parties involved when working in the UK: the Health and Safety at Work etc. Act 1974; the Health and Safety at Work etc. Act 1974 (Application outside Great Britain) Order 2013; the Management of Health and Safety at Work Regulations 1999; the Construction (Design and Management) Regulations 2015 (CDM 2015); the Workplace (Health, Safety and Welfare) Regulations 1992; the Work at Height Regulations 2007; and the Control of Asbestos Regulations 2012.

Under CDM 2015, clients, principal designers, principal contractors, designers and contractors all have specific duties.

The client must make sure that the construction phase does not start unless there are suitable welfare facilities and a construction-phase plan in place. The principal designer has overall responsibility for CDM 2015 compliance during the pre-construction phase.

The principal contractor must plan, manage and monitor the construction phase in liaison with contractors; prepare, develop and implement a written plan and site rules (the initial plan must be completed before the construction phase begins); ensure that suitable welfare facilities are provided from the start and maintained throughout the construction phase; and secure the site.

The contractor must plan, manage and monitor their work and that of their workers and ensure there are adequate welfare facilities for their workers. Where contractors are involved in design work, including for temporary works, they also have duties as designers. When working overseas or in other jurisdictions, advice should be sought regarding any local requirements and obligations. Otherwise, the UK standard is a good model to use to provide a safe and healthy working environment for employees in any location, at home or overseas.

The best time to plan and quantify the detailed compound, offices and welfare requirements is during the early contractor involvement or tender planning stage. It is important to have a detailed solution in place in order to estimate the preliminary cost budget accurately.

Table 3.1 provides a generic list of what may need to be planned, designed and costed. This should be considered carefully and edited to exclude any unnecessary items. Equally, any missing site-specific items should be added.

Table 3.1 Generic list of items possibly requiring attention

Subject	Description	Design output required
Land	Environmental survey: flood risk, tidal impact, earthquakes, subsidence landslides, etc.	Report
	Land survey of site	Survey drawing 1 : 100
	Geotechnical survey of soils	Soils report
	Phasing and location to suit permanent works	Drawings and programme
	Existing utility services survey	CAT survey, CCTV survey, drawings and utility correspondence
Access	Highways entrance/exit and diversions	Design, layout drawing 1 : 100, details 1 : 20 and 1 : 10, and specification
	Wharf/harbour/jetty	Design, layout drawing 1 : 100, details 1 : 20 and 1 : 10, and specification
	On-site haul roads and safe walking routes	Design, layout drawing 1 : 100, details 1 : 20 and 1 : 10, and specification
	Bridges (vehicle and pedestrian)	Design, specification, layout drawing 1 : 100, base (civils) details and steel fabrication drawings

Table 3.1 Continued

Subject	Description	Design output required
Access (*continued*)	Vehicle weighbridge	Design, specification, 1 : 100 layout drawing, base (civils) details and weighbridge drawings
	Wheel wash	Design, specification, 1 : 100 plant layout drawing and base (civils) details
	Airstrip and/or helipad	Design, layout drawing 1 : 100, details 1 : 20 and 1 : 10, and specification
Construction logistics plan	Public access, diversions	Design, layout drawing 1 : 100, details 1 : 20 and 1 : 10, and specification
	Traffic management	Design, layout drawing 1 : 100, details 1 : 20 and 1 : 10, and specification
	Rail branch/siding/platform	Design, layout drawing 1 : 100, details 1 : 20 and 1 : 10, and specification
	Highways licences	Application form and letter of submission
Water	Desalination plant	Design, specification, 1 : 100 plant layout drawing and base (civils) details
	Borehole and treatment plant	Design, specification, 1 : 100 plant layout drawing and base (civils) details
	Metered water supply from existing infrastructure	Design, layout drawing 1 : 100, details 1 : 20 and 1 : 10, and specification
	Tanker/bowser hardstanding and connection point	Design, layout drawing 1 : 100, details 1 : 20 and 1 : 10, and specification
	Water storage tank, pumps and distribution pipework	Design, specification, 1 : 100 plant layout drawing and base (civils) details
	Dust suppression and wheel wash	Design, layout drawing 1 : 100, details 1 : 20 and 1 : 10, and specification
	Trenches, ductwork for buried services	Design, layout drawing 1 : 100, details 1 : 20 and 1 : 10, and specification
	Fire-fighting provisions for compound and building work in progress	Design, layout drawing 1 : 100, details 1 : 20 and 1 : 10, and specification
	Insulation and trace heating to protect pipework	Design, layout drawing 1 : 100, details 1 : 20 and 1 : 10, and specification
Electricity	Metered electricity supply from existing network	Design, layout drawing 1 : 100, details 1 : 20 and 1 : 10, and specification
	Generator and diesel tanks, refuelling logistics (emergency or permanent)	Design, specification, 1 : 100 plant layout drawing and base (civils) details
	Renewable energy: solar or wind with energy storage battery packs	Design, layout drawing 1 : 100 plant layout drawing, details 1 : 20, drawing and base (civils) details

Table 3.1 Continued

Subject	Description	Design output required
Electricity (*continued*)	Electrical distribution power and lighting including 110 V, major plant and so on	Design, layout drawing 1 : 100, details 1 : 20 and 1 : 10, and specification
	Substation/switchgear building	Design, specification, 1 : 100 layout drawing, building details and plant drawings
	Trenches and ductwork for buried services	Design, layout drawing 1 : 100, details 1 : 20 and 1 : 10, and specification
	Construction site electrical regulation testing regime	Specification and method statement
	Environmental impact assessment (for diesel spillage, etc.)	Report
	Security fencing to protect electrical equipment	Design, layout drawing 1 : 100, details 1 : 20 and 1 : 10, and specification
	Portable appliance testing (PAT) regime	Specification and method statement
Gas	Metered gas supply from existing network	Design, layout drawing 1 : 100, details 1 : 20 and 1 : 10, and specification
	Bulk liquid petroleum gas (LPG) tank	Design, specification, 1 : 100 plant layout drawing and base (civils) details
	Gas distribution pipework	Design, layout drawing 1 : 100, details 1 : 20 and 1 : 10, and specification
	Gas valve and meter building	Design, specification, 1 : 100 plant layout drawing and building details
	Trench and ductwork for buried services	Design, layout drawing 1 : 100, details 1 : 20 and 1 : 10, and specification
	Construction site gas regulation testing regime	Specification and method statement
	Environmental impact assessment (for gas leakage, etc.)	Report
	Security fencing for gas equipment	Design, layout drawing 1 : 100, details 1 : 20 and 1 : 10, and specification
	Automatic safety and shut-off systems	Design, layout drawing 1 : 100, details 1 : 20 and 1 : 10, and specification
Voice and data	Aerial/dish mast base	Design, layout drawing 1 : 100, details 1 : 20 and 1 : 10, and specification
	Satellite receiver switched link	Design, layout drawing 1 : 100, details 1 : 20 and 1 : 10, and specification
	Cooper wire, switched link	Design, layout drawing 1 : 100, details 1 : 20 and 1 : 10, and specification
	Fibre-optic, switched link	Design, layout drawing 1 : 100, details 1 : 20 and 1 : 10, and specification

Site compounds and set-up

Table 3.1 Continued

Subject	Description	Design output required
Voice and data (*continued*)	Data and voice wiring network	Design, layout drawing 1:100, details 1:20 and 1:10, and specification
	Cable ducts and draw-pit layout	Design, layout drawing 1:100, details 1:20 and 1:10, and specification
	Voice and data equipment schedule	Schedule
	Internal communication system – radios	Schedule
Security	Fire alarm system	Fire alarm system schematic drawing 1:100 and specification
	Intruder alarm system	Intruder alarm system schematic drawing 1:100 and specification
	CCTV system/webcam systems	CCTV system/webcam system schematic drawing 1:100 and specification
	Biometrics (fingerprint scan or retina scanning)	Biometrics system schematic drawing 1:100 and specification
Wastewater	Outfall connection to existing sewers	Design, drawing 1:100, details 1:20 and 1:10, and specification
	Temporary foul and surface water drainage above and below ground	Design, drawing 1:100, details 1:20 and 1:10, and specification
	Sewage storage/treatment plant	Design, drawing 1:100, details 1:20 and 1:10, and specification
	Land drainage	Design, drawing 1:100, details 1:20 and 1:10, and specification
	Sewage collection and removal from site	Assessment of tanker size and number of visits
	Environmental impact assessment of wastewater	Report
Welfare	Facilities for all site personnel	General arrangement layout drawing 1:100
	Predicted manpower plan	Labour histogram produced from bar chart programme
	Canteen	Detailed room drawings 1:50
	Drying room	Detailed room drawings 1:50
	Changing room	Detailed room drawings 1:50
	Male and female toilets	Detailed room drawings 1:50
	Showers and emergency showers	Detailed room drawings 1:50
	Living accommodation on site	Detailed room drawings 1:50
	Health and first aid	Detailed room drawings 1:50
	Remote welfare facilities about site	Detailed room drawings 1:50

Table 3.1 Continued

Subject	Description	Design output required
Offices	Facilities for all management staff	General arrangement layout drawing 1:100
	Management organogram	Organisation chart
	Subcontractors' offices	Detailed room drawings 1:50
	Main contractor's offices	Detailed room drawings 1:50
	Office furniture layout	Detailed workstation and furniture layout drawing 1:50
	IT office equipment	Detailed IT, printing and office equipment layout drawing 1:50
	Filing system/room	Detailed room drawings 1:50
	Stationery store	Detailed room drawings 1:50
	Cleaners' store	Detailed room drawings 1:50
	Male and female toilets	Detailed room drawings 1:50
	Print room	Detailed room drawings 1:50
	Drawings room	Detailed room drawings 1:50
	Meeting rooms	Detailed room drawings 1:50
Gantry	For congested sites (usually inner city)	Structural steelwork design layout and fabrication drawings
	Highways licence	Application form and letter of submission
Materials	Storage buildings	Design, drawing 1:100, details 1:20 and 1:10, and specification
	Lay-down areas	Design, drawing 1:100, details 1:20 and 1:10, and specification
	Unloading plant	Plant and equipment specifications
	Unloading labour	Labour histogram
	Racking	Design, drawing 1:100, details 1:20 and 1:10, and specification
	Pallets and packing timber	Assessment of quantities required
	Tarpaulins	Assessment of quantities required
Materials handling	Crane	Schedule of lifts (hook time analysis)
	Forklift/telehandler	Assessment of loads and quantities
	Lorry	Assessment of loads and quantities
	Tractor and trailer	Assessment of loads and quantities
Materials testing	Laboratory	Design, drawing 1:100, details 1:20 and 1:10, and specification
	Sample room small items	Design, drawing 1:100, details 1:20 and 1:10, and specification
	Sample testing external storage area	Design, drawing 1:100, details 1:20 and 1:10, and specification
	Samples area external	Design, drawing 1:100, details 1:20 and 1:10, and specification

Table 3.1 Continued

Subject	Description	Design output required
Compound	Temporary soil heaps	Location and design drawing 1 : 100 and specification
	Hardstandings	Design, drawing 1 : 100, details 1 : 20 and 1 : 10, and specification
	Fencing	Design, drawing 1 : 100, details 1 : 20 and 1 : 10, and specification
	Hoarding	Design, drawing 1 : 100, details 1 : 20 and 1 : 10, and specification
	Noise screens	Design, drawing 1 : 100, details 1 : 20 and 1 : 10, and specification
	Gates	Design, drawing 1 : 100, details 1 : 20 and 1 : 10, and specification
	Vehicle entry barrier	Design, drawing 1 : 100, details 1 : 20 and 1 : 10, and specification
	Security hut	Design, drawing 1 : 100, details 1 : 20 and 1 : 10, and specification
	Electronic turnstile	Design, drawing 1 : 100, details 1 : 20 and 1 : 10, and specification
	Lighting	Design, drawing 1 : 100, details 1 : 20 and 1 : 10, and specification
	Car park and cycling racks	Design, drawing 1 : 100, details 1 : 20 and 1 : 10, and specification
	Footpaths	Design, drawing 1 : 100, details 1 : 20 and 1 : 10, and specification
	Signage	Design, drawing 1 : 100, details 1 : 20 and 1 : 10, and specification
	PA system	Design, drawing 1 : 100, details 1 : 20 and 1 : 10, and specification
	Fire stations and assembly	Layout drawing 1 : 100
	Designated smoking areas	Layout drawing 1 : 100
Waste management	Skip standing and segregation area	Design, drawing 1 : 100, details 1 : 20 and 1 : 10, and specification
	Concrete plant 'wash out' area	Design, drawing 1 : 100, details 1 : 20 and 1 : 10, and specification
	Food refuse bins and storage	Design, drawing 1 : 100, details 1 : 20 and 1 : 10, and specification
	Incinerator	Design, drawing 1 : 100, details 1 : 20 and 1 : 10, and specification
Plant maintenance	Refuelling station and fuel storage	Design, drawing 1 : 100, details 1 : 20 and 1 : 10, and specification

Table 3.1 Continued

Subject	Description	Design output required
Plant maintenance (continued)	Recharging station for battery-powered vehicles	Design, drawing 1 : 100, details 1 : 20 and 1 : 10, and specification
	Vehicle repairs and maintenance bay	Design, drawing 1 : 100, details 1 : 20 and 1 : 10, and specification
Workshop buildings	Onsite fabrication shop	Design, drawing 1 : 100, details 1 : 20 and 1 : 10, and specification
	Welding shop	Design, drawing 1 : 100, details 1 : 20 and 1 : 10, and specification
	Painting shop	Design, drawing 1 : 100, details 1 : 20 and 1 : 10, and specification
	Pre-cast concrete factory	Design, drawing 1 : 100, details 1 : 20 and 1 : 10, and specification

A detailed description of issues – including land and access; communications, energy, clean water supply and wastewater disposal; office and welfare accommodation space planning; and materials (unloading, distribution, fabrication, handling, storage and testing) – is provided in the following sections.

3.2. Land and access
3.2.1 Site visit and inspection

A visual feel for the site, its surroundings and accessibility is critical. Sufficient time should be taken to walk the site perimeter and across the central areas. Photographs should be taken or a film made; provided permissions and licences are in place, drones can used to provide an aerial view.

Access routes should also be walked, especially in remote areas away from major roads or in congested town centres where many restrictions can apply (e.g. limited access in length, width, height and axle weight). Look for overhead cables, noise and time limitations. This information should then form the basis of the construction logistics plan.

To prepare for a site visit, contact all relevant authorities and utility companies and request copies of their drawings showing all existing infrastructure (i.e. roads, sewers, water mains, electricity cables, gas mains and communications cables). During the site visit make a point of finding physical proof of items such as utility marker posts, manholes, draw pits, valve access covers, fire hydrants, substations and cable pylons or poles shown on the drawings provided. Also, identify all potential locations for the site compound(s).

3.2.2 Site surveys for topography, ground conditions (geotechnical) and environmental impact

If not already done, arrange for these surveys to be carried out as early as possible; it can take between 4 weeks and 3 months to procure, carry out and publish the survey reports. The analysis, testing, reports and recommendations on their findings are just some of the fundamental prerequisites before being able to begin designing and drawing details of the temporary compound and accommodation. The remaining prerequisites are the works programme, resource histograms, risk assessment and method statements, and general attendance and special attendance requirements of the package contractors. This information should be requested and submitted with the associated quotes as supporting documentation. All tender enquiries for the permanent work packages must therefore be sent out and quotes returned during the same period when survey reports are being prepared. This approach ensures that all the information needed is available to start detailed compound and temporary facilities designs.

3.2.3 Locating the compound

The ideal location for the compound is one that can remain in place for the full duration of the contract, and not interfere with the permanent works or compromise the working space to carry them out.

Where using renewable power such as solar or wind, consider the location that can maximise the potential of the plant (e.g. a south-facing location for solar cells).

Consideration should be paid where work is to be delivered in a residential area. To minimise disruption and noise to residents, battery storage can be used to store energy produced by generators during the day and used to power the site at night, with little to no noise disruption.

Finding the ideal location requires detailed analysis and sketch modelling of the drawings and specifications provided for the permanent works. The timing of the availability of such information depends highly on the two principal methods of procurement

- *Traditional*: the client provides a complete, fully detailed design, specification and bill of quantities for the main contractor/specialist subcontractors to price.
- *Design and build*: the client provides a schematic design and performance specification requiring the main contractor/specialist subcontractors to complete the detailed design, specification, materials take-off and cost plan.

In cases where the detailed design will not be completed pre-tender or pre-guaranteed maximum price, then the following options must be considered

- the use of a piece of land or an existing building close to the site (this removes the risk of the temporary set-up being in the way of the permanent works)
- include for relocating the set-up in the tender cost budget.

On high-security sites such as nuclear facilities, prisons and airports, it is beneficial to have two contractors' compounds. One should be located outside the secure perimeter to make people access, deliveries and storage easier to manage and control. Office space for project management, commercial and administration support staff is best placed in this compound. A second compound should be located inside the secure perimeter to provide welfare facilities for operatives and office space for the production management staff.

Figure 3.1 Site compound for a road project. (Courtesy of Costain)

Lease negotiations for land and premises can take a considerable time. This will involve securing the option to lease contracts and obtaining outline planning approval during the tender or 'early contractor involvement' period. Without these, it would be irresponsible to include assumptions about their availability for use in any tender submissions.

See Figures 3.1 and 3.2 for typical compound layouts for a road project and a small site.

3.2.4 Infrastructure link-up

In well-developed areas there are two other important matters to consider

- the proximity of utilities to connect up to and the length of time required to seek permission of use (i.e. sewers, water mains, electricity mains, gas mains, telecommunications cable network)
- access between the road network and a site for cars and heavy goods vehicles.

Established roads and utilities infrastructure may not exist in remote or underdeveloped areas and so energy and water sources, water treatment, communications and transport links have to be considered from first principles. Provision for designing and constructing the following should be included

- Energy source: power station, substation, overhead transmission cables.
- Water source: pipeline, borehole, river or sea water intake, dam with a reservoir.

Figure 3.2 Compound layout for a small site. (Courtesy of Costain)

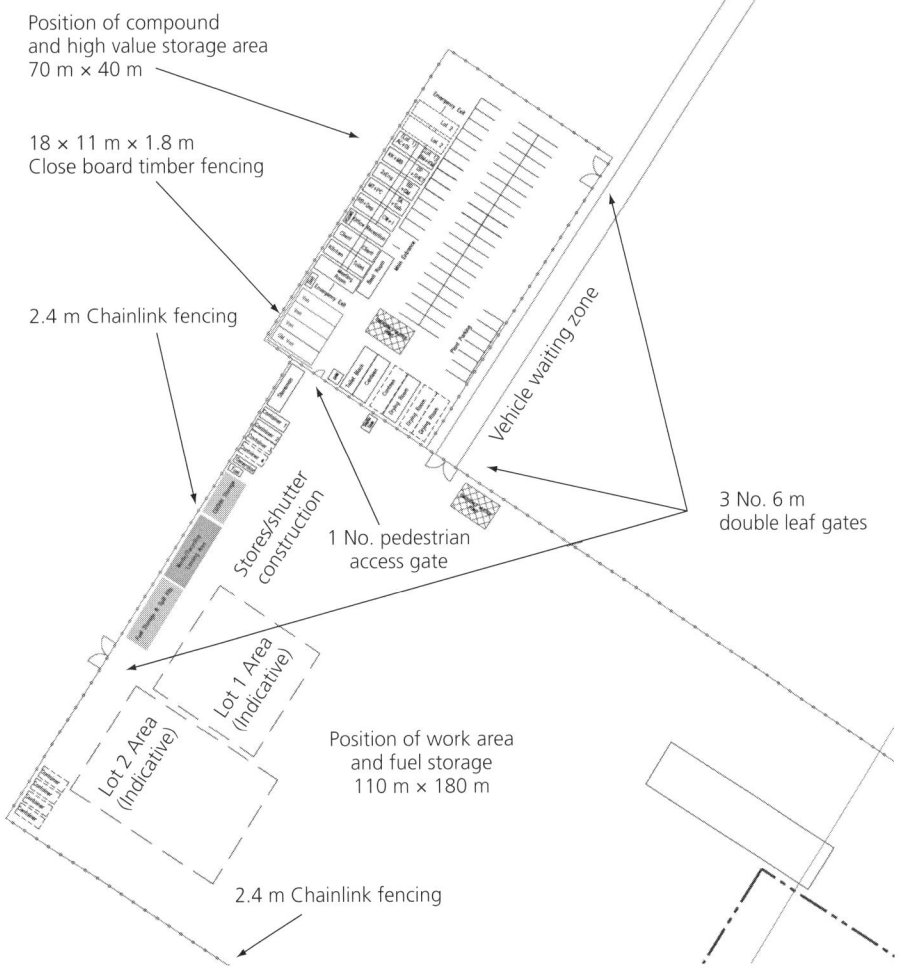

- Water treatment: purification plant for clean water, sewage treatment plant for wastewater.
- Communications: overhead cables, satellite receiver dish, aerial mast.
- Transport links: roadway, airstrip, harbour with quayside, helipad, railway siding, canal.

3.3. Communications, energy, clean water supply and wastewater disposal
3.3.1 Communications
It is common business practice to provide high-speed internet access on site for all site-based personnel. Advances in technology for mobile devices means we now only need a single lightweight hand-held device that can be used as both a laptop and mobile phone.

These devices allow instant access to drawings and servers while on site. For the longer term use, a keyboard, display screen and computing power in the site office, hardwired to the internet via fibre-optic links, is still a necessity.

In the UK, the mobilisation lead time for a new cable internet link, measured from placing an order to commissioning, is on average 16 weeks. However, there is rarely more than 4–6 weeks between signing contracts and being expected to make a start and have a management team based on site.

Two types of IT solution are therefore required

- During the first 16 weeks, use mobile 4G or satellite receiver systems. These provide an instant email and database service. Performance is variable and often frustratingly slow, but is better than no service at all.
- In the period leading up to week 16, the local area cable network (LAN), switchgear and fibre-optic/copper-wired link will be installed and 'go live'. The LAN will be used from week 16 for the remaining duration of the project, linking users to shared servers and equipment such as printers, plotters, scanners, the internet and telephones. Access to the project database and documents will be available at acceptable upload/download speeds.

Before deciding on a system and placing supplier orders, two steps are required. For the first step, a briefing document must be drawn up to pass to the IT system designer. This comprises three parts

- A questionnaire asking for: the site address; project duration, with the start and completion dates; number of users; software required and the type of web-based document-sharing system; and types and number of shared equipment such as printers, plotters, scanners, internet, telephones, projectors and so on.
- A compound layout drawing and a detailed office plan layout drawing showing the positions of the communications hub room, workstations, printers, plotters, scanners, telephones and projectors. Positions of data socket outlets should be drawn on the plan, including some in the meeting room(s). It is prudent to have a few extra socket outlets included for visiting 'hot desk' users.
- A site location map.

The brief is passed to the IT systems designer in the second step so that a fully costed IT system specification and design can be produced. This should always be checked and explained to ensure that the proposals are a correct interpretation of the brief.

3.3.2 Energy

Electricity is the most popular and flexible energy to use. Reliable sources include either: national grid or local distribution cable networks, or hired electricity generators (commonly diesel engine powered). Generators can also be driven by hydrogen gas, gas engine or turbine or even jet engine. The choice of fuel will depend on the safety risk assessment, availability and cost.

Consider innovative technologies and whether renewable power such as solar or wind can be incorporated to support the power needs of the site and reduce the carbon footprint of the site. A hybrid solution will reduce both the carbon dioxide equivalents (CO_2e in kg) and increase the silent-run hours, which in a residential area will show compliance with the organisation's corporate social responsibility. There are systems available that can eliminate noisy generators at night through battery storage, and battery/renewable energy solutions that reduce generator run times further.

3.3.2.1 Estimating the demand for electricity

In estimating demand for electricity, the first step is to produce a schematic electrical system layout which resembles a hierarchical pyramid. Start at the base by naming each temporary building or purpose zone within the compound and list every electrical energy-consuming item within each, along with its energy consumption rating in kilowatts, including the following

- Offices: power socket outlets, light fittings, space heaters, water heaters, domestic appliances, office equipment, IT and communications switchgear, fire alarm, intruder alarm, access control, trace heating.
- Welfare facilities: power socket outlets, light fittings, space heaters, water heaters, domestic appliances, kitchen equipment, fire alarm, intruder alarm, access control, trace heating.
- Compound: external lighting, CCTV system, access control, vehicle entry barrier, weighbridge, wheel wash, battery charging plug-in sockets for large battery-powered moving plant.
- Materials storage shed: power socket outlets, light fittings, space heating, fire alarm, intruder alarm, access control, dehumidifier.
- Materials testing laboratory: power socket outlets, light fittings, space heating, fire alarm, intruder alarm, access control, trace heating, cube curing tank, special test equipment.
- Maintenance workshop: power socket outlets, light fittings, space heating, fire alarm, intruder alarm, access control, welding equipment, hoist, vehicle inspection ramp.
- Fabrication workshop: power socket outlets, light fittings, space heating, fire alarm, intruder alarm, access control, overhead gantry crane, manufacturing plant, welding equipment.
- Items of major plant located in the compound or elsewhere on site, for example, tower cranes, concrete batching/mixing plant, mortar silo mixers, materials or passenger hoists.
- Site distribution to transformers on all floor levels to provide power outlets for 110 V power tools, welding points in plant rooms and so on.
- Site distribution to all floor levels for temporary lighting, fire alarm system and so on.

Using the schematic described above as well as utilisation factors, the total electricity demand can be calculated and used to pick the most economical source of supply.

The second step is to prepare a detailed distribution layout plan showing cable sizes, routing and switchgear locations. In order to do this, the electrical engineer will need the completed detailed compound layout plan and temporary buildings layout plans. Distribution of electricity on site is covered in BS 7375:2010 (BSI, 2010).

To minimise energy usage on site and reduce expenditure, you should request the following from your cabin supplier as a minimum

- passive infrared (PIR) sensors in all rooms
- light-emitting diode (LED) lights
- automatic door closers to prevent loss of heat
- timers on heaters.

3.3.3 Clean water supply

All the water needed on a project for both welfare and construction purposes must be clean and free from impurities (potable is preferable). The daily consumption over the duration of the project should be estimated to establish the required quantity and flow rate of water required. To carry out this estimate, a schematic diagram will be drawn up to show all hot and cold water appliances in welfare facilities and draw-off points for standpipes, mortar silo mixers, concrete mixers, wheel-wash plant and so on. The water treatment, storage and distribution pipework/pumping scheme can then be designed. For high-rise or multi-storey buildings, vertical risers are required with a tap-off point on each floor level. Include an allowance for frost protection by having trace heating and insulation to all pipework.

Once the total quantity and flow rate required have been calculated, the next step is to decide on a water source. Depending on the location, the choices of source will be one of the following

- Connect to the existing water mains supply. Obtain an application form from the relevant water supply company and apply for a new (temporary) metered water supply pipe. They will provide a proposal and quotation for the supply. Once they have received payment for the quoted sum, they will install their pipework from their mains (including a water meter) up to a stop valve located at the site boundary. Water usage is paid for on a rate per cubic metre.
- Import water by bulk tanker. Again, obtain an application form from the relevant water supply company and apply for them to supply water in bulk tankers. They will provide a proposal and quotation for the supply. Once they have received payment for the quoted sum, they will commence deliveries. Water usage is paid for on a rate per cubic metre. A tanker hardstanding area is required with suitable valve chambers to allow the tanker discharge pipes to be connected to the site network.
- Extract water from a river, a borehole or the sea. Water extraction licences and borehole drilling licences have to be applied for and granted before proceeding with this option. Water pumps, purification plant and clean water 'buffer' storage tanks are required. When using sea water, desalination plant is also needed.

- Build a dam and form a reservoir. This option would only be the choice of last resort. Obtaining local planning permission will be necessary before proceeding. Water pumps, purification plant and clean water 'buffer' storage tanks are required.

3.3.4 Wastewater disposal

Foul water from all toilets, sinks, washbasins and canteen equipment has to be collected and treated before it can be discharged back to nature. A system of pipework and manholes needs to be designed and drawn, to connect up all the waste outlets and drain into one of the following

- an existing main foul water sewer system adjacent to the site
- a waste collection tank located above or below ground (this will require frequent emptying by a suction tanker and the waste transported to the nearest sewage works for disposal and treatment)
- temporary sewage treatment works built on site (the last-resort option, which requires local planning permission to be obtained before proceeding).

Similarly, surface water from roofs, guttering, hardstandings, car parks and roads has to be collected by a system of pipework, manholes and petrol interceptors and disposed of by one of the following methods

- connecting to a main storm-water sewer system adjacent to the site
- connection to underground soakaway constructed on site as part of the system
- connection to a catchment and settlement pond constructed on site (if there is a watercourse adjacent, the pond can have an overflow; otherwise the pond is designed large enough to rely on evaporation).

Groundwater arising from the works has to be properly disposed of by a controlled method. This water can be from one of several sources, as listed below

- From ground surface runoff during rainstorms: a system of ditches, land drains and silt traps linked to a soakaway should be designed.
- From groundwater seeping into excavations: this is normally dealt with by excavating sump holes to collect the water and then pumping it into a settlement tank, before pumping it to a soakaway or settlement pond.
- From the dewatering of deep substrata to allow excavation below the water table. A typical dewatering scheme will comprise a series of wellpoints connected via a common header collection pipe to the suction side of a large pump. The pump runs constantly and the discharge water is passed into a segmented settlement tank. The water is then discharged into a soakaway or settlement pond. Water quality should be monitored by daily sampling. Consideration should be given to water sampling at depth.

3.4. Office and welfare accommodation space planning

Detailed work on drawing up plans for site offices, meeting rooms, changing rooms, toilets, drying rooms, canteens and so on can only start once the number of people that have to be catered for and the time they will spend on site has been estimated. A detailed

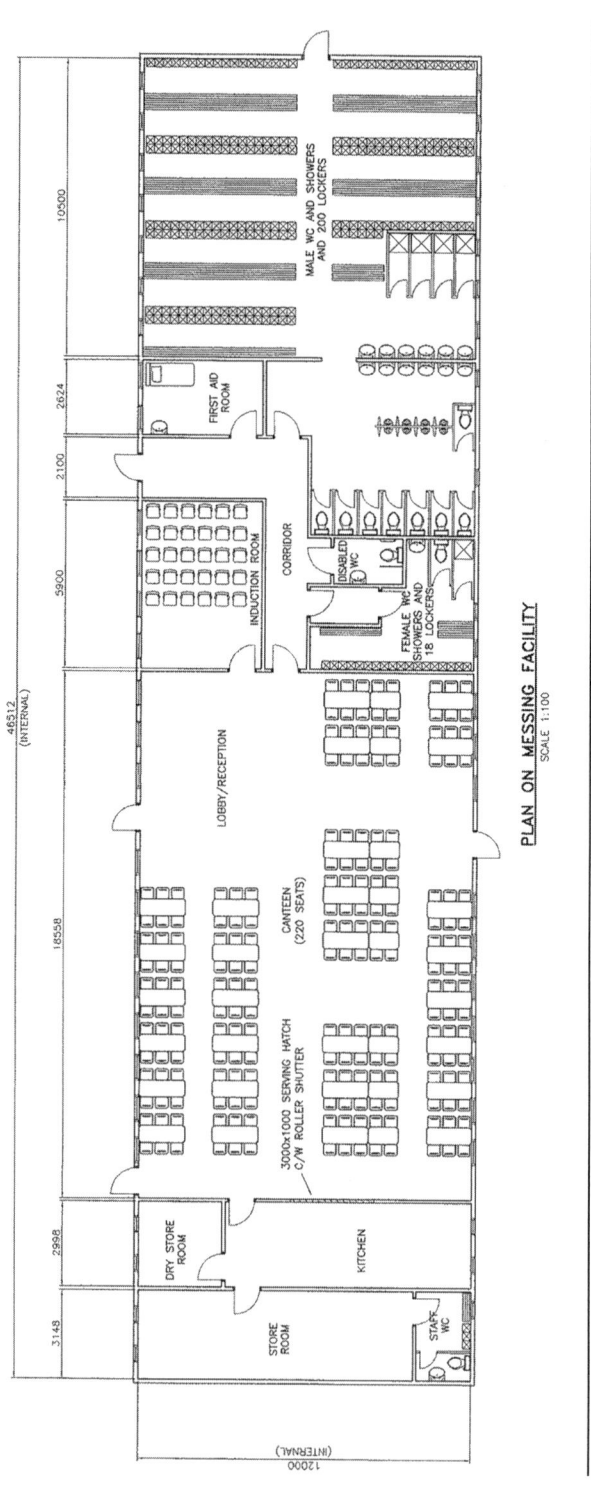

Figure 3.3 Example of layout of site office and welfare facilities. (Courtesy of Costain)

Site compounds and set-up

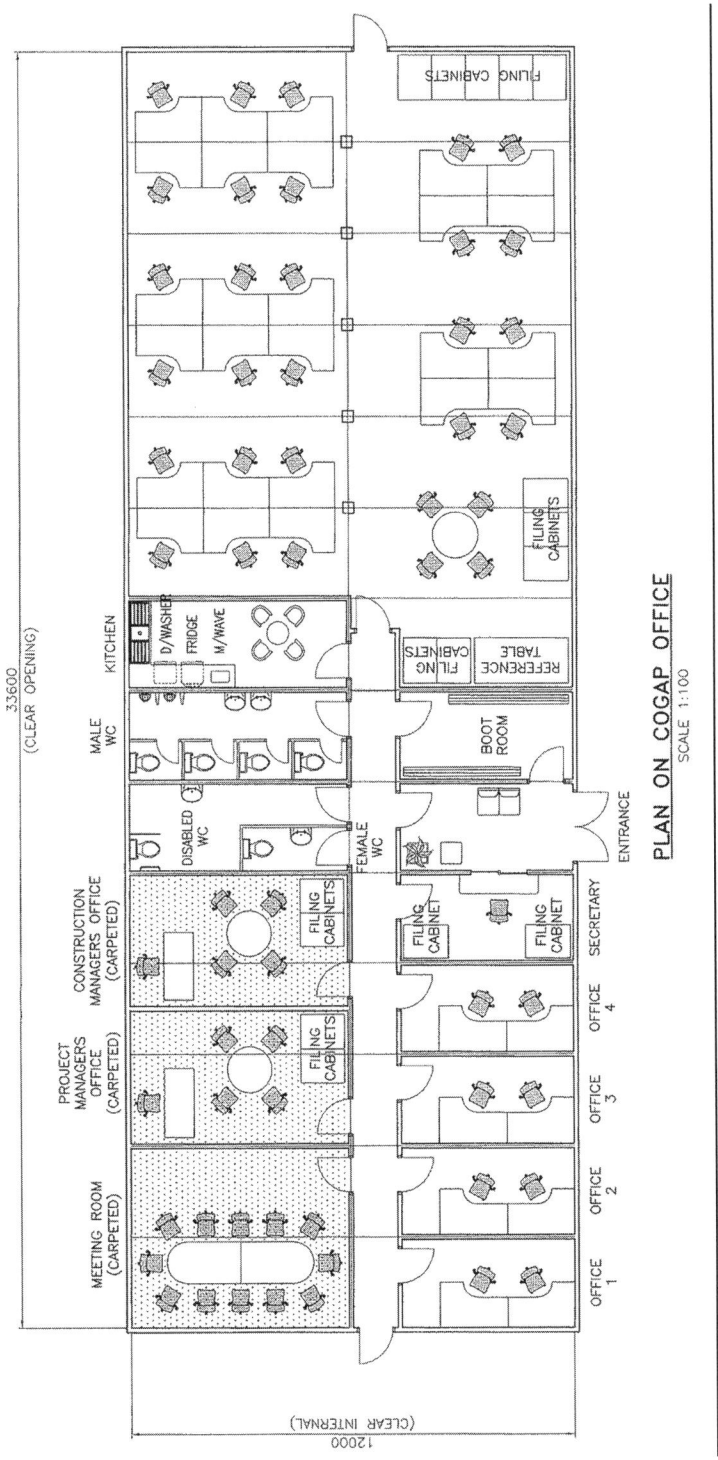

bar chart programme of work and a resource histogram for both management and operatives is therefore needed.

Once these quantities are known, the following must be considered

- Segregation or integration: the need to separate people relative to their job roles, work groups, different employers and for reasons of privacy. Bringing multi-disciplinary teams together in an open-plan environment has become common practice. There will be a requirement to provide different-sized meeting rooms for privacy and confidentiality.
- Productivity and quality: open-plan offices should not be assumed. Construction site offices are notoriously noisy. Many technical and managerial job roles require privacy, relative quietness, wall space to display bar chart programmes, table-top space to lay out two A0 or A1 size drawings side by side, a filing cabinet, a cupboard and a desk. Providing individual offices where needed will improve productivity, quality and staff motivation. Regular team meetings should be used to open up lines of communication.
- Communication and motivation: operatives need a good-quality canteen and welfare facilities. Break times are short and time spent away from the workface must be minimised. Design seating and table layouts that allow sufficient places for all to be served in 'one sitting'. Employees should not have to sit in a van to eat due to lack of space. Equally important are changing, clothes drying and shower rooms. Workers should not be expected to travel to work in their overalls or dirty clothes. The induction/training room needs to be a dedicated space and large enough to accommodate daily new arrivals for safety, health and environment (SHE) inductions and to accommodate all operatives for shift briefings.
- Storage and records: separate and secure rooms with ventilation and temperature control for IT switchgear 'hub' cabinet and document filing.
- Modular or volumetric accommodation system: there are numerous suppliers and systems to choose from. Cost comparisons will need to take into account any foundations required, setting up and dismantling, craneage and transport, and connectivity (i.e. pre-wired and only needing to be 'plugged in').

Deciding on and drawing up a room layout and furniture plan is a fairly straightforward exercise. Most suppliers of modular accommodation have standard generic drawings available for use or to modify into site-specific layouts. In cases where existing buildings are used as temporary facilities, obtain floor plans from the landlord or carry out a dimension survey of all rooms and draw the floor plan. The completed furniture layouts are also used to show small power, data and voice socket outlet positions. Figure 3.3 shows a typical layout of office and welfare facilities.

3.5. Materials (distribution, fabrication, handling, storage, testing and unloading)

There is a huge variety of length, width, height, weight, shape and physical nature of construction materials and components, ranging from lightweight, small items such as screws or nails up to heavyweight, large prefabricated modules.

When considering the choices for deciding on the load size, the method of delivery onto site, storage and movement to the workface, the principal considerations include the following

- Roads, hardstandings and bridges: weight and size restrictions and requirements.
- Mechanical handling plant: self-weight, lifting capacity, availability and utilisation efficiency.
- Offloading areas with safe access and fall-prevention measures if workers have to gain access onto a delivery vehicle.
- Manual handling: maximum weight restrictions.
- Temporary storage areas or buildings: capacity, weight and size limitations and requirements, dry environment.
- Off-site or on-site fabrication.
- Traffic restrictions on the access to site.

Alternative methods and designs for providing a solution for each of these have to be drawn up and fully costed so that they can be compared and allow the most practical, efficient and cost-effective combination to be determined. Main contractors usually rely on specialist suppliers and subcontractors to provide details of their requirements.

3.6. Hoarding and fencing

Site security and public protection comprise one of the most important elements of site set-up. Solid timber hoardings should be designed for wind loading, crowd loading and any suction loads from, for example, passing trains. The Temporary Works Forum (2014) has produced a guide for the management and design of solid hoardings. The principal risks with solid timber hoardings are: failure due to high wind loading and failure due to impact. Hoarding components (posts, rails, etc.) and connections (screws, nails, bolts, etc.) should be appropriately designed for the specific site wind conditions, and the foundations should be adequate to prevent overturning, sliding (relying on friction) and passive failure (if an embedded solution is selected). If it is wished to avoid embedded foundations (perhaps due to the presence of buried services) then kentledge can be used. Concrete blocks, sand bags and so on can be used. However, drums filled with water can have a tendency to leak and the water can evaporate. The hoarding could be in use for several years and at the end of the project the drums will be full of stagnant water, leading to a disposal problem.

Care should be taken when using proprietary solid metal hoardings, and the supplier's assembly and use instructions should be followed explicitly. There are likely to be limitations on their use relating to specific site wind exposure conditions, which need to be checked.

Fencing tends to be more appropriate as demarcation and segregation. Open-mesh types allow the wind to pass through. The supplier's assembly and use instructions should be followed explicitly, especially the limitations on any sheeting, netting or signage that may be attached to the fencing. To prevent these fences 'blowing over' additional kentledge may be required, or if space allows the fences could be triangulated or buttressed, or raking struts with kentledge could be provided. However, the users still need to exercise care because component failure is possible if the system is used in an inappropriate manner or inappropriate location.

REFERENCES

BSI (British Standards Institution) (2010) BS 7375:2010. Distribution of electricity on construction and demolition sites. Code of practice. BSI, London, UK.

Temporary Works Forum (2014) *Hoardings – A Guide to Good Practice*. Temporary Works Forum, London, UK, TWf2012: 01 (revised 2014).

UK Government (1974) Health and Safety at Work etc. Act 1974. Statutory Instrument 1974/37. The Stationery Office, London, UK.

UK Government (1974) Control of Asbestos Regulations 2012. Statutory Instrument 2012/632. The Stationery Office, London, UK.

UK Government (1992) Workplace (Health, Safety and Welfare) Regulations 1992. Statutory Instrument 1992/3004. The Stationery Office, London, UK.

UK Government (1999) Management of Health and Safety at Work Regulations 1999. Statutory Instrument 1999/3242. The Stationery Office, London, UK.

UK Government (2005) Work at Height Regulations 2005. Statutory Instrument 2005/735. The Stationery Office, London, UK.

UK Government (2007) Work at Height (Amendment) Regulations 2007. Statutory Instrument 2007/114. The Stationery Office, London, UK.

UK Government (2013) Health and Safety at Work etc. Act 1974 (Application outside Great Britain) Order 2013. Statutory Instrument 2013/240. The Stationery Office, London, UK.

UK Government (2015) Construction (Design and Management) Regulations 2015. Statutory Instrument 2015/15. The Stationery Office, London, UK.

FURTHER READING

Hall F and Greeno R (2001) *Building Services Handbook*, 5th edn. Butterworth Heinemann, Oxford, UK.

Useful web addresses

Firefly Hybrid Power – manufacturer of hybrid power systems: http://www.fireflyhybridpower.com (accessed 01/08/2018).

Gaia Group – manufacturer of off-grid power supply system (Solatainer): http://www.gaiagroup.com (accessed 01/08/2018).

Health and Safety Executive (HSE) – Health and safety in the construction industry: http://www.hse.gov.uk/construction/index.htm (accessed 01/08/2018).

UK legislation: http://www.legislation.gov.uk (accessed 01/08/2018).

Pallett, Peter F and Filip, Ray
ISBN 978-0-7277-6338-9
https://doi.org/10.1680/twse.63389.057
ICE Publishing: All rights reserved

Chapter 4
Tower crane bases

Ray Filip
Temporary Works Consultant and Training Provider, RKF Consult Ltd

Stuart Marchand
Managing Director, Wentworth House Partnership

Tower cranes are common items of construction plant on many construction sites, especially in towns and cities. They are assembled on site and positioning is critical to ensure adequate site coverage. Precautions need to be taken where multiple cranes are used on a site in order to prevent clashing. The principal items of temporary works associated with tower cranes are bases and mast ties (for very tall cranes which are unable to free-stand). This chapter is concerned with the design and construction of bases.

The number and types of tower cranes to be used on a project should be identified in the planning and tender stage, and sufficient consideration should be given to the following

- Programme – when will the cranes be required and how many cranes will be required to meet the programme?
- Crane positioning – to avoid clashing with the permanent works.
- Crane specification – required crane height, lifting capacity and site coverage.
- Proximity hazards – these could be buried or overhead services, trees neighbouring buildings or property owners who may impose over-sailing restrictions.
- Crane assembly – how the crane will be assembled (transporting crane sections to site and positioning of a mobile crane for assembly) and dismantling the crane after use involve similar issues.
- Ties – tall cranes may be unable to free-stand and the mast may need to be tied to the structure being constructed.
- Foundations – the position of the foundation and the available space on site for the type of foundation selected. Tower crane bases can be large and economic viability has to be considered.

Many contractors do not own their own tower cranes so will hire them from specialist suppliers. The tower crane manufacturer (via the supplier) will provide the foundation loads and the contractor (often the principal contractor but sometimes a specialist subcontractor) will be responsible for the design of the base. The designer should be provided with a comprehensive design brief. The contractor will then construct the base to the details provided by the designer.

As part of the design brief, it is essential that adequate soils information is available for the foundation design. The soils information should extend to a depth below the deepest part of the crane foundation. This could be to a depth below any anticipated depth of piles, and for this situation boreholes will be required to determine the soil profile to the required depth. The design of the base can be carried out by a contractor's in-house temporary works design team, an independent specialist Temporary Works Designer or a permanent works structural/geotechnical engineer. The designer must have the necessary knowledge, skills and experience to comply with the Construction (Design and Management) Regulations 2015 (UK Government, 2015). The designer should carry out a designer's risk assessment and identify and eliminate risks where possible. The foundation design may be an iterative process to determine the most cost-effective and practicable solution for the site. The foundation design should then be checked as per the recommendations in BS 5975:2019 (BSI, 2019) and generally would be classified as design check category 2 or 3 (depending on complexity and location). Where possible and practicable, it can be cost-effective to incorporate the tower crane base into the permanent works or to utilise the permanent works (e.g. core foundations). If this is done then the Permanent Works Designer (and perhaps also the pile designers) should be consulted to ensure the base does not adversely affect the permanent works (similarly, if the tower crane mast is to be tied to an adjacent structure). The permanent works loading will also be required, and load cases with and without the permanent structure should be considered.

4.1. Types of tower crane

There are three main types of tower crane in common use

- Self-erecting (also known as fast erecting) – a compact design of crane that tends to be used for constricted sites. These are quick to erect and can be relocated if necessary. They can also be truck or crawler mounted.
- Horizontal jib (also known as saddle jib, flat top or hammer head) – the most common type of crane used for out-of-town sites. The jib is fixed in the horizontal plane and a trolley with the hook travels along the jib to change the lifting radius. Such cranes are generally assembled and removed from site by a mobile crane.
- Luffing jib – a static crane used where over-sailing is an issue. The hook tends to be fixed on the end of the jib and the jib can be raised and lowered to change the radius. Such cranes are generally assembled and removed from site by a mobile crane.

For very tall structures, certain types of tower crane are capable of 'climbing'. The crane mast can be erected within the structure (e.g. inside a lift shaft) or externally, and the crane can progressively raise itself as construction of the structure continues upwards. Wedges and collars are fixed around the tower and then connected to the structure to transfer loads and provide stability. This is a specialist operation and the permanent works should be checked to ensure it can support the imposed loads from the ties.

4.2. Loading on foundations

It is important that the tower crane manufacturer knows the location of the crane relative to any object or structure that may increase wind loads locally (e.g. tall buildings and the tops of hills or escarpments).

The manufacturer of the tower crane will provide the following loading information

- In service – when wind speeds are sufficiently low (known as 'working wind') that the crane is able to operate.
- Out of service – when the crane is unable to operate due to high winds and the crane is free to weather vane (i.e. the operator releases the slew brake) but may be exposed to maximum wind speeds. Luffing cranes are also required to be left out of service with the jib in the raised position to prevent over-sailing. Some suppliers also provide storm loading from the front and rear of the crane.
- Assembly – additional loads relating to the erection of the crane may be provided by some manufacturers.

The loads will include overturning moments (due to wind, the self-weight of the crane and lifting), vertical loads (due to the self-weight of the crane, the lifting tackle and the load being lifted), horizontal loads and slewing torque when in operation. The foundation designer will design the foundation based on these loadings and should determine the critical combination of loads.

There may be rare occasions when saddle jib cranes have to be locked in position (prevented from slewing, e.g. due to over-sailing issues) in the out-of-service condition. Crane manufacturers should be consulted and a specific set of foundation loads obtained.

In some parts of the world tower cranes and their foundations are required to withstand earthquakes and typhoons or hurricanes. Extreme cold and snow or ice build-up or extreme heat could also become issues in some parts of the world. The crane manufacturer should be consulted.

It is very important that the designer has the competence to understand whether the details provided by the crane supplier are accurate and comprehensive.

4.3. Foundation options

The tower crane is mainly selected by considering its location in relation to the permanent works to be constructed and the required height, lifting capacity, lifting radius, ease of erection and dismantling, economic viability and external restrictions. There are then a number of different foundation options available.

4.3.1 Cruciform base for static crane

The tower crane has a cruciform steelwork base with kentledge blocks around the base that provide stability (Figure 4.1). The crane base is not anchored to the foundation and

Figure 4.1 Diagrams of typical cruciform base arrangements. (Courtesy of RKF Consult Ltd)

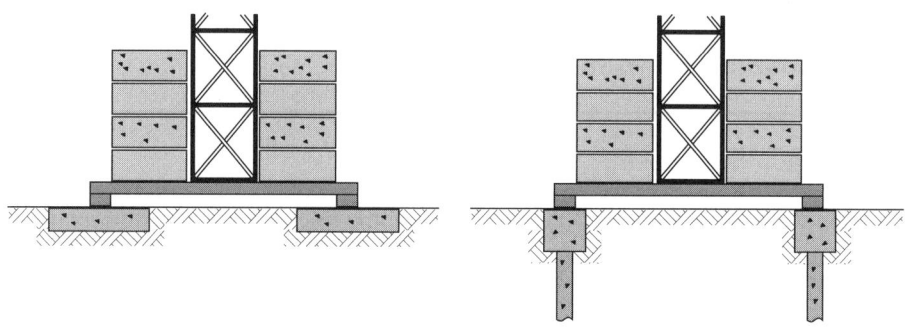

any residual uplift (due to applied overturning moments) is resisted by sufficient kentledge blocks, which are provided by the crane supplier (Figure 4.2). These blocks are installed as the crane is being erected. The foundation is designed mainly for downward compressive loading. The critical loading condition which gives the maximum bearing pressure occurs when the overturning moments are applied about a diagonal axis. All other loads (vertical loads, slewing torque and horizontal load) are also applied to the foundation and must be designed for.

The foundation designer then designs and details the foundation to resist the applied loads without undue settlement. If the loads are small and the ground conditions allow, the crane could be founded on well-compacted structural fill. However, it is more common for the foundations to be a simple reinforced concrete pad designed to spread the load to reduce the applied pressure on the ground. Occasionally, if the structure being constructed has a substantial ground-bearing slab, the crane could be positioned on the slab. If the loads are high and the ground is poor, it is common to provide a pile with cap at each corner and then tie the piles together with a simple reinforced concrete beam. These piles would mainly be designed for vertical compressive loads. The piles could be reinforced concrete or steel sections, and it is of economic benefit to utilise similar piles to those being installed for permanent works. The horizontal load and slewing torque could also be taken by the piles, or the passive resistance of the ground and friction could also be considered. If the passive resistance of the ground is considered then steps must be taken to ensure that the ground is not removed at any stage during the construction (i.e. for drainage or services excavations).

A significant disadvantage of this type of foundation option is the amount of space required. The steel cruciform, kentledge blocks, power supply and fencing (to form an exclusion zone) can typically require an area of 10 m × 10 m. On congested inner city sites where space is at a premium, this type of base might not be a feasible option. However, this type of base does not have to be founded at ground level. It is common for the crane to be positioned on top of structures to free up space at ground level. In this configuration the 'strong points' of a structure can be utilised, whereby the crane will be positioned on a temporary steel frame that has been designed to span between the

Figure 4.2 Typical cruciform base with concrete kentledge blocks showing the amount of space required at ground level. (Courtesy of Wentworth House Partnership)

structure columns or beams. The structure is relied upon to carry the crane loads, and the Permanent Works Designer should confirm this. It is also possible to omit all the crane kentledge by connecting the crane directly to the structure with large tie rods and utilising the self-weight of the structure (Figure 4.3).

4.3.2 On rails

Travelling cranes have a cruciform steelwork base with kentledge blocks (as above) but bogies are attached to each corner. They travel on rail sections (details of which are provided by the crane manufacturer), which sit on compacted ballast (if the soil beneath is good) or a reinforced concrete strip footing, which may also be piled depending on the ground conditions (Figure 4.4). The strip footing is designed as a beam to Eurocode 2 (BSI, 2004a) and is considered to be concentric with the crane bogie, and the effect of

Figure 4.3 Cruciform base with concrete kentledge blocks positioned on top of a structure. Note the temporary steelwork to transfer the compressive crane loads into the columns. (Courtesy of RKF Consult Ltd)

the bogie loads on a single rail will overlap. Bearing pressure and settlement as well as structural failure of the beam need to be considered. Where piles are required, the beam is designed to span between the piles, which are not designed to carry tensile loads (as described above). For a ground-bearing beam this design loading, from the bogie, is considered to be spread evenly over the full width of the beam and the length of the beam to provide safety from ground-bearing failure. The reinforced concrete section is then designed, both longitudinally and transversely.

The crane supplier will also provide details of minimum foundation widths (to allow the rails to be fixed to the foundation), how the rails are to be connected to the foundation and the required tie spacing between foundation beams. The crane track can also be curved in plan, subject to a minimum radius of curvature, as advised by the crane manufacturer.

4.3.3 Expendable base (pad base) for static crane

This option can be used if there is insufficient space at ground level for a cruciform base, as it is only the crane mast that extends above ground level. The steel cruciform and kentledge blocks are removed and the mast is connected to an 'in situ' reinforced concrete base with a sacrificial anchorage system supplied by the manufacturer (Figure 4.5). The manufacturer may supply details of the minimum thickness and concrete strength for the foundation to ensure the anchorage system is sufficiently embedded to transmit the crane loads

Tower crane bases

Figure 4.4 Typical cruciform base on rails. (Courtesy of RKF Consult Ltd)

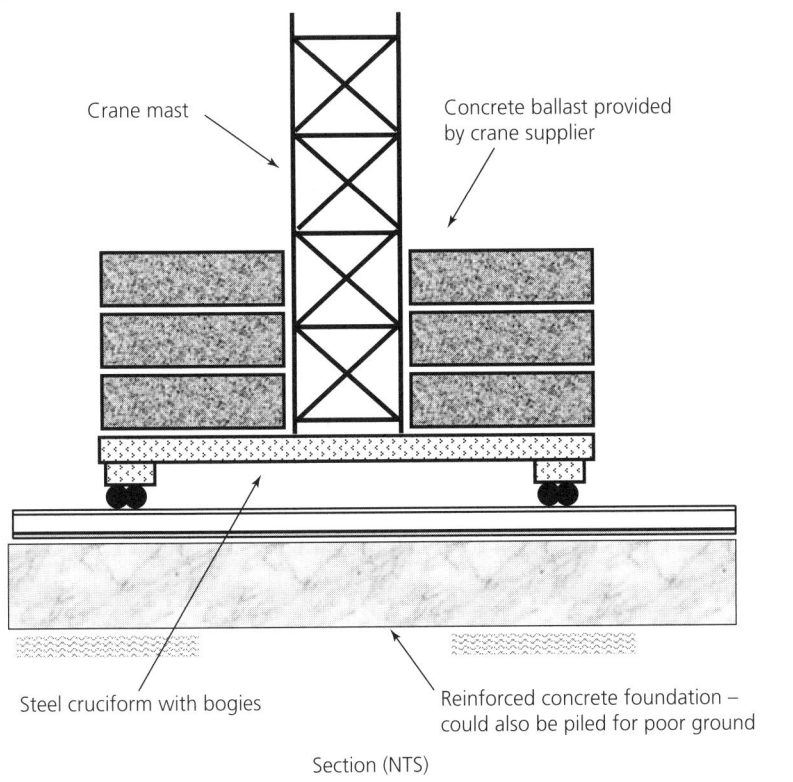

Section (NTS)

(Figure 4.6). The foundation designer will design the foundation for the imposed loads to prevent bearing-capacity failure, overturning, sliding or structural failure. If the loads are relatively low and the soil is relatively strong, the foundation can be designed as ground bearing. However, these bases can be large: 6 m × 6 m in plan and at least 1.5 m depth is not uncommon. Due to the depth of the foundation care should be taken when excavating to prevent ground collapse, especially in poor ground. It is often of economic benefit to

Figure 4.5 Diagrams of typical expendable base arrangements. (Courtesy of RKF Consult Ltd)

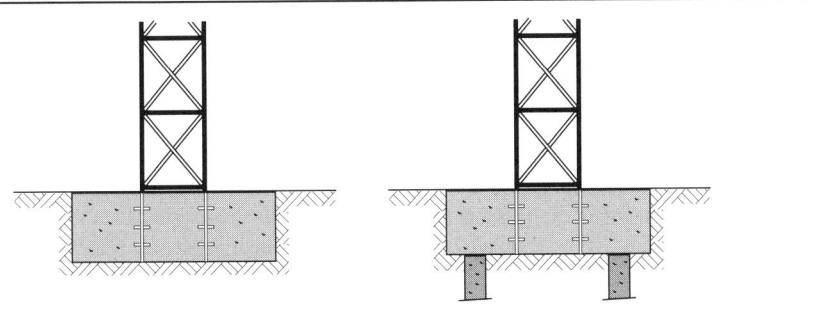

Figure 4.6 Construction of an expendable base. (Courtesy of RKF Consult Ltd)

incorporate the crane foundation in the permanent foundations (e.g. incorporating the crane foundations in a core foundation in a building). Various angles of the jib relative to the foundation need to be considered to find the critical case. Although the case of the jib over the diagonal gives the worst elastic loading on the ground, shape factors applied to the loaded ground area result in this case not necessarily being the most likely to cause ground failure. Due to the depth of the foundation (governed by the minimum anchorage depth provided by the crane manufacturer) it is likely that the foundation will require the minimum reinforcement for the top and bottom mats.

If the crane loads are high and the ground is poor the size of the reinforced concrete base can be very large and result in an uneconomical or impractical solution. The foundation size can be reduced by using piles. Often a single bearing pile is positioned at the corners of the square reinforced concrete foundation. The minimum spacing of the piles should be at least three times the diameter of the pile. The piles should be positioned to be clear of the tower crane anchorage to avoid reinforcement congestion and problems in positioning the anchorage system. The pile-cap should extend a reasonable distance in plan beyond the piles in order to avoid problems associated with punching shear around the pile perimeter. If piles are being installed elsewhere on the project, it is of economic benefit to use the same type of pile for the tower crane foundation. The overturning moments are applied about a diagonal axis, and in this arrangement the piles would have to be designed for compressive and tensile loads. The pile reinforcement is fully anchored into the foundation and, as above, the minimum top and bottom mat reinforcement is generally adequate. If permanent works piles are to be used for the crane foundation, the foundation designer should confirm the loads with the pile designer. In general, the permanent piles are designed for larger compressive loads than from the crane foundation, but not for tensile loads.

4.3.4 Climbing

A tower crane can 'climb', whereby additional sections of mast can be installed as the build progresses upwards. The crane can climb up the outside or the inside (often inside a lift shaft) of a structure. The crane does not have to be founded at ground level, and specialist steel frames are used that transfer the loads from the crane to the structure being constructed. A specialist customised steel frame is provided.

4.4. Foundation design principles

Foundations have previously been designed in accordance with the Construction Industry Research and Information Association (CIRIA) report C654 (Skinner *et al.*, 2006). Crane manufacturers now can provide design information in alignment with Eurocodes, and as a result CIRIA C654 has recently been revised as CIRIA C761 (CIRIA, 2019). Producing CIRIA C761 proved challenging due to the misalignment of the product design code with the general Eurocodes and the different information provided by different manufacturers.

The self-weight of the tower crane and the foundation are taken as 'permanent action'. All other loads are taken as 'variable actions'. There are three main aspects to the design as given in Eurocode 7 (BSI, 2004b)

- Stability – the EQU limit state – applies to pad/gravity bases in relation to overturning.
- Geotechnical capacity – the GEO limit state – applies to pad/gravity bases and piles.
- Structural design – the STR limit state – applies to all bases.

The Eurocodes form a uniform suite of design standards. However, there are inconsistencies that make the rational design of items with an interface difficult. In this particular instance, the interface is between a tower crane, currently designed to product standards FEM 1.001 (FEM, 1998) or DIN 15018-1:1984 (DIN, 1984), and its base, designed to BS EN 1990:2002 (BSI, 2002) to BS EN 1997:2004 (BSI, 2004b). In the future, tower cranes will be designed to BS EN 13001-1:2015, BS EN 13001-2:2014 and BS EN 13001-3-1:2012 (BSI, 2012, 2014, 2015). However, the inconsistencies between tower crane and base design standards will remain.

The FEM (2014) has provided guidelines for applying load factors to tower crane bases, and these have been used in CIRIA C761 (CIRIA, 2019).

4.5. Foundation construction and inspection

The construction of the foundation should be adequately supervised and the designer's assumptions on ground conditions should be confirmed on site. The ground at formation level should not be allowed to deteriorate (e.g. from weathering) prior to the foundation being constructed. The designer's approval should be obtained for any on-site changes to the details.

An exclusion zone should be set up around the crane base, and if there is a potential risk of vehicle impact it is prudent to provide protection around the foundation (e.g. concrete blocks).

If the foundation depth is significant, consideration should be given to how the excavation will be carried out to prevent ground collapse. Consideration should also be given to how the reinforcement will be lifted and placed into position (if prefabricated), or how access and support will be provided to the top mats if constructed in situ.

The crane manufacturer will provide details of construction tolerances. In particular, the anchorage system for an expendable base has to be installed to a very tight tolerance on level.

All tower crane foundations should be regularly monitored and inspected for unexpected settlement, especially differential settlement. If the crane is on tracks the spacing, fixings and end stops should also be inspected. Crane manufacturers will provide specific guidance, and TIN 031 (CPA, 2015) provides some useful checklists.

REFERENCES

BSI (British Standards Institution) (2002) BS EN 1990:2002. Eurocode. Basis of structural design. BSI, London, UK.

BSI (2004a) BS EN 1992:2004. Eurocode 2: Design of concrete structures. BSI, London, UK.

BSI (2004b) BS EN 1997:2004. Eurocode 7: Geotechnical design. General rules. BSI, London, UK.

BSI (2012) BS EN 13001-3-1:2012. Cranes. General design. Limit states and proof competence of steel structure. BSI, London, UK.

BSI (2014) BS EN 13001-2:2014. Crane safety. General design. Load actions. BSI, London, UK.

BSI (2015) BS EN 13001-1:2015. Cranes. General design. General principles and requirements. BSI, London, UK.

BSI (2019) BS 5975:2019. Code of practice for temporary works procedures and the permissible stress design of falsework. BSI, London.

CIRIA (2019) *Tower Crane Foundation and Tie Design*. CIRIA, London, UK, Report C761.

CPA (Construction Plant-hire Association) (2015) *Tower Crane Bases and Ties*. CPA, London, UK, Technical Information Note 031.

DIN (Deutsches Institut für Normung) (1984) DIN 15018-1:1984. Cranes; steel structures; verification and analyses. DIN, Berlin, Germany.

FEM (European Materials Handling Federation) (1998) *Rules for the Design of Hoisting Appliances*. FEM, Brussels, Belgium. FEM 1.001.

FEM (2014) Guidelines for Considering Tower Crane Loads on Supporting Structures. FEM, Brussels, Belgium, Position paper.

Skinner H, Watson T, Dunkley B and Blackmore P (2006) *Tower Crane Stability*. Construction Industry Research and Information Association (CIRIA), London, UK, Report C654.

UK Government (2015) Construction (Design and Management) Regulations 2015. Statutory Instrument 2015/15. The Stationery Office, London, UK.

FURTHER READING

BSI (2006) BS 7121-5:2006. Code of Practice for safe use of cranes. Tower cranes. BSI, London, UK.

CPA (2009) *The Effect of Wind on Tower Cranes in Service*. CPA, London, UK, Technical Information Note 020.

CPA (2009) *Out of Service Wind Speeds*. CPA, London, UK, Technical Information Note 027.

HSE (Health and Safety Executive) (2010) *Tower Crane Incidents Worldwide*. HSE, London, UK, Report RB820.

Useful web addresses

CIRIA (Construction Industry Research and Information Association): https://www.ciria.org (accessed 01/08/2018).

Construction Plant Hire Association, Tower Crane Interest Group: https://www.cpa.uk.net/tower-crane-interest-group-tcig (accessed 01/08/2018).

Chapter 5
Site roads and working platforms

Christopher Tate
Chief Engineer, VolkerFitzpatrick

Except on very compact construction sites, vehicles are likely to be used to transport materials from place to place. Site traffic may comprise road-going vans and delivery lorries, or construction plant ranging from dumpers weighing a few tonnes to tipper trucks of 35 t or more. In order for these vehicles to move safely and efficiently across terrain that might be rough and undulating with poor load-carrying capacity, temporary site roads are needed; these should be distinguished from 'haul routes' used by heavy earth-moving equipment. Site roads generally fulfil only the most basic function of a permanent carriageway and will often be designed for a relatively short working life. They are generally constructed cheaply using unbound granular material or hydraulically bound stabilised soil and must be capable of being simply and rapidly repaired should localised failure or excessive rutting occur during service. Purpose-built working platforms are needed to support heavy construction plant operating in defined locations; for example, tracked piling rigs and cranes and mobile cranes with outriggers. Platforms must be strong enough to ensure that the ground underneath is not overstressed during peak loading, and they should be relatively flat and level so there is no risk of plant toppling during manoeuvring.

5.1. Introduction

Provision for the efficient movement of delivery vehicles, muck-away lorries and construction plant across a site is vital, yet may be given little forethought and a low or non-existent budget. Ground that appears hard and able to support traffic in a hot summer season can quickly be transformed into a quagmire as the autumn and winter rain arrives; production is curtailed and minds become focused on the problem of transporting materials from A to B. Temporary site roads capable of functioning all year round should be central to project planning. If designed appropriately and built before ground conditions are allowed to deteriorate, site roads are cost-effective; they will, of course, have a tangible cost but should more than prove their worth by keeping supply lines open (Figure 5.1).

Working platforms are considered by some to be a luxury. Cases of mobile cranes tilting or toppling during a heavy lift or piling rigs unable to extract casing as a result of poor ground support suggest otherwise. Correctly designed platforms enable lifting or piling operations to be undertaken with confidence at any time of year and whatever the foreseen ground conditions, and have a place on most projects.

Figure 5.1 The need for a properly designed and constructed site road. (Courtesy of Tensar International Ltd)

This chapter deals primarily with the design and construction of temporary roads and working platforms for use by site vehicles and plant. The temporary diversion of public roads within the site is a special case, which is covered briefly.

5.2. Site roads: principles and design
5.2.1 Alignment

In most cases there is little need to be concerned with the finer points of vertical and horizontal alignment. The vertical alignment should as far as is practicable follow the terrain in order to minimise regrading costs. Unbound granular material with a longitudinal gradient of 10% or less can be traversed without difficulty by road-going vehicles; if unavoidable, the gradient could be increased to 15% on straight sections where no stopping/starting is anticipated. In the horizontal plane, sight lines are important on congested sites where it may be necessary for traffic to weave between obstacles; across open fields on a new highway construction project, the shortest convenient route between any two points is the most favoured.

A minimum road width of 3.5 m will accommodate normal site traffic in single file, with local widening (typically $32/R$) on short radius (R) bends to allow for the passage of long or articulated lorries. Passing places 15 m long with 1:5 tapers should be provided at suitable intervals and at blind bends; the length may be reduced to 12 m if site traffic is not expected to include articulated lorries. For two-way working, a 6.5 m wide road will avoid passing vehicles running too close to an edge and causing excessive haunch damage.

5.2.2 Traffic loading

To prevent costly overdesign or premature failure it is essential to make the best estimate of traffic likely to use the road (both during and after its construction). This will involve itemising the weights/volumes of the various building materials involved in the project and assessing the transport needed to supply the imported materials and dispose of any surplus site arisings. The transport (probably mostly comprising vans and lorries) should be categorised by fully laden weight; unladen vehicle movements are usually neglected, as are car movements. On top of this, the day-to-day activities of dedicated construction plant (excavators, dumpers, fork-lift trucks, etc.) should be considered because this type of traffic will, on occasion, also make use of the site roads. The resulting number of movements of vehicles of each type, and the likely distribution of traffic across the site, will enable the magnitude of traffic loading to be calculated for each branch of a site road network.

The damage inflicted on a road by the passage of a vehicle is related to the axle loading. Current knowledge is that there is a fourth-power relationship (i.e. doubling the vehicle axle loading results in 16 times the damage). This effect is taken into account by assigning each type of vehicle likely to be using the site roads an equivalent number of 'standard' axles (a standard axle is 8160 kg or 80 kN). Table 5.1 lists the equivalent numbers of standard axles associated with various types of road vehicle and construction plant operating at full payload (Britpave, 2007). It is clear that a single fully laden lorry will have as much influence on the structural design of the road as a fleet of vans.

In the absence of data for a particular item of wheeled construction plant, its equivalent number of standard axles (sa) can be estimated from the manufacturer's technical

Table 5.1 Equivalent standard axles for typical vehicle types (fully laden)

Vehicle No.	Vehicle description	Gross vehicle weight: t	Equivalent number of standard axles (sa)
V1	6-axle articulated lorry	44	5.15
V2	5-axle articulated lorry	38	4.70
V3	4-axle articulated lorry	35	7.35
V4	3-axle articulated lorry	26	5.65
V5	4-axle draw bar lorry/trailer	35	5.40
V6	4-axle rigid lorry	32	5.35
V7	3-axle rigid lorry	26	4.80
V8	2-axle rigid lorry	17	3.15
V9	Van	5	0.02
V10	Car	2	0.0005
V11	5 t fork-lift	11	0.40
V12	3 t fork-lift	7	0.07
V13	3-axle dump truck	34	11.26
V14	6 t dumper	10	0.39

Data taken from Britpave (2007), courtesy of Alex Lake.

specification using the fourth-power relationship

$$sa = \sum(\text{gross weight on each axle in kg}/8160)^4 \tag{5.1}$$

Tracked plant, such as a small excavator, exerts a relatively low pressure on the ground while travelling and its influence on site road design compared to, say, a fully laden four-axle tipper lorry, is not significant. In cases where the site road is to be used regularly by heavy tracked plant (e.g. large excavators, piling rigs or crawler cranes) the final design will need to be checked using the principles discussed in Section 5.3.

On large civil engineering projects where heavy dump trucks or motorised scrapers are employed for earth moving, it is not normal practice for haul routes to be hardened because the pattern of plant movements may be subject to continual change. Earthworks operations, by their nature, generally take place in the late spring to early autumn period; at this time of year the ground itself is able to provide the necessary support for the vehicles involved (albeit with continual reprofiling of an often deeply rutted surface). Heavy earth-moving plant would therefore not be routinely included in the traffic-loading assessment for site roads.

The product of the anticipated number of movements (n_1, n_2, n_3, ...) of each fully laden type of vehicle (V_1, V_2, V_3, ...) and the equivalent number of standard axles associated with that vehicle (sa_1, sa_2, sa_3, ...) should be summed for all the expected vehicle types

$$sa_{(total)} = n_1 sa_1 + n_2 sa_2 + n_3 sa_3 \ldots \tag{5.2}$$

This will give the total number of standard axles to be carried by each branch of the site road network over its lifetime. Because site roads are likely to have a relatively uneven surface, there will be an element of dynamic loading not usually considered in the design of public roads; to take account of this it is suggested that the calculated total number of standard axles be doubled. The resulting number of standard axles is the design traffic loading

$$sa_{(design)} = 2 sa_{(total)} \tag{5.3}$$

In cases where the site road is merely the foundation layer for a permanent road to be built at a later date, a more simplistic approach to assessing design traffic loading can be adopted. There are published data relating the number of standard axles expected during construction

- the length of permanent road under construction (Powell *et al.*, 1984)
- the size of the development served by the permanent road (BSI, 2001).

5.2.3 Ground conditions

Having established the design traffic loading, the subgrade must be evaluated in order to formulate a design for the site road.

If a comprehensive ground investigation report is available for the site, it may well contain some or all of the information needed. Such information includes

- current subgrade California bearing ratio (CBR) or undrained shear strength – fundamental properties for the thickness design of the road
- plasticity index – as a guide to the equilibrium CBR and the performance of the subgrade when saturated (Powell *et al.*, 1984)
- soil grading – to determine whether the soil can be incorporated in the road structure, either directly as a granular layer or after treatment with hydraulic binders such as lime or cement
- moisture condition value (MCV) and compaction-related data, including remoulded CBR – only important if significant site regrading is needed to achieve a satisfactory road alignment
- soil chemical suite comprising organic matter, sulfate/oxidisable sulfide and pH value – only necessary if stabilisation of the soil is being considered
- an indication of ground permeability (e.g. soil infiltration rate) – to assess whether provision for temporary drainage needs to be made.

Should information be lacking it will be necessary to excavate trial pits at intervals along the proposed route to sample and test the ground to obtain the required data. If time is short, in situ testing may substitute for some of the laboratory tests. For example, a lightweight deflectometer, a Mexe-probe or Transport Research Laboratory (TRL) dynamic cone penetrometer (DCP) could be used to estimate the current CBR (HA, 2009), and a hand vane could be used to measure the undrained shear strength. For a clay soil, shear strength can be converted to CBR using the approximate relationship (Black and Lister, 1979)

$$\text{CBR (\%)} = \text{shear strength (kPa)}/23 \tag{5.4}$$

Basic soakage tests (Garvin, 2016) performed in the trial pits will help decide if the site roads need to be drained or whether natural percolation into the subgrade will suffice.

If soil stabilisation is being considered in lieu of granular material for construction of the road, it is essential that representative bulk soil samples from the trial pits are taken to a specialist laboratory to determine which type of binder is most suitable and the necessary percentage addition. Chemical tests and swelling tests must also be undertaken to ensure there are not elevated levels of undesirable or potentially disruptive substances in the soil (HA, 2007).

5.2.4 Formulating the design

Most site roads are unpaved, at least for the temporary construction condition. They normally comprise unbound granular material, with or without geosynthetics, or soil treated with hydraulic binder (usually with an unbound granular protection course).

5.2.4.1 Unbound granular roads

Elastic analysis can be used for the design of unbound granular site roads, but for temporary works it is simpler to use an empirical method aimed at limiting rut depth. Powell *et al.* (1984) quote an equation (rearranged below) to determine the thickness h

(in mm) of unbound granular material that will limit rut depth to 75 mm for a given traffic loading (sa_{design}) and subgrade CBR (%).

$$h = 190 \log_{10}(sa_{design})CBR^{-0.63} \tag{5.5}$$

This equation was first presented by Giroud and Noiray (1981) and relates to low-permeability subgrade soils (clay/silt); the granular material itself should have a CBR >80%.

As an example, for traffic loading of 1000 standard axles (approximately 100 four-axle tippers with 20 t payloads and dynamic loading allowance) and a subgrade CBR of 2.5% (firm clay), the thickness of unbound granular material (well-graded crushed rock) needed for the temporary road would be

$$h = 190 \times \log_{10}(1000) \times 2.5^{-0.63} = 320 \text{ mm (to the nearest 10 mm)} \tag{5.6}$$

While a rut depth of 75 mm is acceptable for a temporary site road, it is too much if the granular layer is to become the foundation for a permanent road. For the latter case, Powell et al. (1984) present a second equation (rearranged below), where the unbound granular material is type 1 subbase and the rut depth is limited to 40 mm.

$$h = 190 \times [\log_{10}(sa_{design}) + 0.24] \times CBR^{-0.63} \tag{5.7}$$

Using the above example, the thickness of unbound granular material increases to

$$h = 190 \times [\log_{10}(1000) + 0.24] \times 2.5^{-0.63} = 350 \text{ mm (to the nearest 10 mm)} \tag{5.8}$$

The use of both design equations is generally restricted to $sa_{design} \leq 10\,000$. The subgrade CBR used for design purposes should be the lesser of

- the value measured in the trial pits at the time of the ground investigation
- the equilibrium value based on soil type and plasticity index (Powell et al., 1984)
- the value estimated using a lightweight deflectometer, Mexe-probe, TRL DCP or hand vane testing at the time of construction.

Having calculated a suitable thickness for the granular road, the following rationalisation should be adopted.

- If the subgrade is weak and the design CBR falls below 2.5%, formulate an alternative design incorporating a geogrid (see Section 5.2.4.2). As well as saving on the thickness of granular material, the inclusion of a geogrid should give better support for construction plant during placing and compaction.
- For subgrades with a design CBR of 2.5% or more, the calculated thickness of unbound granular material may be adopted, subject to the following provisos
  - Where the design CBR does not exceed 15%, a *minimum* road thickness of 225 mm would be prudent.

Figure 5.2 Summer construction on chalk subgrade – no temporary site road is required

- On non-clay subgrades with a design CBR in the range >15–30%, the road thickness should not be less than 150 mm.
- On subgrades of gravel or weathered rock where the CBR is >30%, it may not be necessary to provide a road structure at all if the ground is relatively free-draining (soil infiltration rate $>10^{-4}$ m/s).
- Subgrades of chalk can be troublesome. Even though low-density chalk can generally be trafficked directly during the drier seasons (Figure 5.2), it may be impassable during winter (particularly if damaged by frost), when the CBR can drop below 2.5%. The site road design should be assessed accordingly. It should be noted that if the site road is also the foundation layer for a permanent public road there may be special requirements in respect of the course thickness and the quality of materials used; the views of the supervising highway authority should be sought.

5.2.4.2 Unbound granular roads incorporating geosynthetics

Geosynthetics include geotextiles (fabrics), geogrids (semi-rigid ribbed meshes) and geocells (honeycombed mattress-like structures). All three types are generally manufactured from polymer-based material (e.g. polypropylene, polyester) and, in the right circumstances, have the potential to reduce the thickness of granular material needed to accommodate any given traffic loading. The overall benefits can be considerable where the subgrade is (very) weak (CBR <2.5%), and may also be worthwhile on stronger ground.

A geotextile covered with compacted unbound granular material to form a road needs the subgrade to deform significantly under wheel loading in order to function as a

Figure 5.3 Interlock of aggregate and geogrid increases lateral restraint (confinement). (Courtesy of Tensar International Ltd)

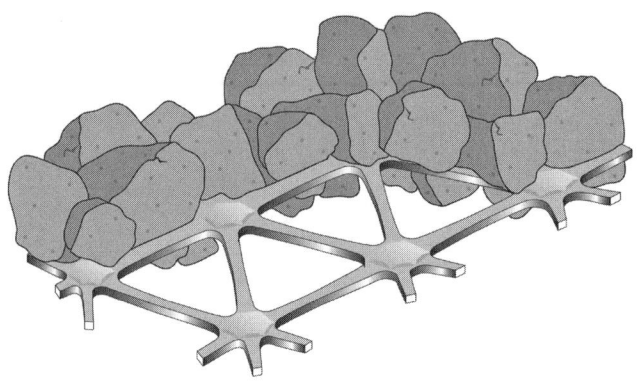

'tensioned membrane' (i.e. a state in which it both confines the granular material and reduces the pressure applied to the subgrade soil). Therefore, a rut must develop before the full contribution of the geotextile occurs; typically the rut depth would be $\gg 75$ mm, which limits the ability of geotextiles to strengthen permanent works. However, a geotextile could prove effective in the special case of a narrow temporary site road where traffic is channelised and a regime for maintenance of rutting is implemented. While a geotextile may not in most circumstances permit a reduction in the thickness of the unbound granular layer, it may confer other benefits. For example, it could prevent the intermixing of the aggregate with a soft cohesive subgrade soil (separation) and filter any soil water rising up into the granular layer on wet sites (filtration). Guidance on designing with geotextiles is available from manufacturers (e.g. Terram, 2017).

By comparison, a horizontal geogrid functions in unpaved road applications by restraining lateral displacement of aggregate subjected to a vertical stress without the inherent need for subgrade deformation; the load-spreading capability of the granular layer is thereby improved and pressure on the subgrade soil reduced. The bearing capacity of the subgrade is also increased as it is relieved of horizontal shear stress at the interface. The net effect may be regarded as a form of mechanical stabilisation (Figures 5.3 and 5.4), which enables the granular layer to be thinned by 30% or more. Design software for site roads (and working platforms) incorporating geogrids is available from a leading manufacturer (Tensar, 2017a). Based on the output from this software, Figure 5.5 compares the thicknesses of granular material needed to accommodate a range of traffic loading with and without TX geogrids. (A minimum road thickness of 150 mm is recommended when TX geogrid is used.) The geogrid can be supplied with a factory-attached geotextile (i.e. as a geocomposite) to provide the combined benefits of confinement, separation and filtration.

Geocells are an extension of the geogrid principle, effectively increasing the depth of the 'ribs' to the full thickness of the granular layer. The quality of aggregate used to fill the

Figure 5.4 Increased restraint (confinement) effected by a geogrid at the base of an unbound granular layer leads to better load spreading and improved subgrade bearing capacity. (Courtesy of Tensar International Ltd)

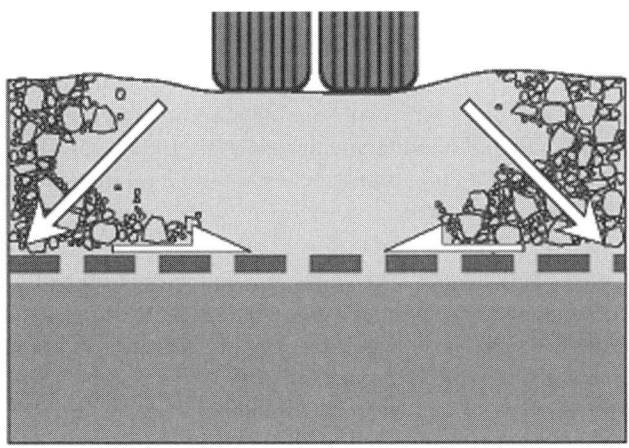

Figure 5.5 Comparison of unbound granular material thicknesses for site roads with and without Tensar TX geogrids (75 mm rut depth)

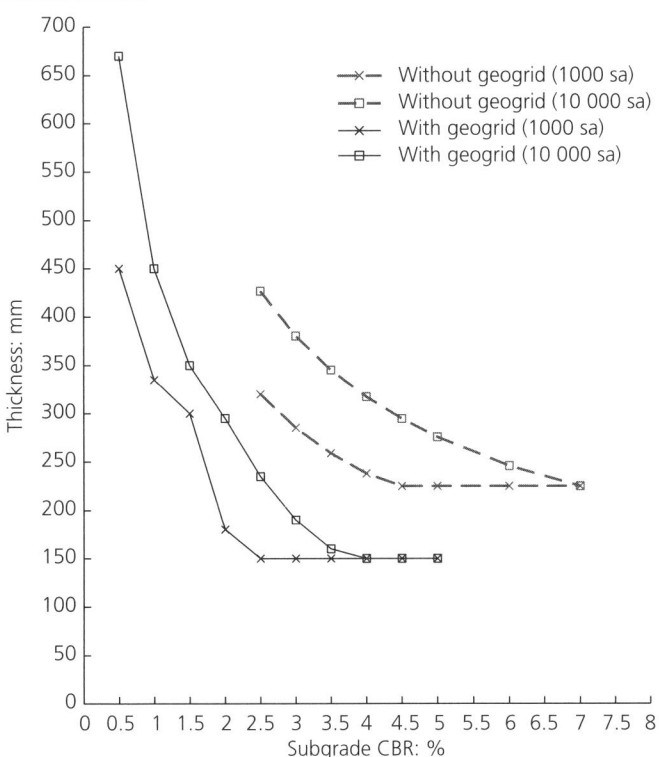

cells may be inferior to that needed with geotextiles or geogrids, thereby recouping some of the extra cost of the geocell structure. Design guidance should be sought from manufacturers.

5.2.4.3 Stabilised soil roads

If the physical and chemical properties of the ground are suitable, forming the road by treating the soil (in situ or ex situ) with one or more hydraulic binders could be an economic alternative to unbound granular material. Details of this technique are discussed fully in Chapter 7 on lime and cement stabilisation. Either of the following two design approaches may be adopted.

- Calculate a suitable thickness of granular material (type 1 subbase) using Equation 5.7 and substitute stabilised soil with a similar stiffness (taken to be 7-day soaked CBR >50%). For temporary roads, a protective granular topping is needed to provide a durable skid-resistant surface course, which will also afford a small measure of frost protection for the stabilised layer over the winter period. The topping should not be thinner than 80 mm (or 2.5 times the D size of the aggregate used) and may be included as part of the design thickness. Stabilised soil should be constructed in layers not thicker than c. 250 mm to ensure that adequate full-depth compaction is achieved by surface rolling; the minimum course thickness should be 150 mm.
- Use a stabilised soil mixture with a higher stiffness than a type 1 subbase and thereby reduce the road thickness needed when compared to unbound granular material. Reference should be made to Chaddock and Atkinson (1997) for the equivalency between a type 1 subbase and a range of lime and cement stabilised mixtures. The minimum course thickness should be 150 mm. A granular topping is still considered desirable, except perhaps where the stabilised mixture has a compressive strength class of C3/4 or higher.

5.3. Working platforms: principles and design

Dedicated platforms are often required at specific locations to support working heavy plant (e.g. a tracked piling rig or crane, or mobile crane outriggers) (Figure 5.6). Platforms may also be needed for falsework to bridge decks and for general scaffolding. In each case, the main function of the platform is to provide reliable support by dissipating the imposed loads sufficiently to avoid yielding of the underlying ground and keep settlement within tolerable limits. A secondary function is to create a working environment that can be kept relatively clean and dry.

For lightly loaded platforms used for storage or supporting falsework on firm ground, a thin concrete blinding layer could suffice (in conjunction with timber sleepers below concentrated loads such as scaffolding standards); a spread of load at 45° through the support is normally assumed for design purposes. At the other extreme, piled foundations might be needed for heavy gantries where the ground is marshy with negligible bearing capacity. Traditional soil mechanics methods can be used for the design of these types of platforms.

Figure 5.6 Granular platforms with geogrids for a tracked sheet piling rig and mobile crane outriggers

5.3.1 Working platforms for tracked plant

Platforms for use by tracked piling rigs and cranes are mostly constructed from unbound granular material (often incorporating horizontal geogrids). Soil stabilisation using hydraulic binders (in situ or ex situ) is an increasingly popular alternative, either as a direct substitute for unbound granular material or to strengthen the subgrade sufficiently to enable a thinner granular platform to serve. Platform thicknesses can vary from a nominal minimum of 300 mm up to 1000 mm or more for the most onerous combinations of loading and subgrade strength.

The frequently used Building Research Establishment (BRE) design method (BRE, 2004) determines the thickness of unbound granular material needed to resist punching shear under the tracks or bearing plates of the operating plant. The method is only applicable where the working platform is significantly stronger than the subgrade and is not appropriate for very soft cohesive subgrades (undrained cohesion <20 kPa). A platform may not be necessary at all where the subgrade comprises stiff clay (undrained cohesion at least 80 kPa). While the BRE method can be used for the thickness design of a platform to support the outriggers of mobile cranes, trafficking by the wheeled vehicles themselves should be assessed in accordance with Section 5.2.

For an unbound granular platform on a very soft cohesive subgrade consideration should be given to improving the strength of the subgrade by treating the soil with one or more hydraulic binders: either using the conventional procedures described in Chapter 7 or, where the subgrade is very soft or marshy to appreciable depth and restricting settlement of the platform is important, adopting deep soil mixing techniques.

Figure 5.7 Comparison of unbound granular material thicknesses for typical piling platform with and without Tensar TX geogrids

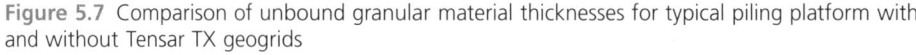

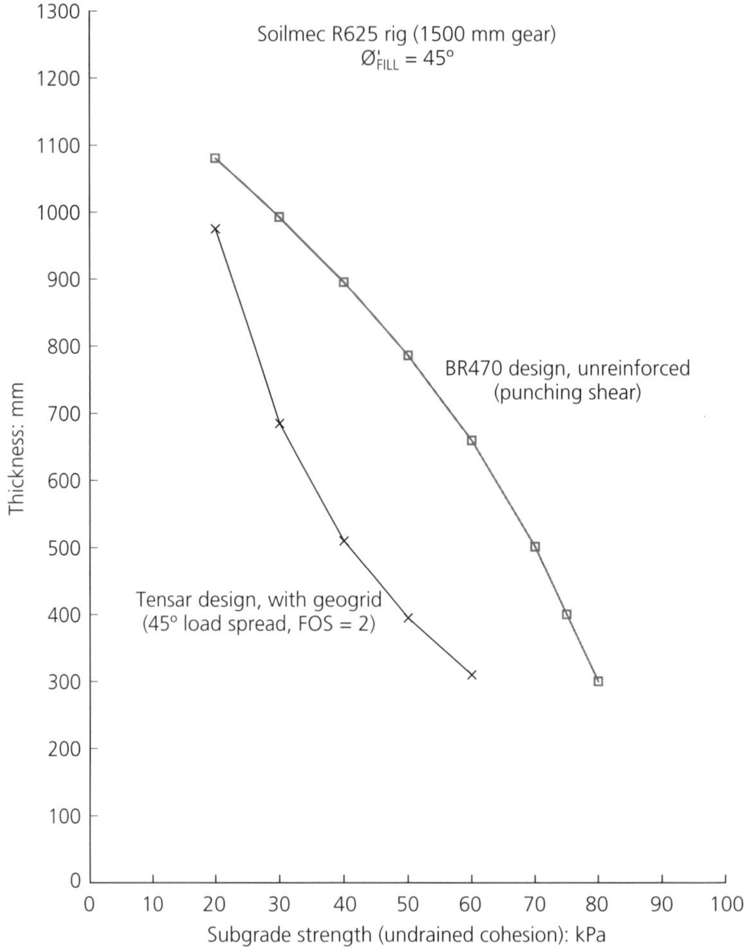

Specialist contractors should be employed for either of these options. Mechanical stabilisation of an unbound granular platform using suitable geogrids is cost-effective in that it can reduce the design thickness of the platform by 30% or more (Figure 5.7); it is also an option where the subgrade has an undrained cohesion as low as 10 kPa, although this would necessitate a specialist design. One design method (Tensar, 2017a) assumes a spread of load through the reinforced granular layer of 45°, the platform being made sufficiently thick to limit the pressure on the subgrade to less than its safe bearing capacity. As with site roads, a geogrid at the base of the granular layer serves to restrain outward displacement of aggregate subjected to a vertical stress, thereby improving load-spreading ability and increasing subgrade bearing capacity. For platforms more than c. 400 mm thick, further geogrids placed horizontally within the body of the granular layer are needed to extend the zone of confinement.

A sound understanding of ground conditions in the environs of the working platform is an integral part of the design process. A programme of ground investigation as outlined for site roads should be undertaken, with emphasis on undrained shear strength rather than CBR as the fundamental design property for cohesive subgrades. The investigation would need to be expanded to include standard penetration testing (to estimate the angle of shearing resistance) should the subgrade prove to be of a granular nature.

It is generally recommended that the gradient of platforms used by tracked cranes and piling rigs with raised masts be restricted to 10% (which includes access ramps), although a level surface is preferable. At the base of a ramp it is good practice to increase the platform thickness by 50%.

5.4. Temporary roads for public use

On some construction projects, particularly highway works, it may be found necessary to close a section of a public road for a period. When no suitable diversion route exists, there may be no option but to build a temporary road within the site for use by public traffic. As the road must be paved to normal highway standards, the route will always be made as short as possible in order to minimise costs.

Road alignment, marking and signage will need to be approved by the local highway authority; temporary lighting may be stipulated at potentially hazardous locations (e.g. short radius bends and junctions).

The thickness design of the temporary road pavement and its foundation can be based on well-established principles relating to traffic loading (number of standard axles) and subgrade strength. However, the intended short life means that the current design charts for highway pavements (HA, 2006) will not necessarily give an economic design, as the minimum thickness limit for the bound layer(s) has been set relatively high.

Reference could be made to earlier design guides for bituminous roads (Powell *et al.*, 1984) and concrete roads (Mayhew and Harding, 1987) where data are provided for traffic loading as low as 100 000 sa.

For even lower traffic loadings, Figure 5.3 in Powell *et al.* (1984) could be extrapolated back to 1000 sa using the following equation to estimate the required thickness of bitumen-bound material, h_{DBM100} (in mm)

$$h_{DBM100} = 190(sa_{temp} \times 10^{-6})^{0.2} \tag{5.9}$$

where sa_{temp} is the traffic loading (in standard axles) expected over the lifetime of the completed road. The bitumen-bound material should comprise a binder course and a surface course; the surface course should have adequate skidding resistance.

The thickness h (in mm) of type 1 subbase needed as the road foundation may be derived from Equation 5.7, where sa_{design} is a measure of the traffic (in standard axles) employed in the building of the road.

For example, consider a temporary public road diversion 200 m long built on a subgrade with a CBR of 6% to carry 5000 sa after construction

$$sa_{design} = 1000 \text{ (minimum recommended input)}; sa_{temp} = 5000 \quad (5.10)$$

$$h = 190 \times [\log_{10}(1000) + 0.24] \times 6^{0.63} = @200 \text{ mm type 1 subbase}$$
$$\text{(to next 10 mm)} \quad (5.11)$$

$$h_{DBM100} = 190 \times (5000 \times 10^6)^{0.2} = 70 \text{ mm DBM100 (to next 10 mm)} \quad (5.12)$$

In practice, it would be unwise to use a subbase thinner than 225 mm as a platform on which to place and compact bitumen-bound material (relaxed to 150 mm should the subgrade CBR exceed 15%). Also, where the thickness of the bound layer is less than 110 mm, there is a risk of premature failure occurring as a consequence of occasional very heavy wheel loading on hot days.

Because temporary roads can be relatively thin they may suffer winter damage when a prolonged period of freezing is experienced and the subgrade is highly frost-susceptible (e.g. chalk). There is no way to mitigate this risk cheaply and the consequences of premature failure during a harsh winter need to be assessed.

5.5. Construction
5.5.1 Materials and maintenance

Granular material for building temporary roads or platforms may be derived from site if suitable deposits of gravel or moderately strong rock occur naturally in the ground; on brownfield sites it should be possible to process demolition arisings to create acceptable granular material. If site-won granular material is in short supply, a design incorporating geogrids is favoured in order to limit the volume of imported material needed; in these circumstances, soil stabilisation also becomes an attractive option.

For convenience, granular materials are classified in accordance with the 'Specification for highway works' (HA, 1998) and should be compacted accordingly. The standard designs for roads (Equations 5.6 and 5.7) require the use of well-graded crushed rock gravel or type 1 subbase. For lower-quality material, for example, type 2 subbase or class 6F2/6F5 capping, an increase in the design thickness is needed. This could be as much as 100% based on material equivalency (BSI, 2001), although realistically a 50% increase is considered more appropriate providing the lower-quality materials have negligible plasticity and are relatively free-draining. No such increase in the design thickness is considered necessary for granular roads incorporating geogrids, although it is important that the maximum particle size of the aggregate remains compatible with the selected geogrid mesh size (Tensar, 2017b). The use of granular material with a rounded particle shape should be avoided (at least near the surface of the road) because it may lack stability under wheel contact pressure.

Where secondary (recycled) aggregates are imported (these will normally be cheaper than primary aggregates) they must have been recovered under an acceptable quality protocol (e.g. WRAP, 2013). Particular consideration should be given to any potential they may have for leaching contaminants and to how they will be disposed of on completion of the works.

For working platforms where the design criteria are different, the angle of shearing resistance (ϕ') of the granular material must be known. This can be determined by a large shear box test in the laboratory but, as the maximum particle size that can be included in the test specimen is 20 mm, the result may underestimate the potential of the material. It is the author's experience that for 75 mm maximum size, well-graded, moderately strong crushed rock or crushed concrete, $\phi' = 45°$ may be safely presumed, provided that the material is non-plastic and relatively free-draining. At the other extreme, naturally occurring sand/gravel deposits with slight plasticity complying with the requirements for type 2 subbase (commonly referred to as 'hoggin') will have $\phi' \sim 35°$.

Site roads and working platforms constructed from unbound granular material must be kept clear of mud and slurry otherwise the desirable free-draining properties will be adversely affected. Any localised deep rutting or haunch damage should be reinstated with fresh material at the earliest opportunity, or progressive failure will result.

Soil stabilisation is a specialist operation and should not be attempted by the inexperienced (see Chapter 7). A strong material (equivalent to or stiffer than unbound granular material) can be formed from most soil types by using an appropriate binder. The binder may be Portland cement, quicklime, fly ash, ground granulated blast furnace slag or combinations of these; alternative proprietary binders are also available. It is important for a road or platform constructed from stabilised soil to be protected against surface abrasion by site traffic, standing water and frost. To this end, a topping of granular material should be provided, except perhaps in cases where the stabilised mixture has a compressive strength class of C3/4 or higher. One disadvantage of stabilised soil construction is that, should premature failure occur, it is not usually practicable to replace the damaged layer with like material. The only option is to dig out the affected area and substitute compacted granular material, which might result in a recurrent 'soft spot'.

On small sites, or for short-term and one-off applications, the placing of proprietary mats could be considered for traffic or plant access over soft ground. Access mats are made variously from heavy-duty flexible plastic composites, aluminium extrusions/plates or timber, and can be bought, or more usually hired, from specialist outlets. Some types are lightweight and capable of being man-handled into position; these can be used to good effect where damage to existing surfacing (e.g. turf) must be avoided. Access mats function by increasing the load contact area; the restraint offered by adjoining linked mats serves to restrict deformation. However, they can prove to be prohibitively expensive for long-term or widespread use.

Inevitably, traffic leaving a construction site will have picked up mud from the site roads or piling platforms. The provision of a wheel-wash facility (or at least a rumble strip) close to the exit will help minimise the spread of mud to adjoining public roads. On larger sites where bulk excavation is taking place, a purpose-made wheel/vehicle wash unit is essential (models are available that recycle the wash water and extract/dewater the suspended solids).

It is important that plant operators and drivers using the site roads and working platforms are aware of the layout and boundaries. Edges should be defined by fencing or timber sleepers; in circumstances where there is an abrupt drop over the edge, a physical restraint should be provided, which might comprise an earth bank or prefabricated barrier units.

5.5.2 Drainage

During inclement weather rainwater could accumulate in unbound granular construction layers, resulting in softening of the underlying stabilised soil (where used) or cohesive subgrade (particularly where the soil infiltration rate is $<10^{-6}$ m/s). Potentially, this could lead to early failure of the site road or working platform.

For site roads, the individual courses and the subgrade should be shaped to a crossfall (5% if feasible, but 2.5% minimum) to shed water, the runoff being channelled away from the edges of the road in drains or temporary shallow open ditches ('grips').

Working platforms may be much wider than site roads and providing a crossfall is not necessarily practicable. It is also desirable that the surface of the platform be relatively level to avoid construction plant tilting. Shaping them to fall in two, or even four, directions towards the edges may be a solution, in conjunction with perimeter drains or 'grips'. Alternatively, for very wide platforms constructed entirely of unbound granular material, shallow land drains could be installed in a herringbone pattern below the platform to channel any water reaching the subgrade to a common outfall; these drains need not be piped, but may simply be slots in the subgrade filled with a free-draining granular drainage medium.

When disposing of water drained from site roads and working platforms it is essential to avoid pollution of aquifers and watercourses. Recourse to silt traps and oil interceptors may be needed and Environment Agency guidance should be sought.

Drainage of surface water from a paved temporary public road would normally be to a filter drain on the low side of the carriageway with an outfall into the existing highway drainage system. In built-up areas, kerbs could be used to direct surface water back to the main highway.

REFERENCES

Black WPM and Lister NW (1979) *The Strength of Clay Fill Subgrades: Its Prediction and Relation to Road Performance*. Transport and Road Research Laboratory (TRRL), Crowthorne, UK, TRRL Laboratory Report 889.

BRE (Building Research Establishment) (2004) *Working Platforms for Tracked Plant*. IHS BRE Press, Bracknell, UK, BR 470.

Britpave (British Cementitious Paving Association) (2007) *Concrete Hardstanding – Design Handbook. Guidelines for the Design of Concrete Hardstandings*, 2nd edn. Britpave, Camberley, UK.

BSI (British Standards Institution) (2001) BS 7533-1:2001. *Pavements constructed with clay, natural stone or concrete pavers. Guide for the structural design of heavy duty pavements constructed of clay pavers or precast concrete paving blocks.* BSI, London, UK.

Chaddock BCJ and Atkinson VM (1997) *Stabilised Sub-bases in Road Foundations: Structural Assessment and Benefits.* Transport Research Laboratory, Wokingham, UK, TRL Report 248.

Garvin S (2016) *Soakaway Design.* IHS BRE Press, Watford, UK, DG 365.

Giroud JP and Noiray L (1981) Geotextile-reinforced unpaved road design. *Proceedings of the American Society of Civil Engineers, Journal of the Geotechnical Engineering Division* **107(9)**: 1233–1254.

HA (Highways Agency) (1998) Specification for highway works. *Manual of Contract Documents for Highway Works*, vol. 1. HMSO, London, UK (amended 2016).

HA (2006) Pavement design. *Design Manual for Roads and Bridges*, vol. 7, section 2, part 3. HMSO, London, UK, HD 26/06.

HA (2007) Treatment of fill and capping materials using either lime or cement or both. *Design Manual for Roads and Bridges*, vol. 4, section 1, part 6. HMSO, London, UK, HA 74/07.

HA (2009) *Design Guidance for Road Pavement Foundations.* HMSO, London, UK, Draft HD 25, Interim Advice Note 73/06 Revision 1.

Mayhew HC and Harding HM (1987) *Thickness Design of Concrete Roads.* Transport and Road Research Laboratory (TRRL), Crowthorne, UK, TRRL Research Report 87.

Powell WD, Potter JF, Mayhew HC and Nunn ME (1984) *The Structural Design of Bituminous Roads.* Transport and Road Research Laboratory, Crowthorne, UK, TRRL Report LR1132.

Tensar (2017a) TensarPave. Version 7.00.10. Tensar International Limited, Blackburn, UK.

Tensar (2017b) *Selection of the Appropriate Tensar TriAx Geogrid.* Tensar Information Bulletin. Tensar International Limited, Blackburn, UK.

Terram (2017) Road & Highways Solutions. Terram, Maldon, UK. See http://www.terram.com/downloads (accessed 01/08/2018).

WRAP (Waste and Resources Action Programme) (2013) *Guidance Notes for the Producers' Compliance Checklist.* WRAP, Banbury, UK.

FURTHER READING

Britpave (British Cementitious Paving Association) (2007) *HBM and Stabilisation 2: The Design and Specification of Residential and Commercial Road Pavements.* Britpave, Camberley, UK.

Jewell RA (1996) *Soil Reinforcement with Geotextiles.* Construction Industry Research and Information Association (CIRIA), London, UK, CIRIA SP123.

WRAP (Waste and Resources Action Programme) (2006) *Guidance on the Use of HBM in Working Platforms.* WRAP, Banbury, UK.

Temporary Works, Second edition

Pallett, Peter F and Filip, Ray
ISBN 978-0-7277-6338-9
https://doi.org/10.1680/twse.63389.087
ICE Publishing: All rights reserved

Chapter 6
Control of groundwater

Neil Smith
Director, Applied Geotechnical Engineering

The main purposes of controlling groundwater are to enable the construction of below-ground works in the dry and to improve the stability of the ground below the sides and base of an excavation. In order to plan and design a control scheme, it is necessary to obtain good-quality information on the stratigraphy, the groundwater regime and the mass permeability of the ground. A number of techniques are available, each being applicable to particular circumstances. The principal parameters governing design are permeability and drawdown. Even with good site investigation, the mass permeability of the ground surrounding a proposed excavation is only approximately known. Sensitivity analysis is therefore an important part of the design process, to give the designer an appreciation of the possible range of flows needed to produce the required outcome. Monitoring of groundwater control schemes is essential to ensure that sufficient drawdown or pressure reduction is achieved and maintained. Environmental aspects must also be considered, to check the quality of the water being extracted from the ground and of the water being discharged from the process. Consents are required, both to abstract the water from the ground and to dispose of it.

6.1. Introduction

Where excavations are required to extend below the groundwater level, it is essential to plan beforehand the means of dealing with the water. There are generally four possible methods

1 Exclude the water by surrounding the area to be excavated with a cut-off wall (there are various ways of doing this; see Chapters 8–11, 13 and 14).
2 Extract sufficient groundwater to depress the level below the excavation surface. (Methods 1 and 2 may be used in combination, where a partial cut-off reduces the volume of water to be extracted.)
3 Reduce the pore water pressure in the soil sufficiently to stabilise the ground around and below the excavation.
4 Excavate below the water (see Chapter 15 on caissons and shafts).

Groundwater control can be achieved by either method 2 or method 3. It is important to realise that low flow does not necessarily mean low pressure and high flow does not necessarily require a high pressure. The techniques for controlling groundwater are

- sump pumping
- wellpointing
- deep wells
- ejectors.

The rarely used technique of electro-osmosis is based on different principles and will not be considered here.

6.2. Techniques
6.2.1 Selection of appropriate technique

Figure 6.1 illustrates the ranges of application of the various different pumped systems for groundwater control. The blurred boundaries are deliberate to emphasise their approximate nature. The particle sizes shown below the permeability axis are a guide only. Note that where there are strata of widely differing permeabilities, it may be necessary to use more than one technique.

6.2.2 Sump pumping: extraction from the surface of the excavation

This is probably the most commonly used system of groundwater control. At its simplest, it consists of a shallow well (such as a perforated 45 gallon drum) placed into a hole excavated below the general base level of an excavation, with a suction pump inlet within

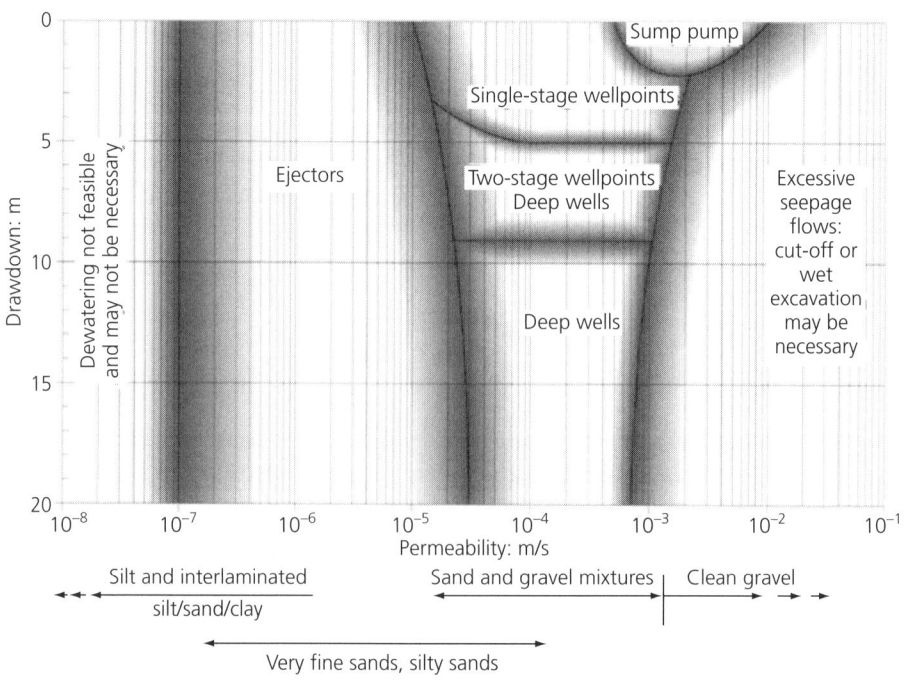

Figure 6.1 Relations between pumped well systems, permeability and drawdown. (Adapted from Roberts and Preene (1994))

the drum. One of the drawbacks of this method is that the water flows towards the excavation and may issue from the lower sides as well as the base, which can reduce the stability of the soil close to the excavation surface. The sump can be fed by a pattern of trenches or French drains leading from the perimeter of the excavation. Sump intakes can be filtered using granular, geotextile wrapping or combined filters.

6.2.3 Deep wells, wellpoints and ejectors: extraction from below the excavation

Wellpoints work by suction pumping, and so the head which they can lift is limited by atmospheric pressure and suction efficiency. Figure 6.2 illustrates the main parts of a wellpoint.

Note that the header pipes, which connect a number of wellpoints to a pump, can obstruct surface activities, as can be seen from Figure 6.3 (in which the tops of wellpoints and pumps can also be seen).

In practice, the achievable drawdown is limited to about 5 m but the use of a second stage of wellpoints can increase the drawdown to around 9 m. In rare instances, multiple levels of wellpoint can be used (Figure 6.4). Wellpoints can be used to enclose an excavation or simply to dewater a length of trench, for example, to install services. To dewater long trenches, sections of wellpoints which extend beyond both ends of the open trench are activated progressively as the work proceeds.

Figure 6.2 The main features of a wellpoint

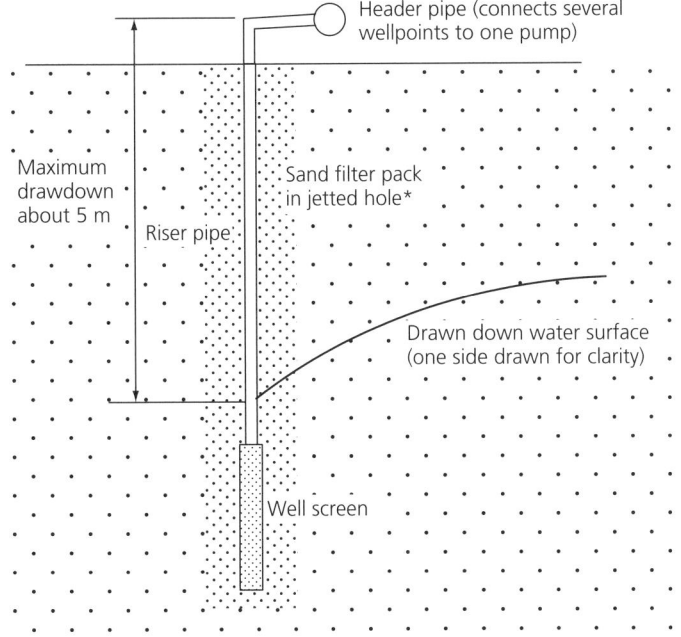

*A filter is essential in stratified soil. May not be needed in well-graded sandy gravel

Figure 6.3 Wellpoint header pipe around excavation. (Courtesy of WJ Groundwater Ltd)

Deep wells are not limited in the drawdown they can achieve because the water is sucked into the bottom of the well and then pushed up the discharge pipe. Pumps operate from within prepared wells which have had lining tubes and surrounding filters previously installed.

Ejectors work on the principle of a nozzle and Venturi pump. Water fed to the base of a hole is forced to flow back up the hole through a nozzle and Venturi to pull groundwater

Figure 6.4 Three stage wellpoint system. Note the rise of the water level below the excavation; a similar rise is present with deep wells

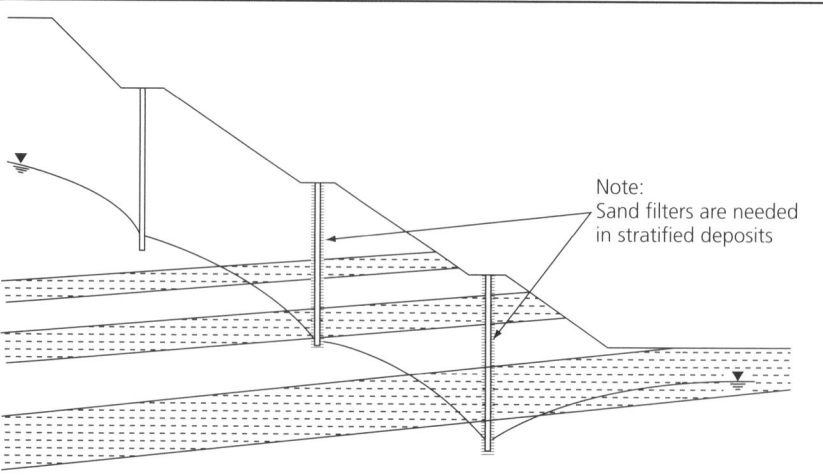

in from below. The practical depth limit on their use is about 50 m. They tend to be used for the control of pore pressure in fine-grained soil. (There is some similarity with the airlift system in that fluid is introduced at the bottom of a discharge pipe (riser) to cause an upward flow.)

6.2.3.1 Development
It is important to 'develop' deep wells. This is, effectively, a cleaning process to remove drilling debris from the hole and also, in suitable ground, wash out fine particles from joints or fissures, thus increasing the effective size of the well. Removal of debris before installing the pump also reduces pump wear. In carbonate rocks such as limestone or chalk, concentrated hydrochloric acid can be added to dissolve slurry and clean fissures. This process releases carbon dioxide in large quantities and should only be carried out by appropriately experienced personnel.

6.2.4 Filters
Whenever groundwater is pumped from a soil, seepage pressures are set up at the point of entry of the water into the well. The seepage pressures can dislodge particles from the soil near the borehole wall so that the flow carries them into the system. This can have deleterious effects both on the system and the ground. Excessive sediment transport will cause wear or clogging to the pump and pipework. Removal of fines from the ground is a common source of settlement of ground in the vicinity of dewatering schemes.

In order to prevent the removal of soil particles, filters are installed. Traditionally, these were made from soil with a grading specified to meet criteria relating the particle size distribution of the filter to that of the aquifer. In some cases, two layers of filter were needed. Filter rules are described in detail in Preene *et al.* (2016).

Woven geotextiles have been developed to provide sophisticated filtering systems. Manufacturers should be consulted for information on the attributes of their particular products.

6.2.5 Hazards
Hazards resulting from the application of the techniques described above include the following

- *Settlement*: when the groundwater level is reduced, the effective stress between soil particles increases. At shallow depth and in compressible soil, this increase can be significant, causing settlement of the ground in the drawdown zone surrounding the system and consequential damage to structures.
- *Removal of fines*: if the filtration system does not function properly, the flow of water can lead to the extraction of fines from the surrounding ground and the creation of voids. In some circumstances, this can lead to catastrophic collapse of the surface.
- *Fouling and clogging*: the effectiveness of wells, wellpoints and ejectors can deteriorate over time because of fouling of the screen by bacterial growth and/or encrustation by chemicals precipitated from the water.

- *Corrosion*: some groundwater (e.g. saline water at sites near the sea) has a chemical composition which may cause corrosion of pipework and pumps.

6.2.6 Recharge

Where settlement is considered to present a hazard, the effects of dewatering can be ameliorated by pumping water back into the ground within the area of drawdown outside the excavation. This will change the groundwater profile during system operation and reduce the drawdown below the surrounding ground.

6.2.7 Monitoring and maintenance

It is essential to confirm that the dewatering system is effective and that it continues to be effective. Plant which is on standby should be regularly checked and run. Groundwater levels must be measured sufficiently frequently over a wide enough area. The flow rate from the system should be checked and collated with groundwater levels. In normal circumstances, the initial pumping capacity used to draw down the water level will be significantly greater than (perhaps twice) that needed to maintain the reduced level. Monitoring the initial stage will show when some of the system can be shut down and put onto standby in case of mechanical or other problems. The occurrence of fouling or clogging is a gradual process and should be monitored by continuing measurement of water levels and flow rates at suitable intervals. In some circumstances it may be necessary to replace wells that have become ineffective. If there are substantial differences between the predicted and actual flow and/or drawdown, these should be assessed carefully.

It may also be necessary to monitor water quality (both the chemistry and the concentration of suspended solids), particularly for discharge or recharge purposes. Where the ground conditions dictate, accurate settlement monitoring should be undertaken; this should begin well before the dewatering is turned on. Note that in some areas close to tidal water, the ground surface can move with the tide. Any natural variation in level could make it very difficult to interpret ground-level data during operation of the groundwater-lowering system.

6.2.8 Consents

It is normally necessary to obtain a license from the Environment Agency to extract more than 20 m^3/day of water from the ground. The abstracted water is classified in a legal sense as trade effluent. As such, a discharge permit is required before the water can be fed to surface water, the sea, watercourses or recharged back into the ground. Note that, even if nothing else is done to the abstracted water, exposure to the atmosphere may cause chemical changes to occur. Advice about obtaining permits from the Environment Agency can be found on its website.

6.3. Investigation for dewatering

The stratigraphy of the area which will be affected by the dewatering can be found from an array of suitably deep boreholes. Samples should be taken to allow the various strata to be classified by testing and their structure be determined by splitting open and describing samples of intact soil. It is necessary to establish the current groundwater level in the vicinity of the site, noting that the water may be flowing and that it may vary with time, for example, responding to tidal fluctuations near the coast or that it may be variably recharged by changes in an adjacent river level.

Direct measurements of permeability can be made in a number of ways, but there are significant drawbacks to most methods. Testing can be carried out in situ during drilling of site investigation boreholes. However, the ground just below the bottom of the borehole may have been disturbed and there may be problems in allowing enough time to establish an accurate rest level. In addition, sediment can be deposited at the bottom of the borehole as a thin low-permeability layer, especially during inflow tests. If the head is reduced too far for a rising head test, the ground around the base of the hole may be disturbed by 'blowing'. Where piezometers are installed, tests can be carried out that are similar to the in-borehole tests. The rest level should have been determined by repeated measurements and the risk of silting or blowing should be minimal. However, tests in boreholes and piezometers suffer from the small volume of ground tested. In most circumstances, it is necessary to carry out a number of tests in different places at different depths.

Permeability tests can be carried out on samples in the laboratory, but sample disturbance may be a problem. Tests which pass a vertical flow through a tube sample may measure a misleadingly low permeability if the sample is anisotropic, as many soils are. To obtain sufficient flow in low-permeability soils, it is generally necessary to apply a much higher hydraulic gradient to the specimen than would occur in practice; this can lead to misleading results. The sample size may not be large enough to include representative fabric (soil structure). Particle size distributions can be used to provide an estimate of permeability in certain soils – uniform sands are most suited to this (see Equation 6.2). In a layered soil, the different soil layers should be separated before taking specimens for particle size distribution testing.

The best way to measure permeability at the scale required for a dewatering scheme design is by a pumping test. This is effectively a small-scale trial of the dewatering system. The test pumping well should be surrounded by a number of monitoring wells so that the drawdown curve can be determined, ideally in two orthogonal directions. These tests are relatively expensive and take between 1 and 2 weeks to conduct. Their major advantage is that the mass permeability of the soil is measured under conditions which are similar to those that will confront the actual dewatering system.

6.4. Analysis and design
6.4.1 Steps in analysis and design
- Draw ground cross-sections to establish the stratigraphy.
- Measure the permeability k and determine a most probable value and a range for all strata.
- Find the equilibrium groundwater level and consider whether this may vary (long-term monitoring data may be available for the locality from the Environment Agency).
- Determine the required drawdown (depth and area).

- Consider the type of system needed (see Figure 6.1).
- Idealise the system: well, slot or combination.
- Estimate the potential flow rate (most probable and range).
- Detail the system (e.g. determine depths and spacings for wellpoints or deep wells).

6.4.2 Permeability

One-dimensional flow in saturated soil is governed by Darcy's law, which states that the velocity of laminar fluid flow through a soil is proportional to the gradient of fluid pressure (head) between the two ends of the element. Turbulent flow in soil is only likely in coarse gravels with high flow. Darcy's law is defined as

$$q = -k\frac{(h_o - h_i)}{l} \tag{6.1}$$

where q is the velocity of flow (m/s), h_o is the head (m) at the point where the fluid emerges from the element, h_i is the head (m) at the point where the fluid enters the element, l is the length (m) of the element and k is the coefficient of proportionality (m/s), referred to as the permeability.

The flow is positive in the direction of negative gradient. Between clay and coarse gravel there is a huge range of permeability, approximately 10^{-10} to 10^{-1} m/s. (An equivalent range of distance is from 1 m to 2.5 times the distance from the Earth to the Moon.) In highly favourable conditions, the most accurate determination of permeability is within a factor of around 3.

The permeability is related to pore size, which in uniform soil is related to the particle size. Hazen's formula can be reliable for some sands and gives a general indication of the order of magnitude of k (in m/s)

$$k = 0.01 D_{10}^2 \tag{6.2}$$

where D_{10} is the sieve size (mm) which allows 10% of the soil by weight to pass through.

Permeability is not a constant, because the pore size decreases if the soil is compressed. This can be important in soft soil (e.g. organic clay), where significant changes in permeability can occur as a result of consolidation, which may be a result of the groundwater lowering.

Soil is not homogeneous. Even in what might appear to be a uniform soil the permeability is almost always greater in the horizontal than the vertical direction. Many soils are layered. Thin clay layers in a sand stratum can dramatically reduce the vertical mass permeability. Glacial soils are very heterogeneous. Predominantly clayey glacial soil can contain pockets, lenses or ribbons of coarse granular material that may provide a conduit to a source of water.

Equation 6.1 shows that the rate of flow is also proportional to the difference in head along the element. In a dewatering scheme this is equivalent to the drawdown, that is, the difference between the equilibrium water level and the lower level required for construction.

Ground cross-sections are seldom uniform and simplifications in geometry may be needed to make analysis possible. The mass permeabilities of the different strata are key parameters in design.

6.4.3 Analysis of pumped well dewatering systems

The flow of water to a dewatering system is three-dimensional. In many circumstances, however, the groundwater regime is simplified to a two-dimensional system using either planar or radial coordinates. Plane flow can be used for lines of wells, for example, dewatering a trench for the installation of services. The idealised equivalent to a row of wells is termed a 'slot'. Radial flow applies in the case of flow to single wells; where a group of wells surrounds an excavation it can be treated as an equivalent single well or as a combination of slots for the sides and wells for the corners.

Figure 6.5 shows the two principal 'ideal' aquifer conditions (confined and unconfined) with partially penetrating wells or slots. Fully penetrating wells, as the name suggests, reach down to the top of the impermeable base layer.

Equations have been derived for volumetric flow rates to wells and slots for confined and unconfined aquifers. As might be expected, the principal variables are the permeability and the required drawdown. In determining the required drawdown, allowance must be made for the increase in head with distance from the well. In the case of a group of wells around an excavation, the head at the centre of the group is higher than the head at the wells themselves (see Figure 6.3).

The expression for radial flow to a fully penetrating well in a confined aquifer is given by the Theim equation

$$Q = \frac{2\pi k D(H - h_w)}{\ln(R_o/r_e)} \qquad (6.3)$$

where Q is the well discharge rate (m³/s), k is the aquifer permeability (m/s), D is the depth of the aquifer (m), H, h_w and R_o are distances of the piezometers from the pumping (m)s, and r_e is the equivalent well radius (m) (see Figure 6.5).

There are similar relationships for a well in an unconfined aquifer and for planar flow to slots in confined and unconfined aquifers. The equations, together with factors for flows from partially penetrating wells and slots, are given in Preene et al. (2016).

The parameters are defined in Figure 6.5. Note that r_e is the *equivalent* well radius. Where a group of wells is distributed around a rectangle $a \times b$, r_e may be taken as $(a + b)/\pi$. If only a single well is being considered, $r_e = r_w$ (i.e. the well radius).

Figure 6.5 Definitions of parameters for flows to wells and slots

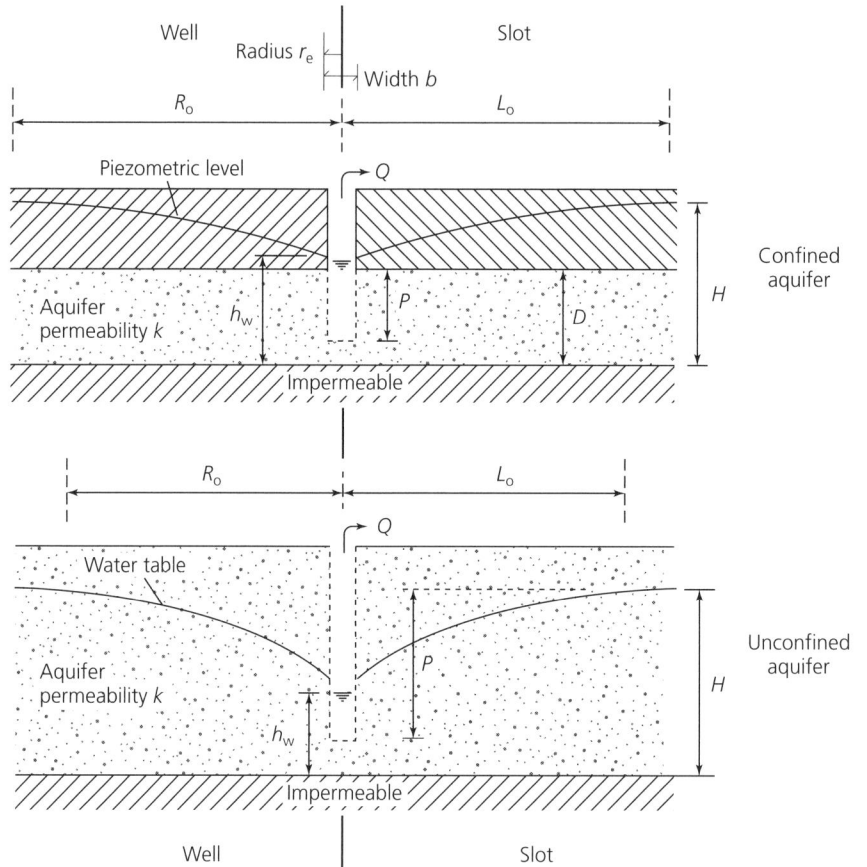

When three-dimensional analysis is required, or for complex two-dimensional geometries, computer programs providing numerical solutions are available. It is, however, advisable to engage the services of a groundwater specialist to make such analyses. Bond (1994) provides advice on the use of geotechnical software.

However sophisticated the analysis, the design of the dewatering system must take account of the uncertainties in both the actual values of permeability and in the structure of the ground affected by the process. It is essential to monitor the effectiveness of the system, ideally starting the process with enough time allowed in the programme to adjust the design if the drawdown is insufficient. When starting a dewatering scheme, it is wise to be aware of where and how quickly additional resources can be obtained (if they are not actually held in reserve). As far as pump capacity is concerned, an immediate need to increase capacity can often be met by use of a standby pump(s), which should be on hand in case of breakdowns.

REFERENCES

Bond A (ed.) (1994) *Validation and Use of Geotechnical Software*. Association of Geotechnical and Geoenvironmental Specialists, Bromley, UK.

Preene M, Roberts TOL and Powrie W (2016) *Groundwater Control – Design and Practice*, 2nd edn. Construction Industry Research and Information Association (CIRIA), London, UK, Report C750.

Roberts TOL and Preene M (1994) Range of application of groundwater control systems. In *Groundwater Problems in Urban Areas* (Wilkinson WB (ed.)). Thomas Telford, London, UK, pp. 415–423.

FURTHER READING

Roberts TOL and Preene M (1994) The design of groundwater control systems using the observational method. *Géotechnique* **44(4)**: 727–734.

Useful web addresses

Environment Agency – information on the regulations surrounding abstraction and disposal of groundwater: https://www.gov.uk/government/organisations/environment-agency (accessed 01/08/2018).

Geotechnical & Geoenvironmental Software Directory – a comprehensive listing of available geotechnical programs, including various groundwater categories (within the geoenvironmental section): http://www.ggsd.com (accessed 01/08/2018).

Pallett, Peter F and Filip, Ray
ISBN 978-0-7277-6338-9
https://doi.org/10.1680/twse.63389.099
ICE Publishing: All rights reserved

Chapter 7
Lime and cement stabilisation

Mike Brice
Associate, Applied Geotechnical Engineering

Neil Smith
Director, Applied Geotechnical Engineering

Weak soils can be improved in strength and workability by the addition of lime and/or cement. Other materials such as pulverised fuel ash (PFA) or ground granulated blast furnace slag can be used, but are not covered in this chapter. However, the principles of application are the same. The use of the process as a temporary measure is generally to provide a sound working surface for construction (although the treated soil has been altered permanently). Lime is usually applied to the soil in the form of quicklime. This has an immediate effect in taking moisture from wet soils and is especially effective in soft, wet, plastic clay. The lime also reduces the plasticity of the clay and so makes it both stronger and more workable. If added in sufficient quantity, it produces a cementing effect with considerable additional increase in strength over time. Cement also reduces the moisture content and plasticity, although less effectively than lime. It does, however, have a greater cementitious effect. It is more suitable for treating lower plasticity or non-plastic soils. The results of adding lime or cement are very soil-specific, so preliminary testing is essential to ensure that the desired effect(s) can be achieved and to find the optimum concentration and type of additive. The effects vary with plasticity and other properties, so that a substantial test programme is needed in variable soils such as glacial till. Blends of lime and cement can be used and other cementitious agents, such as PFA, may be incorporated.

7.1. Introduction

This chapter deals with the use of lime and cement additives to improve the strength of near-surface soils for temporary works construction. In a temporary works context, the addition of lime or cement is most likely to be required in the context of haul road or working platform improvement (see Chapter 5), and in the rendering of wet material suitable for use as a general fill. In some circumstances, such as the enhancement of thick soft deposits, deep methods such as lime mixing or lime columns may be beneficial. These require highly specialised plant and are of a different character to superficial lime and cement mixing. They are not considered further here.

The two additives differ in their actions: lime is of more benefit in the rapid drying of soils and as an additive to improve 'heavy' plastic clay soils. In low plasticity clay soils (plasticity index (PI) < 10%) and in predominantly granular soils lime may be used as a drying agent, but otherwise its effectiveness is limited and cement is of greater benefit. The chemical composition of the soil to be treated is an important factor in determining the most suitable admixtures. The mineralogy of the clay-size fractions may also be relevant, but this would most usually be addressed through plasticity testing.

7.2. Materials and their effects on the soil
7.2.1 Lime

Lime is available in two forms: quicklime and slaked lime. In Europe quicklime is most commonly used for soil treatment, being significantly more effective in drying out a soil by the heat of hydration than slaked lime, and is less dusty to handle. Only quicklime is discussed below.

In the first instance, the addition of lime dries the soil both by absorbing water directly (up to 32% of the dry lime weight) and by driving off water by the exothermic nature of the hydration reaction. As a rule of thumb, the reduction in soil moisture content amounts to approximately 1% per percentage point of added quicklime. This effect is seen in all soils. In cohesive soils there is also a chemical reaction with clay minerals in the soil; the lime will alter the nature of the clay minerals, greatly increasing the plastic limit and thereby reducing the plasticity. Both of these reactions are immediate and are not affected by ambient temperature. They can combine to transform a soft plastic soil into a dry friable material, making it significantly stronger and easier to work. Together they are termed 'soil modification'. The effectiveness of the second reaction is dependent on the nature of the clay minerals within the soil (montmorillonite being more affected than illite and chlorite, for example), and the clay content and mineralogy will also affect the quantity of lime required.

The proportion of lime required is small; 1–3% (by dry weight) of lime to soil is needed in order to effect soil modification. It has been noted that 0.5% lime addition should be adequate for the majority of British plastic clays that have been classified as unacceptable on the basis of excess moisture content (Perry *et al.*, 1996a).

If a higher proportion of lime is added the pH of the soil is elevated to pH > 12 and a cementitious action of the lime on the clay minerals in the soil occurs, leading to an increase in strength with time (typically weeks, although measurable strength increases will still occur after several years). Above 5–8% lime (depending on the original clay content of the soil), the effect is lost and long-term strength begins to decrease. Such high proportions of lime may not be economical in any event. This cementitious reaction is affected by soil temperature, and the reaction will be suspended for as long as the temperature falls below 4–5°C. It also requires the presence of water, and therefore wetting of treated fills may be required if the cementitious effect is to be relied upon. The cementitious action is termed 'soil stabilisation'.

A resting ('mellowing' or 'maturing') period between mixing and compacting the fill is recommended where the stabilisation component of the lime treatment is important. The

Figure 7.1 Typical relationships between strength of treated high-plasticity clay, percentage lime and time after mixing. (Data taken from Bell (1988))

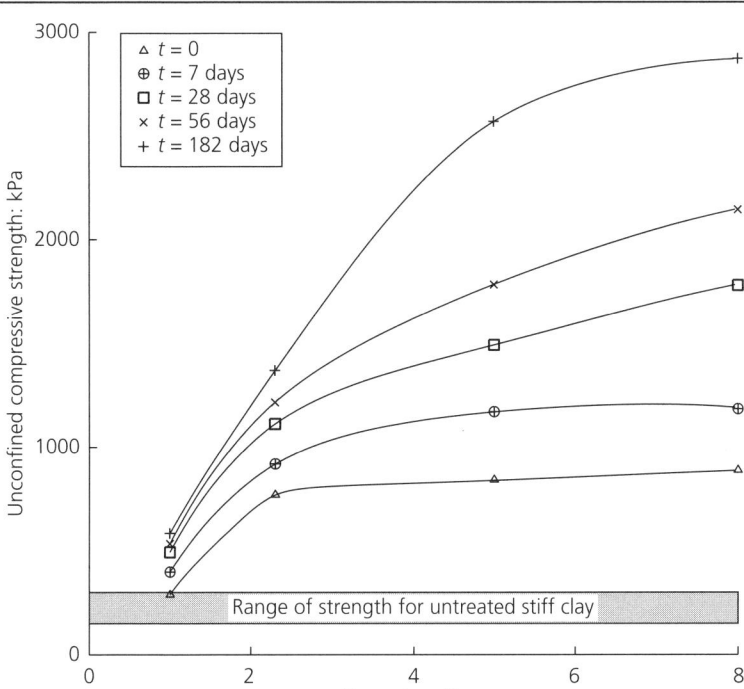

resting period, which is typically 24–72 h for lime-treated soils, allows the lime to hydrate and migrate through the soil, thereby fully effecting the changes in plasticity index and general handling improvements. Following this resting period, the soil is remixed and compacted. If the resting period is too protracted then carbonation of the quicklime becomes appreciable, and less quicklime is available for the stabilisation reaction.

For high-strength end products (such as road-capping layers), compaction to less than 5% air voids is usually required and control of the moisture content is needed to ensure this is achievable.

Figure 7.1 depicts data for the strength of a lime-treated high-plasticity clay from Bell (1988). The flattening of the $t = 0$ line at lime concentrations above 2% clearly shows the limitation of the modifying effect. The subsequent strength gains largely reflect the stabilisation process.

7.2.2 Cement

Cement is used predominantly for soil stabilisation, similar to the cementitious action of excess lime. There is a modest immediate drying effect (soil modification) on first mixing the cement with the soil. This amounts to a 0.3–0.5% reduction in overall moisture

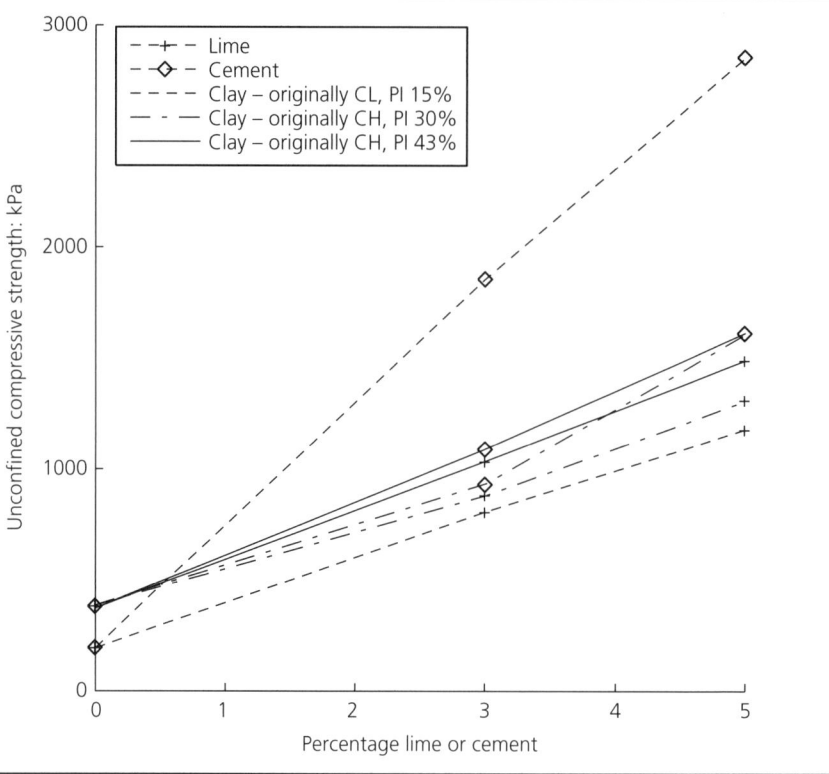

Figure 7.2 Relationships between lime/cement content, compressive strength and original clay plasticity. (Data taken from Christensen (1969))

content per percentage point of cement added, but this is usually viewed as a secondary benefit. The cementitious action is far faster than the equivalent action in the lime-stabilised soil and is not dependent on the presence of clay minerals. The reaction is again suspended below a soil temperature of 5°C, but the period over which the treated soil is susceptible to low temperature is so much shorter than with lime treatment that cement additives might confidently be used closer to the winter season. The faster 'set' of a cement additive means that there is far less leeway in the duration of the resting period. Testing is required to establish the optimum duration of the resting period.

Figure 7.2 presents data from Christensen (1969) for lime- and cement-modified soils of different plasticity, tested for strength at 7 days. It can be seen that there is little difference between the effects of the lime and cement for the high-plasticity soil, but a very marked difference in the results for low-plasticity clay.

7.2.3 Blends

In some cases a combined treatment with a lime/cement blend may be appropriate. In any case it is imperative, in view of the significant costs of lime/cement treatment, that pre-contract laboratory testing is carried out. The best additive to use for a given soil and the proportions in which it is to be added in order to achieve the required performance must be established.

Where it is only required to produce a modest increase in strength, for example, simply to make a site subgrade suitable for traffic, the proportions of additive needed may be very small. There can be problems with the even distribution of such small proportions of additive, so a dry 'inert' bulking medium such as PFA can first be added. (Note: PFA is not inert and will also confer an additional cementitious element to the long-term strength.)

7.3. The performance of treated soils
7.3.1 Soil modification
The addition of lime to a clay soil immediately tends to flatten the compaction curve, increase the optimum moisture content and reduce the maximum dry density for a given compactive effort. It then follows that acceptable fills will be found over a wider range of moisture content. The reduction in maximum dry density is more than compensated for by a gain in strength and stiffness. These changes arise as the clay minerals flocculate, thereby effectively changing the grain size of the clay minerals.

7.3.2 Soil stabilisation
The above effects are not clearly seen with purely cement-treated soils. Soils treated with cement, or soils treated with higher proportions of lime than required for modification alone, undergo stabilisation in which cementitious products form within the soil. The process is much slower for lime-treated soils than for cement-treated soils. The stabilisation process is marked by an increase in strength and stiffness with time.

The changes brought about by the initial soil modification usually render the treated soil frost susceptible. Lime-stabilised soil remains frost susceptible for around 3 months after placement, and frost action during that time would probably disrupt the cementitious bonds developed during the stabilisation process.

Cement-stabilised soils are likely to be less frost susceptible than lime-stabilised ones, due both to the different grain size distribution of the original soil and to the more rapid development of strong cementitious bonds.

In stabilised soils, the lime can undergo an expansive chemical reaction with sulfate chemicals in the soil, potentially leading to heave of the treated soil. Heave magnitudes of 60% of treated layer thickness have been recorded. Such effects are the single main cause of failure of lime- or cement-treated earthworks. The reaction involves the formation of ettringite and thaumasite and is therefore directly equivalent to sulfate attack on concrete. The rapid oxidation of soil sulfides to sulfates during the lime mixing process makes it essential to determine both sulfide and sulfate contents of the soil to be treated during the planning stage.

Organic material in the soil interferes with the stabilisation reaction, tending to reduce the pH locally. An upper limit of 2% organic content for lime-treatable fills is often quoted (Highways Agency, 1991). Above this limit lime addition is taken to be ineffective, although it is recognised that the form of the organic matter is important; Oxford clay with up to 10% disseminated bituminous content has been successfully treated.

Lime-treated soils can lose strength if inundated. Reductions of 30–50% in California bearing ratio (CBR) are quoted (Highways Agency, 1991), so it is important that treated materials are adequately protected and drained.

7.4. Testing

The proportion of additive (lime or cement) is expressed as a percentage of additive per unit dry weight of soil. Testing soil to establish the suitability of lime/cement treatment is highly dependent on the end use of the treated product.

It is not appropriate to apply the results of testing of the original soil to the treated product. The nature of the soil is altered by treatment and new earthworks control criteria need to be derived. The moisture condition value (MCV) test is recognised as being appropriate for the site control of treated soils.

Before the commencement of treatment on site, laboratory testing should be carried out in order to determine (for the treated soil) the following properties

- The sulfate, sulfide and organic contents of the soil to be treated (Highways Agency, 2007; Longworth, 2004). It is important to note that sulfides and sulfates are often leached out of the top metre or so of UK soils. Hence, if soils from deeper than this are to be treated, then it is important that they are adequately sampled and tested. Groundwater should also be tested for sulfates. A limit of 1% total sulfate is usually applied in UK highway works, but it is recognised that lower sulfate concentrations can still result in unacceptable swelling of treated materials and so site-specific testing (relating the sulfate content to heave in an inundation test) should be carried out where soil stabilisation is required (Perry et al., 1996b).
- The relationships between moisture content, MCV and dry density, as well as (if applicable to end use) undrained strength and CBR for different proportions of added lime/cement. The initial consumption of lime (ICL) is the quantity of lime required to bring the soil to a pH of 12.4, and marks the boundary between a soil-modification effect and soil stabilisation. Where soil stabilisation is required (as opposed to simple modification), the ICL marks the lowest lime proportion that needs to be tested. Increments of 0.5% lime are usually tested until the optimum lime proportion is identified. It is important that if a resting period is to be used on site then a similar resting period is allowed in the laboratory (Perry et al., 1996b).
- Heave on inundation, carried out over 14–28 days on a compacted sample in a CBR mould, is important as an indicator of adverse sulfate reactions and should be carried out where soil stabilisation is required.
- Frost susceptibility tests may be appropriate, but these are expensive tests requiring large samples and it may be more appropriate to ensure that treated soils are protected from frost or are not stabilised during the season when frost is a risk.

Quality control testing of the end product is necessary. The testing undertaken would depend on the original intention of the stabilisation process. For example, if the intention is to produce a surface with a given trafficability, then plate-bearing tests or in situ CBR tests would be appropriate. The rate for such testing is commonly of the order of 1 test per 1000 m^2 of treated surface. If, on the other hand, the purpose is to produce a useable bulk fill, then compaction/density testing (to ensure the treated product could be readily compacted to achieve a given standard, e.g. maximum 5% air voids) would be appropriate. A suitable testing rate in this case might be 1 test per 250 m^3 of treated fill.

7.5. Plant

Lime/cement additives are usually mixed into the soil in situ, either for subsequent compaction in place or for subsequent excavation and placement/compaction elsewhere. There may be practical problems on site, with dispersion through the soil of small proportions of lime/cement. In this situation, the additive can be bulked up by the addition of PFA to improve dispersion. This also imparts a cementitious component, leading to a further increase in strength with time.

In addition to storage facilities for the additive, the plant requirements comprise spreading and mixing equipment as well as the preliminary grading and subsequent compaction plant. On economic grounds, agricultural machinery (drag-bars, ploughs, rotovators, etc.) may be suitable for the spreading and mixing on minor works. For best results, however, and to ensure a uniformity of product, specialist spreading/soil pulverisation plant is required (Figure 7.3). More than one pass of the soil pulveriser may be required

Figure 7.3 Addition of lime using an integrated spreader and mixer unit and a simple mixer unit (working on pre-spread lime), both tractor drawn. (Courtesy of Con-Form Contracting)

to produce uniform results. Wetting of the fill may be required in order to achieve specification. Specialist soil pulverisation plant often incorporates a spray bar for this purpose. High mobilisation charges would generally apply to such plant.

A delay between mixing of the lime with the soil and the subsequent compaction allows the quicklime to slake and, depending on the delay, to migrate through and modify the clay soil, facilitating compaction. The duration of this delay, often 1–72 h, depending on the soil and on site constraints, should be modelled during the original testing phase. Following compaction, the cementitious reactions are likely to lead to a gain in strength with time. The time over which the strength increases significantly is termed the 'curing' period.

7.6. Health and safety

The reaction of quicklime with bodily fluids (in sweat, mucus membranes, eyes) is exothermic and the products are caustic. Burns are therefore a significant risk where lime is being handled, and suitable protective clothing together with the intelligent imposition of exclusion zones on the site are necessary. Equipment and procedures for immediate treatment must be set up before work is allowed to start.

REFERENCES

Bell FG (1988) Stabilisation and treatment of clay soils with lime. Part 1 – Basic principles. *Ground Engineering* **21(1)**: 10–15.

Christensen AP (1969) *Cement Modification of Clay Soils*. Portland Cement Association, Skokie, IL, USA, RD002.01S.

Highways Agency (1991) Design and preparation of contract documents. *Design Manual for Roads and Bridges*, vol. 4, section 1, part 1. HMSO, London, UK, HA 44/91 (amended 1995).

Highways Agency (2007) Treatment of fill and capping materials using either lime or cement or both. *Design Manual for Roads and Bridges*, vol. 4, section 1, part 6. HMSO, London, UK, HA 74/07.

Longworth I (2004) Assessment of sulphate-bearing ground for soil stabilisation for built development. *Ground Engineering* **37(5)**: 30–34.

Perry J, MacNeil D and Wilson P (1996a) The uses of lime in ground engineering: a review of work undertaken at the Transport Research Laboratory. In *Lime Stabilisation* (Rogers CDF, Glendinning S and Dixon N (eds)). Thomas Telford, London, UK, pp. 27–45.

Perry J, Snowdon R and Wilson P (1996b) Site investigation for lime stabilisation of highway works. In *Advances in Site Investigation Practice* (Craig C (ed.)). Thomas Telford, London, UK, pp. 85–94.

FURTHER READING

BRE (Building Research Establishment) (2005) *Concrete in Aggressive Ground*. BRE, Watford, UK, BRE Special Digest 1.

BSI (British Standards Institution) (2012) BS EN 13286-47:2012. Unbound and hydraulically bound mixtures. Test method for the determination of California bearing ratio, immediate bearing index and linear swelling. BSI, London, UK.

BSI (2013) BS EN 14227-1:2013. Hydraulically bound mixtures. Specifications. Cement bound granular mixtures. BSI, London, UK.

BSI (2018) BS 1924-1:2018. Hydraulically bound and stabilized materials for civil engineering purposes. Sampling, sample preparation and testing of materials before treatment. BSI, London, UK.

BSI (2018) BS 1924-2:2018. Hydraulically bound and stabilized materials for civil engineering purposes. Sample preparation and testing of materials during and after treatment. BSI, London, UK.

Mitchell J and Jardine FM (2002) *A Guide to Ground Treatment*. Construction Industry Research and Information Association (CIRIA), London, UK, Report C573.

Useful web addresses

British Lime Association: http://www.britishlime.org (accessed 01/08/2018).

Britpave (British Cementitious Paving Association): https://www.britpave.org.uk/default.ink (accessed 01/08/2018).

National Lime Association (USA): http://www.lime.org (accessed 01/08/2018).

Portland Cement Association: http://www.cement.org (accessed 01/08/2018).

Temporary Works, Second edition

Pallett, Peter F and Filip, Ray
ISBN 978-0-7277-6338-9
https://doi.org/10.1680/twse.63389.109
ICE Publishing: All rights reserved

Chapter 8
Jet grouting

John Hislam
Director, Applied Geotechnical Engineering

Neil Smith
Director, Applied Geotechnical Engineering

Jet grouting is a process of increasing the strength and/or decreasing the permeability of soil by wholly or partially eroding a volume of ground and replacing it with grout or a mixture of grout and the disturbed soil. The erosion and grout placement are both effected by high-pressure jetting from drillholes via a tube known as a 'monitor'. Zones of grout can be cylinders created by full rotation of the monitor, segments of cylinders created by limited rotation or panels created without rotation. Closely spaced drillholes are used to make overlapping zones of treated ground to form the desired shape. Any shape is possible provided that the drillholes can be suitably oriented. A number of factors affect the strength of the treated ground, but unconfined compressive strengths of up to 8 or 2 MPa are quite possible in coarse or fine soil, respectively, although a few weeks will be required for the maximum strength to be reached. The permeability of the treated ground can be reduced to 10^{-7} m/s or less. The technique can be used to form retaining structures around excavations or (with subhorizontal drillholes) arches above tunnels under construction or groundwater cut-offs. By treating only cylinders of limited height in a defined depth zone, bottom cut-offs or props between retaining walls can be created. On a small scale, tunnel to shaft break-ins or break-outs can be facilitated or leaky basements sealed.

8.1. Introduction
Jet grouting is a method of ground improvement achieved by replacing the in situ soil with cement grout or by mixing soil and cement grout in situ. It can be used where permeation grouting (filling the voids between soil particles) would be impossible or ineffective. It is carried out from drillholes using radial fluid jets and forms generally cylindrical zones of cemented ground co-axial with the drillhole. High-pressure fluid is used to erode the soil and flush part or all of it to the surface. If only some of the soil is removed, that which remains must be thoroughly mixed with the grout.

Figure 8.1 shows the basic systems of single, double and triple fluids. The pipework which delivers the fluids into the ground is termed (confusingly) the 'monitor'. Figure 8.2 shows an example. The monitor is incorporated into the drillstring. During drilling, the

Figure 8.1 Jet grouting systems

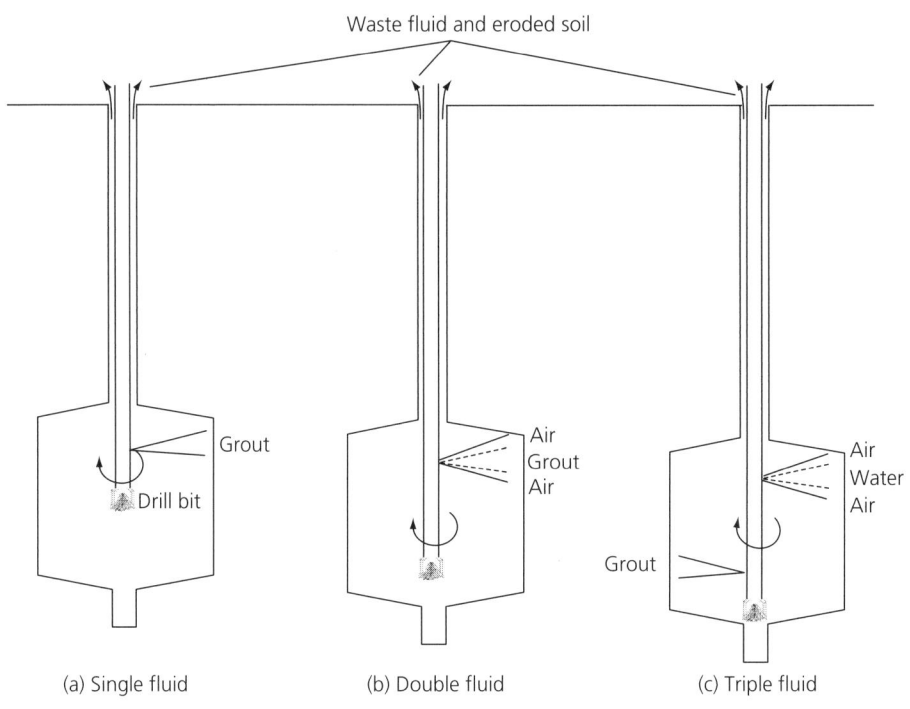

drill flush is emitted through the bit. Usually the jet grout mixture is used as the flushing medium during drilling, as this helps to stabilise the sides of the hole. As for the drillhole cuttings, the excess spoil generated during the jetting process is flushed to the surface via the annular gap between the monitor and the hole.

The technique is not restricted to the creation of vertical columns. It can be used from drillholes of any inclination. The treatment can be used over a limited section of a drillhole to form a disc of improved ground. The treated ground has a relatively high compressive strength and a low permeability. It can, therefore, be used to form arches, gravity structures or tubes to create, for example

- a water cut-off, either around the sides or below the floor of an excavation
- a retaining structure to support the sides of an excavation
- a horizontal low-level prop before excavation
- a supporting arch to maintain ground stability during tunnel construction
- on a small scale, it can be used to stop leakage through basement retaining walls
- a region of low permeability ground to permit tunnel break-out from, or break-in to, shafts.

Coupland (2010), de Wit *et al.* (2007a) and Josifovski *et al.* (2015) provide case histories of interesting projects; many others can be found in the literature. Most applications

Figure 8.2 Monitor and drill bit. (Courtesy of Prof. R. Boulanger, University of California)

entail the construction of cylinders of ground, as the jets are rotated throughout the process of erosion and filling. If rotation is limited or prevented, different geometric forms are possible. The work is covered by an 'execution of special geotechnical work' attachment to Eurocode 7 Part 1 (BSI, 2001).

8.2. Construction methods

There are three systems of jet grouting (see Figure 8.1). The first stage for all is to drill the monitor in to the full depth required. Grouting takes place as the monitor is withdrawn from the hole. The simplest grouting method is the single-fluid system, which uses grout both to erode the soil and to form the cylinder. The double-fluid system uses a combined grout and compressed air jet. The compressed air makes the combined jet much more aggressive. The triple-fluid system separates the erosion and the filling processes, using water instead of grout (plus air) to erode the soil with a grout jet at a lower level.

For the combined fluid jets, the compressed air is forced through an annular nozzle, with the nozzle delivering the water or grout at its centre. The flow of air therefore surrounds the more viscous fluid as it leaves the jet. High jetting velocities are used (up to 300 m/s), requiring fluid pressures of the order of 20–60 MPa and compressed air pressures of about 1.5 MPa. Triplex pumps are used to attain the high pressures required. Extreme care must be observed when the monitor is above ground level, as the jets, especially because they carry particulate matter, can cause damage and injury.

The size of cylinder that can be treated with this method depends to a significant extent on the nature of the soil, as well as the method adopted. While jet grouting can treat virtually all soils (and some very weakly cemented rocks), some materials are more resistant to erosion than others, so the cylinder size that can be treated can vary between strata. Stiff clay is much more resistant than silt or fine sand. Very coarse soil particles require more force to move them than fine grains. The drillhole would typically be about 150 mm diameter. Grouted column diameters from 0.75 m (for the single-fluid method) to 2 m are commonly quoted, and up to 5 m diameter has been claimed. Cylinders for temporary works would normally be used in groups, and the selection of the spacing between drillholes must therefore ensure sufficient overlap of adjacent cylinders. In theory, the depth which can be treated is not limited, but in practice the method has been used to about 50 m below ground.

If the mixing process is not properly effective, lumps of unmixed soil can be trapped within the column (see Stark *et al.*, 2009).

The treatment is generally terminated a short distance below ground level because of the risk of the jetted fluids breaking through to the surface. While the plant used does not need to be large (so that restricted access working is possible), where there is a height constraint columns may have to be constructed in stages. If this is the case, it is obviously important to ensure good overlap between adjacent sections.

8.3. Design principles
8.3.1 Material properties

The principal properties of the treated ground needed for design are its strength and permeability.

The strength varies with the construction technique, for example, the proportion of the in situ soil which is mixed with the grout, how well it is mixed and the nature of the soil itself. Time is also a factor, which may influence the design if the work is on the critical path. Figure 8.3 gives an indication of the strength values that may be achieved, depending on soil type and time after mixing. It can be seen that the full strength of the mixture will take several weeks or more to be reached.

Permeabilities of 10^{-7} m/s and lower have been reported from various sites. Preliminary testing for strength and/or permeability is necessary to ensure that the design and the achieved properties are compatible. Generally, samples of the treated ground are recovered in cored drillholes and tested in the laboratory. Piezometers can be installed in these holes to permit measurement of the in situ permeability.

8.3.2 Design of the treated mass

There are two aspects to design. The requirements for the overall shape and dimensions of the treated ground can be determined using conventional analysis techniques, for

Figure 8.3 Typical strengths of jet grout-soil mixtures

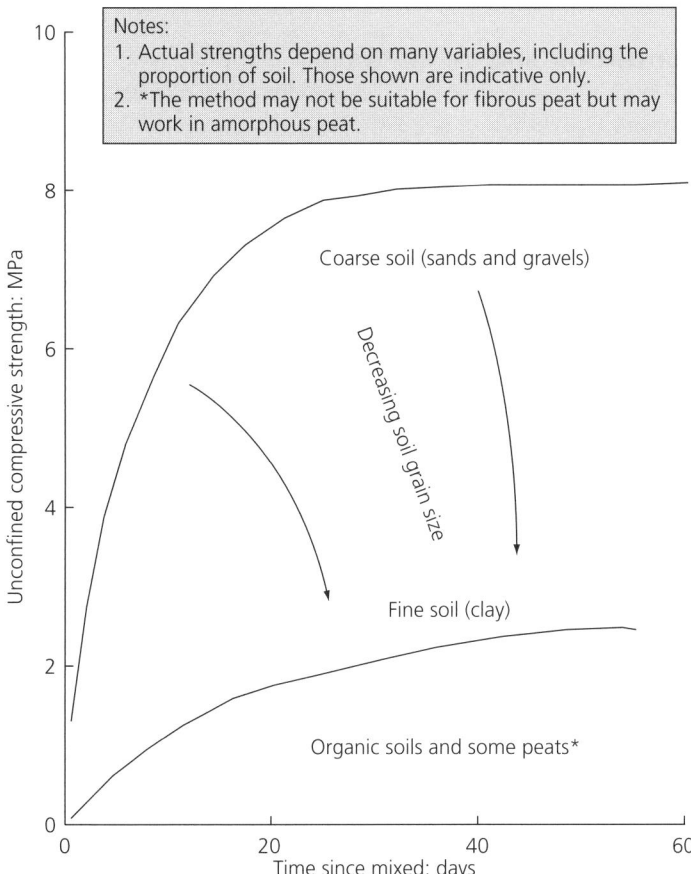

example, by treating a jet-grouted retaining structure as a gravity structure and using normal earth pressure calculations. Where the strength of the grouted mass and the programme time are both critical, due attention must be given to the slow development of strength in some soils. Where a cut-off is required, conventional groundwater flow analysis can be used.

The second aspect of design involves the selection of the construction method, the determination of the column diameter and grout mix requirements, and the monitor rotation and lifting speeds. All of these are empirical; they depend partially on the particular equipment and should therefore be the province of the specialist contractor.

8.3.3 Design verification

Any design of a jet-grouted mass of ground must be verified. Trials are needed in the actual ground conditions. The dimensions of the treated zone require checking; samples need to be taken (by rotary coring methods) for strength and laboratory permeability testing as appropriate. It may also be wise to install piezometers in the cored holes to test the in situ permeability of the mass.

8.4. Monitoring and validation

It is necessary both to monitor the construction process and to validate the adequacy of the treated ground (see de Wit *et al.*, 2007b).

If the location is critical, it may be appropriate to survey long drillholes before grouting. Parameters to be monitored during construction should include the grout density, the fluid pressures and flow rates, and the rate of rotation and withdrawal. When a column has just been completed, its diameter can be checked using callipers. Conventional instrumentation such as inclinometers can be used to monitor the performance of retaining structures. It may be prudent to deploy precise survey techniques and borehole extensometer systems during the grouting process to measure and control heave or settlement.

Preliminary trials can be used to verify the condition of the treated ground by exhumation (see, for example, Stark *et al.*, 2009). Treated ground can be drilled to recover cores for testing strength and permeability. Piezometers can be installed within or around treated ground to confirm its effectiveness in controlling groundwater flow.

8.5. Secondary and side effects
8.5.1 Ground heave and hydrofracture

There is a risk that the process may produce heave of the surface in the area around the treated ground. There are two potential mechanisms. In high-plasticity clays, heave may be caused by swelling resulting from the availability of large quantities of water. Alternatively, excessive fluid pressures may cause hydrofracture, especially if the flushed material is prevented or restricted from escaping directly back up the drillhole to the surface. The restriction can cause a sudden increase in the fluid pressure in the bore. If the fluid pressure becomes higher than the lowest normal pressure in the soil, a fracture in the drillhole wall can occur, which results in an escape of fluid, forcing the ground aside and bringing about surface heave. Some recovery may occur if the pressure is released but this is likely to be uneven and is almost certain to be incomplete.

When used to stabilise very soft clay (having a consistency similar to toothpaste) for a 1 km length of tunnel for the Singapore Mass Rapid Transit scheme, heave of up to 550 mm was recorded (Berry et *al.*, 1987). Average heave above the westbound tunnel (at 15–25 m depth) was 140 mm and that above the eastbound tunnel (at 10–15 m depth) was 70 mm.

8.5.2 Reaction with the ground

The setting cement increases the ground temperature and this can cause the alkaline cement to react with organic material in the soil to produce ammonia. This can be of particular concern where the treated ground surrounds a confined space such as a tunnel. If the groundwater is acidic, for example, in peaty or contaminated ground, the grout mix will need special consideration.

8.5.3 Spoil disposal

The process produces large volumes of waste in the form of slurry consisting of a mixture of water, soil and grout, the largest component of which is water. The volume discharged may be three times the volume of the treated ground. This demands considerable care in collecting the waste at the surface of the drillhole and then directing it to a suitable point for disposal from the site. The consequences of poor control of the waste are a dirty and potentially unsafe site. Control issues regarding drillhole location become important. Off-site disposal of this effluent adds a significant cost to the process.

REFERENCES

Berry GL, Shirlaw JN, Hayata K and Tan SH (1987) A review of grouting techniques used for bored tunnelling with emphasis on the jet grouting method. *Proceedings of the Singapore Mass Rapid Transport Conference*, Institution of Engineers, Singapore.

BSI (British Standards Institution) (2001) BS EN 12716:2001. Execution of special geotechnical works. Jet grouting. BSI, London, UK.

Coupland J (2010) New York's Fulton Street Transit Centre – Dey Street Structural Box. *Proceedings of the DFI/EFFC Conference on Geotechnical Challenges in Urban Regeneration, London, UK*.

de Wit JCM, Bogaards PJ, Langhorst OS et al. (2007a) Design and construction of a metro station in Amsterdam, challenging the limits of jet grouting. *Proceedings of 14th European Conference on Soil Mechanics and Geotechnical Engineering, Madrid, Spain*, pp. 1061–1066.

de Wit JCM, Bogaards PJ, Langhorst OS et al. (2007b) Design and validation of jet grouting for the Central Station, Amsterdam. *Proceedings of 14th European Conference on Soil Mechanics and Geotechnical Engineering, Madrid, Spain*, pp. 1299–1305.

Josifovski J, Susinov B and Markov I (2015) Analysis of soldier pile wall with jet-grouting as retaining system for deep excavation. *Geotechnical Engineering for Infrastructure and Development* January: 3953–3958.

Stark TD, Axtell PJ, Lewis JR et al. (2009) Soil inclusions in jet grout columns. *DFI Journal* **3(1)**: 33–44.

FURTHER READING

Bell AL (ed.) (1994) *Grouting in the Ground*. Thomas Telford, London, UK.

Temporary Works, Second edition

Pallett, Peter F and Filip, Ray
ISBN 978-0-7277-6338-9
https://doi.org/10.1680/twse.63389.117
ICE Publishing: All rights reserved

Chapter 9
Artificial ground freezing

Neil Smith
Director, Applied Geotechnical Engineering

Artificial ground freezing (AGF) is a method for temporarily increasing the strength and decreasing the permeability of ground. As AGF works in all ground types, it is particularly useful in highly variable water-bearing ground when sinking deep shafts and in difficult tunnelling situations. The ground is frozen by passing a cold fluid through pipes buried in the ground. The system creates overlapping cylinders to form a freeze-wall to stabilise the void and exclude groundwater. The moisture content of the soil must normally be above 10% and the rate of flow of groundwater less than about 2 m/day if brine is the coolant or 20 m/day for nitrogen. Recent developments involve initial freezing with nitrogen and then maintaining the freeze with brine. The shape of the frozen ground is limited only by the ability to orientate the freeze tubes. Several weeks are required for the frozen ground to be ready for use if using brine, but nitrogen is much quicker. The system design needs specialist knowledge of the complex properties of frozen ground. The structural design of vertical freeze-wall cylinders can be based on empirical formulae. For other applications, conventional analyses can be used to determine the required strength and dimensions of the frozen ground. Numerical thermal modelling is required to predict freeze-wall growth. Monitoring of the work throughout the entire process is essential.

9.1. Introduction

It is well known to everyone who lives in a climate where freezing temperatures occur that, when frozen, the soil near the surface is generally stronger than when unfrozen. It is also obvious that water ceases to flow when it is frozen. From these two facts, the principal benefits of ground freezing are clear: the soil is strengthened and groundwater flow is prevented. AGF is therefore a specialist process that is used for temporarily preventing groundwater ingress into excavations and, where the ground is unstable, assists in strengthening the ground. Its use is not widespread due to the relatively small number of occasions where the process may be required and, as a result, there are few organisations with real competence in its use. General opinion appears to balk at the up-front cost of employing the process but experience has shown that, in a number of cases, if a value, risk and cost-effectiveness exercise had been carried out first, the adoption of AGF would have brought savings. The fact that AGF operates by modifying the water within the soil pores means that it is much less affected by variations in soil type than, for example, dewatering or grouting, and hence there is less chance that treatment will be ineffective.

The basis of ground freezing is to pass a cold fluid through a pipe buried in the ground and thereby freeze the pore water. This solidifies a column of ground approximately centred on the pipe. By combinations of suitably closely spaced pipes, it follows that the shape of a mass of frozen ground is limited only by the ability to orientate boreholes. It is, therefore, a very flexible technique of ground stabilisation. Provided the soil has a sufficient moisture content and groundwater is not flowing too quickly, any soil can be frozen so that heterogeneous mixtures of sand/clay layers or glacial clay with gravel lenses can be treated effectively.

In normal circumstances, a zone of frozen ground is created around the volume of ground in which construction is required. An array of boreholes is drilled and pipework installed to carry the refrigerating fluid, which can be either brine (slow-freeze fluid circulation) or liquid nitrogen (fast-freeze gas exhaust). The spacing between the boreholes is close enough to ensure that the cylinders of frozen ground around each pipe will touch and overlap within a reasonable time to stop groundwater flow and create a thick enough wall to provide the necessary strength.

When construction has been completed, the fluid circulation or gas exhaust is stopped and the ground begins to thaw. The ground does not necessarily return to its previous condition.

9.2. Construction principles
9.2.1 Methods of ground freezing

The oldest method of freezing the ground uses a solution of chilled brine. Conventional refrigeration plants, normally using ammonia, can reduce the temperature of the brine to as low as $-40°C$, although a temperature of $-25°C$ is likely to be suitable in most circumstances. The cooling fluid is then pumped around the circuit to freeze the ground surrounding the buried pipes. The process requires various components of plant, including a compressor, a condenser, an evaporator, a pump and distribution pipes.

Considerable time is required for a brine-based system to bring the ground to a condition in which it is sufficiently frozen for work to begin. The stages are drilling boreholes and installation of pipework, chilling the brine and building the freeze-wall. This process may take several months and the formation of ice cylinders will take at least a month.

The distribution system would typically consist of a header pipe (or ring main) and freeze tubes within the ground. The freeze tubes can be connected to the header in series or in parallel but, for risk reduction, it is normal that there should be at least two parallel circuits to ensure that if a pipe leaks the whole system does not have to be shut down. Adjacent freeze tubes should be connected to different parallel circuits to ensure that the worst spacing between pipes is double the installed spacing. This needs to be considered in the design factor of safety.

More recently, the ability to deliver liquid nitrogen to site in reasonably large quantities and to store it in vacuum-insulated tanks has provided an alternative to brine. The system does not require refrigeration plant or pumps to circulate the fluid. The nitrogen

Figure 9.1 Schematic diagram of a brine system

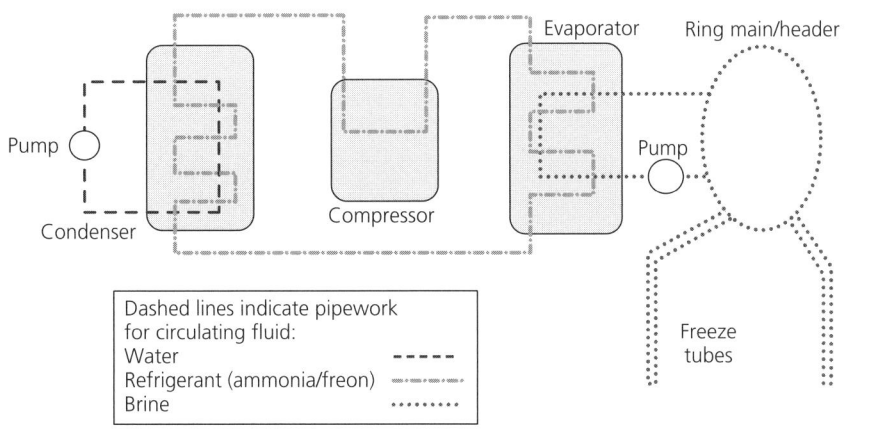

vaporises at −196°C at a pressure sufficient to drive the fluid through the circuit without the aid of pumps. At the end of the process, it is vented to the atmosphere rather than recirculated. The very low temperature of the fluid reduces the freeze-wall formation time from several weeks to a few days. The high temperature differentials make it more appropriate to connect the freeze tubes in parallel rather than in series.

The loss of product and consequent maintenance costs may make the use of nitrogen uneconomical if the ground must be kept frozen for an extended period. It is, however, possible to use nitrogen initially to produce the freeze-wall rapidly, and then to replace it with a brine system for longer-term freezing (Viggiani and Casini, 2015).

Schematic diagrams of the two systems are shown in Figures 9.1 and 9.2. Figure 9.3 shows a system installed in preparation for shaft construction.

Figure 9.2 Schematic diagram of a liquid nitrogen system

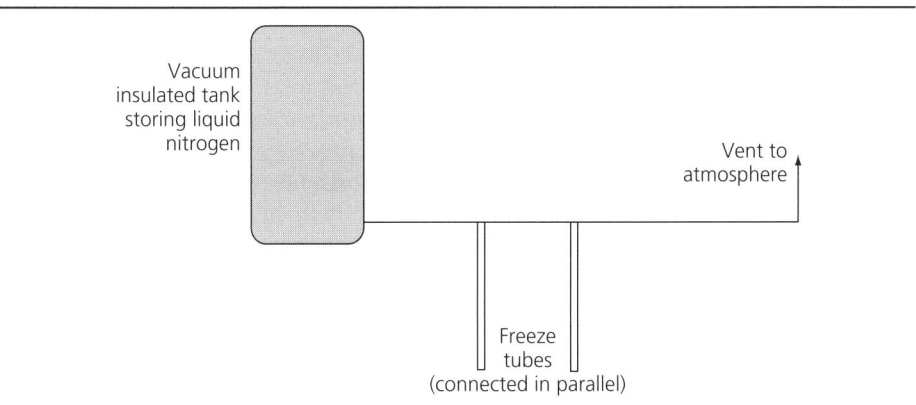

Figure 9.3 Freezing system for shaft construction through 30 m of superficial deposits to bedrock at Belmont, NSW, Australia. (Courtesy of the British Drilling and Freezing Company Limited)

9.2.2 Advantages and disadvantages

In variable ground, freezing will be effective on nearly all soils, although the rate of freezing will differ between, for example, clay (slower) and gravel (quicker). Another significant advantage of freezing is that, given suitable conditions, the frozen ground will encircle service pipes, making it possible to reduce the disruption and re-routing of services often required by other forms of ground treatment. It is possible to use the technique on a small scale, for example, to plug leaks in basements while a permanent seal is formed.

However, there may be problems in achieving freeze-wall formation if the groundwater is saline or contains other dissolved salts, or if the ground is contaminated with hydrocarbons. Below a moisture content of about 10% there may be insufficient ice in the body of the ground to form a competent retaining structure. Flowing groundwater can also prevent freeze-wall formation, which can be a particular issue where adjacent older structures are present. The ground may heave during freezing; movement is dominantly perpendicular to the freeze pipes so that the use of horizontal freeze pipes, particularly with clay, must be carefully considered. Consolidation occurring on thawing can exceed the heave movement. The structure of peat or organic soil may be broken down by freezing.

9.3. Design principles
9.3.1 Introduction
There are two principal aspects to the design of a ground freezing scheme.

- Calculations of the thermal behaviour of the ground are needed to ensure that the columns of frozen ground around each freeze tube meet to provide a continuous freeze-wall which is sufficiently thick. This is a specialist subject that requires the knowledge and experience of the specialist contractor's staff. The text in the following section is, therefore, limited to describing the main parameters involved and some aspects of the behaviour of frozen soil.
- The structural design of the freeze-wall must ensure that it has sufficient strength to support the ground around the intended excavation and that its deformation will be tolerable; frozen ground exhibits significant creep characteristics.

Time is a further factor to be considered in design. If the time available for freeze-wall formation is short, then nitrogen may be preferred to brine or the spacing between freeze tubes may be reduced. Overcapacity should be built into the design as a protection against the malfunction of any elements of the system, or simply a less effective system than anticipated.

9.3.2 Thermal behaviour
9.3.2.1 Thermal parameters

There are similarities between the flow of heat through the ground and the flow of water. In the latter case flow is governed by the permeability and the head gradient, whereas for heat flow the parameters are thermal conductivity (K) and temperature gradient. The thermal conductivity is the amount of heat per unit time flowing through a unit area under a unit temperature gradient. The temperature is usually expressed in degrees Kelvin, resulting in the slightly confusing situation that the units of thermal conductivity K are W/m K where the second 'K' refers to the temperature (degrees Kelvin). The range of thermal conductivity values for different soil types is far less than the range of permeability. Organic soil has a K value of about 0.25 W/m K; other soils have a K value in the range between about 0.6 and 4 W/m K; values for rocks range up to about 7 W/m K. For comparison, the K value of wood is about 0.1–0.2 W/m K and that of iron is 80 W/m K. Thermal conductivity is not constant with temperature, however. It increases with reducing temperature, and this effect may need to be considered for nitrogen freezing.

For thermal design, the heat capacity is measured per unit volume (not mass) of material and is the quantity of heat transfer required for a unit change in temperature of a unit volume of the ground. It is denoted by C and has units of J/m^3 K, where K is degrees Kelvin. The range of C is also small for a range of soils, roughly 1.8–2.5 MJ/m^3 K. It decreases with decreasing temperature, tending towards zero at a temperature of absolute zero.

The formation of an individual cylinder of ice around a freeze pipe follows a declining growth rate due to the insulating effect of the ice as it forms. In practice, cylinders of 1.0 m diameter form after about a month; it is rare to assume that cylinders will exceed 1.2 m diameter.

9.3.2.2 Engineering properties of frozen soil

The enhanced strength and stiffness of frozen (compared to unfrozen) soil is largely due to the strength of the ice. However, frozen soil is not simply a combination of ice and soil particles. It also contains unfrozen water and air, so the composite material is very complex, as is its behaviour. Ice itself exhibits significant creep characteristics at low shear stress levels; this is manifested in the flow of glacier ice. In soil, the creep of the ice transfers stress to the intergranular (effective) stress between particles; if this eventually exceeds the frictional capacity of the soil, failure will occur.

As examples of the variability of the strength of frozen soil, low strain peak shear strengths of frozen Ottawa sand have been found to range from about 5 to 8 MPa, increasing with confining pressures, which ranged from about 1.4 to 6.9 MPa. Strength is also heavily dependent on strain rate, however. The strength of frozen silt has been shown to vary by a factor of 3 for strain rates varying from 10^{-6} to 10^{-3} s^{-1}. While the variations in strength can be seen to be very significant, it is also clear that the strength of the frozen material is very much higher than strength values associated with unfrozen soil.

The two aspects of design must proceed in tandem. The engineer designing the freezing system must be provided with the strength and stiffness requirements for the frozen ground. It is also important that the designer knows the duration for which the ground is to retain its enhanced strength and stiffness.

9.3.3 Structural design

Many applications of ground freezing are for the construction of shafts. Regardless of the plan shape of the shaft, it is normal to form a circular freeze-wall which envelopes the shaft. The use of a circular wall ensures a uniform compressive stress in the horizontal plane. The required freeze-wall thickness may be determined using empirical formulae (Harris, 1995) or complex finite element programs (Haasnoot, 2010).

9.3.4 Practical aspects of design

Other applications include the construction of temporary arch supports for tunnel construction in unstable ground and stabilisation of localised zones for tunnel break-in to (or break-out from) shafts and cross-passage construction. A relatively frequent application of the technique is to 'rescue' tunnelling projects that have run into difficulty. Freezing has been used in instances where dewatering and/or grouting has been ineffective.

Tunnel projects by their very nature have many restraints on working space which directly influence positions for larger drilling rigs and the orientation possible for the freezing pipe drilling. These aspects were tested on the Storebælt undersea tunnels in Denmark (Kofoed and Doran, 1995) for the most difficult cross-passages and for junctioning the tunnels. Figure 9.4 illustrates some aspects of this problem. The section shows freeze pipes and notional ice cylinders protecting the crown of the excavation. Ideally, the drilling would be parallel to the excavation to provide a simple pattern, but in this case it is not possible because there is no space for the drilling rig to operate in the crown of the tunnel. This is a very common problem when planning freeze patterns associated with tunnelling works. One critical aspect of freezing works is the directional control during drilling, and for this reason many authors consider that the maximum

Figure 9.4 Cross-passage freeze works, Storebælt tunnels, Denmark. (Courtesy of Kofoed and Doran (1995))

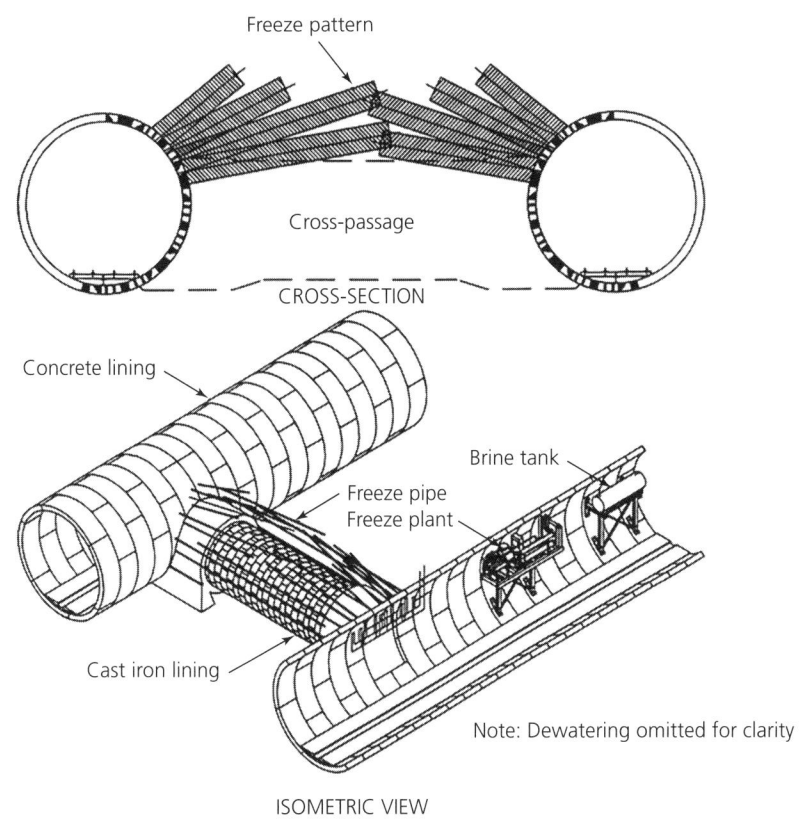

practical length for freeze drilling is 25 m. Longer lengths are sometimes used but must be surveyed and actual freeze pipe spacing checked. Figure 9.5 shows a typical drilling rig and drilled pipes at the bottom of a shaft in Cairo, Egypt. The double row of freeze pipes forms a conical freeze that extends back onto the intact tunnel in order to rescue the tunnel-boring machine.

Another infrequent use is to form a temporarily strengthened roadway for heavy plant to cross weak ground. In uses other than to facilitate shaft construction, the project engineer may use a conventional analytical approach to determine the required properties and dimensions of the frozen zone. This then provides the criteria for the design of the freezing system and freeze tube layout.

9.4. The effects of freezing and thawing

It is important to realise that the soil will not necessarily return to its previous state after it has been subjected to a cycle of freezing and thawing. The structure of some clays can be disrupted by the freezing process, generating fissures and causing an increase in the

Figure 9.5 Drilling rig and freeze pipes at bottom of shaft, Cairo, 2010. (Courtesy of David Hartwell)

permeability (Harris, 1995). The ratio of post- to pre-frozen permeability may be as much as two orders of magnitude.

When water freezes and turns to ice its volume increases by approximately 9%. In some soils, however, especially low-plasticity clays and silts, much larger ground movements may be caused by frost heave. This occurs when lenses of frozen water are formed by water that migrates in response to suctions which occur in the ground during freezing. The localised increases in volume can also cause significant increases in pressure on adjacent structures such as foundations or retaining walls. With often crowded modern urban infrastructure it is not uncommon for a new tunnel to pass immediately under critical infrastructure such as rail tunnels or major sewers. In these situations monitoring and management of heave is critical.

Ground settlement or deformation can occur on thawing, with the potential to create voids around the newly completed structure. In such circumstances, it may be necessary to use compensation or 'back-wall' grouting to fill the voids and ensure the long-term integrity of the structure.

Various researchers have investigated the effect of the freeze–thaw cycle on soil strength. Some have found that the strength of clay increases following the cycle, where other studies report a decrease in strength. If the post-freezing strength of the soil is a matter of concern, it is advisable to carry out testing in advance of the works.

The thickness of the freeze-wall will continue to increase until the rate of heat extraction is balanced by the supply from the ground. The void excavated within the freeze-wall will generally be maintained at a temperature at which men and machines can work effectively. There will, therefore, be some water generated in the void by melting. In addition, where the void has to be ventilated and the air drawn from the surface is warm and humid, substantial condensation can take place. Where sprayed concrete linings are used it is necessary to make allowance for the effect of the freeze on the strength development of the concrete where it is sprayed directly onto frozen ground. It is common to allow a sacrificial first layer of sprayed concrete.

9.5. Monitoring

It may be necessary to survey the drilled holes prior to installation of the freeze tubes. Significant deviations from the design hole location may result in gaps in a freeze-wall. Accurate hole drilling is especially important for deep shafts and when precision is required (e.g. the rescue of a tunnelling machine).

Once the freezing process has been started, the ground temperature should be monitored at suitable locations throughout the length of dedicated freeze tubes to ensure that the plant is functioning correctly and to check that the ground is freezing as predicted. Separate monitoring holes with an array of temperature sensors are normally used; ideally these should monitor critical points but are also best drilled at a different angle to the freeze tubes so that the sensors are at varying distances from the freeze pipes and the development of the ice cylinder can be confirmed. Temperature monitoring must continue through the whole period of freezing until the ground has thawed.

Ground levels must be monitored to check for heave and post-thaw settlement. For schemes where the frozen zone is required to control groundwater, piezometers are required to measure groundwater conditions.

Acknowledgements

The author would like to acknowledge the assistance received from Dr Alan Auld when writing this chapter for the first edition and from Mr David Hartwell for the second edition of this book.

REFERENCES

Haasnoot J (2010) Large scale ground freezing in the Netherlands. *Proceedings of the DFI/ EFFC Conference on Geotechnical Challenges in Urban Regeneration, London, UK.*

Harris JS (1995) *Ground Freezing in Practice*. Thomas Telford, London, UK.

Kofoed N and Doran SR (1995) Storebælt Eastern Railway Tunnel, Denmark – Ground freezing for tunnels and cross passages. *Proceedings of the XI ECSMFE, Copenhagen, Denmark.*

Viggiani GMB and Casini F (2015) Artificial ground freezing: from applications and case studies to fundamental research. *Geotechnical Engineering for Infrastructure and Development* January: 65–92.

FURTHER READING

Auld FA, Belton J and Allenby D (2015) Application of artificial ground freezing. *Geotechnical Engineering for Infrastructure and Development* January: 901–906.

BGFS (British Ground Freezing Society) (1995) Technical Memoranda on the Ground Freezing Process. BGFS, Nottingham, UK.
- TM1: Harris JS and Wills AJ. Introduction to artificial ground freezing.
- TM2: Harris JS. AGF processes.
- TM3: Harris JS. Site investigation for AGF works.
- TM4: Jones RH. Control of ground movements in AGF works.
- TM5: Harvey SJ and Wills AJ. Value, risk and cost effectiveness in AGF works.
- TM6: Harris JS and Bell MJ. Shaft freezing.
- TM7: Harris JS. Tunnel freezing.
- TM8: Auld FA. Casting concrete against frozen ground.

BTS/ICE (British Tunnelling Society/Institution of Civil Engineers) (2010) Section 4 – Ground stabilisation processes. In *Specification for Tunnelling*. Thomas Telford, London, UK, pp. 116–118.

Useful web addresses

Tunnels and Tunnelling – information on contractors and consultants specialising in ground freezing: http://www.tunnels-directory.co.uk (accessed 01/08/2018).

Temporary Works, Second edition

Pallett, Peter F and Filip, Ray
ISBN 978-0-7277-6338-9
https://doi.org/10.1680/twse.63389.127
ICE Publishing: All rights reserved

Chapter 10
Slope stability in temporary excavations

Neil Smith
Director, Applied Geotechnical Engineering

Most construction sites require excavation, and the stability of the sides must be considered in relation to safety and economy. Where possible, it is usually most economical to form excavations in open cut. The essence is to determine the steepest practicable slope. Several factors influence slope behaviour: (*a*) the nature and variability of the soil; (*b*) the groundwater conditions; (*c*) the loading close to the crest; (*d*) the time the excavation will remain open; and (*e*) the consequences of failure. Factors (*a*) and (*b*) require adequate information from site investigation, and (*c*)–(*e*) require good liaison with the construction planning team. Construction is generally straightforward, although it must be coordinated with other processes (especially groundwater control). Slopes in rock require similar considerations, but the behaviour of rock slopes tends to be dominated by the discontinuities in the rock, as the blocks of intact rock are comparatively strong. The important characteristics of rock slope discontinuities are their inclination relative to the cut slope (noting that there may be more than one set of discontinuities), frequency, persistence, openness and smoothness. Safety is a paramount consideration: unsupported trenches are examples of steep cut slopes and collapses are a major cause of accidents on construction sites.

10.1. Introduction
This chapter gives guidance on the design and construction of temporary slopes for excavations on construction sites. Construction of below-ground works within an unsupported excavation has generally been viewed as the cheapest option. However, the use of battered (as opposed to vertical) slopes requires excavation of significantly larger amounts of material and the extra excavation can bring substantial additional cost if the material has to be taken off site, especially if any of it is contaminated. The use of sloping sides may also affect the choice of plant, for example, if cranes have to be sited outside the excavation and so need a longer reach. Figure 10.1 is a decision diagram showing the steps involved in determining whether excavating to a batter is a viable option for the project.

Provided that the safety and financial risks associated with slope movement are properly controlled, relatively simple approaches to the design of temporary slopes on construction sites can be acceptable. For large temporary slopes, sophisticated investigation, analysis and monitoring may be appropriate (Kovacevic *et al.*, 2007).

127

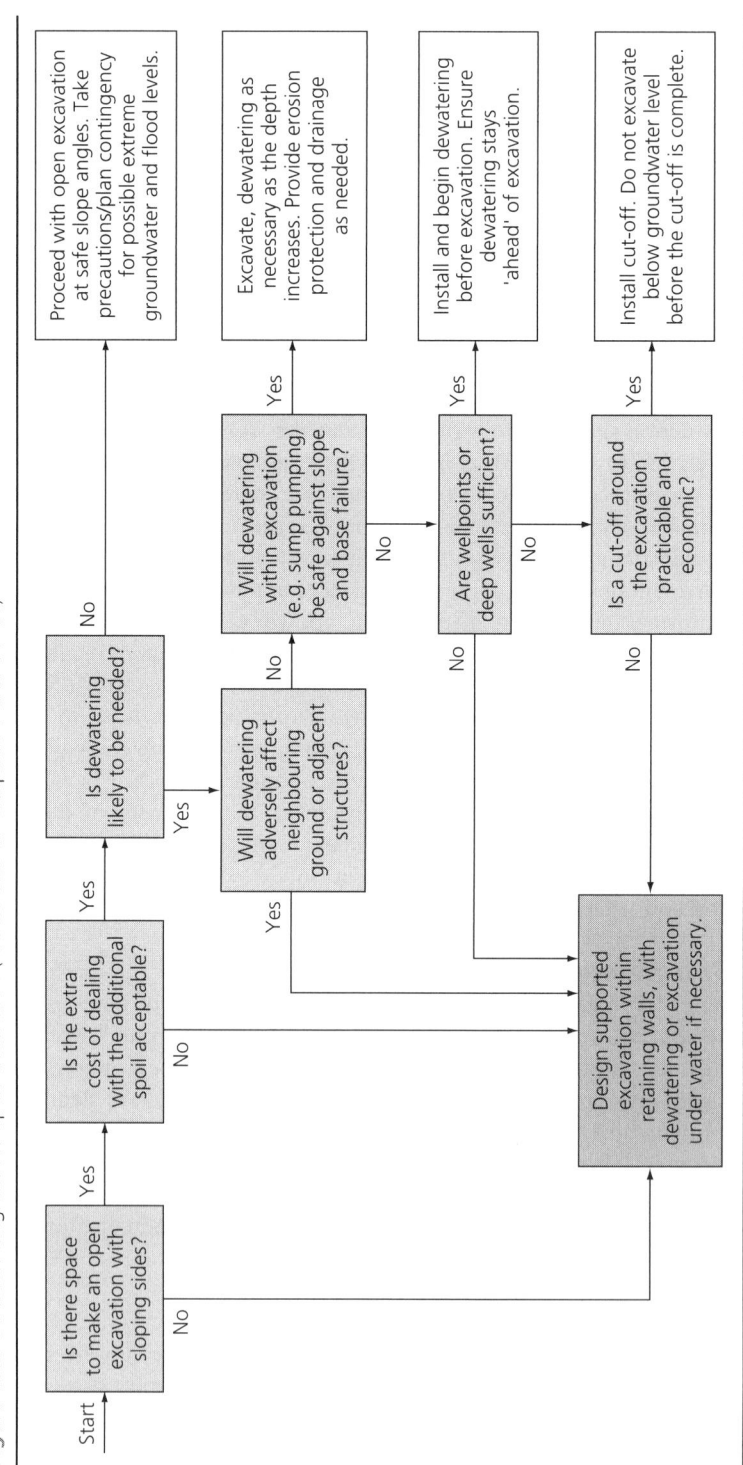

Figure 10.1 Decision diagram for open excavation (relates also to Chapters 6 and 11–14)

In addition to the nature of the ground and the groundwater regime, time is an important parameter for temporary slope design; the shorter the design life of the slope, the better. The design must also take account of the consequences of movement occurring before the slope fails. In this regard, it must be remembered that soil can tolerate much larger pre-failure strains than most buildings or hard-paved surfaces. Failure of adjacent structures may therefore take place before the slope itself fails, and the sensitivity of adjacent structures must therefore be considered as part of the slope design.

Eurocode 7 (BSI, 2004) gives general guidance on slope design in Section 11. Clause 2.2(1)P requires that both short-term and long-term design situations shall be considered. In this context, the distinction between short and long term depends on the nature of the ground as well as the design life of the slope. In coarse soil, long-term conditions are established quickly and there is generally no requirement to consider short-term conditions. In fine soil, initial stability is governed by undrained parameters; an indefinitely long period is required for fully drained (long-term) conditions to be established. Tomlinson (2001) presents information from a number of excavations in London Clay; the period prior to slippage ranged from 1 day to no failure after 4 months. The design life of temporary slopes in clay lies somewhere between fully undrained and fully drained conditions. Local knowledge of stability problems may be very useful. A walkover survey of the site and surroundings should be carried out before any design work is commenced.

10.2. The consequences of failure

In most circumstances, if it is possible to carry out an excavation within open cut slopes, it is also possible to arrange construction in order to minimise both the chance of failure and the consequences, should ground movement occur. The following relatively simple matters need to be considered at an early stage of project planning

1. Try to keep excavations away from the site property boundary to avoid a risk to neighbours or the public (and party-wall problems).
2. Ensure there are no critical services close to the crest or toe of the slope.
3. Plan drainage or surface-covering measures to minimise rainwater infiltration to the ground beneath the slope, including the elimination of ponded water close behind the crest.
4. Avoid routing haul roads or locating heavy plant just above the crest of a slope.
5. Avoid stockpiling spoil near the crest.
6. Where relatively steep slopes are needed, make an initial cut to a flat angle and then cut to the steepest slope in short lengths (as short as is practicable and less than the slope height). Do the work, then reinstate the slope to a flat angle before excavating the next bay. Consider constructing in 'hit and miss' bays.
7. Keep the period for which the slope is open down to a minimum, for example, by using a two-stage approach as in (6) above.

10.3. Construction principles

The excavation of temporary slopes in soil is a straightforward process. For excavation in rock, it is necessary to determine whether the rock can be ripped or whether blasting will be required. The assessment of 'excavatability' (also referred to as 'rippability') has been examined by a number of authors, for example, Pettifer and Fookes (1994). Monitoring of the slope after excavation is an essential part of the process.

10.4. Design principles – some simple fundamentals
10.4.1 The nature of fine- and coarse-grained soil

It is widely appreciated in the construction industry that excavated slopes in fine-grained (clay) soils may stand for a considerable period at a steep angle, even vertical, but that slopes in coarse soil (sand and gravel) will quickly ravel to an 'angle of repose'. This difference in behaviour is not the result of any fundamental clay property that is different from sand but derives from the surface tension of the pore water, which creates high capillary (suction) pressures within the pores of fine-grained clay soil. Due to the difference in grain size (and therefore also pore size), clay can support suction pressures of the order of 1000 times that of gravel. (These suction pressures allow a clay to remain saturated at levels tens of metres above the level at which the hydrostatic pressure is zero – the phreatic surface, or 'water table').

When an excavation is cut into clay, the surrounding ground experiences lateral and vertical stress relief. In simple terms, this produces a tendency for the soil to expand and hence generates a corresponding suction in the pore water while the effective (intergranular) stress between the soil particles, which controls strength, remains largely unaffected. An example of the extent of suction within a clay slope is given in Figure 10.2, which is a prediction by Kovacevic *et al.* (2007) for a major excavation at Heathrow Terminal 5. Over time, groundwater flows towards the zones of high pore suction and an equilibration process takes place which reduces the suction and decreases the stability of the clay slope. In a homogeneous clay of low permeability, this process is slow, so that a relatively steep slope may remain stable for a substantial time (months or years).

The suction pressure that can be maintained in a clean gravel is negligible and equilibration is rapid so that, on excavation, the effective stress (and thereby the shear

Figure 10.2 Predicted contours of pore pressure within the London Clay at Heathrow Terminal 5. Note the extent of the suction zone and the high suction pressures. (Reproduced from Kovacevic et al. (2007))

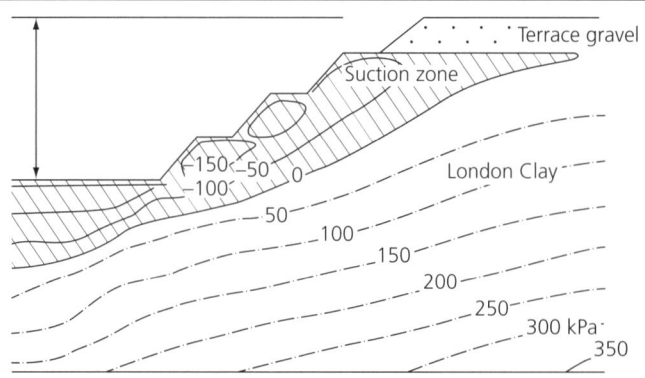

strength) within the soil reduces; a steeply cut slope will quickly degrade to the angle of repose of the soil.

In real soils, gravels and sands will generally contain some fine grains, which may be sufficient to allow small suctions to be maintained for short periods. This leads to slopes that may stand longer than expected at steeper angles than the angle of repose, but this additional stability may only persist for a short time. Another possible factor which may maintain a steep slope in coarse soil is the deposition of salts at grain contacts when groundwater evaporates. Reliance on these factors is risky unless supported by good local experience.

On the other hand, real clays (especially glacially deposited clays) contain discontinuities such as fissures or thin bands, pockets or ribbons of coarser soils. The discontinuities or occurrences of coarser soil allow suction pressures to fall more rapidly, and the result is that the slope is likely to be less safe than predicted (or safe for a shorter time) and there may be a higher risk of small failures within a larger slope (see Kovacevic *et al.*, 2007).

10.4.2 Effects of groundwater
If groundwater is introduced into a previously dry soil, its principal effect is to reduce the effective stress between the soil particles. Where the groundwater is static, the effective stress in coarse-grained soil can reduce to as little as around half the dry value. In clay soils initially supporting high suctions, far greater reductions are possible. If groundwater is flowing in an unfavourable direction (e.g. towards the surface of a cut slope), the water can reduce the effective stress further still. The unit weight of the saturated soil is also higher due to the replacement of air by water in the soil voids. Both of these effects reduce the stability of a soil slope.

In silts, sands and more permeable soils, flows can be established quickly such that excavation side slopes can be rapidly eroded by groundwater issuing into the excavation. This leads to severe problems on site. The solution is (*a*) to ensure that the groundwater regime is properly known from investigation and monitoring prior to the design of temporary works; and (*b*) to install suitable groundwater control measures before excavation starts (see Chapter 6 on control of groundwater).

In rocks, adverse groundwater conditions reduce the stress across discontinuities and hence the friction available to resist movement. They also impose lateral forces on steeply inclined discontinuity planes.

The design can incorporate drainage measures such as a lateral drain above the slope crest to direct surface water runoff away from the slope. Exposed surfaces on or above the slope can be covered by impermeable sheeting to conduct water to drain away.

10.4.3 Geotechnical categories
While it is not specifically stated, Eurocode 7 (BSI, 2004) is principally directed to ensuring that failure does not occur. However, in some circumstances of temporary slope design it may be accepted that there is a relatively high possibility of a failure.

Eurocode 7, Clause 2.1(14–21) defines three geotechnical categories (1, 2 and 3) as described in the following sections.

10.4.3.1 Category 1
Clause 2.1(8)P of Eurocode 7 states that 'small earthworks [shall be identified] for which it is possible to ensure that the minimum requirements [see UK National Annex (BSI, 2014)] will be satisfied by experience and qualitative geotechnical investigations, with negligible risk'. Referring to this clause, table NA.1 of the UK National Annex states that 'Minimum requirements are not given in this National Annex and should be agreed where appropriate with the client and other relevant authorities'.

10.4.3.2 Category 2
Clause 2.1(17) of Eurocode 7 defines category 2 works as having 'no exceptional risk or difficult soil or loading conditions'. Most temporary slopes on construction sites would fall into this category, but if the consequences of failure can be reduced sufficiently they could be treated as category 1.

10.4.3.3 Category 3
Clause 2.1(20) of Eurocode 7 states that this category includes slopes which fall outside the other two categories. Clause 2.1(21) states that this category 'should normally include alternative provisions and rules to those in this [Eurocode 7] standard'. The notes to this clause give the examples of 'unusual or exceptionally difficult ground or loading conditions'.

Temporary slopes classified as category 3 are outside the scope of this guidance, and specialist advice should be sought at an early stage in the planning of the project.

10.4.3.4 Categorisation example
Figure 10.3 is a photograph of a small slip which occurred approximately 2 weeks after the toe of an existing cutting was removed to give access for construction work. A steep (c. 80°) slope about 1.5 m high (visible at the right-hand side of the photograph) was cut into a stiff clay to make an access road, on which the two men are standing. At the time of the photograph, fill had been placed in front of the slipped mass to restrain further movement and to re-establish the access road. The slipped clay was subsequently replaced with coarse granular fill. The slip that occurred could be considered a category 1 event, because the consequences of the failure were trivial and easily accommodated by the contractor within a contingency.

However, the boundary of the site was very close to the slope crest (at the fence in Figure 10.3) and the designer's main concern was to avoid any ground movement that would affect the neighbouring property. The risk of such a failure classifies as a category 2 event. It would therefore be possible in principle for the designer to accept a lower factor of safety against a small failure such as the one which did occur, but to require a higher factor against any failure that would reach the crest of the slope. This example indicates that it is possible for the same slope to have more than one category.

Figure 10.3 Small slip caused by temporary removal of the toe of a 100-year-old cutting in stiff clay. Some fill has been placed in front of the slip to stabilise the ground and to re-establish the access road on which the men are standing

10.4.3.5 Factor of safety
The appropriate factor of safety (F) must be determined on a site-specific basis, taking all of the influencing elements into account.

10.4.4 Design for soil slopes of 'small' size
Figure 10.4 provides suggested temporary cut slope angles for different materials. Here the designer should ensure that all interested parties understand and accept the risks involved in the selected slope angle(s). Simple slip surface analyses as referred to in Section 10.4.5.2 may be used in support of the designer's judgement.

10.4.5 Design for soil slopes of 'medium' size
10.4.5.1 Coarse-grained soil
In coarse-grained soil, if the groundwater level is above the excavation bottom, digging should not be started until groundwater control measures have been put in place either to cut off the supply of groundwater to the excavation or to reduce its level to below the proposed excavation surface. (In deep excavations, it may be possible to dewater and dig in stages; see Chapter 6.) Stable slopes can then be formed at the angle of repose, less a margin for a factor of safety. If the consequences of a failure are trivial then a contractor may opt to take a risk and cut the slope to a steeper angle, having made contingencies for the disruption and cost of a failure. Table 10.1 is adapted from BS 8002:2015 (BSI, 2015) and lists critical friction angles (ϕ'_{crit}) for sands and gravels formed of silica. (Most

Figure 10.4 Suggested temporary cut slope gradients for different materials

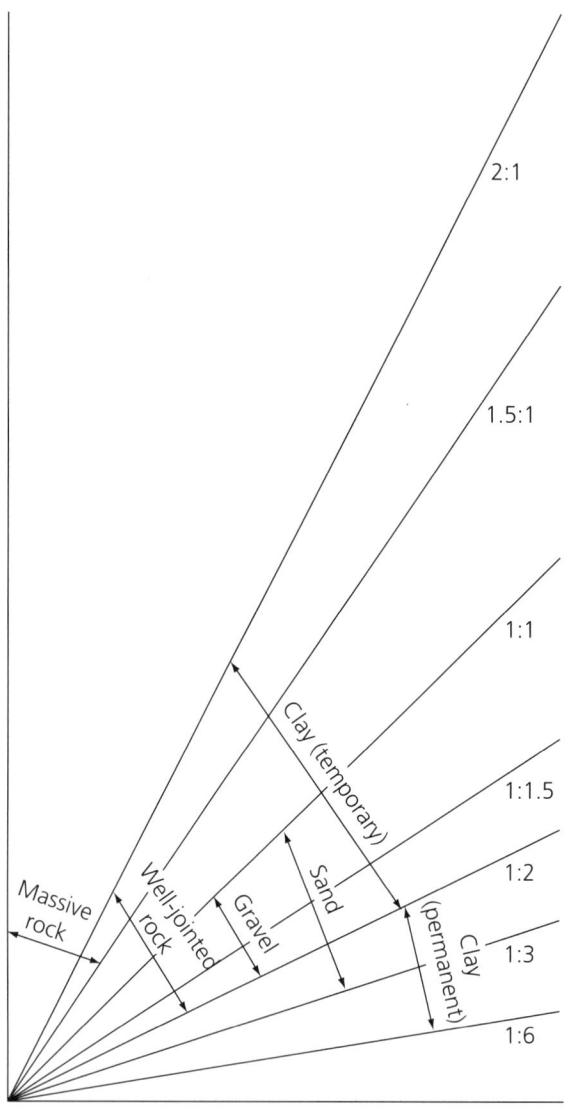

common sands are siliceous, but it should be remembered that different friction angles would apply to sands of different mineralogy, e.g. carbonate or mica.)

As an example, the ϕ'_{crit} value for a well-graded sub-angular sand and gravel may be taken as 36°. The factor of safety to be used on slopes cut into coarse-grained soil will depend on the circumstances. Where the consequences of slope movement are small (category 1), $F = 1.0$ may be acceptable, provided that the value of ϕ'_{crit} is reasonably conservative. For longer-term slopes or where failure is to be avoided (category 2), a

Table 10.1 Critical angle of friction $\phi'_{crit} = 30° + A + B$ for siliceous sand and gravels

A – Angularity[a]	A: °
Rounded	0
Sub-angular	2
Angular	4

B – Grading of soil[b]	B: °
Uniform	0
Moderate grading	2
Well graded	4

[a] Angularity is estimated from a visual description of the soil.
[b] Grading can be determined from grading curve by use of: uniformity coefficient = D_{60}/D_{10}, where D_{10} and D_{60} are particle sizes such that in the sample 10% of the material is finer than D_{10} and 60% is finer than D_{60}.

Grading	Uniformity coefficient
Uniform	<2
Moderate grading	2 to 6
Well graded	>6

A step-graded soil should be treated as uniform or moderately graded soil according to the grading of the finer fraction.
Intermediate values of A and B may be obtained by interpolation.

Data taken from BS 8002:2015, 'Code of practice for earth retaining structures' (BSI, 2015).

value $F = 1.25$ may be prudent. The factor of safety is applied to $\tan(\phi'_{crit})$; if the slope angle is β, then

$$F = \frac{\tan(\phi'_{crit})}{\tan \beta} \tag{10.1}$$

Hence, for $\phi'_{crit} = 36°$, $F = 1.25$ would give a slope (β) of 30° (1 : 1.73).

It may, of course, be possible to cut slopes to angles steeper than ϕ'_{crit}, provided that some failures can be accepted. Factors such as the presence of some fine soil combined with slight cementing of the soil grains can give a considerable increase in the stable slope angle over that which theory would predict.

For the case with the water table at the ground surface, stable angles are approximately half that of dry slopes.

10.4.5.2 Fine-grained soil

The analysis of temporary slopes in fine-grained soil is much more complex. Despite the use of sophisticated methods to back-analyse two failed temporary slopes, Kovacevic *et al.* (2004) found that 'actual times to failure are still difficult to predict'. The presence of undetected inhomogeneities in the ground can hasten the loss of suction and cause failure within a few days of cutting the slope, and sometimes during excavation. The designer of a temporary slope in fine-grained soil on all but very major projects is likely to have to work with incomplete information about the stratigraphy and relatively poor-quality data on soil properties. The design must therefore start by minimising the consequences of ground movement.

Conventional limit equilibrium analyses may be used to analyse slopes. A problem with these analyses is that the failure mechanism must be assumed (or guessed) in order to perform the analysis. If the mechanism is wrong, the analysis is likely to give misleading results. A recent development that can help with the determination of the critical ultimate limit state mechanism is discontinuity layout optimisation (Smith and Gilbert, 2007). However, the early approaches of Bishop (1954) (circular surfaces) and Janbu (1954) (general surfaces) are still in wide use. An advantage of the assumption of a circular plane of sliding is that a family of potential surfaces can be defined in terms of a grid of centres and a range of radii. The most critical of these can then be found.

In uniform clay soil, the critical slip surface is likely to be moderately deep and to emerge somewhere close to or a little beyond the toe of the slope. Where a weak layer is present, the most critical surface found in a circular analysis will be tangential to the weak layer, but a distinctly non-circular surface will probably be the most critical. In some cases, non-circular surfaces can be simplified to an active wedge at the upslope side, a passive wedge at the downslope side and a polygon sliding on the weak layer between the two wedges.

In view of the uncertainties involved, the analysis is best viewed as a tool to guide the judgement of the designer in the selection of the most appropriate slope to adopt. Whatever method is used, sensitivity analyses are essential to give an appreciation of the possible effects of differences in shear strength. It is normal to use undrained soil parameters in the analysis of the short-term stability of clay slopes, but to allow for some loss of strength, which is dependent on the design life of the slope and the nature of the material. A range of strength values should be used to examine the effects of softening (loss of suction) on stability.

10.4.6 Design of rock slopes

Rock slopes of category 1 or 2 will generally be designed using relatively simple kinematic considerations. The process essentially involves determining whether there are discontinuities or combinations of discontinuities that can delimit a block of rock able to slide or topple into the excavation. If there are no such discontinuities, then the chance of failure at this scale of project will be very small, because failure must occur through the intact rock. If blocks with the potential to fail do exist, then the friction on the discontinuities must be assessed and the factor of safety for the block must be calculated. If weak discontinuities slope into the excavation, then it may be necessary to cut the slope at the same angle as (or flatter than) the discontinuities.

If the factor of safety is insufficient, it may be improved by installing rock bolts to increase the restraint. If the friction is reduced because of the groundwater conditions, then dewatering will enhance the factor of safety.

10.5. Monitoring

There are two main principles in the construction of temporary slopes. As mentioned in Sections 10.2 and 10.3, one is to minimise the angle, length and open period of the slope and the other is to monitor the slope. Monitoring systems must be properly designed, bearing in mind that it is no use gathering data that cannot be analysed and interpreted. It is possible to acquire so much information that important trends cannot easily be discerned.

All monitoring systems should include regular visual inspection of the slopes, especially the areas just behind the slope crest and at the toe. Substantial slope movements can be preceded by smaller movements which are manifested in cracking behind the crest (tension cracks) or bulging of the ground near the toe. If such movement is seen, there is often time to carry out some stabilisation work, such as placing backfill at the toe before a major failure takes place. The ground should also be checked for indications of groundwater emerging from the slope; this may simply be dampness or softening, rather than flowing water.

A wide range of monitoring systems is available, from very simple to very sophisticated. At the simple end of the spectrum, a row or rows of ranging poles arranged in straight lines along and above the slope can be visually monitored on a daily basis. It should be noted, however, that the predominant component of movement of the ground at the top of a slipping mass will be vertical, so level measurement is important. Targets fixed to pegs can be surveyed precisely, either manually or using automated instruments. In-ground instrumentation can take the form of slip indicators, inclinometers or piezometers.

Slip indicators are essentially simple tubes installed in boreholes drilled through the zone of likely movement into stable ground below. Slippage is evidenced by a closure of the tube at the surface of shearing; however, these instruments do not provide information on the amount of movement.

There are two types of inclinometer. The early instruments consisted of tubes grouted into boreholes (taking care to try to match the stiffness of the installation to that of the ground). Like slip indicators, these must extend into fully stable ground at depth in order to provide a datum for lateral movement. A 'torpedo' is lowered down the tube, using locating grooves to orient readings of the tilt of the torpedo at (normally) 0.5 m intervals down the tube. In that way, a continuous profile of the tube is obtained from each monitoring visit. More recently, inclinometers have been developed to measure the tilt of devices that are permanently installed in the ground. This allows remote reading and the distribution of data through the internet.

Piezometers measure the pressure in the groundwater at a particular point. If the instrument is not located suitably, it will not give useful information and could mislead.

Some types are capable of measuring negative (below atmospheric) pressure and can therefore measure suction in the soil mass. Measurement of suction in a temporary slope would normally only be used on category 3 slopes.

Monitoring systems for movement or pore pressure can be electronically logged and remotely accessed. They can be automated to warn when trigger values (e.g. an amount or rate of movement) are reached. However, such systems should not replace careful and frequent visual observation of the slope condition.

It should be noted that it is generally less practicable to install instruments at or close to the toe of the slope, as this is the main area of construction activity.

Finally, the contractor and designer (and any other concerned parties) must agree trigger levels and contingencies to stabilise a moving slope. The simplest and quickest approach to movement is often to place new fill in front of the slipped or slipping mass (see Figure 10.3). This may be only an emergency measure to arrest movement and to allow work on redesign of the slope. Careful excavation of failed fine-grained soil and replacement by coarse granular material such as hardcore can be a sufficient remedy. A further alternative may be the installation of soil nails, which is effectively a way of reinforcing the soil mass. Nails are usually drilled and grouted in place, although driven nails can be used. Unlike ground anchors, they are not pre-tensioned and require further movement of the ground to develop tensile resistance. While costlier than replacement, they may allow the reinstatement of a steep slope required for construction.

REFERENCES

Bishop AW (1954) The use of the slip circle in the stability analysis of slopes. *Géotechnique* **5(1)**: 7–17.

BSI (British Standards Institution) (2004) BS EN 1997-1:2004. Eurocode 7. Geotechnical design. General rules. BSI, London, UK.

BSI (2014) NA + A1:2014 to BS EN 1997-1:2004 + A1:2013. UK National Annex to Eurocode 7. Geotechnical design. General rules. BSI, London, UK.

BSI (2015) BS 8002:2015. Code of practice for earth retaining structures. BSI, London, UK.

Janbu N (1954) Application of composite slip surfaces for stability analyses. *Proceedings of the European Conference on the Stability of Earth Slopes, Stockholm, Sweden*, vol. 3, pp. 43–49.

Kovacevic N, Hight DW and Potts DM (2004) Temporary slope stability in London Clay – back analyses of two case histories. *Advances in Geotechnical Engineering* **3**: 1–14.

Kovacevic N, Hight DW and Potts DM (2007) Predicting the stand-up time of temporary London Clay slopes at Terminal 5, Heathrow Airport. *Géotechnique* **57(1)**: 63–74.

Pettifer GS and Fookes PG (1994) A revision of the graphical method for assessing the excavatability of rock. *Quarterly Journal of Engineering Geology* **27(2)**: 145–164.

Smith CC and Gilbert M (2007) Application of discontinuity layout optimization to plane plasticity problems. *Proceedings of the Royal Society A: Mathematical, Physical and Engineering Sciences* **463(2086)**: 2461–2484.

Tomlinson MJ (2001) *Foundation Design and Construction*, 7th edn. Pearson Education, London, UK.

FURTHER READING

Bromhead E (1992) *The Stability of Slopes,* 2nd edn. Taylor & Francis, London, UK.
Hoek E and Bray JW (1981) *Rock Slope Engineering.* E & FN Spon, London, UK.

Useful web addresses

Geotechnical & Geoenvironmental Software Directory – a comprehensive listing of available geotechnical programs, including soil slope stability and rock slope stability: http://www.ggsd.com (accessed 01/08/2018).

Pallett, Peter F and Filip, Ray
ISBN 978-0-7277-6338-9
https://doi.org/10.1680/twse.63389.141
ICE Publishing: All rights reserved

Chapter 11
Sheet piling

Ray Filip
Temporary Works Consultant and Training Provider, RKF Consult Ltd

Interlocking steel sheet piles are profiled steel sections which are installed vertically into the ground by a variety of methods to support the ground and exclude water when excavating or to carry applied loads. They are used in temporary works to support the ground and to exclude water while a structure or service is installed below ground level. They are generally extracted to be reused; however, they may be incorporated into the permanent works, such as in sealed basement walls where they may also be used to carry vertical loads from the building or structure. Sheet piles can also be used as anchorages and in conjunction with other types of retaining walls. They may be used with tubular or I-section piles to form a high-modulus combination wall. 'Cofferdam' is a term given to a closed sheet pile 'box' used for deep retention schemes and which may include provisions for excluding water. Isolated steel I- or H-sections or steel tubes can be driven into the ground to form bearing piles. The design concept of sheet piling is unusual in that it utilises the shear strength of the ground to support itself; the element being supported is therefore also part of the supporting mechanism.

11.1. Major alternatives
Alternatives to sheet piling include the following

- Contiguous or secant piling can be used in place of sheets to form temporary and permanent retaining walls or cofferdams.
- Diaphragm walling can form very substantial temporary and permanent retaining walls.
- Concrete caissons are an alternative to cofferdams.
- Concrete bearing piles (driven pre-cast or augered and cast in situ) which can carry compressive, tensile and horizontal loads.
- Post-and-plank trench boxes are appropriate for minor retaining structures (see Chapter 12 on trenching).
- 'Top down' is an alternative to traditional basement construction using temporary frames.

11.2. Types of steel sheet piling
The main types of steel sheet piles are: straight web, U-profile, Z-profile and combination wall (Figure 11.1) and each individual element interlocks with adjacent elements

Figure 11.1 Examples of a U-profile, Z-profile, combination wall, straight web and box pile

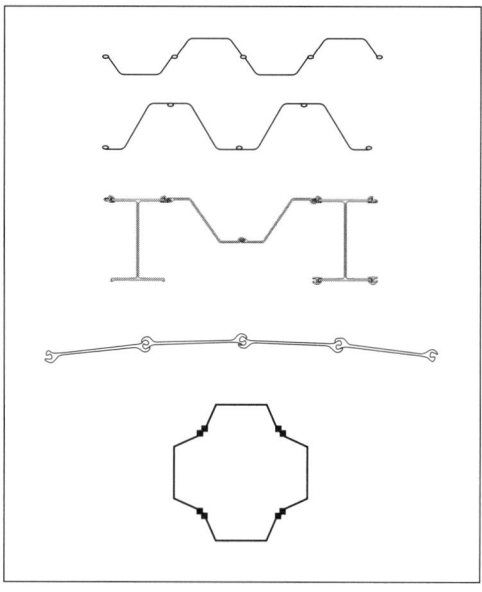

(whereas the majority of lighter trench sheets overlap rather than interlock). Straight web sheets are intended for the construction of circular structures where they work in hoop tension. U- and Z-profile hot-rolled sheets are the most common and can be used in straight lines or can also be installed to a gentle radius. Combination walls are for high-modulus work (e.g. for marine work) and comprise steel I-beams or large-diameter steel tubes with U- or Z-profile sheets between. Where a line of piles is required to turn a corner or form a junction, special corner sections are available (Figure 11.2).

Sheet piles are available as hot-rolled sections in standard steel grades varying from S240GP to S460GP to BS EN 10248-1:1996 (BSI, 1996a) and cold-formed sections in standard steel grades from S235JRC to S355JRC to BS EN 10249-1:1996 (BSI, 1996b). From grade 270 to 355 the increase in stress is 30% but the increase in procurement cost is approximately 2% (also improved driveability). Specialist products such as marine-grade steels are also available. For further information on the available range of products, rolling tolerances, corrosion protection and achievable radii, see the ArcelorMittal *Piling Handbook* (2016) or similar information from other manufacturers.

11.3. Installing sheet piles

Piles can be installed individually (pitch and drive) or installed progressively using the panel technique (panel driving) in combination with piling gates for improved installation accuracy, driveability and stability. With combination walls, the heavier element (I-section or tube) is installed first using piling gates, with double Z-sheets or triple U-sheets filling the gap between to allow for installation tolerances. A number of common installation techniques are used.

Figure 11.2 Examples of corner sections. (Reproduced from ArcelorMittal *Piling Handbook*, 8th edn (2008), courtesy of ArcelorMittal Ltd)

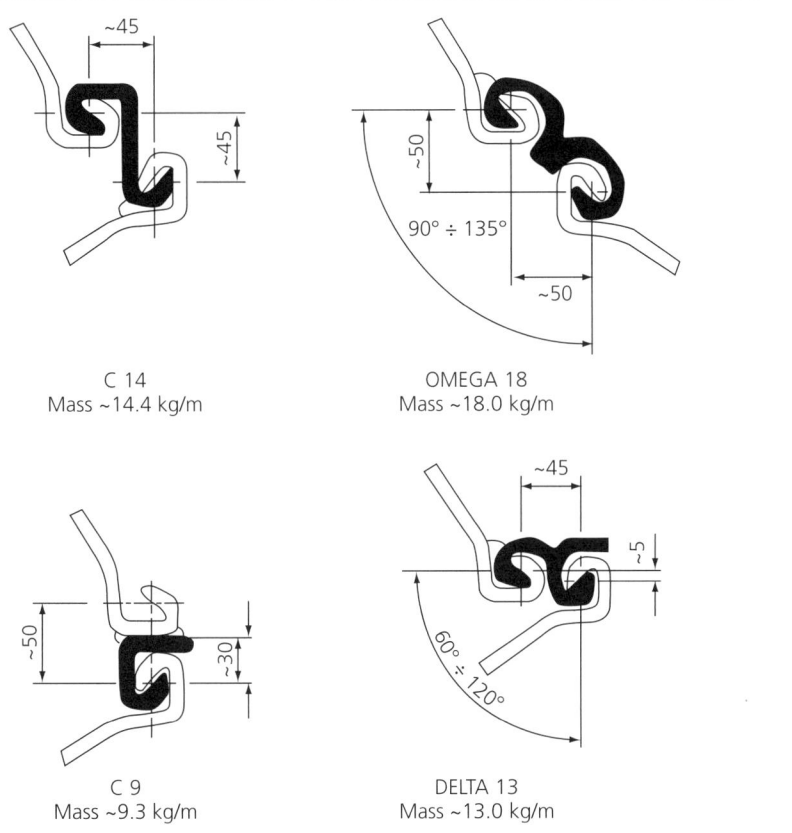

11.3.1 Open trench
Relatively short and light sheet piles (and trench sheets) may be installed by placing them into a pre-excavated trench and then backfilling, or by pushing them into relatively soft ground using an excavator bucket.

11.3.2 Dig and push
Sheets are progressively pushed into the ground using an excavator bucket as the excavation progresses. The sheets will not be significantly 'toed' in, hence multiple frames are required. This technique is generally used in soft ground conditions and accuracy is poor unless guides are used.

11.3.3 Vibrators
Vibrators are suitable for use in predominantly granular soils and soft to firm silts or clays. Small versions known as 'excavator-mounted vibrators' are attached to a standard site excavator, whereas larger (more powerful) versions are attached to a piling rig or freely suspended from a crane. Spinning eccentric weights generate the vibration (they

spin in opposite directions to cancel out the horizontal components leaving the vertical component), which is used to liquefy the soil and reduce resistance rather than drive the pile. With rig-mounted vibrators a crowding force is often provided to concurrently push the piles, and the self-weight of the vibrator is utilised for crane-suspended versions. Vibrators are not very suitable for use in stiff clay.

11.3.4 Percussive or impact hammers

These are suitable for most soil types and can often work where hard driving or subsurface obstructions are likely to be encountered. Most are suspended by cranes, and a weight is dropped onto the top of the pile. The most basic form of impact is known as a 'drop hammer'; the weight is dropped onto the top of the pile and must then be lifted for the next blow. Although drop hammers are still used, impact force is today provided by compression of diesel, air or (most efficiently) hydraulic impulse (75–95% efficiency compared to 40–60% for diesel).

Hammers can be single-acting (where the power is used to lift the drop weight so it can fall under gravity) or double-acting (using the impulse to accelerate the drop weight downwards as well as to lift it back up for the next blow). Double-acting hammers provide a much faster impact rate (120 blows/min), but single-acting hammers have their place in highly cohesive or tough ground conditions. Hammers generate significant noise (even when acoustically shrouded) and develop vibration in hard driving conditions.

Noise and vibration are significant issues, and local authorities can set limits under the 1974 Control of Pollution Act 1974 (UK Government, 1974).

11.3.5 Hydraulic pushing

Hydraulic pushing (commonly known as 'silent piling') is used in cohesive soils. Piles are pushed into the ground using a steady pushing force from powerful hydraulic jacks. Equipment with jacking forces of 2000 kN is available. Reaction to the pushing force is achieved by pulling on adjacent piles, which utilise the skin friction of the soils. A key element to the process is installing the first few piles, which can be achieved by using a reaction frame with kentledge or self-weight and the crowd force of a rig. These methods are vibration-free and quiet (the principal noise is from the engines), and thus they are suitable in built-up or sensitive areas. In rock, a pilot drill precedes the pushing process (see Figure 11.3).

11.3.6 Driving aids

Methods are available to aid the installation process, as described in the following sections.

11.3.6.1 Pre-augering

A small auger is screwed into the ground at the sheet pile clutch interlock positions. This loosens the dense ground locally and probes for buried obstructions. For retention schemes, pre-augering below formation level should be avoided as it reduces passive resistance.

Figure 11.3 Examples of (a) a hydraulic pusher and (b) a hydraulic impact hammer. (Courtesy of Dawson Construction Plant Ltd)

(a)

(b)

11.3.6.2 Water jetting

A lance is connected to the pile and water is forced under pressure to the toe of the sheet. The water softens clay or loosens gravel as the pile is forced downwards. Water jetting should be used with care to prevent fines being washed out around the installation area or pressure build-up in the vicinity causing heave to adjacent foundations or services. Jetting to the bottom of the pile wall may form a scour pathway.

Piles are often installed by tracked machines such as leader rigs or with piling equipment suspended from crawler cranes. Consideration must be given to the working platform beneath the machinery (see Chapter 5 on site roads and working platforms).

11.3.7 Water control

Clutched sheet pile walls become less permeable with time as fines will fill the interlocks. Some water seepage is inevitable through the clutches, and minor seepage can be controlled with internal sump pumping. Sealants can be applied into the clutches to aid with water control. The type of proprietary product to be used depends on the degree of moisture penetration permitted. Sealants include the following

- Bituminous sealants: applied pre-driving and mainly used for temporary sheet piling.
- Hydrophilic or resin-based sealants: applied pre-driving. They are more effective than bituminous sealants and are mostly suitable for permanent works, but can be damaged during installation, particularly by heat generated by hard vibratory installation.
- Seam welding clutches: applied after installation, and provide the highest degree of watertightness. They are predominantly used for permanent sealed basements.

See also Chapter 6 on control of groundwater.

11.3.8 Extracting piles

Piles can be extracted by a vibrator, which can be rig-mounted or crane-suspended. When crane-suspended, a significant line pull will be necessary as the vibrator will only loosen the soil and the crane must pull the pile out. If the sheet piles have been in the ground for some time or have been driven into hard ground, some blows with an impact hammer may be necessary to break any interlock or ground friction. 'Silent' piling equipment can also extract piles. Purpose-made hydraulic extractors are available that use spreader beams on the ground for reaction to the jacking force. Extractors up to 10 000 kN capacity are available, which exceeds the tensile strength of most sheet piles.

11.3.9 Cofferdams

Cofferdams are essentially a sheet pile 'box' formed to support the ground and possibly to exclude water from the excavation (Figure 11.4). Cofferdams are conventionally single skin, but for very deep excavations double-skin walls are used which are formed from two parallel lines of piles filled with rock and tied together with tie rods. These are generally designed as a gravity-retaining structure. Corner sections (see Figure 11.2) are used to form a continuous wall in square or rectangular cofferdams. Circular cofferdams are generally installed using circular piling gates by the panel-driving method, although leader rigs can also be used.

For all cofferdams (and retaining walls in general), internal framing or external anchorage can be used. The design and detailing of a cofferdam and its framing or anchorage

Sheet piling

Figure 11.4 Sheet pile excavation with heavy-duty proprietary framing. (Courtesy of Groundforce Shorco Ltd)

requires careful consideration by a competent designer with experience of the construction process that will occur within the cofferdam. The sequence of installation and removal must be considered in the design to identify the critical loads. It is not unusual for the critical-load case to occur during the removal process rather than installation.

It is difficult and generally not economical to exclude 100% of the water, and so sump pumps are used to pump water out of the cofferdam. Dewatering can be used to control water ingress but can be expensive and has significant risks associated with it (see Chapter 6 on control of groundwater). Water may also enter the cofferdam over the top of sheet piles (when positioned in a river or sea) or from beneath. Care should be taken to avoid base failure by piping or heave. The sheet piles can be driven to a greater depth to penetrate an impervious soil layer (cohesive material) as a cut-off typically around 2 m. If it is not possible to utilise a cut-off, an underwater concrete plug can be formed in the base of the cofferdam and the water pumped out once the concrete has hardened.

The supports to cofferdams can be provided by ground ties and anchorage, fabricated steel internal framing or heavy-duty proprietary hydraulic framing (hydraulic struts with capacities of 2500 kN and more are available). Further information is available in the

Construction Industry Research and Information Association (CIRIA) publication SP95 (Williams and Waite, 1993).

11.4. Eurocode 7

A variety of design methodology and guidance documents have been developed such as those by ArcelorMittal (2016) and CIRIA publications SP95 (Williams and Waite, 1993), SP104 (CIRIA, 1995) and C760 (Gaba *et al.*, 2017). The most recent codes are Eurocode 7 (BSI, 2004a) and Eurocode 3 (BSI, 2003) and the associated UK National Annexes (BSI, 2006, 2014). Eurocode 7 is a comprehensive limit state design code based on the use of partial factors with three design approaches (geotechnical application classification categories 1, 2 and 3) in terms of increasing perceived risk, with design procedures altered accordingly. In the basis of the design, loads (called 'actions' in the Eurocodes) and load combinations should be taken from Eurocode 0 (BSI, 2002) and Eurocode 1 (BSI, 2004b). Section 2.4.2 of Eurocode 7 (BSI, 2004a) lists the items that should be considered in geotechnical design, with consideration being given to actions occurring jointly and time effects. The code requires that: 'for each geotechnical design situation it shall be verified that no relevant limit state, as defined in Eurocode 0 (BSI, 2002), is exceeded'.

Limit states should be verified by one of the following methods

- use of calculation
- adoption of prescriptive measures such as the use of conservative rules, attention to specification and control of materials, workmanship, protection and maintenance procedures
- experimental models and load tests
- observational methods.

Section 2.2 of Eurocode 7 requires that both short- and long-term design situations are considered. Typical ultimate limit states from section 2.4.7.1 of Eurocode 7 are

- loss of equilibrium (EQU) of the structure or ground, considered as a rigid body (this limit state is often critical to sizing structural elements)
- internal structural failure or excessive deformation (STR)
- failure or excessive deformation of the ground (GEO)
- loss of equilibrium due to uplift (buoyancy) by water pressure or other vertical actions (UPL)
- hydraulic heave, internal erosion and piping caused by hydraulic gradients (HYD).

Partial factors are defined in annex A of Eurocode 7 (BSI, 2014), with the factors being increased for abnormal risk situations or reduced for less severe cases, temporary structures or transient design situations. A partial factor of 1.0 should be used for accidental situations.

Section 2.4.7.3.4 of Eurocode 7 provides details of three design approaches for the various limit states. Section 4 of Eurocode 7 states that: 'to ensure the safety and quality of a structure, the following shall be undertaken, as appropriate'.

- The construction processes and workmanship shall be supervised; this would include identifying differences in the actual ground and water conditions compared to those assumed in the design. Any differences in the method of construction from those assumed in the design, should be reported.
- The performance of the structure shall be monitored during and after construction to validate assumptions made in the design.
- The structure shall be adequately maintained to ensure safety and serviceability.

For flexible retaining walls supported by anchors or struts, the magnitude and distribution of earth pressures, internal structural forces and bending moments depend to a great extent on the relative stiffness of the structure and the stiffness, stress and strength of the ground. For problems of ground–structure interaction, analyses should use stress–strain relationships for ground and structural materials and stress states in the ground that are sufficiently representative, for the limit state considered, to give a safe result.

Sections 7–9 of Eurocode 7 provide guidance on the design of piled foundations, anchorages and retaining structures (where diagrams of modes of failure are shown). Eurocode 7 does not provide specific guidance on the supervision, monitoring and maintenance of retaining structures. This specific guidance can be found in BS EN 12063 (BSI, 1999) for sheet pile walls and BS EN 12699 (BSI, 2015) for displacement piles. Specifiers may use *ICE Specification for Piling and Embedded Retaining Walls* (ICE, 2016) for guidance.

Issues in Eurocode 7 that remain to be resolved include tie bar and 'dead-man' anchorage systems and factoring water pressures.

11.5. Design

The first and perhaps the most important stage in the design process is to interpret the site investigation information and establish a geotechnical model for analysis. Most site investigations are commissioned for the permanent structure, and it is not uncommon for additional investigation to be required to aid the temporary works design. Eurocode 7 Part 2 (BSI, 2004a) now gives guidance and rules for the extent of the site investigation required for embedded retaining walls and bearing piles.

There are two philosophies of design that allow the inherent shear strength of the soil to be utilised to help the sheet piling support it

1. The soil–structure interaction (SSI) approach, which requires the elastic properties of both the soil and the structure to obtain the stresses by iterative analysis.
2. The limit-equilibrium (LE) approach, which sizes the elements and penetration such that the whole is in equilibrium when the maximum stresses are mobilised.

Although the SSI approach is a very precise analysis, it uses the elastic properties of soil, which can only be determined to a very imprecise level. Historically, therefore, the LE approach has been the common method of calculation. It permits very simple pile designs to be carried out manually; whether the use of the SSI or the LE approach provides the best solution is best determined by use of specialist software.

A design should aim to be both safe and economical, and the structure classification should determine the expertise employed in the design. Codes and standards that should be followed have recently changed; British Standards were effectively withdrawn and ceased to be maintained on 31 March 2010, to be replaced by Eurocodes. There are associated European Standards effectively in place to be used by the designer in conjunction with the Eurocodes, but the designer should be aware that, although limit state methods and analysis are common in both present and former codes of practice, the application and values of partial factors and the resulting factors of Safety are different.

The elements described in the following sections need to be considered in the design of sheet pile walls, cofferdams and so on and their supports.

11.5.1 Stresses due to applied loads

Bending and axial stresses generated by lateral earth and water pressures surcharges, applied loads (e.g. impact) and vertical loads should be checked against the stress in the steel in accordance with Eurocode 3 (BSI, 2003). Pressures can be different if the wall is retaining an excavation or if backfilling is being compacted behind the wall. Variations in soil profile and properties should be considered. When using U-profile sheets in cantilever, the designer should be satisfied that clutch interlock and ground cohesion/friction is sufficient to overcome the possibility of 'neutral axis clutch slip' or apply a factor of 0.8 to the modulus (see Section 11.5.16 and the UK National Annex to Eurocode 3 (BSI, 2003, 2006a)). As a general rule, clutch interlock and ground cohesion/friction in typical UK soils are usually sufficient to overcome neutral axis clutch.

Timescale should also be considered in cohesive materials; in short-term usage undrained (total stress) and in the medium to longer term drained (effective stress) design are appropriate. The pressures on the retaining wall will generally increase with time.

11.5.2 Limit equilibrium methodology

This considers overall stability to prevent rotation of the wall or forward displacement (sliding). Destabilising forces are due to lateral earth, water and surcharge pressures (known as 'active pressures'). Stabilising forces are due to the embedment of the wall in the soil (known as 'passive resistance') and restraint by framing or anchorage. Note that passive resistance, being a reactive force, can act in any direction to provide a stabilising mechanism. The sheets should sufficiently penetrate below formation level to maintain stability, even if multiple levels of framing or anchorage are provided. The stability of the surrounding soil must also be considered to ensure there is no possibility of slip failures forming. Stability due to heave, piping or excessive ingress of water must also be considered.

11.5.3 The installation and removal sequence

This should be considered carefully to identify the critical design case. It is not unusual for the critical stresses in the sheets, framing or anchors to occur during the backfilling and removal process, rather than during installation and excavation. The most onerous water table conditions should be considered.

11.5.4 Deflection

Deflection should be limited to prevent damage to surrounding roads, buildings or services caused by settlement or ground movement. Deflection limits should be established and a monitoring regime established if necessary. Readings should be taken on a regular basis during excavation and also once excavation has been completed. Trigger levels should be established, such as

- green – low movement; continue taking readings
- amber – increased movement; increase frequency of readings
- red (maximum designed deflection) – significant movement; thorough investigation of the causes and remedial measures to be considered.

Deflections as well as the required penetration of cantilevered sheet pile walls can be significant for retained heights over about 4.0 m (unless very heavy piles are used).

11.5.5 Corrosion protection

Corrosion protection is mainly a consideration for permanent works and not temporary works. Cathodic systems are sometimes required for marine applications.

11.5.6 Workmanship, construction tolerances

Accidental over-dig at formation level (a maximum of 0.5 m or 10% of retained height), softening of the formation in cohesive soils (usually the top 1 m of formation on the passive side) and accidental impact should be considered. Cofferdams in water should include flooding valves and measures should be taken if water ingress becomes excessive.

11.5.7 Framing and anchor support loads

These are affected by the stiffness of the structural members involved (stiffer members attract load) and the soil–structure interaction (the SSI design can give loads as much as 85% higher than those obtained with the LE approach). Timber, steel tubes or steel structural sections are common materials for framing, and occasionally reinforced concrete is used for walings. Useful guidance and detailing is given in CIRIA SP95 (Williams and Waite, 1993). Design should be undertaken by a site-experienced engineer. Steelwork walings tend to be I-sections, sometimes as battened pairs, with tubular or UC struts. The struts can rake down onto concrete thrust blocks. Long horizontal struts (where the effective length and self-weight bending are significant) can be supported by king posts. Web stiffeners should be used at waler–strut junctions and are installed at points of concentrated load; shear plates or welded/bolted connections are used to carry shearing components of reactions from raking members or corner braces.

Torsion on the waling beam should be considered due to the vertical reaction imposed from raking props, unless the vertical reaction can be carried by corbels. Self-weight bending and impact loading (in the least favourable position) should be allowed for. Prevention of progressive collapse from impact is usually achieved by back-analysing, assuming any one strut is removed and limit state conditions. The design of steel framing was developed from BS 449-2:1969 (BSI, 1969), BS 5950:1990 (BSI, 1990) and Eurocode 3 Part 5 (BSI, 2003), with BS 449-2:1969 and Eurocode 3 being used.

11.5.8 Tension cracks
Allowance should be made for tension cracks in clay behind the wall filling with water.

11.5.9 Thermal effects
These can be significant in large struts and should be considered in the design.

11.5.10 Releasing strut and anchor loads
These should be released gently by incorporating sand boxes, hydraulic jacks or striking packs.

11.5.11 Circular cell structures
Circular steel or reinforced concrete walings are designed for hoop-stresses.

11.5.12 Twin-wall cofferdams and other gravity structures
Well-drained granular fill is essential and the bearing resistance of the ground should be considered (soft soil formations should be avoided). A width/height ratio of >0.8 is required to prevent shear failure of the filling.

11.5.13 Site constraints
On confined sites, horizontal framing can be designed to carry live loading such as plant and material storage.

11.5.14 Driveability
Driveability refers to the ability of sheets to be installed into the ground without buckling, damage or end resistance, or wall adhesion preventing penetration. It should be noted that in all soils other than those of low strength the calculated structural requirements of the sheet piles will almost invariably be less onerous than the strength needed for driveability. The ArcelorMittal *Piling Handbook* (2016) provides the calculation method for suitable sections relative to items such as the standard penetration test (SPT) N values, and soil descriptions, length and steel grade of pile and installation methods, but tends to give overly optimistic findings for driveability. Generous factors are advised in the calculations.

11.5.15 Design sequence
The design of a sheet pile wall for LE analysis should follow the sequence of steps as follows

- Cross-sections of the soil profile should be established and moderately conservative soil parameters established from available data such as the site investigation report or from guidance values given in documents such as CIRIA SP95 (Williams and Waite, 1993) or CIRIA C760 (Gaba *et al.*, 2017). Groundwater, river levels and tidal ranges should be considered (generally

worst-case scenarios). A certain degree of engineering judgement may be necessary with regard to the reliability of drainage behind the wall.
- A decision should be taken if skin friction and skin adhesion are to be considered. If there is any doubt, they should be ignored.
- Consider the timescale and whether total stress or effective stress conditions apply to each stratum.
- Partial factors of safety are applied to the previously determined moderately conservative soil parameters to derive design parameters.
- Establish the coefficients of active (K_a) and passive (K_p) earth pressure loading, together with water pressure on both sides of the wall.
- Use coefficients and the Rankine formula to develop active and passive pressure diagrams. Over-dig, tension cracks filling with water on the active side and soil softening on the passive side should be considered. The active and passive pressure diagrams should then be combined to give a design analysis model.
- Supports may be introduced at appropriate levels using a framing/anchorage arrangement to allow the work to be carried out as easily as possible and limit deflections.
- Forces and moments are resolved to test the equilibrium of the total pressure diagram in the ultimate-limit condition. This is the maximum load situation and will give the required penetration (adding a length to permit reverse passive reaction). Partial factors are applied to variable and applied loads in accordance with Eurocode 7; the worst case for Design Approach 1 (UK National Annex to Eurocode 7 (BSI, 2014)) combinations 1 or 2 determines the required pile length with the design effect moment and forces.
- The depth of penetration determines the length of pile required.
- The maximum moment multiplied by a factor plus the yield strength of the pile steel will give the minimum required pile section modulus and therefore the pile section designation for structural adequacy. The pile section modulus may need to be increased to account for 'driveability' (see Section 11.5.14). After checking driveability, the chosen section is verified for structural adequacy in accordance with Eurocode 5 (BSI, 2004c) and the associated UK National Annex (BSI, 2006b), after taking into account effects of corrosion on the section properties, including a check on the section classification and the effects of that. Piles can be verified for bending strength using plastic section properties if remaining in class 2, but if the section drops to class 4, then buckling and shear may determine the section. It is best to adopt sections remaining in class 3 or higher for structural walls.
- If using the LE method, multiply the forces put into the supports by a factor F_T to allow for the interaction of the soil and piles. This gives the design forces for anchors or frames.
- Back-analyse the frames to test for any one strut or tie accidentally removed for the serviceability limit state, to check that progressive collapse will not occur.
- Maximum potential deflections and ground settlements may be estimated using SSI or finite-element methods of analysis. Alternatively, the appendix to CIRIA C760 (Gaba et al., 2017) shows deflections and settlements from extensive site studies for use with the LE method.

11.5.16 Partial factors that may be considered for use in sheet piling design

Refer to Eurocode 7 (BSI, 2004a) and the associated UK National Annex (BSI, 2014) for the factors given therein. The following have been found to be appropriate for use in limit state design to date, but do not mix factors from more than one source

- $F_s = 1.2$, a divider applied to the tangent of moderately conservative angle of friction and moderately conservative drained cohesion values to give design tan ϕ and C' when the soil is under effective conditions (1.25 in the UK National Annex).
- $F_s = 1.5$, a divider applied to moderately conservative undrained cohesion values to give design cohesion when the soil is under total conditions (1.4 in the UK National Annex).
- $F_{pen} = 1.2$, a multiplier applied to the calculated depth from the net pressure diagram, from the point of zero net pressure to the bottom of pile, to add a finite length for F_3 (the stabilising passive force at bottom of pile). This is applicable only to a cantilever or fixed earth configuration, and only for finding the actual length of the pile to be driven.
- $F_m = 1.2$, a multiplier applied to the calculated moment in all steel members and piles to compensate for the difference in design stresses between the old factored yield and the new yield stress, with partial factors as in this list.
- $F_t = 1.85$, a multiplier applied to calculate loads from sheet piles onto all rigid frame members to compensate for the difference between the results obtained with the LE and SSI methods.
- $F_\beta = 0.8$, a divider applied to the required bending capacity of U-piles to compensate for the clutches being on the neutral axis. (Note that there are research results which appear to disprove that this reduction is necessary in practice.)

11.5.17 Other allowances suggested in the design
- Accidental over-dig: design rules are given in Eurocode 7 (BSI, 2004a). Consider 10% of the retained height or 0.5 m, whichever is the lesser.
- Softening of the top 1 m of ground after excavation: applies only in cohesive soil on the passive side of the pile. Usually taken as zero cohesion at formation level to full cohesion at 1 m depth.

11.6. Inspection and maintenance

Excavations and cofferdams should be inspected as per the requirements of the Construction (Design and Management) Regulations 2015 (UK Government, 2015, regulations 22, 23 and 24). Many excavations and cofferdams are in use for a significant period of time and a regular maintenance regime should be established. This may include checking and replacing timber wedges, remedial measures to seal interlocks, adequate means of access/egress and adequate edge protection. For management issues, see Chapter 2 and BS 5975 (BSI, 2019).

11.7. Plastic sheet piles

Plastic sheet piles are lightweight and are often used for permanent solutions where relatively short piles are required in relatively soft grounds. Typical projects could include river banks and canals or cut-off walls where strength is not a primary consideration (but stiffer plastic combination-type walls and king post walls are possible). Similar installation methods to steel sheet piling can be used (the choice of equipment will be more limited), and steel strengtheners (called 'mandrels') can be used to aid installation in denser or stiffer soils. Hard driving can generate heat, causing damage to the piles, and very cold temperatures can also cause problems. Plastic sheet piles have good longevity properties as they do not rot or rust.

REFERENCES

ArcelorMittal (2008) *Piling Handbook*, 8th edn. ArcelorMittal, Luxembourg.

ArcelorMittal (2016) *Piling Handbook*, 9th edn. ArcelorMittal, Luxembourg.

BSI (British Standards Institution) (1969) BS 449:1989-2:1969. Specification for the use of structural steel in building. Metric units. BSI, London, UK (withdrawn).

BSI (1990) BS 5950:1990. Structural use of steelwork in building. BSI, London, UK (withdrawn).

BSI (1996a) BS EN 10248-1:1996. Hot rolled sheet piling of non-alloy steels. Technical delivery conditions. BSI, London, UK.

BSI (1996b) BS EN 10249-1:1996. Cold formed sheet piling of non alloy steels. Technical delivery conditions. BSI, London, UK.

BSI (1999) BS EN 12063:1999. Execution of special geotechnical work. Sheet pile walls. BSI, London, UK.

BSI (2002) BS EN 1990:2002. Eurocode: Basis of structural design. BSI, London, UK.

BSI (2003) BS EN 1993:2003. Eurocode 3: Design of steel structures. BSI, London, UK.

BSI (2004a) BS EN 1997-1:2004. Eurocode 7: Geotechnical design. General rules. BSI, London, UK.

BSI (2004b) BS EN 1991:2004. Eurocode 1: Actions on structures. BSI, London, UK.

BSI (2004c) BS EN 1995:2004. Eurocode 5: Design of timber structures. BSI, London, UK.

BSI (2006a) NA to BS EN 1993-3-1:2006. UK National Annex to Eurocode 3: Design of steel structures. Towers, masts and chimneys. Towers and masts. BSI, London, UK.

BSI (2006b) NA to BS EN 1995-2:2004:2006. UK National Annex to Eurocode 5: Design of timber structures. Bridges. BSI, London, UK.

BSI (2014) NA + A1:2014 to BS EN 1997-1:2004 + A1:2013. UK National Annex to Eurocode 7: Geotechnical design. General rules. BSI, London, UK.

BSI (2015) BS EN 12699:2015. Execution of special geotechnical works. Displacement piles. BSI, London, UK.

BSI (2019) BS 5975:2019. Code of practice for temporary works procedures and the permissible stress design of falsework. BSI, London, UK.

CIRIA (Construction Industry Research and Information Association) (1995) *Remedial Treatment for Contaminated Land*. CIRIA, London, UK, SP104.

Gaba A, Hardy S, Doughty L, Powrie W and Selemetas D (2017) *Guidance on Embedded Retaining Wall Design*. CIRIA, London, UK, C760.

ICE (Institution of Civil Engineers) (2016) *ICE Specification for Piling and Embedded Retaining Walls*, 3rd edn. ICE Publishing, London, UK.

UK Government (1974) Control of Pollution Act 1974. Statutory Instrument 1974/40. HMSO, London, UK.

UK Government (2015) Construction (Design and Management) Regulations 2015. Statutory Instrument 2015/15. The Stationery Office, London, UK.

Williams B and Waite D (1993) *The Design and Construction of Sheet-Piled Cofferdams.* CIRIA, London, UK, SP95.

FURTHER READING

ArcelorMittal (2004) *Steel Sheet Piles. Installation.* ArcelorMittal, Luxembourg.

ArcelorMittal (2014) *Impervious Steel Sheet Pile Walls. Design & Practical Approach.* ArcelorMittal, Luxembourg.

ArcelorMittal (2016) *Steel Foundations Solutions. General Catalogue.* ArcelorMittal, Luxembourg.

BSI (British Standards Institution) (2010) BS 6349-2:2010. Maritime works. Code of practice for design of quay walls, jetties and dolphins. BSI, London, UK.

BSI (2015) BS 8002:2015. Code of practice for earth retaining structures. BSI, London, UK.

BSI (2015) BS 5930:2015. Code of practice for ground investigations. BSI, London, UK.

Byfield M and Mawer R (2001) *The Development of Section Modulus in Larssen U-shaped Sheet Piles.* University of Cranfield, Cranfield, UK.

Dawson R (2001) Steel to replace concrete. *Proceedings of the Institution of Civil Engineers – Geotechnical Engineering* **149(4)**: 205–207.

Day RA and Potts DH (1989) *Comparison of Design Methods for Propped Sheet Pile Walls.* Steel Construction Institute, Ascot, UK, Publication 077.

Driscoll R, Scott P and Powell J (2009) *EC7 – Implications for UK Practice.* CIRIA, London, UK, Report C641.

Filip RK (2004) Recent advances in quiet and vibration-less steel pile installation. *Proceedings of the 30th Annual Conference on Deep Foundations, Chicago, IL, USA*. Deep Foundations Institute, Hawthorne, NJ, USA, pp. 1–11.

Gaba A, Hardy S, Doughty L, Powrie W and Selemetas D (2017) *Guidance on Embedded Retaining Wall Design.* CIRIA, London, UK, C760.

ICE (Institution of Civil Engineers) (1996) *The Observational Method in Geotechnical Engineering.* Thomas Telford, London, UK.

Packshaw S (1962) Cofferdams. *Proceedings of Institution of Civil Engineers* **21(2)**: 367–398.

Padfield CJ (1984) *Design of Retaining Walls Embedded in Stiff Clay.* CIRIA, London, UK.

Rowe PW (1955) A theoretical and experimental analysis of sheet pile walls. *Proceedings of Institution of Civil Engineers* **4(1)**: 32–69.

Rowe PW (1957) Sheet pile walls in clay. *Proceedings of Institution of Civil Engineers* **7(3)**: 629–654.

Symons IF, Little JA, McNulty TA, Carder DR and Williams SGO (1987) *Behaviour of a Temporary Anchored Sheet Pile Wall on A1(M) at Hatfield.* TRL, London, UK, TRL Research Report 99.

ThyssenKrupp GfT Bautechnik (2010) *Sheet Piling Handbook*, 3rd edn. ThyssenKrupp, Essen, Germany.

Yau JHW and McNicholl DP (1990) Failure of a temporary sheet pile wall: case study. *Proceedings of the Seminar on Failures in Geotechnical Engineering, Hong Kong.*

Useful web addresses

ArcelorMittal Ltd: http://corporate.arcelormittal.com (accessed 01/08/2018).

CIRIA (Construction Industry Research and Information Association): www.ciria.org (accessed 01/08/2018).

Dawson Construction Plant Ltd: http://www.dcpuk.com (accessed 01/08/2018).

Groundforce Shorco Ltd: https://www.vpgroundforce.com/gb (accessed 01/08/2018).

Mabey Hire Ltd: https://www.mabey.com/uk (accessed 01/08/2018).

MGF Ltd: https://www.mgf.ltd.uk (accessed 01/08/2018).

Steel Piling Group: https://www.steelpilinggroup.org (accessed 01/08/2018).

T.H.E. Plastic Piling Company: https://www.plasticpiling.co.uk (accessed 01/08/2018).

ThyssenKrupp: https://www.thyssenkrupp-infrastructure.com/index-2.html (accessed 01/08/2018).

Temporary Works, Second edition

Pallett, Peter F and Filip, Ray
ISBN 978-0-7277-6338-9
https://doi.org/10.1680/twse.63389.159
ICE Publishing: All rights reserved

Chapter 12
Trenching

Ray Filip
Temporary Works Consultant and Training Provider, RKF Consult Ltd

Each year a number of site operatives are killed in relatively shallow trenches, and the first question that should always be asked is: Do operatives have to enter the trench or can an alternative method of work be identified?

A trench is defined as an excavation whose length greatly exceeds its width and depth. It is generally considered that a shallow excavation is less than 6 m deep (most trenches) and a deep excavation is greater than 6 m deep (specialist advice should be sought). Trenches can be excavated by hand or mechanical digger to allow a service, pipeline or foundation to be installed or, when backfilled with granular fill, used to improve drainage. A trench may have battered sides when site circumstances allow, or temporary shoring may be installed (traditional timbering, sheets and frames, proprietary trench boxes) to protect workers entering the trench. Controlling water ingress into a trench is often a major issue (see Chapter 6 on control of groundwater). A safe system of work and adequate planning is necessary, especially when workers are required to enter a trench, and the trenching operation can also have a detrimental effect on the surroundings when appropriate management control measures are not provided or inappropriate solutions are used. Good health and safety practice is critical to trenching operations. Principal designers have a duty to get involved to ensure that designers can demonstrate that trenching operations can be carried out safely, especially where site operatives could become trapped by collapsing soil.

12.1. Introduction
12.1.1 Major alternatives

Trenchless techniques (to minimise surface disruption) include methods such as micro-tunnelling, auger/thrust boring, pipe ramming, impact moling, directional drilling, mole ploughing and rock boring. Trenchless techniques might not be possible or practicable because soils are unsuitable or very large diameter or very long lengths of pipework are to be installed, or due to cost limitations, site-specific restrictions and the non-availability of specialist machinery and labour (Watson, 1987).

Other alternatives to trenching include traditional timber headings or tunnelling techniques for larger diameter services, the installation of services above ground to avoid having to excavate a trench, and specialist techniques such as soil nailing or ground freezing.

12.1.2 Soils

Soils can be classified by particle size (broadly cohesion-less or cohesive), compactness and structure with a separate classification for organic soils (soil classifications are provided in various documents such as BS 5975:2019 (BSI, 2019) and BS 8002 (BSI, 2015) (see also Eurocode 7 (BSI, 2004a)). Accurate soil descriptions are essential to allow soil design parameters to be determined. Designers will summarise and approximate actual soil conditions to form an idealised model for design, and these approximations have to be confirmed by monitoring the soils being excavated.

Granular soils are sands and gravels that have an angle of repose. This angle is dependent on the particle shape, particle roughness, compaction and grading. A well-graded material has a variety of sized particles and these tend to pack together, giving higher angles of repose. Poorly graded materials have a predominance of a single-size soil and tend to have lower angles of repose. Granular soils have high permeability, and close sheeting will generally be required for deeper trenches or when a high water table is encountered, in order to prevent collapse, significant ground movement and fine materials being washed out. If the groundwater pressure is not relieved (dewatering) or released through any gaps in the support scheme, there will be a large increase in the pressure on the support scheme.

Cohesive soils are bound together due to the effects of water between the fine particles in clays and silts (this is known as 'cohesion'). When excavated, clay soils can stand near vertical without support for a period of time. The cohesion in silts will vary with moisture content, and they are less stable than clays. The permeability is low (particularly in clays) and drainage occurs over a significant time period; hence extensive dewatering is often not required in cohesive soils (sump pumping is often adequate) unless bands of sands or coarse silts (known as 'partings') are present or the soil is heavily fissured. Even though these soils can be relatively stable in the short term, a risk assessment should determine whether support is necessary (considering fissures, surcharges, adjacent works, timescale, etc.).

Rock has cemented particles and, even though the rock mass itself may be very strong, inclined bedding planes and fractures can lead to instability of individual blocks. Heavily weathered rock may need to be considered as a very dense cohesionless material. Excavating rock can require heavy machinery or specialist techniques such as blasting.

When groundwater is encountered, the following items should be observed: the level at which the water was encountered, the rate of inflow and rise of water, and the final level that the water reaches.

12.1.3 Battering trenches

Any vertically sided trench is rarely totally acceptable for any appreciable length of time. Collapse has to be avoided, which means battering or benching the sides to a safe slope, or supports should be provided. An appropriate risk assessment or stability analysis (see Chapter 10 on slope stability) is required to ensure a batter allows a safe system of work to be provided. When designing a battered trench (which may be stepped) the actual site conditions of surcharges, groundwater, excavated depth, space available, possible deterioration of the soil with exposure and timescale should be considered. Virtually all soils can be excavated to a safe angle, but the presence of water will detrimentally affect the safe angle and dewatering is likely to be required to improve stability. Additional protective measures such as netting or fencing may be used to prevent any loose material falling on operatives. Blinding or sheeting may be used as protection from weathering, and berms and ditches dug to cut off surface water. In layered soils, different batter angles may be appropriate for each of the strata.

Regular checks should be carried out looking for potential signs of instability or degradation. Warning signs of rotational slip failures are: bulging at the toe and tension cracks appearing at the top of the batter. If these signs are noted then the batter can be regraded, weight added to the toe of the slope or the drainage of the batter improved.

In dry conditions over a short duration, granular soils have an internal angle of friction (φ) that is typically in the range $c.$ 25–50° and batters may be safely cut at an angle slightly less than φ. When groundwater is present, the safe angle will be reduced. In the longer term (months), the angle will need to be less than the effective internal angle of friction (φ'). Cohesive soils can be cut near vertical in the very short term (up to a week), but drainage occurs and collapse will inevitably follow. Where men are required to work in a confined excavation, the sides will need to be safely battered or sheeting and propping or boxes used. Safe batter angles are listed in many publications (e.g. Irvine and Smith, 1983).

12.1.4 Risks, planning and construction

Designers must carry out an appropriate risk assessment and design out risk where possible (principle of prevention); any residual risks should be communicated so they can be managed effectively on site. Risk assessments and method statements are required on site. Some common considerations when planning and designing a trenching operation include the following

- Knowledge of the site: history, topography, previous slippage, voids, flooding, location of existing services, foundations, existing soil and groundwater conditions.
- Can the trench be battered or is shoring required and, if so, which type of shoring is suitable? If supports are required, is the necessary equipment readily available on site or will it be hired (delivery and lead-in times need to be considered)?
- Reasons for trench: the item to be installed will determine the depth and width of the trench (including working space) and the equipment to be used.
- Soil, water and surcharge pressures must be considered on any shoring to prevent collapse and overloading of shoring equipment. These pressures will vary depending on the type of soil, depth of trench, head of water and magnitude of surcharge. Soil pressures (particularly in cohesive soils) will also vary with time. Out-of-balance pressures should be considered on sloping ground or when high surcharges exist on one side of the trench. Allowance should be made for variations in the soil and groundwater conditions (or tidal variations).

- Limit deflections and ground movements to an acceptable amount (note adjacent structures and services). Is heave of the base likely?
- Consider the installation and removal sequence and allow for construction tolerances.
- Allow for accidental impact on structural members (in particular props), accidental over-dig and provision for softening of the formation level.
- Is groundwater a problem and might pumping be required? Dewatering may draw in fines and cause boiling and settlement in the vicinity.
- Access, egress, edge protection (plus stop logs for plant and vehicles) and confined spaces measures must be provided.
- Consider undermining of adjacent structures or ground movement, which may damage services.
- For permanent schemes or very long-term temporary schemes, an allowance for deterioration of structural members (e.g. corrosion of steel) should be made.
- Can the excavated material be used as backfill?

Thermal and seismic effects are generally not considered.

Some common considerations for constructing and monitoring a trenching operation include the following

- Operatives and supervisors should be briefed, trained, competent and follow a safe pre-instructed method of work. Emergency procedures should be communicated and correct personal protective equipment worn.
- Ground and water conditions should be monitored and any changes from the design assumptions should be communicated to the designer.
- Shoring equipment should be inspected and damaged or faulty items disregarded. Once installed, measures should be taken to prevent damage due to accidental impact (e.g. from an excavator bucket).
- Existing services should be located by hand digging or vacuum excavation (avoid using mechanical diggers).
- How will the trench be protected from the possible inrush of surface runoff (sandbags, bunds or ditches may be considered)?
- Monitor the trench carefully if excessive vibration is generated from nearby operations or large surcharges are positioned near the trench, as well as during periods of bad weather.
- Trenches should be inspected as per the requirements of the Construction (Design and Management) Regulations 2015 (UK Government, 2015; regulations 22–24). Some contractors and clients (e.g. National Grid) are introducing an 'excavation tag', similar to the 'scaffold tag' system that has been in existence for many years.
- How is the trench backfilled and compacted to prevent settlement?
- How will the support system be removed? A detailed removal sequence should be provided and followed on site; simply stating that removal is the reverse of installation is not adequate.

Some contractors and utility companies have standard solutions for relatively shallow low-risk trenching operations. These predesigned solutions indicate sizes for trench sheets, walers, struts, traffic clearances and surcharge limitations. Appropriate risk assessments and method statements are provided by a competent person, and operatives must have been appropriately trained. Such solutions are very useful for short-duration or repetitive 'reactive maintenance' contracts (repairing a damaged service).

12.2. Techniques

Over a very short period of time and for a relatively shallow depth most soils will stand near vertical, allowing sufficient time for certain shoring systems to be installed. If the sides do need to be supported (as a result of a risk assessment being carried out) due to the nature of the soil, longer-term use, sensitive surroundings or for deeper trenches, then a trench support scheme will be required. A competent person should make the decision (by risk assessment and consideration of site-specific conditions) if any of the types of equipment described below are suitable and, if so, which type to use. Work should not be permitted outside the protected area. All equipment should be assembled and used as per the manufacturer's instructions, and consideration should be given to a safe method of installation and a safe method of removal.

12.2.1 Traditional timbering

'Timbering' is the term given to traditional support techniques using timbers. In the 1970s, regulations were introduced that forbade operatives entering inadequately supported trenches. As a result, many traditional methods of timbering trenching became impracticable and hydraulic support systems were developed. However, the traditional timbering techniques have been developed and are still in use, particularly in the utility sector. Many of these solutions are for relatively low-risk narrow trenches, shafts and headings. These methods are versatile and the equipment easily handled, but they are labour intensive and rely on good workmanship. Many companies have standardised solutions for use in a variety of circumstances, and these have been designed to satisfy a range of conditions likely to be experienced on site. The design principles rely on soil arching (the full overburden pressure can be reduced) and have been proved satisfactory by experience. The Timber Research and Development Association (TRADA) publication *Timber in Excavations* (1990) provides design examples, but as this dates from 1990 the examples require updating to current design standards. The recommendation is that timber quality should be at least SC4 (C24).

12.2.2 Trench and drag boxes

Trench and drag boxes (Figures 12.1 and 12.2) are generally considered to be suitable as a shield for low-risk situations in relatively stable dry ground (capable of standing near vertical for a short period of time) or rock. A trench box can be placed in a pre-excavated trench and moved either by raising it or moving it forward using an excavator. A drag box can be dragged along a wider pre-dug trench using an excavator. Both versions require at least two struts for stability, and there are limitations on the maximum diameter of pipe that can be installed (maximum distance from the toe of the box to the bottom strut).

Figure 12.1 Trench boxes. (Courtesy of Groundforce Shorco Ltd)

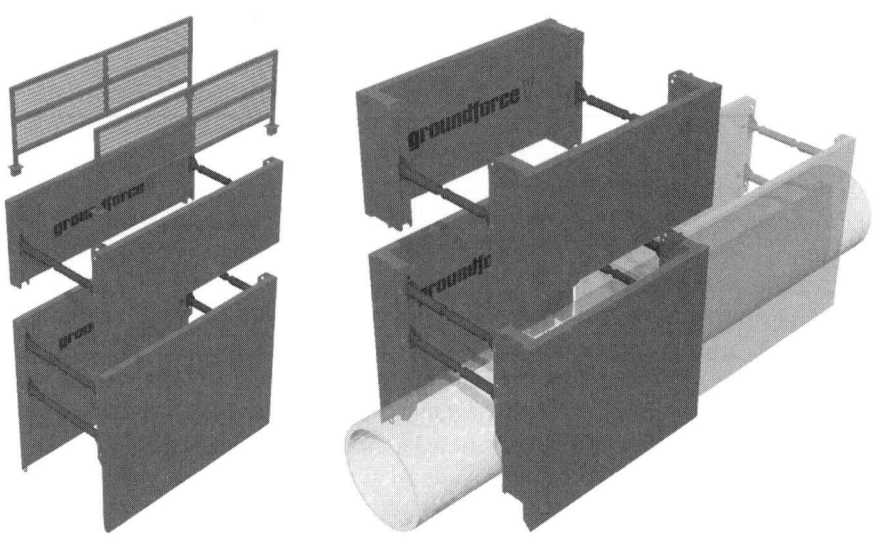

Drag boxes are suitable for relatively shallow trenches (up to about 4 m depth). Trench boxes can be assembled in lifts to achieve significant depths (over 6 m is not uncommon). Trench sheets can be placed at the ends of the boxes to provide additional protection to the open ends. With both versions the pre-excavated trench is wider than the box to allow the box to be installed and moved. Trench boxes do not support the sides of the excavation, but act as a shield to protect the workforce (even when backfill is placed between the box and the surrounding ground there is a period of time when the ground has to be self-supporting). There is a risk of the ground collapsing onto the box, affecting operatives, plant and services on the surface around the box, whereas those inside the box are shielded.

Excavation should not be permitted beneath the box (known as 'flying' a box), as friction between the box and the surrounding ground cannot be relied upon to support the box. There is an alternative installation method for a trench box known as 'dig and push' (Figure 12.3). With this method the soil is supported throughout installation; although this reduces the risk of significant ground movement, it is slower.

The utility sector will often use small plastic trench boxes. They are suitable for small and shallow excavations (up to around 1.2 m depth and around 1.2 m × 1.2 m plan dimensions) in pavements that are required for repairing or replacing utility pipes. They are lightweight so they can be easily handled, and provide protection to the operative during the work.

12.2.3 Vertical shores

Vertical shores are an intermittent system comprising a pair of vertical aluminium rails and a pair of hydraulic struts. They are installed into a pre-dug trench in dry stable

Figure 12.2 Trench box on site (edge protection not installed). Note backfilling between the box and ground, dewatering and excavated spoil kept away from edge of trench. (Courtesy of Groundforce Shorco Ltd)

Figure 12.3 'Dig and push' sequence for installing a trench box. (Courtesy of Groundforce Shorco Ltd)

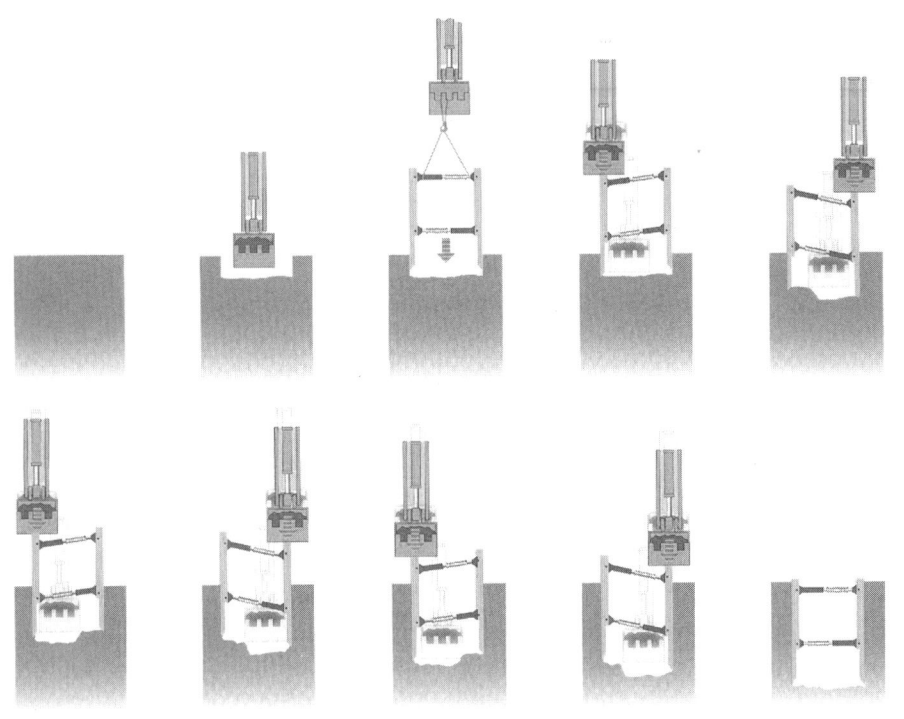

ground, which can stand unsupported vertically for a period of time. They are suitable for relatively narrow and shallow trenches. The spacing of the rails is determined by the ground conditions, and is typically 0.5–1.2 m. The shores are hinged and designed to be manually installed from outside the trench, being activated by a hand pump to a working pressure indicated by a gauge. Only small-diameter pipes can be installed below the bottom strut, which is typically $c.$ 0.5 m from the base of the trench. Vertical shores can be used between services, but there is a risk of ground collapse between the shores. For this reason, vertical shores are limited in their application.

12.2.4 Trench sheets and hydraulic waling frames

Steel trench sheets and hydraulic waling frames (Figure 12.4) are the most adaptable of systems and hence the most widely used. They are generally used in poorer ground, for deeper trenches where surcharges are higher or where ground movement may need to be limited and some design is likely. The trench sheets (or light sheet piles) provide continuous support and may rely on penetration below the formation level for passive resistance. The sheets can be pushed in by an excavator or driven using an excavator-mounted vibrator or small piling rig (the driving method depends on the length of the sheet, the ground conditions and the site limitations). Plastic sheets may also be used if the ground is softer and relatively short sheeting for shallow trenches is required.

Figure 12.4 Sheets with steel waling beams and props. (Courtesy of Groundforce Shorco Ltd)

A simple guide system (baulk timbers or steel beams) should be used to ensure the sheets are driven near vertical and to maintain a straight line. When site conditions allow, the sheets can be partially installed into a pre-excavated 'lead trench' which is then backfilled. Multiple levels of horizontal waling beams with hydraulic, adjustable struts (or timber walings and struts) are used, with the spacing of the struts suiting the pipe lengths to be installed. The trench depth is limited by the length of sheets that can be installed by light machinery, especially with interlocking (rather than overlapping) sheets, which require additional height for 'clutching' the sheets together. The slenderness of the sheet can also preclude driving into dense or stiff soils. These systems can also be installed using the 'dig and push' method.

Small gaps can be left in the sheeted wall to allow for services crossing the trench and, if relatively stable ground is encountered, intermittent sheeting can be used. This is the principal reason why the horizontal waling system is so adaptable and popular.

Similarly, manholes can be constructed in a square- or rectangular-sheeted excavation using proprietary hydraulic supports known as 'manhole braces', whereby all four walers in a frame are hydraulically driven off a common manifold (timbers can also be used). Gaps can be left in the sheets or some sheets left above the formation level to allow pipework to be installed.

12.2.5 Post and plank vertical H-sections

Post and plank (Figure 12.5) vertical steel H-sections are pre-driven by an excavator or piling rig or installed into augered holes at $c.$ 1–3 m centres with horizontal planks (timber, pre-cast concrete, in situ concrete or steel sheets) which span between the flanges of the vertical H-section. When placed into augered holes, the bottom of the H-section is often held in concrete. The planks are installed and wedged in position, one beneath the other, as the excavation progresses downwards. When using timber, steel or pre-cast planks, the soil behind the planks has to be excavated slightly larger to accommodate the planks, with sand backfill placed behind the planks to fill the gap, but some ground movement is inevitable. In situ concrete planks accommodate construction tolerances and are cast against the excavated face; backfilling is, therefore, not required. This method has the advantages that the main structural supports are installed prior to excavation and can be easily altered, and gaps can be left to accommodate services. However, it can be time-consuming, labour-intensive and some ground movement is possible.

A variant uses light steel sections spanning horizontally between the flanges of the posts with trench sheets pushed down progressively, and this dramatically reduces the labour requirement. A proprietary version known as a 'side rail system' comprises posts, sliding panels (each $c.$ 1.5–2.5 m deep) and struts, which form a continuous ground support and are installed by excavator. This system can be used for widths up to 7 m and depths in excess of 7 m. The equipment is relatively large and heavy and heavy excavators are required for the installation. For deeper trenches with large diameter pipes, an in situ concrete blinding is cast at the base and lower levels of struts can be moved upwards or removed. Additional support such as trench sheets may be required to provide additional support at each end.

Proprietary equipment is designed for the most robust loadings envisaged, and in all but obviously excessive conditions the need for detailed calculations is unlikely. Where it is felt wise to make calculations, empirical methods as given in the following section may suffice.

12.3. Design to CIRIA 97 trenching practice

The Construction Industry Research and Information Association (CIRIA) Report 97, *Trenching Practice* (Irvine and Smith, 1983), was last revised in 1992; however, it is still an authoritative guide to this type of work, even though some items require revision to current standards (e.g. the rule 'obligatory support at deeper than 1.2 m' is still quoted). The calculation of earth pressures is still relatively empirical as soils are a naturally variable material.

Supports for trenches often have several levels of support; in such cases traditional soil mechanics theory such as the triangular Coulomb pressure diagram have been found not

Figure 12.5 (a) Post and plank excavation and (b) side rail system trench. (Courtesy of Groundforce Shorco Ltd)

(a)

(b)

to apply (because these support systems are relatively flexible and there is a degree of soil arching and displacement which leads to a redistribution of pressures). From full-scale tests, Terzaghi and Peck (1996) derived empirical trapezoidal pressure diagrams for calculating maximum strut loads. From further experience, these pressure diagrams were developed to incorporate rules for the design of walings and sheets. These are incorporated into simple charts in the CIRIA Report 97 (Irvine and Smith, 1983), which was prepared assuming the following conditions

- Dry conditions (assumes water is dealt with by dewatering), as water pressure has a great effect on soil behaviour.
- Trenches are up to 6 m deep, with supports (walings and struts). Experience has shown that in a wide range of soils the strut loads are similar up to this depth; specialist advice must be sought and an experienced person must design the trench support scheme for trenches greater than 6 m deep.
- Short-, medium- and long-term trenches in granular soils and mixed soils but only short- to medium-term trenches in clays. Pressure increases with time in clays, so different parameters are used in long-term trenches in clay.
- Steel sheets with a minimum section modulus z of 35 cm^3/m (or timber poling boards of 32 mm thickness) which can be driven using non-specialist driving techniques. There is no minimum 'toe in' of the sheets quoted and the maximum cantilever of the sheeting should not exceed 500 mm.
- Timbers are a minimum grade of SC4 (C24) and steel walers are a minimum grade of S275 to BS EN 10025:2004 (previously grade 43A) (BSI, 2004b). Adjustable props meet regulations as described in BS 4074:2000 (BSI, 2000).
- The slope across the trench does not exceed 1:4.
- Surcharge is limited to 10 kN/m^2 and deflection is not considered a major problem.
- Not to be used in soft clays (undrained cohesion <30 kN/m^2) or saturated silt.
- There is no heave or boiling at the base of the trench.
- The supports are be installed tight against the sides of the trench to ensure soil arching takes place and the load on the waling is relieved.

The effective depth of a trench is the actual depth, but an adjacent batter is considered to increase the effective depth. Timber waling and trench sheets are designed for loads equivalent to 50% of the pressure diagram to allow for arching. Steel walings, being stiffer, are designed for the full pressure diagram.

The chart in Figure 12.6 is for use in granular soils, mixed soils and short-term trenches in clay. It uses a rectangular pressure diagram and the pressure is calculated as

$$\text{pressure} = 3.07H + 1.76 \text{ (kN/m}^2\text{)} \tag{12.1}$$

A second case is provided for medium-term trenches in clay as in this case the earth pressure increases with time. Unless the water table is lowered by dewatering the full hydrostatic pressure diagram should be used. The pressures due to soil and due to surcharge are calculated as

$$\text{pressure due to soil} = 0.65 \times K_a \times \gamma \times H = 0.65 \times 0.27 \times 17.5 \times H = 3.07H \tag{12.2}$$

$$\text{pressure due to surcharge} = 0.65 \times K_a \times S = 0.65 \times 0.27 \times 10 = 1.76 \tag{12.3}$$

Figure 12.6 Example design chart from CIRIA Report 97 (Irvine and Smith, 1983). (Reproduced courtesy of CIRIA)

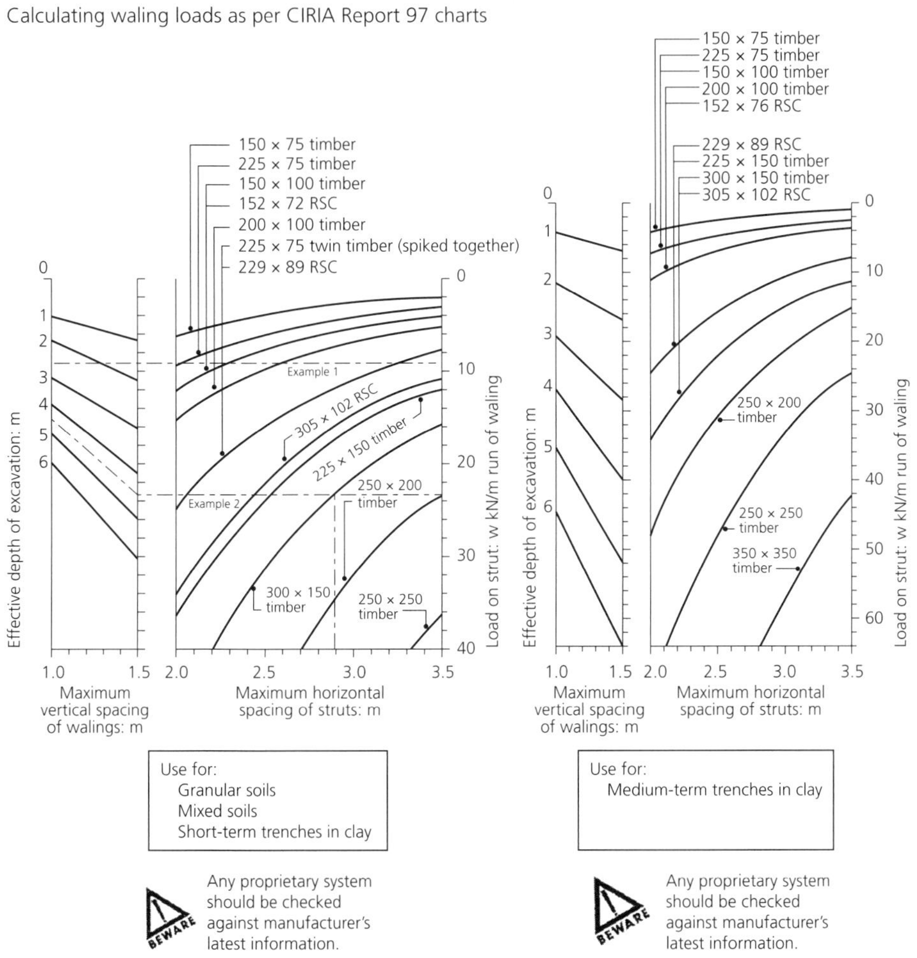

where $\gamma = 17.5 \text{ kN/m}^3$ is the density of soil, H is the effective depth of the excavation (m) and $S = 10 \text{ kN/m}^2$ is the surcharge. The coefficient of active earth pressure is defined as $K_a = 1 - \sin \phi$, where $\phi = 35°$ is the soil friction angle, and $K_a = 0.65$ approximates the maximum triangular pressure to rectangular pressure.

For medium-term trenches in clay, an alternative chart is provided based on similar principles to those mentioned above. The formula for the uppermost 1 m is as above; below this depth it increases to

$$\text{pressure} = 6.56H + 3.75 \text{ (kN/m}^2\text{)} \tag{12.4}$$

12.4. Controlling water

Water can enter a trench from a number of sources: ingress of groundwater when excavating below the water table, surface runoff, excavating through land drains or ditches, damage to nearby water mains or artesian water pressure. Water-control measures are necessary to keep the base of the excavation dry and prevent excessive soil softening (by blinding) so that work can progress safely and efficiently and to avoid the base of excavations in sandy soils boiling during dewatering from inside a sheeted trench.

Some of the associated risks in trenching are

- lowering the water table may lead to fine soils being drawn out of the soil, leading to settlement and possible damage to surrounding buildings, services and roads
- a reduction in near-surface bearing pressures due to high water table
- difficulty in achieving compaction densities if soils are too wet
- fine soils exhibit short- and long-term characteristics as drainage occurs
- heavy rain can cause high surface runoff rates
- when excavating above artesian water pressure (where an impervious soil overlies a granular soil containing pressurised water) the weight of the remaining impervious soil may not be sufficient to resist the upward water pressure and this can cause the base of the trench to heave
- potential contamination of rivers, streams and other watercourses (bunds and silt busters) and discharge licences may be required
- battered trenches become increasingly unstable when water inflow occurs.

Site investigation data should be studied to determine the groundwater level and whether artesian water pressure may be encountered. The permeability of the soils should also be estimated to determine the rate at which groundwater will flow into the trench. Trial holes can be a simple way of crudely estimating inflow rates and the stability of soils. It should also be noted that soils tend to be laid down in layers; vertical and horizontal permeability will therefore differ.

When determining dewatering techniques, professional advice should be sought. For further details, see Chapter 6 on control of groundwater.

REFERENCES

BSI (British Standards Institution) (2000) BS 4074:2000. Specification for steel trench struts. BSI, London, UK.
BSI (2004a) BS EN 1997-1:2004. Eurocode 7: Geotechnical design. General rules. BSI, London, UK.
BSI (2004b) BS EN 10025-1:2004. Hot rolled products of structural steels. General technical delivery conditions. BSI, London, UK.
BSI (2015) BS 8002:2015. Code of practice for earth retaining structures. BSI, London, UK.
BSI (2019) BS 5975:2019. Code of practice for temporary works procedures and the permissible stress design of falsework. BSI, London, UK.
Irvine DJ and Smith RJH (1983) *Trenching Practice,* 2nd edn. CIRIA, London, UK, R97.
Terzaghi K and Peck R (1996) *Soil Mechanics in Engineering Practice*. Wiley, London, UK.

TRADA (Timber Research and Development Association) (1990) *Timber in Excavations*, 3rd edn. Thomas Telford, London, UK.

UK Government (2015) Construction (Design and Management) Regulations 2015. HSE, London, UK.

Watson TJ (1987) *Trenchless Construction for Underground Services*. CIRIA, London, UK, TN127D.

FURTHER READING

ArcelorMittal (2016) *Piling Handbook*, 9th edn. ArcelorMittal, Luxembourg.

BSI (British Standards Institution) (2002) BS EN 13331-1:2002. Trench lining systems. Product specifications. BSI, London, UK.

BSI (2015) BS 5930:2015. Code of practice for ground investigations. BSI, London, UK.

CPA (Construction Plant-hire Association) (2001) *Selection of Proprietary Shoring Equipment and the Use of Chains to Support Shoring Equipment*. CPA, London, UK, Guidance STIG0201.

CPA (2004) *Risk Assessments for Shoring and Piling Operations*. CPA, London, UK, Guidance STIG0403.

CPA (2016) *Management of Shoring in Excavations*, Parts 1 and 2 – revision 1. CPA, London, UK.

Department for Transport (2000) *Safety at Street Works and Road Works. A Code of Practice*. Stationery Office, London, UK.

HSE (Health and Safety Executive) (1997) *Safe Work in Confined Spaces*. HSE, London, UK, Guidance INDG258.

HSE (1997) *Safety in Excavations*. HSE, London, UK, Guidance CIS8 (Revision 1).

HSE (1999) *Health & Safety in Excavations*. HSE, London, UK, Guidance HSG185.

HSE (2000) *Avoiding Danger from Underground Services*. HSE, London, UK, Guidance HSG47.

HSE (2009) *ACoP Safe Work in Confined Spaces*. HSE, London, UK, Guidance L101.

HSE (2012) *Excavation: What You Need to Know*. HSE, London, UK, Guidance CIS64.

HSE (2013) *Avoidance of Danger from Overhead Electric Power Lines*, 4th edn. HSE, London, UK, Guidance GS6.

Mackay EB (1986) *Proprietary Trench Support Systems*. CIRIA, London, UK, TN95.

Preene M (2000) *Groundwater Control – Design and Practice*. CIRIA, London, UK, C515.

Sommerville SH (1986) *Control of Groundwater for Temporary Works*. CIRIA, London, UK, R113.

ThyssenKrupp GfT Bautechnik (2010) *Sheet Piling Handbook*, 3rd edn. ThyssenKrupp, Essen, Germany.

UK Government (1991) New Road and Streetworks Act 1991. Statutory Instrument 1991/22. HMSO, London, UK.

Useful web addresses

CIRIA (Construction Industry Research and Information Association): www.ciria.org (accessed 01/08/2018).

Construction Plant-hire Association: https://www.cpa.uk.net (accessed 01/08/2018).

Dawson Construction Plant Ltd: http://www.dcpuk.com (accessed 01/08/2018).

Groundforce Shorco Ltd: https://www.vpgroundforce.com/gb (accessed 01/08/2018).

Mabey Hire Ltd: https://www.mabey.com/uk (accessed 01/08/2018).
MGF Ltd: https://www.mgf.ltd.uk (accessed 01/08/2018).
RMD Kwikform: https://www.rmdkwikform.com (accessed 01/08/2018).
ThyssenKrupp: https://www.thyssenkrupp-infrastructure.com/index-2.html (accessed 01/08/2018).

Pallett, Peter F and Filip, Ray
ISBN 978-0-7277-6338-9
https://doi.org/10.1680/twse.63389.175
ICE Publishing: All rights reserved

Chapter 13
Diaphragm walls

Chris Robinson
Design Manager, Cementation Skanska Limited

Andrew Bell
Chief Engineer, Cementation Skanska Limited

Diaphragm walls (also known as slurry walls) can be used to form cofferdams, quay walls, foundation elements and embedded retaining walls (e.g. to construct tanks, shafts, basement excavations or cut and cover tunnels). Their use is becoming more prevalent in congested urban environments where construction below ground is becoming ever more common due to the low availability of land for development, and the associated high land prices, and to the development of deep urban infrastructure. The use of diaphragm walls becomes increasingly economically attractive when they can be used to form part of the permanent works. This chapter describes common uses of diaphragm walls, their method of construction (including aspects of additional temporary works required to facilitate construction), a summary of typical construction details which can aid construction of a high-quality finished product and a description of the most significant design considerations.

13.1. Introduction

Diaphragm walling refers to the in situ construction of vertical walls by means of deep trenches. Stability of the trench excavation is maintained by the use of a support fluid; this is most commonly a bentonite suspension, although polymer support fluids can also be used. Diaphragm walls are constructed in discrete panels, typically ranging in length from 2.8 to 7.0 m, using purpose-built grabs or milling machines. The panels are constructed in the required sequence (see Section 13.3.1) abutting each other to form a continuous structural wall. T-shaped panels or counterforts can be constructed where very high bending capacity elements are required, typically to minimise propping requirements. A variety of details are available to form the ends (stop ends) of each panel, which will generally incorporate a water bar, and can include details to facilitate the transfer of shear forces between panels.

13.2. Applications

Although an expensive form of temporary works that are generally used where other forms of temporary ground support are not practical, diaphragm walls become very economical when they can be used to form part of the permanent works.

In addition to providing ground support and water exclusion, diaphragm walls can also be utilised to carry vertical loads. Where the magnitude of vertical loading exceeds the available wall capacity (based on an embedment length required to maintain wall stability), the panels can be taken deeper to carry the vertical load. Discrete isolated panels known as 'barrettes' can be constructed to carry vertical loads alone. These might be adopted where diaphragm walls are being constructed elsewhere on a scheme and it would be uneconomical or impractical to mobilise additional plant to construct conventional bearing piles, or where the founding level is so deep as to preclude these from being adopted.

13.3. Construction methods and plant
13.3.1 Planning

The construction methods and plant selected will depend on the characteristics of each project. These will include consideration of the ground conditions (including groundwater chemistry which may impact on support-fluid performance), the depth of the walls to be constructed, access restrictions, headroom restrictions, any working time restrictions, party wall issues (which may impact on panel dimensions), panel trench stability, and so on.

Figure 13.1 depicts the typical sequence of operations for diaphragm wall construction. After construction of the first diaphragm wall panel (opening panel), further panels are constructed to suit the specific site conditions and constraints. Panel sequencing is typically ordered such that a minimal number of stop ends (peel-off, reusable type stop ends) are in the ground at any one time. This is not a constraint where non-reusable stop ends (e.g. pre-cast concrete stop ends) are used.

As well as diaphragm walls being a temporary works solution in their own right, their construction can require some significant additional temporary works to ensure successful execution. These include guide walls, support fluid, temporary anchorages and/or temporary props, jet-grouted props at depth, working platforms for tracked plant (see Chapter 5), slope stability assessment (see Chapter 10), temporary screening to protect the public and personnel, protection of watercourses and adjacent services, and foundations for wet and dry silos (for support fluid).

Where ground movements are critical, additional precautionary measures may be adopted, which can vary from limiting panel lengths to preliminary underpinning of particularly sensitive adjacent structures.

As noted by Fernie *et al.* (2001), ancillary works (broadly speaking all operations prior to bulk basement excavation) can cause a significant proportion (up to around 80%) of recorded settlements of adjacent structures. Caution should therefore be taken to ensure that any precautionary measures do not cause more problems than they are intended to mitigate.

13.3.2 Site preparation

Demolition works may be needed in advance of the commencement of diaphragm walling operations. Often this will entail excavation or grubbing out of existing

Diaphragm walls

Figure 13.1 Typical construction sequence. (Courtesy of Cementation Skanska Limited)

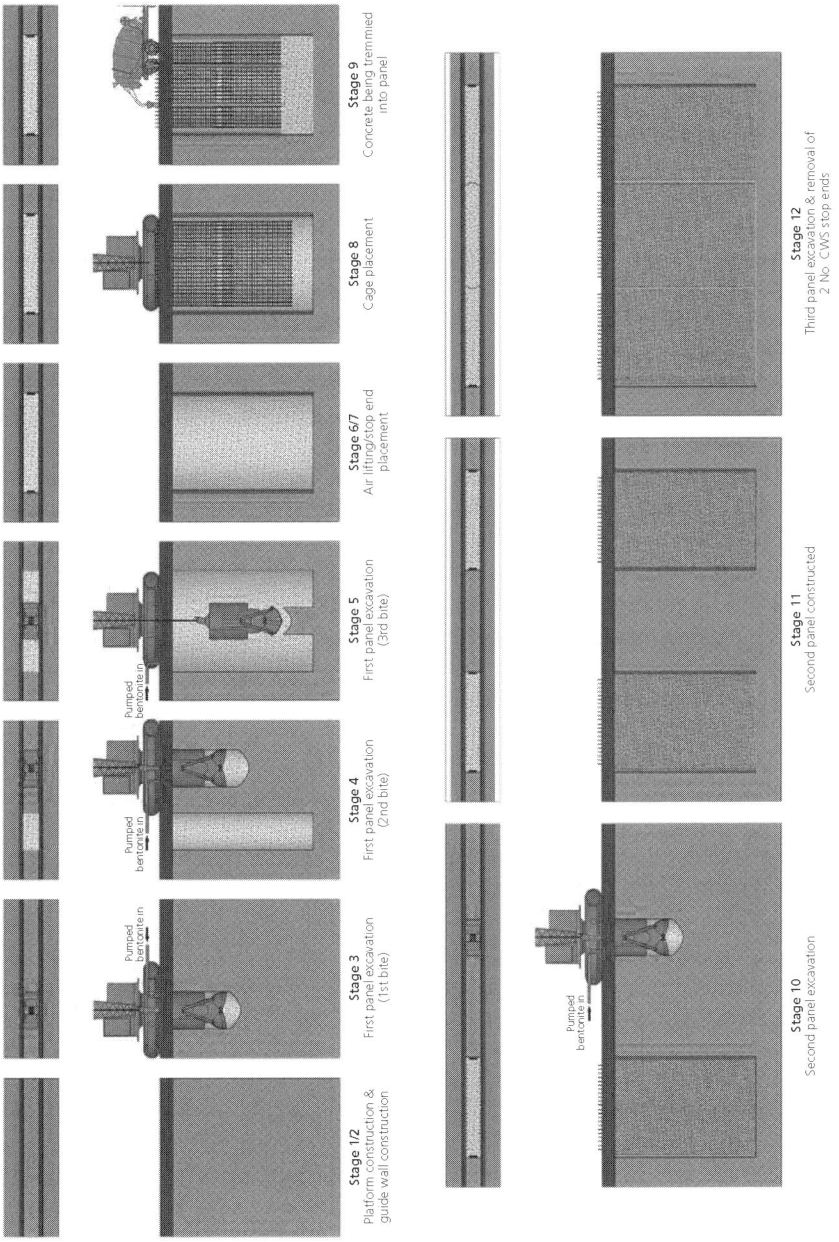

foundations. Where this occurs along the line of the wall it is good practice to ensure the backfill to these areas is selected not only to give adequate support to tracked plant but also to avoid high overbreak occurring. This would adversely affect the finished surface of the diaphragm wall, which may require trimming back if the overbreak is in excess of the specification limits. Suitable backfill materials often used are stiff clays or clay-bound hogging. Properly placed and compacted, these materials will give a relatively smooth finish to the wall over the backfill depth and potentially aid wall verticality. This is also important to allow the support fluid to perform well and to minimise fluid loss and the risk of temporary panel instability.

13.3.3 Working platforms
As with any other site where tracked plant (predominantly piling rigs and cranes) is to be employed, a suitable working platform should be provided. Reference should be made to Chapter 5 for full details of working platform design. As well as supporting the tracked plant, the platform may be required to support significant loads from support fluid silos and tanks.

13.3.4 Guide walls
Guide wall construction will usually follow on from construction of the working platform. The guide wall is required to undertake two main functions: it facilitates provision of tighter tolerances and dimensional control of the diaphragm wall panel excavation at commencement level, and also enables the bentonite support fluid to be controlled. The guide wall will also need to provide support for keying off stop ends and the panel reinforcement cages via trapping beams.

Care should be taken to ensure that the guide wall itself remains stable with the diaphragm walling plant operating in close proximity to it. This may be achieved through provision of suitable backfill or strutting. Once the panel has been constructed and cured for a day or so, it will provide the required support to the guide wall in lieu of the original backfill.

A trench stability calculation should be undertaken to ensure the panel excavation is sufficiently stable. The method described by Huder (1972) is often adopted for this analysis.

It may be necessary or prudent to design the guide wall to span laterally while supporting vertical loads (e.g. trapping off the panel reinforcement) in case the ground beneath it collapses during panel construction.

13.3.5 Support fluid
The most common support fluid used for diaphragm wall construction is a bentonite suspension. Most commercially available bentonite powders tend to be a blend of bentonite and polymers. The exact properties and combinations will depend on the required properties. Most support-fluid properties can be adjusted by the addition of the right polymer or other chemical additives. Detailed discussions with suppliers are important prior to commencement of diaphragm walling works to ensure that the correct materials are procured for the methods to be employed.

In principle, there is no reason why polymers cannot be used alone without bentonite minerals. There is quite a large body of experience of polymer support for diaphragm wall construction in North America and Asia, but very limited experience within the UK.

The support fluid will require cleaning or de-sanding throughout the construction process to ensure that the required properties are maintained.

Prior to commencing panel excavation, it is normal practice to undertake compliance testing of the fresh bentonite to ensure it has the required properties as defined by the contractor's method statement and/or the project specification. Typical compliance values for sodium activated bentonite support fluid and polyacrylamide polymer support fluid are given in *ICE Specification for Piling and Embedded Retaining Walls* (ICE SPERW), section C20.6 (ICE, 2017).

During panel excavation, compliance testing is undertaken at a regular predefined frequency as described in ICE SPERW, section B20/6. Further compliance testing should be undertaken upon completion of panel excavation and prior to concreting the panel. It is particularly important to ensure the bentonite within the panel is sufficiently de-sanded prior to placing concrete.

The size of the bentonite farm (the area required for mixing, cleaning and storing the support fluid) will depend on the scope of the works, the number of rigs used, the maximum panel volume and the nature of the ground supported.

Deep panels in fine granular materials will require far more fluid cleaning than if in stiff cohesive ground.

It should be noted that one of the major costs of diaphragm walls is the establishment and removal of the bentonite farm. Diaphragm walling, therefore, tends to be less economical for small areas of wall compared to other solutions. There are, however, certain circumstances which will preclude the adoption of alternatives (e.g. secant piled walls), such as the depth of excavation required and watertightness.

A suitable area of the site should be identified for location of the bentonite farm (or other support-fluid mixing and storage area). The bentonite farm will occupy a relatively large proportion of a typical site; Figure 13.2 depicts a typical set-up.

A check should be undertaken to ensure that the bearing capacity of the materials forming the foundations for the dry and wet silos is adequate. It is common for a reinforced concrete slab to be provided to ensure sufficient stability of the bentonite farm components, which may require to be bolted down to withstand wind loading when empty.

13.3.6 Reinforcement cages

Reinforcement cages for diaphragm wall construction can be supplied prefabricated or site fixed. Site fixing of cages has certain advantages (if sufficient space is available) because cages can be fabricated to fit the excavated panel widths, typically up to around

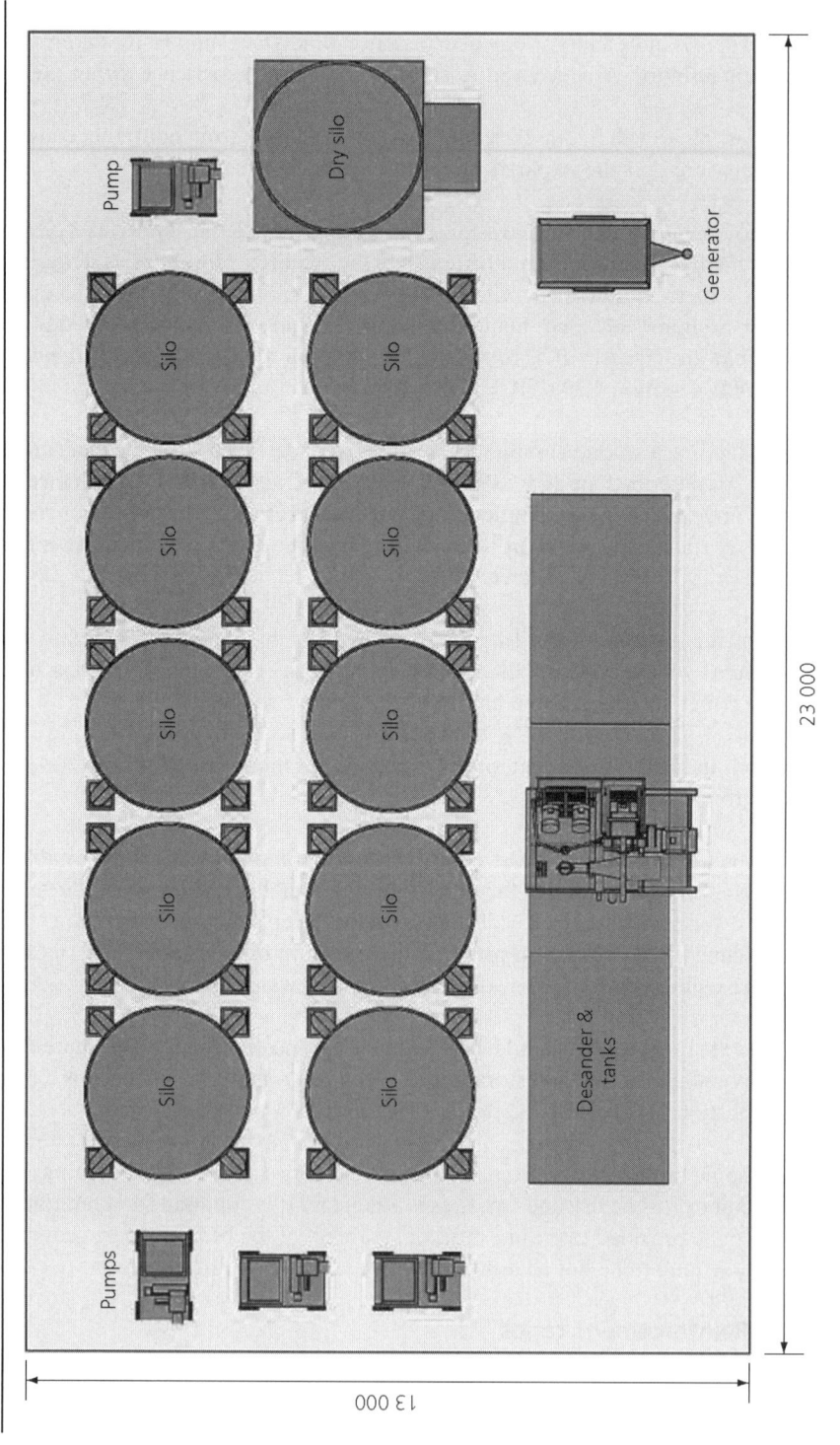

Figure 13.2 Typical diaphragm wall bentonite farm set-up. (Courtesy of Cementation Skanska Limited)

Figure 13.3 Tandem lifting of a reinforcement cage. (Courtesy of Cementation Skanska Limited)

7 m. The maximum cage width that can easily be transported by road in the UK is 2.8 m. Multiple cages can be provided within a wide panel, but this may not work where full reinforcement continuity is required within a panel.

Lifting of diaphragm wall reinforcement cages will almost certainly require a tandem lift (Figure 13.3) to ensure that the cage is not damaged when lifting from the horizontal to the vertical plane. This is classified as a complex lift and requires greater consideration than more conventional lifting operations. See Chapter 20 for further guidance on lifting operations.

Where cage sections are to be spliced, the splice detail should ensure that the reinforcement configuration is as free from bar congestion as possible. This will assist construction of a high-quality finished product. Reinforcement bar congestion is often the primary cause of poor-quality concrete, particularly at the near face of the panel. The splice should ensure that operatives are not required to place their hands within the cage where serious injury could be incurred should the cage sections move relative to each other.

Careful checking of diaphragm wall cages is required, particularly when there are a large number of relatively complex details, for example, box-outs for pull-out bars (e.g. Kwikastip or Startabox), starter couplers and so on. BS EN 1538:2010 (BSI, 2010) specifies tolerances for reinforcing elements (including couplers) both in the plane of and perpendicular to the plane of wall (note that new tolerances are given in ICE SPERW (ICE, 2017)). Failure to address these areas sufficiently can lead to support-fluid inclusions remaining within the completed panel.

13.3.7 Concreting

Diaphragm wall panels are concreted in an identical manner to bored piles constructed under support fluid or which have groundwater in the bore. Sectional tremie pipes are employed which typically have a diameter of 200–300 mm. The diameter of tremie pipe will depend on individual contractor's equipment and the depth of tremie pipe required (i.e. the depth of panel or barrette to be concreted).

The reinforcement cage detailing should fit the dimensions of pipe to be used and the number of tremie positions required to ensure a high-quality finished product. A short panel (around 2.8–4.5 m long) would typically require a single tremie location, whereas longer panels (around 6.5–7.5 m long) would require two positions. Corner panels will require particular consideration.

Concreting of diaphragm wall panels can be significantly more complex than for bearing piles. Single diaphragm wall panels may have multiple tremie locations for concrete placement and, perhaps more significantly, the concreting operation can take many hours due to the relatively large volumes of concrete that need to be placed. When very large concrete pours are required, a prudent measure is to ensure that the concrete supplier has a standby batching plant available. To comply with ICE SPERW (ICE, 2017), tremie pipes must be maintained at levels to ensure that they have a minimum penetration of 3 m. The tremie pipe must not be withdrawn from the panel concrete until the concreting operation has been completed. In addition, a sufficient head of concrete must be maintained within the tremie pipe to ensure that this exceeds the support-fluid pressure.

To attain the highest possible quality finished product, it is essential that panel construction is completed in the shortest practicable time without unnecessary delays.

For guidance on the performance of fresh concrete and its method of placement using tremie methods in deep foundation elements, reference should be made to the EFFC/DFI (2016) *Best Practice Guide to Tremie Concrete for Deep Foundations.*

13.3.8 Grabs (rope and hydraulic)

In general, moderately deep diaphragm wall panels (up to around 40–50 m deep) will be excavated using a conventional grab. Grabs can take the form of either traditional rope grabs or, more recently, hydraulic grabs (Figures 13.4 and 13.5).

Figure 13.4 Rope grab. (Courtesy of Cementation Skanska Limited)

Rope grabs are more basic than the more modern hydraulic grabs. They consist of a short body, at the base of which is a clamshell bucket. The grab has one set of ropes to carry the weight of the grab (and spoil when the clamshell bucket is full) and a second set to close the clamshell bucket and cut the panel. The grab is generally supported by a crawler crane. Although rope grabs are old technology, they still have a place in modern construction. For example, as a conventional crawler crane is used in conjunction with the rope grab, the system lends itself to low-headroom scenarios where a short jib crawler crane is required. Alternative low-headroom hydraulic systems are commercially available, but these may be significantly more costly. However, as with consideration of all alternative systems, the system to be adopted also needs to be compatible with the site-specific ground and project conditions such as programme requirements and so on.

Hydraulic grabs have a much higher closing force than rope grabs, and this can improve production rates or enable them to be used to excavate panels in stronger ground. The grab can be supported on a Kelly bar or, more recently, a rotator itself supported by a modified crawler crane unit. Rotators are available for 180° and 360° plan movement. The rotator allows a high degree of flexibility of panel orientation, which is particularly important on tight sites and at corner panel locations. The use of a rotator in

Figure 13.5 Hydraulic grab. (Courtesy of Cementation Skanska Limited)

conjunction with steering teeth on the grab clamshell bucket allows the grab to be steered according to ground conditions to maintain the panel verticality within tight specification tolerances. Hydraulic grabs are also fitted with electronic inclinometers that enable the grab operator to see a graphical representation of panel geometry and verticality in real time. The data enable the grab operator to react quickly to maintain panel verticality.

13.3.9 Hydrofraise/hydromill/trenchcutter

For diaphragm walls constructed within relatively strong strata, or for deep diaphragm walls, the use of hydromills becomes necessary (Figure 13.6). A hydromill is essentially a reverse circulation trench cutter. This cuts the panel rather than digging it (as is done with grabs). Hydromills can have counter-rotating cutting wheels or cutting chains, depending on the manufacturer and the ground conditions.

Deeper diaphragm wall panels require a longer cycle time to lower a grab into the panel, take a bite from the base of the panel, be lifted from the panel and deposit the spoil.

Figure 13.6 Hydromill. (Courtesy of Cementation Skanska Limited)

Reverse circulation hydromills, on the other hand, continuously cut the ground, which then goes into suspension within the support slurry. This is circulated from the panel through de-sanding equipment and returned into the panel. The hydromill thereby negates the requirement for the cutting tool to be continuously lowered and lifted from the panel. The use of hydromills is costlier than the use of grabs; however, they will usually be employed where grabs cannot be used to progress panel excavation due to the strength of the ground or where the greater production rates outweigh the greater plant costs.

As stated, the adoption of particular plant or techniques needs to be compatible with the ground conditions. The use of hydromills within cohesive materials can be problematic in clays with a relatively high plasticity, such as London Clay. The mills can become blocked by the clay and their efficiency dramatically reduced. One solution would be to employ a hydraulic grab to excavate the upper section of a panel (e.g. to the base of London Clay) and then employ a hydromill to continue panel excavation to the final depth.

As with hydraulic grabs, hydromills are fitted with electronic instrumentation to enable the panel excavation to be undertaken to achieve the required tolerances.

Figure 13.7 Peel-off type stop end. (Courtesy of Cementation Skanska Limited)

13.3.10 Panel joints (stop ends)
Stop ends are used to perform two principal functions: to temporarily house a water bar, and to provide a profiled end to the diaphragm wall panels to facilitate shear transfer between adjacent panels.

Stop ends most often used in modern practice take the form of either reusable 'peel-off' profiled steel plate (e.g. CWS (continuous water stop) stop ends; see Figure 13.7) or pre-cast concrete profiles.

When using a hydromill it may be possible to grind a slotted profile into the end(s) of the adjacent panel(s) to provide both a shear key and a significantly more tortuous path for groundwater.

If water ingress is not critical, then stop ends may be omitted entirely.

13.4. Design
13.4.1 Scope
This section is not intended to provide full guidance on how to design embedded retaining walls. This subject is covered in detail within the various codes of practice and best practice

guides referenced at the end of this chapter and the next. In particular, the reader is referred to the Construction Industry Research and Information Association (CIRIA) Report C760 (Gaba *et al.*, 2017), BS 8002:2015 (BSI, 2015) and Eurocode 7 (BSI, 2004b).

The aim for this section is to provide a commentary on those aspects specific to the design of diaphragm walls.

13.4.2 Geotechnical model

Prior to undertaking any stability analyses for embedded retaining wall design, it is imperative that a representative ground model can be produced. As a general rule of thumb, the site investigation should ideally extend to a depth of around 50% greater than the deepest part of the temporary works structure.

The appropriateness of specific investigation techniques should be assessed in the light of guidance given by geotechnical design codes of practice. Reference should be made to Eurocode 7 (BSI, 2004b) for the derivation of geotechnical data for site characterisation and design purposes.

The site investigation should also provide the designer with other relevant information that may have an impact on the design, including

- adjacent structures – magnitude and level of permanent and variable surcharge loading
- adjacent structures – any limitations on ground-borne vibrations which may influence the construction technique(s) to be adopted
- adjacent highways – magnitude of variable surcharge loading
- information on adjacent services – for example, potential for water main rupture and magnitude of tolerable ground-borne vibrations
- buried infrastructure and impact of ground movements (e.g. London Underground Ltd (LUL) assets).

During construction of the embedded retaining wall any significant variation in the ground or groundwater conditions should be communicated to the design team without delay. It is imperative that the contracting organisation has a sufficient understanding of the design to decide what constitutes a significant variation. The wall designer should provide a wall manual in accordance with ICE SPERW (ICE, 2017). Depending on the design route adopted for the contract, the intention of ICE SPERW is to provide sufficient information to facilitate communication of pertinent design and construction information between parties.

13.4.3 Embedded retaining wall design

Most embedded retaining wall design in current practice is undertaken adopting the factor on strength method described in BS 8002:2015 (BSI, 2015) and Eurocode 7 (BSI, 2004b). In certain circumstances practitioners may prefer to adopt alternative design methods, such as Potts and Burland's (1983) method based on specific experience of embedded retaining wall design in stiff clays, where this is not precluded by the project specification.

The design of embedded retaining walls will often involve both undrained (total stress) and drained (effective stress) design. This will of course depend on the geological materials present at the particular site under consideration. Where the retaining wall is required to provide ground support or groundwater exclusion for a longer time than undrained behaviour can be assumed, effective stress analysis will be necessary.

Most practitioners adopt a limit state design approach when designing embedded retaining walls. The limit states considered are the ultimate limit state (ULS) and the serviceability limit state (SLS). The ULS involves consideration of the safety of people and the safety of the structure itself, for example, instability of the structure, failure by rupture of the structure or any part of it or excessive deformation of the structure such that adjacent structures reach their ULS. The SLS describes the consideration of conditions relating to the performance of the structure under normal operating conditions, for example, deformation of the structure or deformation of the ground supported by the structure.

Two sets of analysis calculations are generally performed when considering the ULS design of embedded retaining walls. The first ULS analysis assesses the required wall toe depth to maintain stability for the relevant earth pressures, applied loads and specified factors of safety (partial factors on resistances, actions, soil mobilisation, etc.). The second set of calculations determines the reinforcement requirements based on the ULS bending moments and shear forces derived from the first analysis.

Project specifications (as well as design codes and guidance documents) usually require the design of the retaining wall to allow for unplanned excavation, usually 10% of the retained height up to a maximum of 0.5 m. While this can be considered to be best practice in most circumstances, it may be possible to make some economies to the design by gaining the agreement of all relevant parties that this design requirement be removed. If this is the case, sufficient control measures must be implemented during the bulk excavation operations and this design constraint must be clearly communicated to all parties. Most design analyses are generally undertaken using some form of soil–structure interaction (SSI) analysis. SSI analyses can be further divided into subgrade reaction/ pseudo finite-element methods and finite-element/finite-difference methods. When considering SLS calculations, these preclude the use of limit equilibrium methods and will require some form of SSI to be undertaken.

The selection of analysis method should be compatible with the problem and the available design input information. There is clearly little point opting for the more complex SSI analysis method if there is little accurate information on design soil parameters. Modern complex analyses can appear deceptively easy to run and can lull the uninitiated into a false sense of security regarding the appropriateness of their analyses.

The relative size of the embedded retaining wall element is generally governed by the magnitude of the ULS bending moment that it is required to resist. SLS considerations are often involved in determining the temporary support requirements (e.g. propping or waling beam stiffnesses and locations) rather than in consideration of the wall element size.

Fernie and Suckling (1996) report that embedded retaining walls in stiff UK soils will generally exhibit excavation-induced lateral movements of around 0.15% of the retained height. The relationship with wall stiffness is reported as being more relevant to cantilever and single-propped systems than for multiple-propped embedded retaining walls.

Bolton et al. (2010) further postulated that wall displacements are unavoidable regardless of the embedded retaining wall element constructed (within the bounds of conventional element sizes and stiffnesses). The wall displacements are controlled to a much greater degree by the wall support system adopted (temporary propping and bottom-up construction versus top-down construction using permanent floor slabs, and even the construction of propping/lateral support elements by pre-excavation techniques).

13.4.4 Vertical capacity of walls

As described in Section 13.2, embedded retaining walls can also be required to carry vertical loads. The embedded retaining wall can be constructed to a deeper toe than required for wall stability alone. Any additional depth of wall can be constructed from plain concrete (i.e. constructed unreinforced), subject to the net vertical loading being compressive in nature. Clearly the additional depth would need to be suitably reinforced to carry any tensile loads (e.g. hydrostatic uplift forces). Due to the high stiffness of diaphragm walls, both active and passive faces below dredge level will contribute to the vertical capacity of the wall (note this may not be possible with significantly more flexible walls such as sheet piles).

Where temporary support to the diaphragm wall is provided through ground anchorages, the vertical stability of the wall due to the increased vertical loading on the wall from the anchorages should be checked. In most cases, the increased vertical load will not require any increase in wall embedment, but the design must demonstrate sufficient vertical capacity of the wall.

13.4.5 Reinforcement design

Reinforcement requirements should be determined by adopting analysis methods compatible with the stability analysis (e.g. Eurocode 2 (BSI, 2004a) for Eurocode 7 (BSI, 2004b) geotechnical design).

Depending on the nature of the diaphragm wall there may be an opportunity to refine the reinforcement design according to the distribution of bending moment and shear force magnitude with depth. Bars can potentially be curtailed or bar sizes changed across splice zones to provide the most economical reinforcement configuration.

The detailed design of diaphragm wall reinforcement cages must consider the potential impact of cage congestion. BS EN 1538:2010 (BSI, 2010) requires that the horizontal clear space between single vertical bars or groups of vertical bars parallel to the face of the panel should be at least 100 mm. For horizontal shear reinforcement, BS EN 1538:2010 recommends a minimum vertical spacing of 200 mm.

The detailed design of the reinforcement cages should also consider how the cages are to be fabricated. For example, such detailed design should consider if closed or open shear links will facilitate easier (and potentially safer) fabrication. While it may seem that such considerations are going to the extremes of detail, they can have a significant impact on the overall quality of the wall, particularly where the reinforcement cages have a large number of complex details (coupled box-outs, pull-out starter bars, etc.).

13.4.6 Watertightness and wall toe level for groundwater cut-off

The client's expectations with regard to watertightness must be suitably managed throughout the design and procurement process. If the wall is to remain as a permanent structure, a useful reference is the ICE document *Reducing the Risk of Leaking Substructure: A Clients' Guide* (ICE, 2009) for risk-based guidance on basement waterproofing.

Where water-bearing strata are retained by the embedded retaining wall, it is often a requirement to take the toe of the retaining wall to a depth where it will provide a vertical cut-off by effectively sealing in to a low-permeability stratum. This relies on the presence of such a suitable stratum within a sensible depth of the toe level, which is designed for wall stability.

Where no low-permeability stratum is present within a depth considered to be economically or practically viable, the required toe level of the wall can be assessed by undertaking a flow net analysis in conjunction with an assessment of flow rates into the excavation which are considered to be tolerable. A sensitivity analysis should be undertaken to assess the likely envelope of inflow rates based on potential variations in permeability (both vertical and horizontal), which in turn depend on the degree of anisotropy of the soil. Greater guidance on groundwater control is given in Chapter 6.

13.4.7 Observation method

Where the embedded retaining wall is governed by temporary works considerations, it may be appropriate to adopt a risk-based design approach (i.e. the observational approach) to realise savings to the project. It is imperative that the project is set up from a very early stage to accommodate all the requirements such an approach requires if it is to be executed successfully. The adoption of the observational approach requires sufficient instrumentation to be in place and be actively monitored, together with an agreed set of trigger/action levels and a comprehensive suite of contingency and emergency plans.

Regardless of whether the observational method is to be adopted, sufficient appropriate instrumentation should be deployed and monitoring undertaken to ensure that design assumptions and construction details can be validated and risk managed appropriately. Further discussion relating to instrumentation is given in Chapter 14, Section 14.4.3.

Often, savings through the adoption of the observational approach arise from a reduced construction programme through revised propping requirements and excavation sequencing.

Reference should be made to CIRIA Report 185 (Nicholson *et al.*, 1999) for comprehensive guidance on the subject of the observational approach.

REFERENCES

Bolton M, Lam SY and Vardanega PJ (2010) Predicting and controlling ground movements around deep excavations. In *Geotechnical Aspects of Underground Construction in Soft Ground* (Mair RJ and Taylor RN (eds)). Balkema, Rotterdam, the Netherlands.

BSI (British Standards Institution) (2004a) BS EN 1992-1-1:2004 + A1:2014. Eurocode 2: Design of concrete structures. General rules and rules for buildings. BSI, London, UK.

BSI (2004b) BS EN 1997-1:2004 + A1:2013. Eurocode 7: Geotechnical design. General rules. BSI, London, UK.

BSI (2010) BS EN 1538:2010 + A1:2015. Execution of special geotechnical works. Diaphragm walls. BSI, London, UK.

BSI (2015) BS 8002:2015. Code of practice for earth-retaining structures. BSI, London, UK.

EFFC/DFI (European Federation of Foundation Contractors/Deep Foundations Institute) (2016) *Best Practice Guide to Tremie Concrete for Deep Foundations*, 1st edn. EFFC/DFI, Bromley, UK/Hawthorne, NJ, USA.

Fernie R and Suckling T (1996) Simplified approach for estimating lateral wall movement of embedded walls in UK ground. In *Geotechnical Aspects of Underground Construction in Soft Ground* (Mair RJ and Taylor RN (eds)). Balkema, Rotterdam, the Netherlands.

Fernie R, Shaw SM, Dickson RA *et al.* (2001) Movement and deep basement provision at Knightsbridge Crown Court, Harrods, London. In *Response of Buildings to Excavation-induced Ground Movements. Proceedings of the International Conference held at Imperial College, London, UK* (Jardine FM (ed.)). CIRIA, London, UK, SP199.

Gaba A, Hardy S, Doughty L, Powrie W and Selemetas D (2017) *Guidance on Embedded Retaining Wall Design*. CIRIA, London, UK, C760.

Huder H (1972) Stability of bentonite slurry trenches with some experience in Swiss practice. *Proceedings of 5th European Conference on Soil Mechanics and Foundation Engineering: Structure Subjected to Lateral Forces*. Sociedad Espanola Mecanica, Madrid, Spain, pp. 517–522.

ICE (Institution of Civil Engineers) (2009) *Reducing the Risk of Leaking Substructure. A Clients' Guide*. ICE, London, UK.

ICE (2017) *ICE Specification for Piling and Embedded Retaining Walls*, 3rd edn. ICE Publishing, London, UK.

Nicholson D, Tse CM and Penny C (1999) *The Observational Method in Ground Engineering: Principles and Applications*. CIRIA, London, UK, R185.

Potts DM and Burland JB (1983) *A Parametric Study of the Stability of Embedded Earth Retaining Structures*. Transport and Road Research Laboratory, Wokingham, UK, Supplementary Report 813.

FURTHER READING

BSI (British Standards Institution) (1997) BS 8110-1:1997. Structural use of concrete. Code of practice for design and construction. BSI, London, UK.

BSI (2009) BS 8102:2009. Code of practice for protection of below ground structures against water from the ground. BSI, London, UK.

BSI (2015) BS 8004:2015. Code of practice for foundations. BSI, London, UK.

Powrie W and Batten M (2000) *Prop Loads in Large Braced Excavations*. CIRIA, London, UK, PR77.

Puller MJ (1994) The waterproofness of structural diaphragm walls. *Proceedings of the Institution of Civil Engineers – Geotechnical Engineering* **107(1)**: 47–57.

Puller MJ (2003) *Deep Excavations: A Practical Manual*, 2nd edn. Thomas Telford, London, UK.

Twine D and Roscoe H (1999) *Temporary Propping of Deep Excavations – Guidance on Design*. CIRIA, London, UK, C517.

Useful web addresses

Deep Foundations Institute: http://www.dfi.org (accessed 01/08/2018).

European Federation of Foundation Contractors: https://www.effc.org (accessed 01/08/2018).

Federation of Piling Specialists: https://www.fps.org.uk (accessed 01/08/2018).

Temporary Works, Second edition

Pallett, Peter F and Filip, Ray
ISBN 978-0-7277-6338-9
https://doi.org/10.1680/twse.63389.193
ICE Publishing: All rights reserved

Chapter 14
Contiguous and secant piled walls

Chris Robinson
Design Manager, Cementation Skanska Limited

Andrew Bell
Chief Engineer, Cementation Skanska Limited

Contiguous and secant piled walls can be used to form embedded retaining walls, for example, to construct tanks, shafts, tunnel portals, road/rail cuttings, basement excavations, cut and cover tunnels and other forms of underground structures. They are commonly adopted where alternative methods cannot be successfully implemented because of the nature of the ground or performance requirements. In common with diaphragm walls, the economics of contiguous and secant piled walls become increasingly attractive when they form part of the permanent works. This chapter includes a description of common uses of contiguous and secant walls and their method of construction (including aspects of additional temporary works required to facilitate construction), a summary of typical construction details which can aid construction of a high-quality finished product, and a description of common design approaches and design considerations.

14.1. Introduction

Contiguous and secant piled walls are two common forms of embedded retaining walls. The most significant difference between these two forms of retaining wall is that contiguous piles are designed to be totally independent reinforced elements with, typically, 150 mm between adjacent pile elements at commencing level, whereas secant piles are designed to form an interlocking wall. Secant piles can therefore retain both ground and groundwater (subject to construction tolerances, as discussed in Section 14.3).

14.1.1 Contiguous pile wall
Contiguous piled walls are suitable where the groundwater table is below excavation level, as groundwater is free to flow through the interstices between pile elements. Permanent works applications for contiguous piled walls therefore require an additional reinforced concrete lining to secure exposed soil and resist long-term groundwater pressures. Figure 14.1 depicts a typical contiguous piled wall.

14.1.2 Secant pile wall: hard/soft or hard/firm
Where short-term water retention is required this system, consisting of interlocking bored piles, offers the most cost-effective and rapid solution. Female (or primary) piles

Figure 14.1 Contiguous piled wall. (Courtesy of Cementation Skanska Limited)

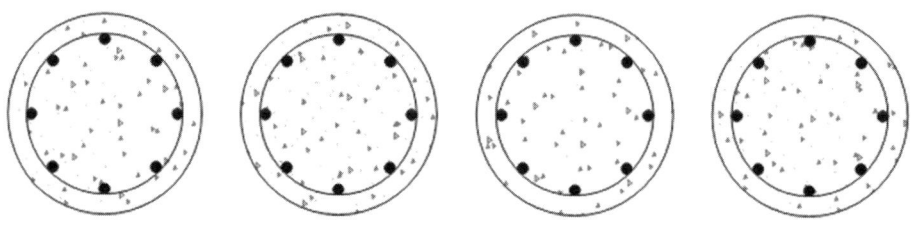

Figure 14.2 Hard/soft or hard/firm secant piled wall. (Courtesy of Cementation Skanska Limited)

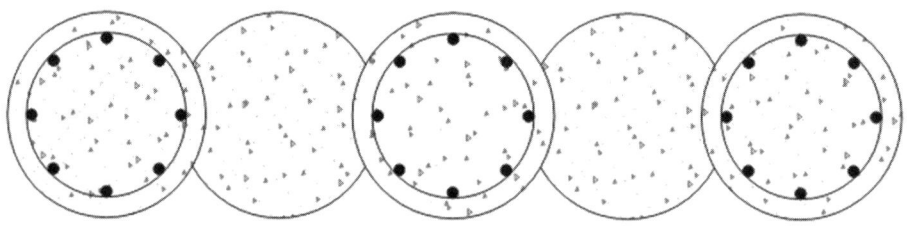

are constructed first using a 'soft' cement–bentonite mix (commonly 1 N/mm^2) or 'firm' concrete (commonly 10 N/mm^2). Male (or secondary) piles, formed in structural reinforced concrete, are then installed between (and cutting into) the female piles, with a typical interlock of 150–250 mm. These walls may need a reinforced concrete lining for permanent works applications, depending on the particular watertightness requirements of the project. An illustration of a typical hard/soft or hard/firm secant piled wall is provided in Figure 14.2.

14.1.3 Secant wall: hard/hard

Hard/hard wall construction is very similar to that for a hard/firm wall except that the primary piles are constructed in higher-strength concrete and may be reinforced by reinforcement cages or steel beam or column sections. Heavy-duty rotary piling rigs, using tools fitted with specially designed cutting heads, are necessary to cut the secondary piles. As structural concrete is used throughout, there may be no need to provide a

Figure 14.3 Hard/hard secant piled wall. (Courtesy of Cementation Skanska Limited)

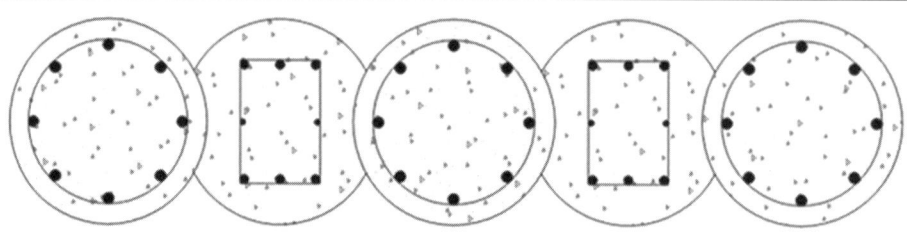

lining wall. The end product provides a fully concreted face and can be an effective alternative to diaphragm wall construction. Figure 14.3 depicts a typical hard/hard secant piled wall.

14.2. Applications

Contiguous and secant piled walls can be adopted to provide solutions to a variety of constructed situations. They tend to be adopted where the general site conditions, depth of excavation, ground conditions or watertightness requirements (note: secant piled walls only) preclude the adoption of alternative (potentially cheaper) methods, for example, construction in open-cut excavations, king post walls or sheet piling (see Chapter 11).

In addition to providing ground support (and a groundwater exclusion function in the case of secant piling), contiguous and secant piled walls can also be used to carry vertical loads (temporary and/or permanent vertical loads). Where the magnitude of vertical loading exceeds the available wall capacity (based on an embedment length required to maintain wall stability), the wall piles can be constructed to a greater depth than required by consideration of wall stability alone in order to satisfy design criteria relating to factors of safety. This additional depth of pile can be constructed from plain unreinforced concrete, subject to the superstructure loading being compressive in nature. The additional depth of excavation would clearly need to be suitably reinforced to carry any tensile loads (e.g. hydrostatic uplift forces). When designing embedded retaining walls to withstand applied vertical loads, it is usual to consider only that part of the wall below the deepest excavation level.

Where particularly onerous ground conditions are present, for example, very loose fine sands, a secant piled retaining wall may represent the preferred solution, even if groundwater is not present above the excavation level. The nature of such ground may pose an unacceptable risk of ground loss through insufficient arching capability of the ground, which may in turn pose a risk to the stability and serviceability of adjacent structures or services or ground to be supported.

14.3. Construction methods and plant
14.3.1 General considerations

Most forms of replacement piling can be employed to construct contiguous and secant walls. The plant employed will have certain constraints or limitations on the physical dimensions of piles that can be constructed. As discussed in Chapter 13 on diaphragm walls, the construction methods and plant employed need to be determined through consideration of each individual project's particular characteristics. This will include consideration of aspects such as site-specific ground conditions, length of wall piles to be constructed, watertightness requirements, access restrictions, headroom restrictions, any working time restrictions and party wall issues.

The drawing sequence presented in Figures 14.4 and 14.5 shows the typical sequence of site operations for contiguous pile wall and secant pile wall construction, respectively. Pile construction needs to be sequenced to ensure that construction of any pile does not adversely affect any previously constructed piles. For secant walls, the sequence needs

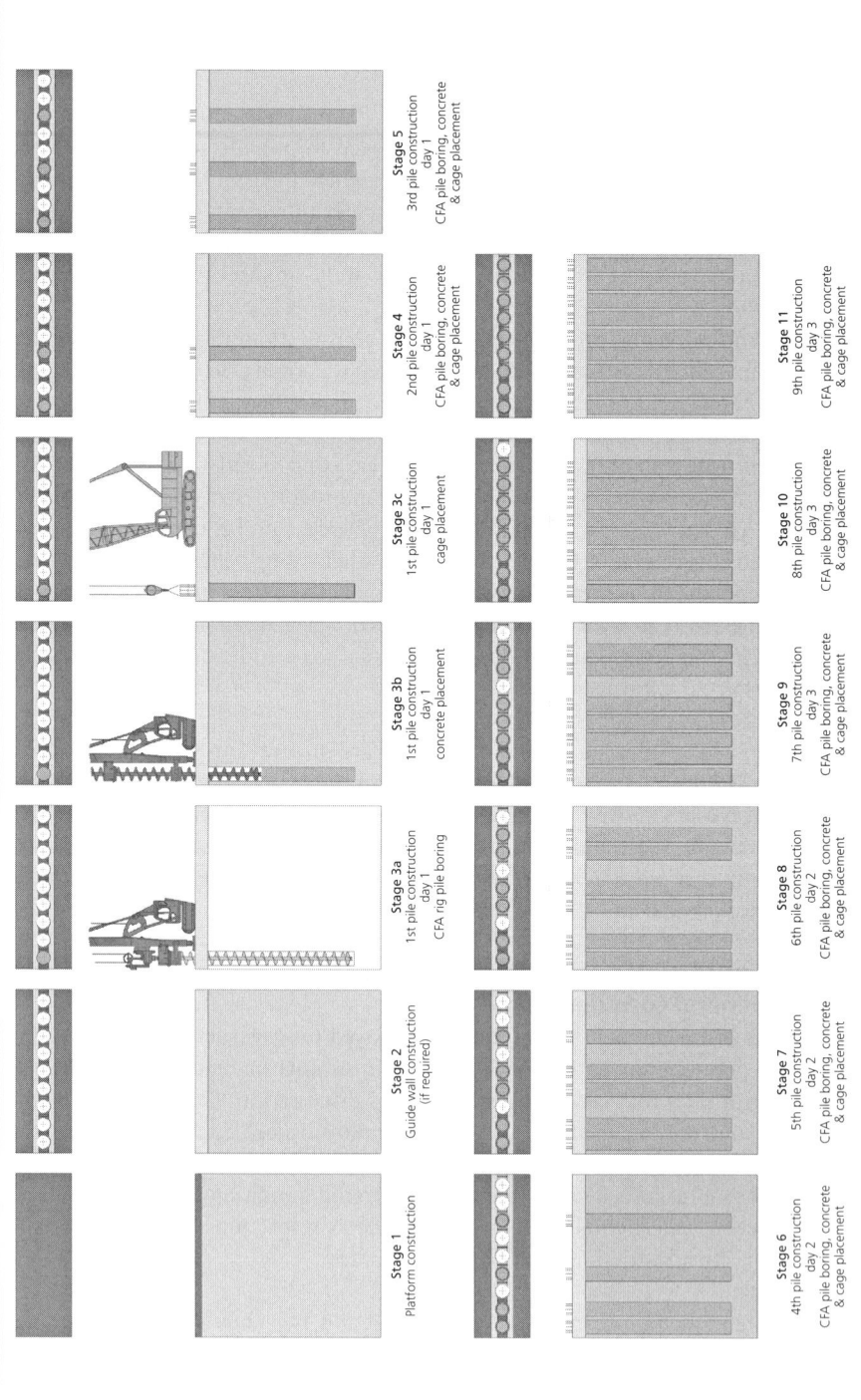

Figure 14.4 Contiguous piled wall and indicative sequence of construction. (Courtesy of Cementation Skanska Limited)

Figure 14.5 Secant piled wall and indicative sequence of construction. (Courtesy of Cementation Skanska Limited)

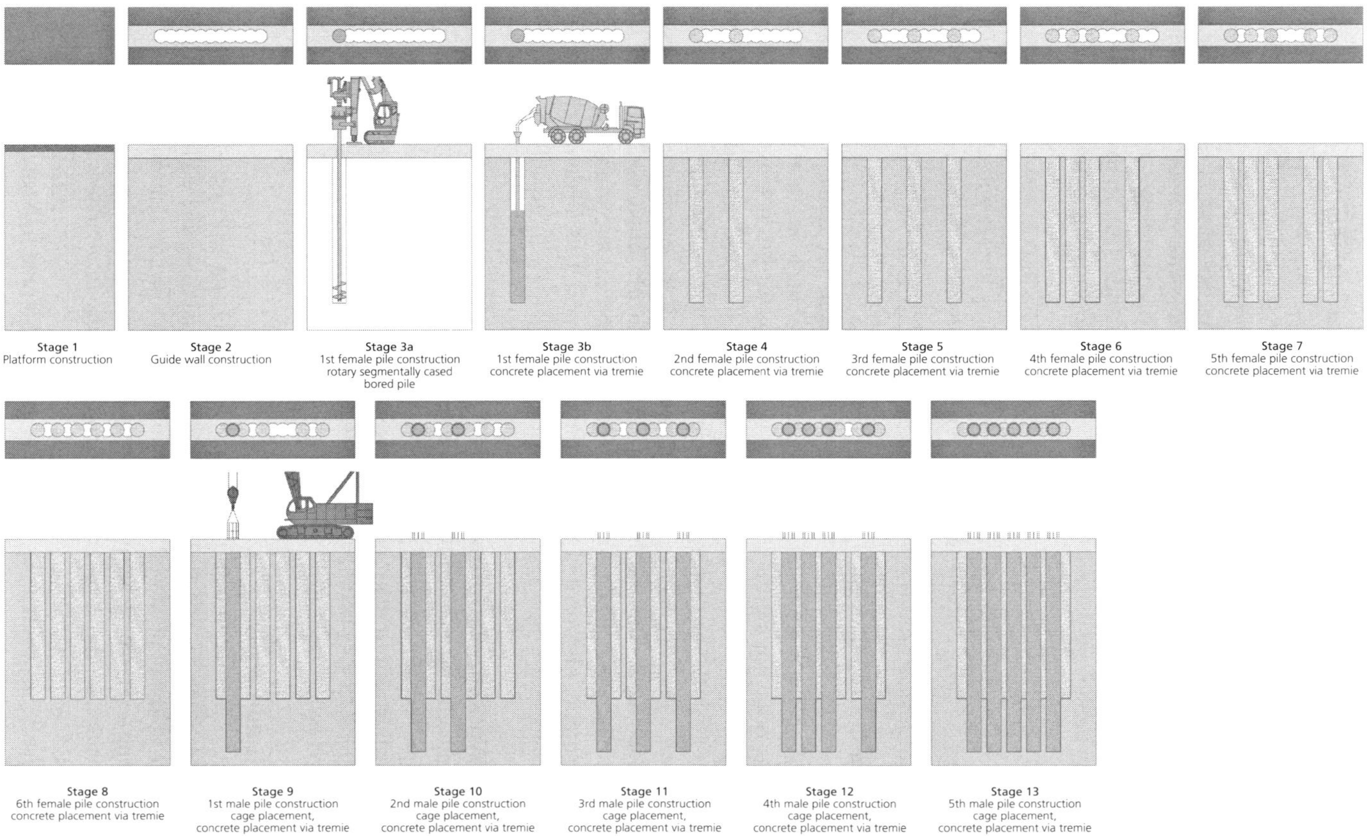

also to take account of the strength gain of the female pile concrete (female pile concrete can potentially crack when male piles are cut into them if the concrete is too strong, which may have a significant impact on the wall watertightness).

As well as being a temporary works solution in their own right, both contiguous and secant piled wall construction can require some considerable elements of additional temporary works for their successful execution. As described in Chapter 13 on diaphragm walls, these include aspects such as guide walls, support fluids, temporary anchorages and/or temporary props, working platforms for tracked plant (see Chapters 5 and 7), consideration of slope stability (see Chapter 10), temporary screening to protect the public and personnel, protection of watercourses and protection of adjacent services.

Precautionary measures may be adopted such as using long temporary casings, limiting the number of piles constructed along sensitive elevations within one shift, or underpinning of particularly sensitive adjacent structures.

14.3.2 Site preparation
Many sites require demolition works to be undertaken in advance of wall construction operations; this will often entail excavation or grubbing out of existing foundations. Where this occurs along the line of the wall it is good practice to ensure the backfill to these areas is selected not only to give adequate support to tracked plant but also to mitigate against high overbreak occurring and adversely affecting the finished surface of the piled wall. This may require trimming back if the overbreak is in excess of the specification limits. Suitable backfill materials that are often used are stiff clays or clay-bound hogging. These materials, properly placed and compacted, will give a relatively smooth high-quality finish to the wall over the backfill depth and can aid wall verticality.

14.3.3 Working platforms
As with any other site where tracked plant (predominantly piling rigs and cranes) is to be employed, a suitable working platform should be provided to ensure that an adequate factor of safety against platform failure is maintained under the range of loading which will be applied by the plant operating on it. Refer to Chapter 5 for full details of working platform design.

14.3.4 Guide walls
Guide wall construction will usually follow on from construction of the working platform, although this is not always the case. It is usual for guide walls to be constructed in advance of secant piled wall construction to enable greater dimensional control at the commencing level to be achieved. Generally a limit of positional tolerance of ± 25 mm at commencement level can be achieved through the use of a guide wall.

When constructing contiguous piled walls, guide walls tend to be used less often because construction tolerances are usually less critical (contiguous piled walls are not designed for groundwater exclusion and therefore do not have any watertightness performance criteria to satisfy). Without a guide wall, a plan positional tolerance of ± 75 mm is typically specified.

Unlike contiguous piled wall construction, secant piled wall construction will almost invariably require the construction of a scalloped guide wall in advance of piling operations. The very nature of secant piled walls requires the piles to interlock, usually for watertightness.

14.3.5 Pile construction techniques

Different pile construction techniques will have different limits to attainable verticality. Most rotary piling techniques will typically have a verticality limit of $1:75$. Stiff double-walled continuous flight auger (CFA) strings may be able to achieve verticality limits of the order of $1:75$ to $1:125$, whereas cased CFA and large-diameter segmentally cased piled walls may be able to achieve a pile verticality of the order of $1:150$ to $1:200$. Specific guidance should be sought from specialist contractors on a case-by-case basis.

A large range of piling techniques can be adopted for construction of contiguous and secant walls, although practical considerations may limit the use of some of these for the construction of interlocking secant walls.

A non-exhaustive list of applicable piling techniques includes CFA, cased CFA, segmental flight auger (SFA), large-diameter rotary bored piling with conventional slip casings, large-diameter rotary bore piling with segmental casing, mini piled case/auger and drilled pile systems (e.g. Symmetrix).

Some of the above piling techniques (e.g. rotary bore piling, case and auger piling) can be employed by adopting a short length of temporary casing (although care needs to be taken when assessing the pile spacing for cased and uncased diameters) where the ground is self-supporting. Others may require full-length casing in unstable ground or the use of a bore support fluid, while CFA and SFA provide full bore support without the need for temporary casings or support fluid.

14.3.6 Reinforcement cages

Reinforcement cages for wall piles are typically fabricated using at least six bar cages. There are exceptions to this general rule, but the design must take account of the potential for unfavourable cage orientation within the pile bore.

The number of longitudinal reinforcing bars forming the pile cage needs to be compatible with the pile diameter and minimum reinforcement requirements. In most instances, the reinforcement requirements of wall piles will satisfy minimum reinforcement requirements without explicit consideration.

Female piles to secant piled walls are not normally reinforced. However, when designing the wall as a hard/hard wall, the female pile will be constructed from a structural concrete mix (with a specified characteristic strength) and will be suitably detailed to facilitate subsequent male pile construction. The requirement for the male piles to cut into the female pile clearly imposes constraints on the amount of reinforcement that can be accommodated within the female pile. Notwithstanding this limitation, reinforcement of the female pile can provide the additional wall capacity (in terms of bending resistance), which may otherwise require larger-diameter male piles or the adoption of alternative construction methods.

It is possible to vary the pile reinforcement within fabricated cages to the structural requirements at different levels within the pile. Bars can potentially be curtailed or changes in bar sizes made across cage splice zones. It should be noted that it is not generally possible to economise pile cages to the same degree as within diaphragm wall cages; this is partly due to the requirements to consider unfavourable cage orientation within the pile bore and the design of the pile as a column element.

Where cage lengths are to be spliced, the splice detail should be carefully detailed to limit bar congestion. This will assist in maintaining a high-quality finished product. Reinforcement bar congestion is often the primary cause of poor-quality concrete at the exposed face of the pile. As with detailing of reinforcement cages for diaphragm wall panels, the cage splices should be detailed to ensure that operatives are not required to place their hands within the cage, where serious injury could be incurred should the cage sections move relative to each other.

It may be possible to introduce couplers and box-outs within pile reinforcement cages. However, it is the authors' experience that a high degree of redundancy (in terms of the number of couplers fabricated into the reinforcement cage) is usually required to ensure that sufficient couplers can be located when the box-out is exposed. Pile reinforcement cages have a tendency to twist within the pile bore. This is particularly the case where pile cages are plunged within wet concrete, but can also occur where cages are hung or suspended within the empty pile bore with concrete subsequently being placed via a tremie pipe. Another consideration in pile box-out design is to ensure that there is sufficient space for concrete to flow between all the reinforcing elements and to ensure that the box-out (typically constructed with polystyrene or Styrofoam to facilitate exposure of the couplers) does not cause buoyancy instability of the pile cage.

14.3.7 Concreting

As described in Section 14.3.5, a variety of piling techniques can be adopted. Where piles are constructed adopting CFA or SFA techniques, concrete will be introduced from the toe of the pile and brought up to the level of the piling platform. Reinforcement cages are installed following the concreting operation.

For open bore piling techniques (e.g. large-diameter rotary bored piling, cased and auger piling), concreting of the pile (or grouting for small-diameter piles) will usually be undertaken following installation of the reinforcement cage. For open bore techniques, sectional tremie pipes are generally employed that typically have a diameter of 200–300 mm. The diameter of the tremie pipe will depend on the individual contractor's equipment and the depth of the tremie pipe. Consideration of the diameter of the reinforcing cage is also important in order to ensure the tremie pipe can easily be inserted and removed without causing damage to, or snagging on, the cage.

For guidance on the performance of fresh concrete and its method of placement using tremie methods in deep foundation elements, reference should be made to the *Best Practice Guide to Tremie Concrete for Deep Foundations* (EFFC/DFI, 2016).

14.4. Design

14.4.1 Scope

This section on design is not intended to provide full guidance on how to design embedded retaining walls. This subject is covered in detail within the various codes of practice and best practice guidance listed at the end of this chapter. In particular, the reader is referred to the Construction Industry Research and Information Association (CIRIA) publication C760 (Gaba *et al.*, 2017), BS 8002:2015 (BSI, 2015) and Eurocode 7 (BSI, 2004b).

The aim for this section is to provide commentary on specific design aspects of contiguous and secant piled walls. High-level guidance on the aspects most relevant to the design of embedded retaining walls is given in Chapter 13, Section 13.4.3.

14.4.2 Temporary support for lateral wall stability

Contiguous and secant walls may be designed as cantilever walls. This will depend on the depth of excavation, the depth to which the wall piles can be economically installed, the prevailing ground conditions, and the magnitude of tolerable wall displacement and associated ground movement.

Where the various factors are unfavourable, some form of temporary support will be required. This may take the form of temporary berms (generally to facilitate installation of another temporary support measure), temporary props or temporary ground anchorages. Each of these measures will require suitable design. Should temporary ground anchorages be adopted, this may require a way-leave if the anchorages are to be installed beyond the site boundary. It will also be necessary to undertake a design check to ensure that vertical stability is maintained, as the anchorages will impart a downward vertical load onto the embedded retaining wall.

Temporary ground anchorages and temporary props will generally require a waling beam to span along the wall to distribute the load between the wall and the anchors/props. Often an in situ capping beam will be constructed to be used in place of a waling beam. Waling beams for ground anchors can be more complex, particularly if they are fabricated from steel. A twin parallel flange channel arrangement is often adopted which allows the anchorage to pass between the two channel sections at the required angle of declination (often 30–45° below horizontal). An alternative solution for ground anchorages is to construct an in situ head block detail (Figure 14.6) which spans adjacent reinforced piles.

14.4.3 Instrumentation

Where wall deflections are a critical element of the design (most particularly where the observation method is being employed), inclinometer reservation tubes can be installed within the reinforced piles. A suitably sized thin-walled steel tube (with a sealed end) is attached to the reinforcement cage (in sections with threaded couplings if spliced cages

Figure 14.6 In situ anchor head block. (Courtesy of Cementation Skanska Limited)

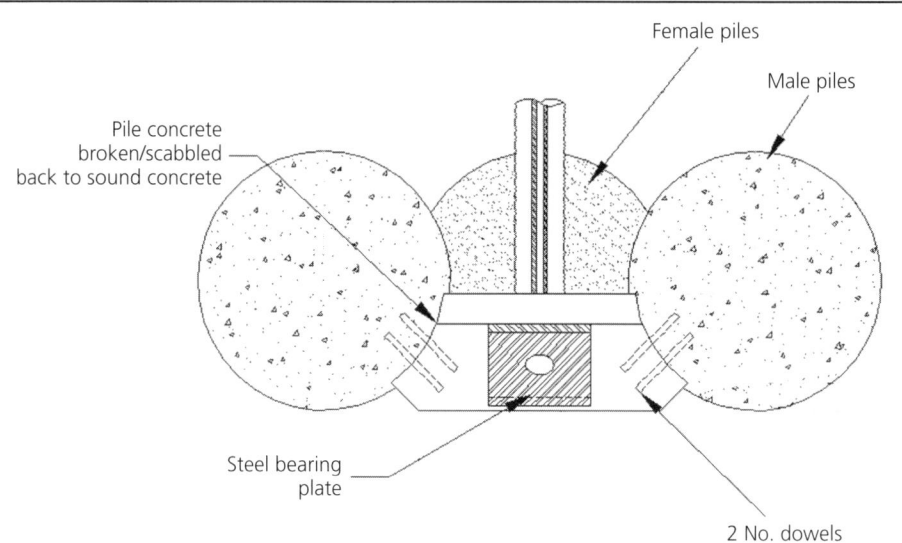

are required). The reservation tube facilitates subsequent installation (grouting in) of inclinometer tubes, which generally take the form of plastic ducts with orthogonal guides for the inclinometer torpedo. It is usual for the inclinometer duct installation to be undertaken outside the piling contract.

Alternative methods of monitoring wall deflections are available such as cast-in-place inclinometers; these are more limiting by their very nature, however, allowing monitoring of deflections only at the levels the instruments are installed (compared to a full deflection profile for conventional inclinometer instruments). Cast-in-place inclinometers are particularly useful in situations where access is difficult or impossible.

Other forms of instrumentation can be installed within wall piles, generally fixed to reinforcement cages (e.g. vibrating wire strain gauges or fibre-optic strain gauges to monitor the state of stress within the wall piles). Again this is most commonly associated with the observational method, for example, when assessing whether additional temporary support measures may be omitted. For full details on the observational method, refer to CIRIA R185 (Nicholson et al., 1999).

14.4.4 Reinforcement design

Reinforcement requirements should be determined adopting analysis methods compatible with the stability analysis, for example, Eurocode 2 (BSI, 2004a) for structural concrete design and Eurocode 7 (BSI, 2004b) for geotechnical design.

Care needs to be taken to ensure that the reinforcement design is compatible with the method of pile construction proposed. For example, plunging cages within CFA/SFA piles has practical limitations on the depth attainable for a given cage weight and rigidity. A maximum practicable reinforcement cage plunge depth of around 15 m is often adopted. Although it is possible to adequately reinforce these pile types to deeper levels, specific guidance from specialist piling contractors should be sought.

There may be opportunity to refine the reinforcement design according to the distribution of bending moment and shear force magnitude with depth. Bars can potentially be curtailed or bar sizes changed across splice zones to provide the most economical configuration.

When designing reinforcement cages, the recommendations of BS EN 1536:2010 + A1:2015 (BSI, 2010) regarding bar spacing should be applied to avoid cage congestion and the potentially negative effect on construction quality.

The reinforcement cage design should take into account any requirements for handling/lifting the cages to ensure that they are sufficiently robust. This may require the inclusion of suitable lifting bands and so on. Where cages are required to be spliced, sufficient attention to detail should be given to the method used to form the splice such that operatives are not required to put their hands within the cage during the splicing operation.

14.4.5 Watertightness and wall toe level for groundwater cut-off (secant walls only)

Secant walls must be designed to ensure they maintain interlock to the required level. It should be noted that whatever degree of interlock is designed for secant walls cannot be guaranteed to provide a watertightness greater than BS 8102:2009 (BSI, 2009) Grade 1. Section B1.9 in *ICE Specification for Piling and Embedded Retaining Walls*, 3rd edition (ICE, 2016) provides information on watertightness assessments.

While it is usual to maintain male–female pile interlock to below the dredge level, there may be opportunity in certain circumstances to found the female piles above the dredge level if a suitable low-permeability stratum is present. It is important to clearly communicate any risks that may be present in such an arrangement, and have a suitable contractual arrangement in place covering risks associated with watertightness and the costs of any potential remedial works.

Where water-bearing strata are retained by the secant wall, it is often a requirement to take the toe of the retaining wall to a depth where it will provide a vertical cut-off by effectively sealing in to a low-permeability stratum. This relies on the presence of such a suitable stratum within a sensible depth of the toe level which is designed for wall stability. In addition, a check should be undertaken to ascertain the level to which male–female pile interlock can be maintained.

Where there is no low-permeability stratum within a depth considered to be economically or practically viable, the required toe level of the secant wall can be assessed by undertaking a flow net analysis in conjunction with an assessment of flow rates into the excavation which are considered to be tolerable. A sensitivity analysis should be undertaken to assess the likely envelope of inflow rates based on potential variations in permeability (both vertical and horizontal), which in turn will depend on the degree of anisotropy of the soil. Guidance on groundwater control is given in Chapter 6.

14.4.6 Vertical capacity of walls

Embedded retaining walls can also be required to carry vertical loads. The embedded retaining wall can be constructed to a deeper toe than required for wall stability alone. Any additional depth of wall can be constructed from plain concrete (i.e. constructed unreinforced), subject to the net vertical loading being compressive in nature. Clearly, the additional depth would need to be suitably reinforced to carry any tensile loads (e.g. hydrostatic uplift forces). When designing embedded retaining walls to withstand vertical loads, it is usual to consider only that part of the wall below the deepest excavation level (i.e. the fully embedded length of the wall).

Where temporary lateral support to the wall is provided through ground anchorages, the vertical stability of the wall should be checked due to the increased vertical loading on the wall from the anchorages.

14.4.7 Other aspects

CIRIA C760 (Gaba et al., 2017) recommends that embedded retaining wall design makes an allowance for unplanned excavation of 10% of the design retained height up to 0.5 m, although this requirement may be reduced or omitted entirely subject to the agreement of all interested parties and the implementation of sufficient control measures to ensure that the design constraints are fully realised.

Where cohesive materials are present at the dredge level, consideration should be given to the potential for these materials to degrade and soften, usually just within the top 1 m, thus potentially adversely affecting the passive resistance afforded to the retaining wall.

The design surcharge load (generally a minimum surcharge of 10 kN/m^2) should be allowed to act on the active side of the retaining wall.

As well as considering the required wall toe level to maintain lateral and vertical stability, certain ground conditions may require the wall toe to be taken deeper by consideration of more global factors. One such scenario is where an embedded retaining wall is to be constructed within deep soft cohesive deposits. The potential for basal failure of the excavation, whereby the soft cohesive materials affect flow around the toe of the wall, should be assessed in such a situation. Piping failure may need to be avoided and potential slip planes may need to be intercepted (see Chapter 10).

REFERENCES

BSI (British Standards Institution) (2004a) BS EN 1992-1-1:2004 + A1:2014. Eurocode 2: Design of concrete structures. General rules and rules for buildings. BSI, London, UK.

BSI (2004b) BS EN 1997-1:2004 + A1:2013. Eurocode 7: Geotechnical design. General rules. BSI, London, UK.

BSI (2009) BS 8102:2009. Code of practice for protection of below ground structures against water from the ground. BSI, London, UK.

BSI (2010) BS EN 1536:2010 + A1:2015. *Execution of special geotechnical works. Bored piles.* BSI, London, UK.

BSI (2015) BS 8002:2015. *Code of practice for earth retaining structures.* BSI, London, UK.

EFFC/DFI (European Federation of Foundation Contractors/Deep Foundations Institute) (2016) *Best Practice Guide to Tremie Concrete for Deep Foundations*, 1st edn. EFFC/DFI, Bromley, UK/Hawthorne, NJ, USA.

Gaba A, Hardy S, Doughty L, Powrie W and Selemetas D (2017) *Guidance on Embedded Retaining Wall Design.* CIRIA, London, UK, C760.

ICE (Institution of Civil Engineers) (2016) *ICE Specification for Piling and Embedded Retaining Walls*, 3rd edn. ICE Publishing, London, UK.

Nicholson D, Tse CM and Penny C (1999) *The Observational Method in Ground Engineering: Principles and Applications.* CIRIA, London, UK, R185.

FURTHER READING

BSI (British Standards Institution) (1997) BS 8110-1:1997. *Structural use of concrete. Code of practice for design and construction.* BSI, London, UK.

BSI (2015) BS 8004:2015. *Code of practice for foundations.* BSI, London, UK.

ICE (Institution of Civil Engineers) (2009) *Reducing the Risk of Leaking Substructure. A Clients' Guide.* ICE, London, UK.

ICE (2016) *ICE Specification for Piling and Embedded Retaining Walls*, 3rd edn. ICE Publishing, London, UK.

Powrie W and Batten M (2000) *Prop Loads in Large Braced Excavations.* CIRIA, London, UK, PR77.

Puller MJ (2003) *Deep Excavations: A Practical Manual*, 2nd edn. Thomas Telford, London, UK.

Twine D and Roscoe H (1999) *Temporary Propping of Deep Excavations – Guidance on Design.* CIRIA, London, UK, C517.

Useful web addresses

Deep Foundations Institute: http://www.dfi.org (accessed 01/08/2018).

European Federation of Foundation Contractors: https://www.effc.org (accessed 01/08/2018).

Federation of Piling Specialists: https://www.fps.org.uk (accessed 01/08/2018).

Temporary Works, Second edition

Pallett, Peter F and Filip, Ray
ISBN 978-0-7277-6338-9
https://doi.org/10.1680/twse.63389.207
ICE Publishing: All rights reserved

Chapter 15
Caissons and shafts

Andrew Smith
Retired

This chapter deals with the construction of shafts and caissons using pre-cast segmental linings together with the use of sprayed concrete lining in shaft sinking operations.

Caissons in the marine environment are not covered in this chapter. Construction methods used, including generally available types of shaft linings, other materials and construction plant commonly used are all discussed, and a brief description is given of specialist processes that can be used to provide ground stability when required. Comments on design considerations for the construction of shafts and caissons using pre-cast segments are also included.

15.1. Introduction

Shaft sinking in the UK is generally carried out using circular pre-cast segmental linings, which were originally developed after the Second World War as an alternative to the original and more expensive cast-iron linings. They are typically used to provide access for tunnelling operations, where they can then be converted to permanent access chambers. More recently, they have become increasingly used to form storage chambers, pumping stations and so on, where they can offer a cost-effective solution to more expensive alternatives such as in situ construction within a piled cofferdam. The advantages are that the permanent works materials are, in effect, used as the temporary works during construction and also that the construction 'footprint' is kept to a minimum (an important consideration in urban areas).

Specification clauses for shaft construction and break-outs from shafts can be found in the British Tunnelling Society *Specification for Tunnelling* (BTS/ICE, 2010).

For small-diameter shafts (up to 4 m diameter), full-circle segmental rings are available, but high unit weights need to be considered when specifying the construction plant required.

As circular structures, segmental shafts normally require no additional bracing during construction because all the ground loads are evenly distributed to produce only compressive loads in the lining. Another advantage is that, properly constructed, the risk of settlement is kept to a minimum; the construction process should ensure that the exposed ground is supported with the minimum of delay and there are no large temporary working spaces to be backfilled on completion.

Pre-cast shaft linings were originally designed to mirror the earlier cast-iron linings so that they had a rib and recessed panel appearance. As such, they had a limited use in the context of depths exceeding around 20 m, and are no longer readily available. These linings have now been replaced with solid units, normally 1000 mm wide, in standard diameter ranges from 4 to 25 m, with different manufacturers having their own bolting systems. The individual segments are cast to very exacting tolerances to ensure accurate alignment between components. The design of these units normally incorporates a sealing system utilising hydrophilic strips or rubber gaskets, the latter being factory fitted. Increasingly, these segments are being manufactured using fibre reinforcement, which has a beneficial effect in terms of fire resistance and also makes it easier to break out openings and attach fixings. Although the majority of manufacturers produce a standard range of products to a set design suitable for the vast majority of projects, they will also provide a design service and special manufacture of linings to suit more demanding or specific design conditions. In addition, as part of their service manufacturers are able to provide specialist items such as corbel units to accommodate landing slabs and also complete roof slabs designed to the engineer's loading requirements.

A more recent development, often used in conjunction with segmental shaft construction, is the use of a sprayed concrete lining (SCL). SCL is used to replace the segmental lining and can be used in conjunction with reinforcement and/or lattice arches or, alternatively (and now more commonly), with fibre reinforcement. Typically, segmental ring construction might be used for the upper levels of the shaft where ground conditions and surface features make this appropriate with a change to SCL in the underlying more stable strata. For permanent works, use of a secondary in situ lining may be required to create a smooth finish. SCL can be used in conjunction with either sprayed-on or sheet-membrane waterproofing. One considerable advantage of a SCL is that openings and so on required in the shaft lining can be formed by the use of additional framing reinforcement but without the need to introduce expensive temporary works support while the portal structures are constructed. On the other hand, openings in shafts built with segmental rings generally require considerable temporary support during construction (particularly at deep levels), as the inherent compressive strength of a shaft ring is lost once an opening is formed in it, until the permanent works have been completed.

15.2. Major alternatives
Alternatives to shaft sinking methods not covered in this chapter include

- sheet piled cofferdams (see Chapter 11)
- diaphragm walls (see Chapter 13)
- contiguous and secant walls (see Chapter 14).

15.3. Common methods of construction
Shaft sinking methods are broadly subdivided into two categories: underpinning and caisson sinking. For more details see the British Tunnelling Society publication *Tunnel Lining Design Guide* (BTS, 2004).

15.3.1 Underpinning

Underpinning involves the excavation and erection of each ring of the segmental lining beneath the previously constructed ring. As each ring is completed cementitious material is injected behind the lining to fill any voids and secure it in the ground, ready to support the next ring, which is bolted up from beneath. Different manufacturers have their own bolting systems for doing this. This method is normally used in firm self-supporting ground or where ground treatment processes have created stable ground conditions. However, it can also be used to recover a situation where a shaft being sunk as a caisson has become stuck, although additional ground stabilisation processes will probably be required in this situation.

The initial segmental ring is placed in a pre-dug excavation and keyed in to a concrete collar cast around it, normally by inserting dowels through the grout holes. It is vital during this process that this initial ring is built within the correct tolerance and fully supported, as any settlement during the casting of the collar could have serious repercussions in keeping the rest of the shaft vertically aligned. Once the first ring of the shaft has been fixed in the ground, it is common to fix plumbing brackets around the top of the ring to check verticality as the shaft is sunk.

The grouting process requires the base of each ring to be sealed. There are geotextile hoses available that can be fixed behind the ring before it is built; after building, the hose is inflated with grout to seal the annulus before void grouting begins. It is, however, more common to push excavated material in under the ring once it has been built to achieve the same result: so-called 'fluffing up'. When using this method, care must be taken not to damage the seals. If an excavator is being used it is possible to fit a purpose-made blade for this purpose. It is also good practice to form small voids in the previously grouted annulus up to the grout holes of the ring above to release any trapped air as grouting takes place.

Excavation of shafts constructed by underpinning is commonly carried out by 360° excavators, initially working from the surface and then lowered into the excavation as it becomes deeper. Alternatively, there is a range of pole grabs available (some telescopic) that can be attached to excavators and used from the surface. With the advent of zero tail swing models it is possible to get machines into all but the smallest shafts. It is very important to accurately trim the excavation to the correct profile; this avoids overbreak and excessive grout use.

Traditionally carried out by hand, adherence to current hand–arm vibration syndrome (HAVS) regulations can make this a time-consuming process. One method of overcoming this is to reverse the bucket on the shaft excavator so that it can dig upwards to assist the trimming process (Figure 15.1). Segments are usually placed by crane using specially manufactured underpinning frames, supplied by the segment manufacturers. Grouting normally uses bagged cementitious material supplied shrink-wrapped and on pallets for weather protection and ease of handling by forklift. It can be mixed and pumped in special composite units driven by compressed air. The segment manufacturers generally provide threaded grout sockets in their segments, and it is important to check that the grout gun nozzle is compatible with the fittings supplied.

Figure 15.1 Underpinning showing trimming. (Courtesy of Joseph Gallagher Ltd)

As well as underpinning using segments, the same basic process can be used with SCL methods (Figure 15.2). Once the shaft has been excavated for the predetermined length, the SCL is applied using either a hand-held nozzle or a robot sprayer. For most operations, this material is supplied ready mixed and either discharged directly into the pump or held in a re-mixer on site. This material can also be supplied ready mixed using bulk road tankers which discharge into static silos set up on site. As with most operations involving SCL, the material is supplied retarded and an accelerator is added at the nozzle. Reinforcement can be provided in the form of mesh or prefabricated arches, but it is becoming increasingly common to use fibre-reinforced concrete, which speeds up the process considerably. Openings typically use steel reinforcement locally, and can be formed incrementally as the excavation proceeds without the need for temporary support. If required, sprayed waterproof or sheet membranes can be incorporated in the lining, normally by sandwiching them between two separate layers of SCL.

15.3.2 Caisson sinking

Caissons can be round, square or rectangular but must be of a regular cross-section with no protrusions, which would cause drag and lead to the danger of lock-up. The majority of caissons are circular, however.

Caisson sinking typically involves constructing the first one or two rings of the shaft at ground level within a substantial reinforced concrete collar using a special cutting ring at the leading edge. Some manufacturers can supply the rings for caisson sinking with an external bolting system which eliminates the needs for bolt pockets within the shaft and

Figure 15.2 Oruga shotcreter and tunnel opening. (Courtesy of Meyco-BASF and Joseph Gallagher Ltd)

also the need to gain access to the inside of the shaft during ring-building operations. As with underpinning, it is vital that the initial rings are built accurately and held in position while the collar is concreted. These rings are surrounded by polystyrene sheets before concreting the collar to create a sleeve through which the shaft can slide. Sacrificial jacking bases are also positioned around the perimeter onto which the vertical shaft jacks are then fixed after concreting. The number of jacks will depend on the diameter of the shaft and the depth to be sunk. For shaft sizes up to around 10 m diameter most segment suppliers manufacture their own pre-cast cutting edges. Over this size it is necessary to use a fabricated steel unit which must be designed to suit the sizes and fixing patterns of the rings to be used. For larger diameters and demanding ground conditions it is essential to have a steel unit that can be welded on site to increase rigidity and prevent shaft distortion during sinking.

The cutting edge (Figure 15.3) must provide an overcut to the rings to be used so that an annulus is formed as the shaft sinks, enabling a lubricant to be introduced. This annulus is typically of the order 50 mm. There are a number of products on the market suitable for this operation. The caisson is sunk by excavating from within and then letting the shaft sink in a controlled manner, almost always by the use of vertical hydraulic jacks, typically of 40 t maximum capacity, positioned around the collar. The size, and hence weight, of this collar must be sufficient to counteract the anticipated jacking loads required. As the shaft sinks further, rings are added at the surface. Specially designed working cages are needed for this operation (Figure 15.4).

Figure 15.3 Steel cutting edge (during trail erection) ready to receive pre-cast units. (Courtesy of PL Manufacturing Ltd)

Figure 15.4 Building pre-cast segments from a working cage. (Courtesy and copyright *NATM Magazine*, 2009)

The annulus created by the cutting edge is kept filled with a thixotropic material such as bentonite or one of a range of synthetic products currently available to support the excavated ground and to minimise friction. On completion of sinking, this material is replaced by the injection of cementitious grout in one operation to lock the caisson into position and to replace the lubricant with solid material to minimise settlement. During sinking, a constant check must be kept on the verticality and square of the shaft, and corrections made on the jacks to keep it within tolerance.

Once a caisson becomes badly out of alignment the consequences can be severe, including getting it stuck and/or segment damage. In this regard, careful attention should be paid to the lubrication process, particularly where there is a risk of ground coming onto the caisson. In addition, a careful analysis of the ground conditions should include a determination of the likelihood of large obstructions such as boulders blocking the cutting edge. There is far less danger of caissons becoming stuck in fine-grained homogeneous soils than in, say, gravels or boulder clay, where alternative methods might be more appropriate.

Caisson excavation can be carried out 'dry' or 'wet' depending on ground conditions. If the ground is naturally stable or has been stabilised by, for example, dewatering (see Chapter 6), excavation can be carried out from the surface or from within the shaft using the excavation plant described for underpinning above. If the conditions are unstable and/or waterlogged, or where the hydrostatic conditions could cause the base of the excavation to 'blow', excavation must be carried out with the shaft flooded to the prevailing hydrostatic level. In these circumstances the excavation plant normally used is either an excavator-mounted pole grab, where special telescopic models can reach depths of around 20 m, or a rope-operated digging grab mounted on a crawler crane. With the latter, the digging ability is governed by the hardness of the material and the submerged weight of the grab. In hard material it is possible to add weights to the grab; if this is not successful, measures such as pre-augering or the use of chisels suspended from the shaft crane must be considered.

It is becoming more common to use caisson sinking, even in stable ground, because the method eliminates the need for the trimming process required when underpinning. The method also minimises the need for personnel to be in the shaft, as the ring building takes place at the surface.

The same plant is used for mixing and pumping the lubricant as for the final grouting, and is described in Section 15.3.1 on underpinning.

With wet caissons it is normally necessary to seal the base with the shaft submerged. This is because dewatering, once the shaft has reached its depth, might cause the base to heave or 'blow' under hydrostatic pressure. Even if dewatering is a possibility, sealing the base in wet ground conditions can be extremely difficult. The depth of the so-called concrete 'plug' must be sufficient to provide enough resistance to the hydrostatic uplift in conjunction with the weight of the shaft rings and the weight of the collar. The latter is normally attached to the shaft, once sunk to its final position, by fixing dowel bars

Figure 15.5 Jacks pushing pre-cast units down, with a clam-shell excavator. (Courtesy of Joseph Gallagher Ltd and Specialist Plant Associates Ltd)

through the top rings into the collar designed to provide the shear resistance required. The concrete plug is placed by tremie methods, almost always using concrete pumps. To provide a key it is usual to install recessed panel rings in the plug location or, alternatively, corbel rings. Segment manufacturers usually supply these as part of their shaft segment range. The plug must be left in place for a minimum of 5 days to cure before dewatering begins. Preparation of the surface can then commence, usually by placing a regulating blinding, to allow construction of the structural base above. Some engineers like to incorporate a dowelled connection between the plug and the structural base in the design to ensure against any possibility of separation along the boundary.

Current practice in the UK for calculating temporary resistance to uplift normally ignores any resistance provided by grouting the annulus or the shear resistance of the ground at the base, and usually assumes a groundwater level at ground level. On top of this, a safety factor of the order of 1.05 is typically applied.

Where the base of the shaft is founded in stable ground or where the base has been rendered temporarily stable by dewatering/pumping or another ground stabilisation process, as an alternative to a deep plug it is possible to provide uplift resistance by under-reaming. The shaft is first stabilised by normal annulus grouting and the cutting edge is usually removed, which in the case of a steel fabricated unit can be reused. The base excavation is then under-reamed, using temporary supports if required, to extend it beyond the shaft footprint. Once the reinforced concrete base has been cast, the passive resistance of the undisturbed ground above the toe is mobilised to counteract uplift.

Circular caissons formed of pre-cast segments can be sunk incorporating an external concrete surround or 'jacket'. As each ring, or possibly pair of rings, are added it is surrounded by in situ concrete using a steel shutter prior to further sinking. This concrete is usually fibre reinforced but can incorporate mesh reinforcement and is typically 250–500 mm thick. This process requires modification of the shaft jacks to give them additional overhang so that they only bear on the pre-cast segments and not the jacket (see Figure 15.5).

The advantages of this system are as follows

- It increases the weight of the shaft and thus can reduce the depth of the anti-flotation plug required, which may be desirable to avoid sinking the shaft into more challenging ground conditions.
- It adds an additional measure to ensure watertightness.
- It adds rigidity to the shaft to counteract any tendency for the shaft to distort.
- It can be used to incorporate vertical injection tubes to the cutting edge area that can be used for jetting during sinking or final grouting operations.
- Most importantly, it enables any portals required at the shaft base for follow-on tunnelling operations to be constructed without the need for internal temporary bracing. The pre-cast linings at the portal positions can be broken out while leaving the intact jacket to take the circumferential hoop loads and maintain the structural integrity and watertightness of the shaft during portal construction. On completion, exit seals can be fitted to the portals prior to the launch of the tunnelling equipment, which, if correctly tooled up, can then cut through the fibre-reinforced concrete lining provided by the jacket during the launch procedures.

There are a number of ground stabilisation processes that can be used to aid shaft sinking and to reduce construction risks; these are discussed in detail in Chapters 6, 8 and 9. It is worth bearing in mind that if the shaft construction involves excavating, moving and disposing of large amounts of saturated material, particularly in urban areas, it may be prudent to consider ground stabilisation on environmental grounds to lessen the impact (see Chapter 7). Likewise, if the shaft is to be sunk using sump pumping to control groundwater, the issues of silt separation and discharge facilities should be seriously considered; very exacting standards are normally demanded from licensing authorities before such discharges can be accepted into surface water disposal systems. If deep well dewatering is being considered, there are issues to be addressed with regard to abstraction and discharge licenses.

The use of such processes needs to be considered and decided upon at the construction planning stage, as installation is more difficult to achieve once construction has started, is likely to be less effective and can be very disruptive and costly.

15.3.3 Pre-cast roof slabs

Most shafts require some form of roof slab for the completed structure. As an alternative to costly in situ construction, often requiring expensive temporary formwork support, most shaft segment manufacturers will provide a pre-cast solution as part of the service they offer. This can also have the benefit of time savings, as the manufacture takes place off site, with the installation itself normally taking 1–2 days. The manufacturer will typically design the slab to the engineer's requirements as part of this service. However, the whole process normally takes around 8–10 weeks, so early planning for this option is advisable.

15.4. Principles of design

The design of a shaft, the method of sinking and the selection of the lining depend on many factors. Shaft projects can vary from small, simple schemes to large and complex systems, and the final use will affect the design.

The most important factor that influences shaft design is the type of ground and whether it is unstable or competent and self-standing when excavated. The presence of groundwater exacerbates the unstable ground conditions and imposes hydrostatic pressures that increase linearly with depth in both unstable and competent strata.

Once ground conditions are known from site investigations, the basic parameters for excavation and muck removal, groundwater control, ground stability control and lining installation can be evaluated and the type of shaft lining chosen. Calculation of active pressures from the ground on to the walls is covered in Chapter 11. These figures are also used to give upward pressures to check the stability of the base in both the temporary and the permanent condition. Note that it is the situation before the base has been installed that is the critical condition, and that particular dangers are created by the hydrostatic forces.

Structural design of pre-cast units follows standard reinforced concrete or fibre-reinforced design as appropriate, and is a service provided by the supplier.

Design of plan-circular structures in the ground is by using hoop compression, suitably factored to make allowance for non-uniformity of ground loading, and allows a very light efficient structure to be used. Openings in hoop compression attract very large concentrated loads around their edges.

It is always prudent to design out any plan shape with intrusions or a non-uniform cross-section, not only for the high stresses at angles but because of the danger of such shapes becoming stuck when being sunk as a caisson.

REFERENCES

BTS (British Tunnelling Society) (2004) *Tunnel Lining Design Guide*. Thomas Telford, London, UK.
BTS/ICE (British Tunnelling Society/Institution of Civil Engineers) (2010) *Specification for Tunnelling*. Thomas Telford, London, UK.

FURTHER READING

BSI (British Standards Institution) (2011) BS 6164:2011. Code of practice for health and safety in tunnelling in the construction industry. BSI, London, UK.

Pallett, Peter F and Filip, Ray
ISBN 978-0-7277-6338-9
https://doi.org/10.1680/twse.63389.217
ICE Publishing: All rights reserved

Chapter 16
Bearing piles

Ray Filip
Temporary Works Consultant and Training Provider, RKF Consult Ltd

Piles are generally relatively long and slender structural members that can be used as an alternative to large pad foundations or large strip footings. They are typically used for the following purposes

- in compression to transmit vertical loads through soils of low bearing capacity and high compressibility to deeper, stronger, less compressible soils or rocks
- in tension to resist uplift forces
- to resist horizontal or inclined loads (vertical or raking piles)
- to resist lateral earth and water pressures in addition to vertical loads (vertical piles) (see Chapter 14 on contiguous and secant piles and Chapter 11 on sheet piling).

Bearing piles can be classified as

- Large displacement piles – solid sections or hollow steel sections with a closed end; they are typically formed of steel tubes, timber or pre-cast reinforced concrete.
- Small displacement piles – piles of small cross-section; they are typically steel H-sections (universal bearing piles), open-ended steel tubes, steel sheet piles or 'screw'-type piles.
- Replacement piles – piles formed by removing soil by boring (e.g. continuous flight auger piling) and then placing concrete into a lined or unlined hole (depending on whether ground requires support); alternatively, the lining may be extracted as the concrete is placed. A reinforcement cage is generally placed into the wet concrete.

There are numerous varieties of piling techniques, machinery and material combinations to suit the various ground conditions, loads to be resisted and site constraints that can be encountered. This chapter can only highlight some, and further advice should be sought from specialist piling contractors.

16.1. Introduction
For temporary works, bearing piles can typically be used

- beneath tower crane bases and large mobile crane outriggers (rather than using very large spreader pads)
- beneath overhead gantry cranes (rather than using large strip footings)
- beneath large site cabin gantries, concrete-batching plants or large storage silos
- beneath heavily loaded falsework or scaffolding
- to support large props during refurbishment work or façade retention schemes
- beneath slide tracks for bridge slides
- to provide reaction when jacking structures or as anchor piles for reaction systems for pile testing
- to resist hydrostatic uplift when constructing basements beneath the water table (often in the permanent condition the weight of the structure will resist flotation)
- sheet piles can be used in the temporary condition for cofferdams or basement excavations and can then be used to provide vertical support to the new structure
- contiguous and secant piles can be used as temporary or permanent retaining walls and temporary vertical loads from cranes and so on can be applied to them – for very deep excavations a diaphragm wall is common
- inclined or vertical steel piles can be used to provide temporary protection (from accidental craft impact) to cofferdams in river
- beneath temporary jetties for access to constructing bridge foundations in a river or for bringing materials or plant onto shore
- to test other piles
- king posts to support long props across an excavation or for providing stability to reinforcement cages (see Chapter 32)
- temporary and permanent steel H-section plunge columns are common for 'top down' construction
- as an alternative to traditional underpinning when a structure is to be undermined by subsequent construction
- to provide support to temporary slabs, gantries and so on to facilitate access for larger plant required to construct the permanent works.

In extreme cases bearing piles can be used for the provision of general site working conditions in very weak or soft ground conditions. Both driven (steel and reinforced concrete) and bored (reinforced concrete) piles are used. Driven piling provides the advantage of instant load-carrying capability, this factor outweighing the usually more acceptable environmental advantages of bored piles. The process of driving piles also effectively load tests the piles, and the bearing capacity of the pile can be determined from the energy required to drive the pile into the ground. While timber and plastic piles can be included here, it is more usual that steel or reinforced concrete piles are adopted (depending on material availability).

The resistance of bearing piles can be determined from

- calculations using bearing capacity theory and design guidance
- full-scale field loading tests
- wave equations (e.g. using the GRLWEAP simulation software)
- pile driving formulae (e.g. the Hiley formula or the Danish formula).

16.2. Types and installation
16.2.1 Use of bearing piles in temporary works
The type of temporary pile selected is often chosen to be the same as the type of permanent works pile in order to save on time and mobilisation costs. However, the position of the temporary piles should be carefully considered to avoid clashes with structures and services and interference with adjacent piles. Noise, vibration, plant access and working space may also become issues.

The type and magnitude of loading on the temporary piles may not be the same as for the permanent piles, and this will need to be considered. A common example is when using piles for a tower crane base (see Chapter 4). When using a cruciform base with piles, the crane supplier will provide sufficient kentledge on the crane such that the main loading on the piles is compressive. However, when using an expendable base, the pile loading can be compressive or tensile, and the piles need to be designed for both load cases. Piles that are also laterally loaded will be designed for significant bending moments, as well as for any compressive and tensile loads that may be applied.

Designers should be aware of site ground conditions, loads to be supported, types of piles and equipment available, preferred pile cut-off levels, site restrictions (noise, vibrations, headroom, settlement limits, etc.) and installation tolerances. The designer should also consider whether the temporary works piles are to be sacrificial, can be incorporated in the permanent works or have to be removed. Reinforced concrete piles can be difficult and expensive to remove, whereas steel bearing piles can be easier to remove if this is necessary.

If installation tolerances (typical tolerances are ± 75 mm in position and $1:75$ in verticality) are important, piling guides or piling gates can be used to improve accuracy.

Designers should be aware of the range of piles and their suitability in certain ground conditions, as well as the limitations of the equipment required for their installation. Some considerations for common pile types are listed below.

- *Bored continuous flight auger* – spoil is removed with minimal ground disturbance, and the technique is relatively quiet, fast and economical, producing piles typically of 300–1200 mm in diameter, up to around 30 m deep and with 20 m reinforcement cage (although not all diameter–length combinations are compatible). The augers can be used for low-headroom work and enlarged pile heads, but capacities are lower. They are suitable for soft ground, where drilling support fluids or casings might otherwise be required. Concrete is placed through the base of the auger string as it is withdrawn and then the cage is installed.
- *Bored rotary* – spoil is removed with minimal ground disturbance and the technique is relatively quiet. It is suitable for most soil types as it offers the highest load-bearing capacity, but it can be relatively slow, with large and heavy machinery being used for installation. Pile diameters are in the range 300–3000 mm, and the base can be under-reamed to give a diameter in excess of 5000 mm. Depths of up to 75 m can be achieved and piles can be raked (up to a maximum angle of around 15°). Chisels can be used for obstructions and can be socketed in rocks. The

technique can be used for some limited-headroom applications. In good cohesive soils open bores may be acceptable, but in collapsing soils it may be necessary to use casings and bentonite suspension or polymers. The reinforcement cage is installed into the open bore and concrete placed (although short cages may be plunged into fresh concrete). This technique is also used for plunge columns and king posts.

- *Driven pre-cast concrete* – pre-cast concrete piles are top driven using a crane or rig-mounted percussive hammers. It is quick to install and is immediately ready to carry load, but capacities are lower. The technique is suitable where a soft soil overlies the harder bearing layer, for granular soils and soft weathered rocks, or where spoil removal is to be avoided. Piles are manufactured in sections off-site and then transported. Pile sizes can be up to 400 mm^2. Pile sections are joined together and progressively driven up to depths in excess of 50 m. This technique can be an economical option where relatively short piles are required. Noise, vibration and ground heave can be issues.
- *Driven steel piles* – H-sections, sheet piles, closed tubes or open tubes are driven with percussive hammers, vibrated (noise and vibration can be problematic) or pressed using 'silent piling' technique. No spoil is removed. The end bearing can be low with H-piles, sheets and open tubes, so these are often used in cohesive soils (where shaft adhesion will be higher). Closed tubes and smaller-diameter open tubes, which may plug (a dense soil plug can form inside the pile), can also rely on end bearing. The technique can also be used in dense granular soils, where hard driving can be expected. The piles have the advantage of being immediately ready to be loaded, can be load tested by the installation process, can work well in compression, tension and lateral loading, and can be relatively easily extracted and reused (if necessary). Tubes in excess of 3000 mm diameter can be installed in lengths of over 30 m, and the length can be increased by welding sections together. Sheet piles can be linked to form box piles (see Chapter 11).
- *Mini piles* – generally less than 300 mm diameter, these can be bottom-driven tubes or reinforced concrete or bored piles. They are often used inside buildings where access is difficult or in limited headroom to carry relatively small loads from props or for underpinning.
- *Bar or strand anchors* – 100–300 mm in diameter and up to 50 m long, these can be used in a wide variety of soils to temporarily carry tensile loads from flotation or to temporarily support retaining walls. When grouped closely together they can act as piles in compression.
- *Specialist grouting* – can be used to improve poor soils so that they can support temporary loads from plant, site traffic or temporary structures such as cabins or silos.

The designer of piles for temporary works purposes should discuss the options available with an experienced and competent specialist piling contractor.

16.2.2 Installation methods

With all piling disciplines, health and safety issues are paramount, as often large and heavy equipment and materials are being used. Many piling rigs have rotating elements from which operatives need to be suitably segregated and the rigs need to move around sites safely to carry out piling operations, so segregation zones are required. Adequate working platforms (see Chapter 5) and segregation between plant and operatives is imperative. Overhead and buried services need to be identified and designers need to be aware of plant limitations (headroom and proximity to buildings). Piles and rigs can be adversely affected by nearby excavations, slopes, ramps, overhead and buried services, and so on.

Modern pile monitoring, testing and instrumentation techniques during installation allow better quality and load-bearing assurance for all types of piling. However, in situ concrete piles will need to be integrity tested to ensure the as-placed concrete quality meets the specification requirements. All piles can be load tested for compressive, tensile and lateral loads to give a better understanding of the behaviour of the pile under loading and to rationalise further design (subject to logistical constraints).

Noise and vibration from large equipment may be an issue when working near sensitive structures, services or railways (noise can equally be an issue in residential areas), and ground heave can be an issue when driving large diameter piles. Relatively light handheld equipment, including adapted breakers and impact moles, are used to install scaffold pole-size steel piles in difficult access conditions; this technique is particularly useful for the provision of access and working platforms and safety fences on embankments and cuttings. Screw piles, installed by rotation, are very useful for lighter loads for similar duties as above.

Driven steel piles can also be removed relatively easily. They have inherent tensile capacity and can be pulled out using a crane-suspended vibrator, hydraulic pressing equipment (pulling rather than pushing) or purpose-made hydraulic pile extractors.

16.3. Design principles
16.3.1 Ground parameters
As with permanent works, adequate site investigation information is a necessary requirement for efficient (and, therefore, economical) pile design. The site investigation should be taken to below the depth of the deepest piles. It is highly desirable that the pile designer has input to the scope of such investigatory work in order to be provided with pertinent information on the ground conditions and thus derive design parameters. There is a variety of methods for determining the design soil parameters with depth. These take into account the amount of data available, its location in relation to pile positions, the degree of uncertainty and variations in the data.

The ultimate compressive load that any pile can carry is the sum of the resistance of the base and the resistance of the shaft (tensile piles provide shaft resistance only and will also need to be anchored to the structure applying the tensile load). Base resistance is the product of the area of the base and the ultimate bearing capacity at the base level, while the shaft resistance is the product of the shaft perimeter and the average resistance per unit area (also known as 'skin friction'). The skin friction is fully mobilised at relatively small displacements (typically around 1% of the shaft diameter, which is around 5–10 mm), while relatively large displacements are required to fully mobilise base resistance (typically around 10% of the base diameter). Factors of safety are applied to the ultimate pile resistance to limit potential settlement; the factors applied may be larger for end bearing than for shaft resistance.

Driven piles and bored piles can be designed by calculation, whereby a penetration depth is usually specified and the depth is then confirmed by full-scale testing on site. However, driven piles can also be designed using empirical driving formulae (e.g. the Hiley formula – see below), whereby a resistance or 'set' may be specified. This is based on laws governing the dynamic impact of bodies. A designer can equate the energy of the piling hammer to the work done in overcoming ground resistance. Allowances are made for losses, friction, elastic contraction and inertia of the pile. The specified 'set' will be the penetration of the pile into the ground per hammer blow. 'Overdriving' (excessive driving) should be avoided as this is likely to damage the head of the pile and possibly the hammer. Hammer manufacturers often quote that practical refusal of the pile is '10 blows per 25 mm penetration'. Perhaps the reason the pile has to be driven very hard is that the hammer is too small or the pile is not stiff enough for the underlying ground conditions.

A re-drive test should be carried out 24 h after initial drive and the new results should be equal to or less than those from the previous test. Figure 16.1 shows a typical pile drive chart. The pile set is given per hammer blow.

The Hiley formulae for resistance is

$$R = \frac{Wh\eta}{S = 0.5c} \qquad (16.1)$$

where R is the ultimate driving resistance (t), W is the weight of ram or moving parts (t), h is the fall height of the ram, η is the blow efficiency (the ratio of the pile weight to the falling weight), S is the set (in./blow) and c is the temporary compression, which is made up of a large number of variables.

Figure 16.1 Impact pile driving

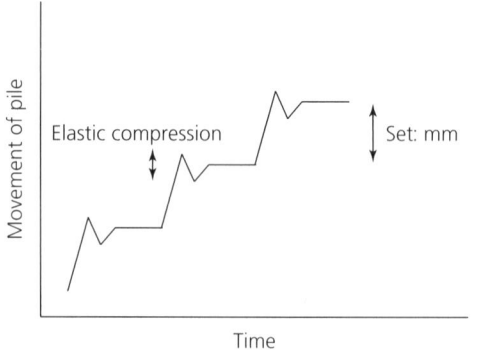

It is interesting to note that the physical characteristics of the soil do not appear in the pile driving formulae, and the formulae do not take into account long-term effects such as re-moulding, group effects and negative skin friction.

The design of bored piles follows fairly well-understood principles that rely on the calculation, or estimation, of the combination of the shaft friction and the end bearing.

In clays the shaft adhesion can be derived from the product of the average undrained shear strength and the pile shaft surface area, the undrained shear strength being empirically factored to realise a match with many years of practical feedback from pile testing. The average shaft adhesion is given by

$$\text{average shaft adhesion} = \alpha C_u \tag{16.2}$$

where C_u is the undrained shear strength and α is a shaft adhesion factor, which can vary from typically 1.5 for soft sensitive clays to 0.2 for very stiff clays. The value of α can also be affected by factors such as material being dragged downwards during installation, the method of installation and the soil around the pile being re-moulded during drilling. Adhesion factors are given by Tomlinson (1994), and typically in London Clay α is taken as 0.45.

The base resistance is derived from the product of a bearing capacity factor (for isolated piles, usually 9), the pile base area and the undrained shear strength at the pile toe.

Undrained shear strength values at depth (for base resistance and shaft adhesion) are often taken from laboratory test results (such as triaxial tests). End bearing can be significantly increased by enlarging the base of the pile by 'under-reaming'.

In granular soils the shaft friction is derived from a factored product of the average effective overburden pressure over the shaft length and the soil's characteristic or average angle of friction between the pile shaft and soil. The factors that are used are essentially based on the method of pile installation; driven piles have larger values than bored piles. Much empirical work has been carried out on rocks (particularly weak rocks), as reviewed, for example, by Tomlinson and Woodward (2008), and specialists should be consulted in order to assist with such design. Chalk is a somewhat special case in that pile design has become accepted as being dependent on the factored effective overburden pressure rather than empirical data from standard penetration tests (Lord *et al.*, 2002, 2003).

Where end bearing can be located on rock, the ArcelorMittal *Piling Handbook* 9th edition (2016) provides some useful capacities.

A group of piles behaves differently to a single pile as the pressure bulb created by a group extends deeper than that for a single pile due to inter-reaction. Group behaviour of temporary piles should be considered the same as for permanent piles. Tomlinson (1994) gives useful guidance on pile groups.

Figure 16.2 Cross-bracing of piles for lateral loading and effective length

16.3.2 Load factors

For temporary works, bearing piles can be designed using lower factors (compared to permanent works) as the piles will be loaded for a relatively short duration. However, it is usual for these lower factors to be justified by on-site testing. In order to produce economical designs, the observational method is also often adopted, whereby trigger action levels are agreed and contingency plans are prepared.

If piles are required to carry lateral loads, raking piles can be used or the piles can be designed to carry a relatively significant bending moment. For structures such as temporary jetties it is also commonplace to cross-brace the structure in order to limit the lateral loading in the piles (Figure 16.2). The durability of temporary works piles is not generally considered to be an issue due to the short-term nature of their use.

16.3.3 Loadings

Lateral loading conditions can arise from impact loading from either berthing vessels or operational impacts in the case of marine works or land-based construction plant. In addition, wind and/or water flow (possibly with floating debris impact or damming loadings) must be included in possible load cases. These loadings are described in greater detail for jetties and load platforms in Chapter 17 and for crane bases in Chapter 4. Tomlinson (1994) gives useful guidance on laterally loaded piles. Guidance can also be found in BS 5975:2019 (BSI, 2019), which is pertinent to all temporary works.

16.3.4 Analytical process

The philosophy behind the analysis is covered in detail in sources such as: Eurocode 7 (BSI, 2004), BS 8004:2015 (BSI, 2015), *ICE Specification for Piling and Embedded Retaining Walls* (ICE, 2016), the Steel Construction Institute's *Steel Bearing Piles* (SCI, 1989) and Tomlinson and Woodward (2008).

16.3.5 Installation tolerances and site constraints

When designing temporary piles and any items relying on the piles, pile designers need to take into account installation tolerances and site constraints. Some common considerations include the following

- Positional accuracy and lack of verticality can lead to eccentric loading and a reduction in available construction space for the permanent works. These can be significant issues when constructing temporary retaining walls for basements or if pre-cast units are to be used in conjunction with the piles. To improve installation accuracy concrete guide walls (see Chapter 14) and steel piling gates can be used.
- Limitations on how close different piles can be installed to existing structures have to be considered both at ground level and at height (building ridge lines and balconies can clash with piling rig masts).
- Piling rigs need to be able to access a site and traverse safely around it. Working platforms will be required (see Chapter 5), and extensive storage areas for items such as fuel storage, auger strings, casings, reinforcement cages may also be needed (unless items are being delivered as and when required).
- Piling operations often require significant attendances. Cranes will be required to lift heavy items, although many piling rigs have winches that can be used for lifting. Excavators will be required for removing spoil, and access for concrete delivery also needs to be considered.
- Noise and vibration needs to be considered in built-up environments or when working near to services or infrastructure such as railways.
- When working inside buildings or under bridges, headroom as well as access becomes an issue. Machinery and techniques are available for working in low headroom, as is equipment that can work in tight spaces.

REFERENCES

ArcelorMittal (2016) *Piling Handbook*, 9th edn. ArcelorMittal, Luxembourg.

BSI (British Standards Institution) (2004) BS EN 1997-1:2004 + A1:2013. Eurocode 7: Geotechnical design. General rules. BSI, London, UK.

BSI (2015) BS 8004:2015. Code of practice for foundations. BSI, London, UK.

BSI (2019) BS 5975:2019. Code of practice for temporary works procedures and the permissible stress design of falsework. BSI, London, UK.

ICE (2016) *ICE Specification for Piling and Embedded Retaining Walls*, 3rd edn. ICE Publishing, London, UK.

Lord JA, Clayton CRI and Mortimore RN (2002) *Engineering in Chalk*. Construction Industry Research & Information Association (CIRIA), London, UK, C574.

Lord A, Hayward T and Clayton CRI (2003) *Shaft Friction of CFA Piles in Chalk*. CIRIA, London, UK, PR86.

SCI (Steel Construction Institute) (1989) *Steel Bearing Piles*. SCI, Ascot, UK, P156.

Tomlinson MJ (1994) *Pile Design and Construction Practice*, 4th edn. E & FN Spon, London, UK.

Tomlinson MJ and Woodward J (2008) *Pile Design and Construction Practice*, 5th edn. E & FN Spon, London, UK.

FURTHER READING

AGS (Association of Geotechnical and Geoenvironmental Specialists) (2006) *Guidelines for Good Practice in Site Investigation*. AGS, Bromley, UK.

Atkinson JH (1993) *Introduction to the Mechanics of Soils and Foundations: Through Critical State Soil Mechanics*. McGraw Hill, New York, NY.

BSI (British Standards Institution) (2000) BS EN 206-1:2000. Concrete. Specification, performance, production and conformity. BSI, London, UK.

BSI (2000) BS EN 1536:2000. Execution of special geotechnical work. Bored piles. BSI, London, UK.

BSI (2004) BS EN 1992-1-1:2004 + A1:2014. Eurocode 2: Design of concrete structures. General rules and rules for buildings. BSI, London, UK.

BSI (2005) BS EN 12794:2005. Precast concrete products. Foundation piles. BSI, London, UK.

BSI (2007) BS EN 1993-5:2007. Eurocode 3: Design of steel structures. Piling. BSI, London, UK.

BSI (2007) BS EN 1993-7:2007. Eurocode 7: Geotechnical design. Ground investigation and testing. BSI, London, UK.

BSI (2009) BS 5228-1:2009 + A1:2014. Code of practice for noise and vibration control on construction and open sites. Noise. BSI, London, UK.

BSI (2009) BS 5228-2:2009 + A1:2014. Code of practice for noise and vibration control on construction and open sites. Vibration. BSI, London, UK.

BSI (2015) BS EN 12699:2015. Execution of special geotechnical works. Displacement piles. BSI, London, UK.

Fleming WGK, Randolph M, Weltman A and Elson K (2009) *Piling Engineering*, 3rd edn. Taylor & Francis, London, UK.

FPS (Federation of Piling Specialists) (2006) *Handbook on Pile Testing*. FPS, Bromley, UK.

Gaba A, Hardy S, Doughty L, Powrie W and Selemetas D (2017) *Guidance on Embedded Retaining Wall Design*. CIRIA, London, UK, C760.

Healy PR and Weltman AJ (1980) *Survey of Problems Associated with the Installation of Displacement Piles*. CIRIA, London, UK, PG8.

ICE (Institution of Civil Engineers) (2016) *ICE Specification for Piling and Embedded Retaining Walls*, 3rd edn. ICE Publishing, London, UK.

Jardine R, Chow F, Overy R and Standing J (2005) *ICP Design Methods for Driven Piles in Sands and Clays*. Thomas Telford, London, UK.

Tomlinson MJ (1995) *Foundation Design and Construction*, 6th edn. Longman Scientific, London, UK.

Turner MJ (1997) *Integrity Testing in Piling Practice*. CIRIA, London, UK, R144.

Useful web addresses

Deep Foundations Institute: http://www.dfi.org (accessed 01/08/2018).

Federation of Piling Specialists: https://www.fps.org.uk (accessed 01/08/2018).
Geosolve Ltd – provider of CAD packages: http://www.geosolve.co.uk (accessed 01/08/2018).
Steel Construction Institute – bearing pile guide and other steel guides: http://www.steel-sci.org (accessed 01/08/2018).
Steel Piling Group: https://www.steelpilinggroup.org (accessed 01/08/2018).

Temporary Works, Second edition

Pallett, Peter F and Filip, Ray
ISBN 978-0-7277-6338-9
https://doi.org/10.1680/twse.63389.229
ICE Publishing: All rights reserved

Chapter 17
Jetties and plant platforms

Paul Boddy
Technical Director, Interserve Construction Limited

This chapter explores the options available to a site designer when a jetty or platform is required. It highlights the materials and design approaches possible and the importance of obtaining good information about the expected loadings. The chapter also highlights the importance of good communication between the site team and the designer.

17.1. Introduction
In marine and river temporary works terms, a plant platform is a structure constructed at the edge of a watercourse to allow the loading and offloading of materials, plant and personnel from a vessel. A plant platform would be necessary where the existing bank is not considered strong enough to withstand the plant loadings.

If the river bed profile is such that a barge cannot approach the bank, the plant platform may need to extend out further to where the river is deeper. If this is the case, the structure becomes a jetty (Figure 17.1).

Plant platforms are also used to gain high-level access in dry sites, for example, at street level for craneage and trucks to stand above a building basement under construction. In structural terms, these can equate to jetties but without being subjected to the mooring and environmental loadings as described in Sections 17.5.2.3 and 17.5.3.

Three types of structure could be considered when specifying a plant platform or jetty – solid structures, open jetty structures and floating jetties – as described in the following sections.

17.2. Solid structure
This is a totally enclosed structure which will not allow water flow through it. This might be the most straightforward to design and install, but its size may be governed by restrictions imposed by controlling bodies, such as the Environment Agency in the UK, due to the possibility of it impeding river flow.

17.2.1 Mass fill gravity platform or jetty structure
This is the simplest method and generally consists of mass fill rock (rip rap). This can, however, result in large structures being constructed, as the sides will need to be battered.

229

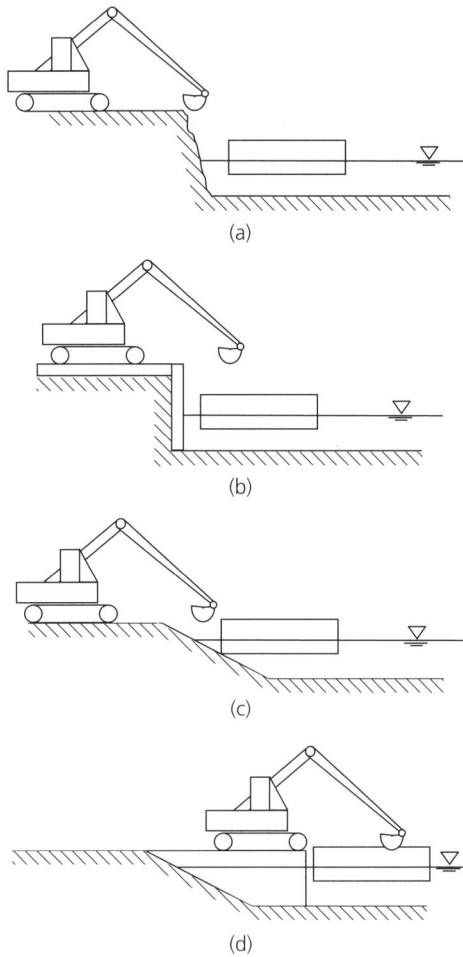

Figure 17.1 Situations for jetty and plant platform: (a) barge can approach the bank, but the bank is weak; (b) plant platform installed; (c) barge cannot approach river bank; (d) jetty installed

Compaction is difficult and the side angles can therefore be relatively shallow. Ensuring that the entire structure is removed when the works are completed can also be difficult.

The stone could be installed by floating plant such as a pontoon and excavator working with a hopper. To ensure the stone structure meshes together, the machine driver would need to choose the stone sizes to ensure they interlocked as far as practically possible. Issues for access would have to be considered, as described in Chapter 18.

The alternative is to build the structure progressively outwards by tipping stone at the riverbank; lorries would then back onto the completed section to place stone for successive sections.

Figure 17.2 Sheet pile platform (a) and jetty (b)

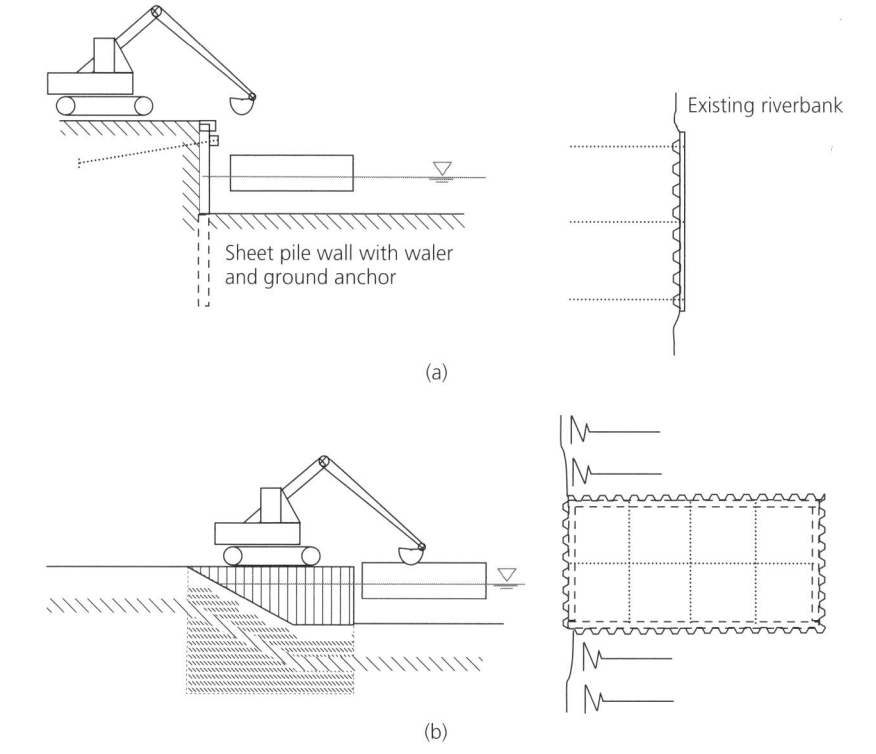

17.2.2 Sheet piled platform or jetty structure

This would be constructed in a similar way to a cofferdam, with either a line of piles installed at the riverside to form a plant platform or a three-sided box being installed for a jetty structure (Figure 17.2). See Chapter 11 for details of sheet piling. To complete the jetty, it would be filled with stone to provide the running surface.

In both cases, ties and walers should be provided to prevent lateral movement. For long sheet piled jetties it is prudent to form compartments with regular sheet pile cross-walls.

17.3. Open jetty structure

For longer structures an open structure may be more appropriate, as this means that river flow underneath is not impeded. This would consist of a series of piers either bearing on the river bed or driven into the bed (Figure 17.3).

The decking design will depend on the use of the jetty and therefore what loading it will need to resist. Steel plate is a durable option but can become slippery when wet; a timber decking (timber sleepers or Ekki mats) may be more appropriate. As the structure is temporary, the main structure will almost certainly be in steel as concrete will require time to cure and be difficult to remove afterwards. Proprietary systems such as steel

Figure 17.3 Open jetty general arrangement

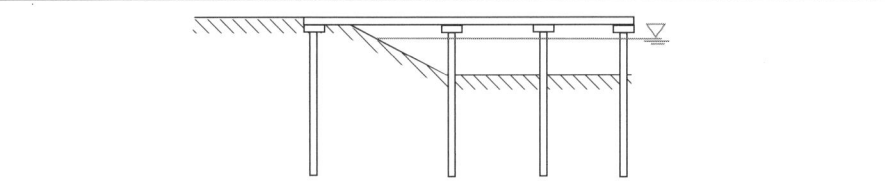

Figure 17.4 Typical span arrangements

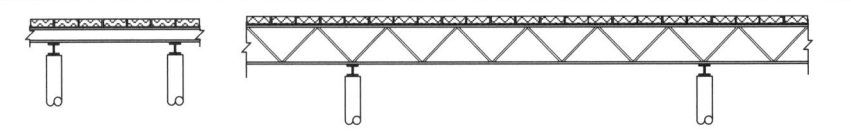

soldiers may also be considered, as they can be hired and returned when the job has been completed.

The span sections will almost certainly be steel and can either be a beam/column section or a truss section. Beam/column sections are more economical for short spans, but for longer spans a truss section may be more feasible (Figure 17.4).

As spans increase the beam/column-section size will increase considerably to resist not only static moments but also deflections due to dynamic loading. Deflections under loading need to be controlled to ensure that the platform does not become unstable.

Lateral stability also becomes an issue at long spans, and bracing will therefore be necessary. For shorter spans, column sections provide better resistance against torsion and impact (although these are structurally less efficient than beam sections).

The pier sections will be chosen for simplicity of installation, and so will be H-sections or circular sections, which can be installed using plant and removed afterwards. To resist lateral forces, raking piles should also be considered, as these require smaller sections than if the vertical piles have to resist all the lateral loads (Figure 17.5). For information on the installation and design of temporary bearing piles, refer to Chapter 16.

17.4. Floating jetties

Floating jetties (Figure 17.6) can be constructed from a variety of materials such as Unifloat pontoons, plastic modular pontoons such as jet floats, and proprietary systems such as floating walkways (more frequently found in marinas). To ensure the jetty is kept in the same location, guide piles can be driven and fastened to the pontoon. Clearly, the fixing should ensure the pontoon can freely rise and fall with the water levels. To assist with access to the bank, the first section can articulate. Refer to Chapter 18 for further details of floating plant.

Figure 17.5 Typical cross-section with raking piles

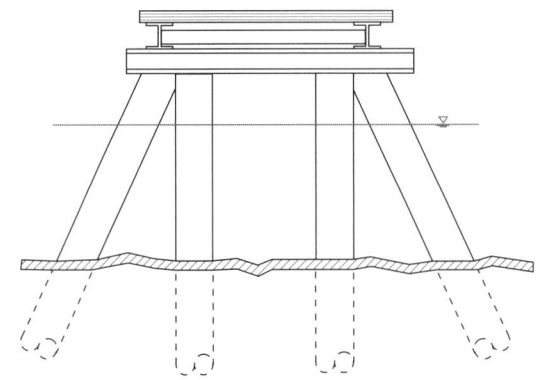

Figure 17.6 Floating jetty with guide piles

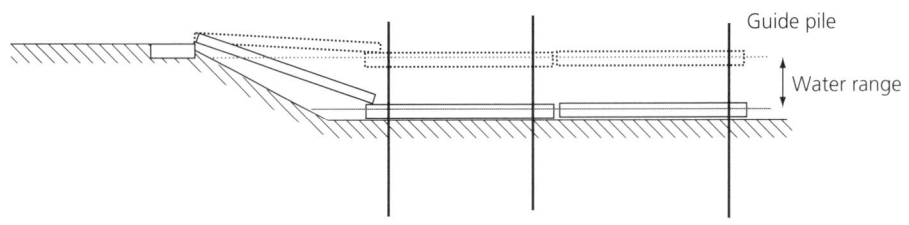

17.4.1 Mooring points

The jetty will be required to hold the vessel in place while it is being loaded and unloaded. Bollards therefore need to be installed on the top to tether the boat or barge. The spacing of these should ensure the barge can be held in place without a pendulum effect occurring (Figure 17.7).

The bollard and connection should be designed to prevent it breaking off under the force from a moving boat. The designer may also wish to consider the design case where the vessel has set sail, forgetting to untether.

17.4.2 Connections

All sections for a working platform or jetty will typically be large, heavy and generally difficult to connect, particularly if being installed by a crane on a pontoon. All connections should therefore be large, simple and with plenty of redundancy should it be found that not all the designed connections can be installed. If welding is to be considered, then

Figure 17.7 Mooring points arrangement to prevent pendulum effect

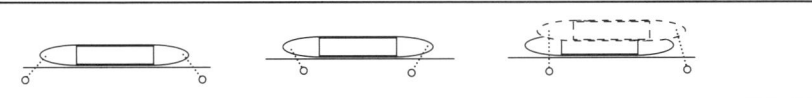

Figure 17.8 Guide lug details

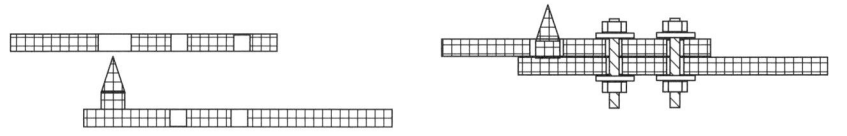

ensure that the specified weld is larger in length and size than that required in order to allow for difficulties in maintaining weld consistency. This is particularly the case if welding is required to be done underwater by divers, who will be relying on touch and feel to undertake their work (assume a throat strength of 110 N/mm^2). It may also be prudent to provide lugs to hold two sections together while the connection is being undertaken (Figure 17.8).

17.4.3 Consents
Any works being undertaken in or around a watercourse in the UK require Temporary Works Consent from the Environment Agency, which may issue conditions on the use and design of a loading platform or jetty. Such conditions will be issued to ensure that any pollution risk is kept to a minimum and that the structure does not impede river flows, particularly in flood conditions.

17.4.4 Interface with site team
As with temporary works, the initial concept may actually begin with the site team who will have a good idea of what is required and (just as important) what materials are available in terms of steel sections.

A prudent Temporary Works Designer will therefore ask for a sketch from the site team with initial sizes based on what is available and where connections are required. The designer can then back-calculate to confirm with the site team whether the sections work and, if not, enquire as to what else is available.

17.4.5 Installation of piles
As with all piling, the critical issue is the initial setting-out to ensure that the pile is located where it is required and within the specified tolerance. This is particularly challenging with raking piles, where the setting-out relates the pile to where it will be on the river bed. A piling gate should be used to fix the location and angle of the pile at deck-connection level, which can be hung over the side of a barge or pontoon. Piling gates are discussed in Chapter 11 and the process in Chapter 16. Refer to Chapter 18 for when piling is to be from floating plant.

17.5. Loadings
When designing a loading platform or jetty, the loads to be imposed on it must be considered. At the beginning of the design process the designer and the site supervisor should agree what plant and materials are to be transported over the structure in order to determine the possible loadings.

The loadings that the structure will be subjected to include

- self-weight
- imposed load from plant
- environmental loads (wind and water)

as described in the following sections.

17.5.1 Self-weight
The self-weight should be considered in the design of the member as for any other structure, but will generally not be the significant part of the loading.

17.5.2 Plant
17.5.2.1 Tracked plant
Tracked plant typically includes excavators, crawler cranes or piling rigs. The loading will be from the tracks. The loading from excavators and crawler cranes clearly varies depending on the working radius and the load being lifted. This will typically induce a trapezoidal load distribution under the tracks. The designer may then wish to convert this load using a rectangular or Meyerhof distribution (Figure 17.9).

For a crawler crane, the pressure diagrams should be available from the crane supplier and will be based on crane type and counterweight, boom length and maximum load at working radius.

For the excavator this information may not be as readily available; some judgement therefore may be necessary to determine the track loads. The designer should consider the three load cases of working over the front, over the side and diagonally. Plant may need to be controlled in its operation to ensure that it does not overload the platform or jetty.

A piling rig will also induce track loads onto the jetty. A Federation of Piling Specialists contractor will provide a pressure diagram similar to that of a crawler crane (e.g. Chapter 5). The designer must review the criteria on which the pressure diagram is based to ensure it is as required, in particular that the dynamic load is factored up.

Figure 17.9 Trapezoidal and Meyerhof rectangular distributions

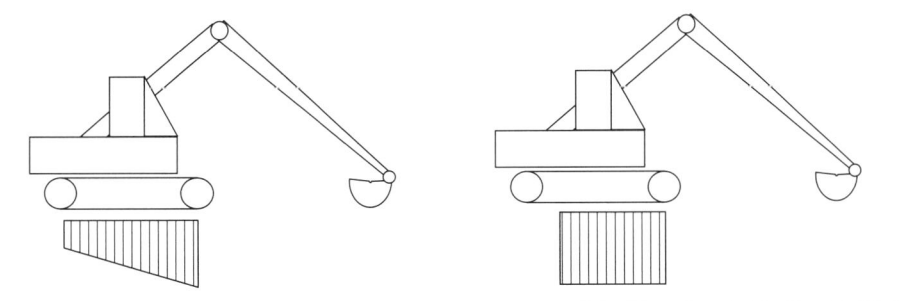

17.5.2.2 Wheeled plant
Axle weights should be confirmed with the supplier or, in the absence of this, a credible reference should be sought. For a solid cofferdam structure, the Construction Industry Research and Information Association (CIRIA) report C580 (Gaba et al., 2003), which gives recommendations on road-going vehicle loadings, may be applicable. In addition, Highways Agency document BD21/01 (HA, 2001) provides axle loadings for critical vehicles expected on UK bridges. Note that dynamic loadings from wheeled plant will comprise a larger percentage of dead load than will that from tracked machines.

17.5.2.3 Craft
The jetty should be designed for a nominal impact loading from craft coming alongside the structure and should allow for some accidental loadings (see BS 6349-1-1:2013, BS 6349-1-3:2012 and BS 6349-1-4:2013 (BSI, 2012, 2013a,b) for assistance). Fendering may need to be fitted to the structure or, if the loadings are too high, it will need to be free-standing.

17.5.3 Environmental loadings
17.5.3.1 Wind loadings
These can be derived from current wind codes such as BS EN 1991-04 (BSI, 2005). The designer may also consider using the BS 5975:2019 (BSI, 2019) temporary works code, which provides further relaxations on loadings due to the time of year. In all wind calculations, the most challenging part is to decide on the pressure coefficient (c_{pe}) value to use for the shape of the jetty where engineering judgement can play a critical role.

17.5.3.2 River flows and wave actions
Generally, these are far more significant than wind loading; river flows can be calculated using basic Bernoulli formulae. The pressure q (N/m^2) for a given flow velocity v is defined

$$q = 500v^2 \tag{17.1}$$

An allowance should be made for the shape of the structure and a friction coefficient c_w value given accordingly to derive the force on the structure. Further information can be found in BS 6349-1-1:2013, BS 6349-1-3:2012 and BS 6349-1-4:2013 (BSI, 2012, 2013a,b), which also provide guidelines on wave actions.

Where flood conditions are possible, consideration should be given to the loading from the damming effect of trapped debris against the structure. If flow is restricted, this loading will be considerable.

17.6. Analysis
The analysis of platforms and jetties can be conducted using traditional methods of analysis; only specific aspects are covered here. Once the forces are known and the materials have been chosen the design is generally straightforward. The designer must, however, consider the factors of safety being applied to the loadings and be satisfied that

an allowance for dynamic loading has been taken into account, particularly with regard to excavators or wheeled plant that may need to brake when on the structure.

REFERENCES

BSI (British Standards Institution) (2005) BS EN 1991-1-4:2005 + A1:2010. Eurocode 1. Actions on structures. General Actions, Wind Actions. BSI, London, UK.

BSI (2012) BS 6349-1-3:2012. Maritime works. General. Code of practice for geotechnical design. BSI, London, UK.

BSI (2013a) BS 6349-1-1:2013. Maritime works. General. Code of practice for planning and design for operations. BSI, London, UK.

BSI (2013b) BS 6349-1-4:2013. Maritime works. General. Code of practice for materials. BSI, London, UK.

BSI (2019) BS 5975:2019. Code of practice for temporary works procedures and the permissible stress design of falsework. BSI, London, UK.

Gaba AR, Simpson B, Powrie W and Beadman DR (2003) *Embedded Retaining Walls – Guidance for Economic Design*. Construction Industry Research and Information Association (CIRIA), London, UK, C580.

HA (Highways Agency) (2001) *The Assessment of Highway Bridges and Structures* (includes correction dated August 2001). HMSO, London, UK, BD21/01.

FURTHER READING

Blake LS (2004) *Civil Engineers Reference Book*, 4th edn. Butterworth-Heinemann, Oxford, UK.

Ehrlich LA (1982) *Breakwaters, Jetties and Groynes: A Design Guide*. Sea Grant Institute, New York, NY, USA.

Elson WK (1984) *Design of Laterally Loaded Piles*. CIRIA, London, UK, R103.

Williams BP and Waite D (1993) *The Design and Construction of Sheet-Piled Cofferdams*. CIRIA, London, UK, SP95.

Useful web addresses

Pontoonworks: http://www.pontoonworks.co.uk (accessed 01/08/2018).

Temporary Works Forum: http://www.twforum.org.uk (accessed 01/08/2018).

Chapter 18
Floating plant

Paul Boddy
Technical Director, Interserve Construction Limited

This chapter explores the range of equipment types available when it is necessary to go afloat. It highlights the importance of good planning, in terms of both using the plant and in ensuring the vessel can actually be used at the location. This chapter gives advice on the necessary calculations and checks that need to be undertaken for a given scenario, and stresses the importance of these checks being carried out by an engineer or naval architect who has both technical competence and practical experience.

18.1. Introduction

There are many situations in the field of construction where conventional plant cannot be used, and to undertake works the contractor may need to go afloat. This would typically be for works being undertaken either at the edge of a watercourse or actually in it.

The need to go afloat occurs for two main reasons as follows

- The works involved are too remote from the edge of a watercourse for plant to physically reach. Examples of this may be the driving of piles for bridge piers in a wide river or undertaking works to an inlet structure in a reservoir.
- The works are located at the edge of a watercourse. For example, sheet piling is required but there is no easy access for conventional plant such as a piling rig. This situation may arise if the topography is too steep to allow plant to approach the bank, or the ground simply cannot support the weight of the proposed plant. The distance travelled to reach the site may also be great, making it cheaper to procure floating plant rather than to provide a haul road.

Other non-engineering reasons why access cannot be gained to a river bank include where there are issues with third parties not permitting access across their land. This can be commonplace with inland waterways, where there is history of conflict between landowners and the waterway. There may also be environmental factors that may prohibit the use of plant in sensitive areas (e.g. sites of special scientific interest).

As with all engineering projects, and in particular marine engineering projects, the access to the site is key, and the planning engineer should ensure that all options are considered both in terms of practicality and cost before the decision about access is made. In many circumstances, gaining access to a site can be more expensive than actually carrying out the works themselves.

Once the decision has been made to go afloat, further logistical issues may arise, as marine plant is generally heavy and will have to be lifted into the watercourse at a convenient location. This may itself require further temporary works, such as constructing jetties or strengthening river banks to accommodate a heavy crane. Alternatively, plant may need to be floated from a remote location. If the latter is viable, then the planner should consider the time required for the plant to be moved from the ingress location to the work site.

In either situation, the planning engineer must also consider possible obstructions such as locks or bridges which would impede the movement of plant, and also whether there is adequate water depth for the plant being considered. Tidal and river flows should also be considered, together with what contingencies are necessary should the river be in spate. The time of year should also be considered.

If the plant needs to be supported from a canal or river bed then other environmental considerations need to be taken into account, such as whether disturbing the river bed will damage habitats. Canal beds are historically lined with puddle clay to prevent water loss, and this could be punctured by plant.

The use of marine plant provides challenges for both the operator and design engineer alike to ensure it is used effectively and safely. The planning engineer should not underestimate the amount of work required not just in the use of floating plant itself but also in its transportation to and from and set-up at the work site.

18.2. Types and uses
18.2.1 Pontoons and barges
These are normally rectangular 'tanks' with flat tops which allow plant such as excavators and crawler cranes to be used safely and to carry out their duties as if they were on dry land (Figure 18.1).

A barge is generally a single structure which is large in dimensions, meaning it will remain in the water and be towed from location to location by a tug boat or similar. In most circumstances, the plant on the barge will remain permanently on board.

A pontoon is typically modular and is designed to be dismantled and moved from one location to another, typically by road, and then craned into the water along with the necessary plant.

Clearly the primary concern for the use of barges or pontoons with plant is to ensure the vessel is large enough to be stable while the given plant is undertaking its specified duties. Each load case should be considered by an appropriately trained design engineer or naval architect to ensure the barge or pontoon remains stable.

Figure 18.1 Typical barge with crane and ancillary plant

It is also essential that the site team understands the limitations of the plant use on the barge or pontoon, and does not stray outside of the boundaries set. Any required changes to site operations should be referred back to the designer to confirm it is safe to proceed.

As well as cranes, excavators can also be used on barges for such activities as dredging or grading of river banks or for positioning of bank protection such as rip rap (large stonework). An excavator working is more dynamic than a crane and the pontoon can experience more roll, which can make working difficult. In these circumstances, the barge may be fitted with 'spud legs', which are generally circular tubes running through the deck of the pontoon. These can be lowered so they embed into the river bed to provide extra stability (Figure 18.2).

Care does need to be taken when using spud legs purely for stability if the watercourse is subject to wave action or the pontoon is operating during an ebb tide. The reason for this is that in either scenario an air gap could appear below the pontoon, meaning its weight is transferred onto the spud legs rather than being supported by the water. If the pontoon has not been designed for this scenario it could either become unstable or individual elements could fail due to the change in load path.

If any of the above scenarios could occur a jack-up barge may be more appropriate (see Section 18.2.3).

Figure 18.2 Modular pontoon with spud stabilisers

Where access is required but a barge cannot reach the location (e.g. the river is not wide enough or there is an obstruction such as a canal or river lock, which can be as narrow as 7 ft) a modular pontoon may be a more appropriate solution. The Unifloat pontoon is the classic example (Hathrell, 1968). It is a steel box approximately 5 m long, 2.5 m wide and 1.2 or 1.8 m deep, and can be connected to others to provide a range of different-sized working platforms depending on the requirements. It was developed for the British Army and had to be able to

- work with a Bailey bridge (another example of army engineering)
- be connected together in the water
- be transported by a 3 t truck.

There are now numerous copies of the original concept; an example is shown in Figure 18.2. These pontoons are used very frequently on inland waters for dredging works, lifting operations and other typical construction activities. Whatever the activity, the ensemble still needs to be checked for stability and the pontoon size adapted if necessary. In addition, the connection details should be considered to ensure that the load can be transferred safely from one unit to another.

The decision of whether to use a barge or a pontoon will depend greatly on the plant size being proposed and the access restrictions at the location where the floating craft is required. The barge shown in Figure 18.1 could only be transported by water because of its size and would also be restricted by any constrictions along the route. It could be used anywhere on the River Thames within London and could be transported to other estuarine locations in the UK and abroad. However, it could not be used upstream of Teddington Lock as the barge width is considerably wider than that of the locks at that location. On the other hand, a modular pontoon could be transported via road to an ingress point and constructed on the water.

Barges tend to be more robust in construction, and therefore are able to accommodate larger payloads and undertake larger marine projects.

Clearly, if considering a pontoon, the planning engineer must ensure that a suitable location is available to launch the pontoon, which may require vehicular and crane access. As already stated, this will require considerable additional temporary works (see Chapter 17). Pontoon sections enable flexibility in the shape of the working platform, allowing non-standard shapes to be developed and thus optimising the size of the pontoon for the required works. This is particularly useful in activities such as inland dredging where the width of the watercourse can be limited.

18.2.2 Lightweight modular systems

For locations where only lightweight access is required and/or there is limited crane access, there are further options that can be considered. One example is Jetfloat, which is a system of modular plastic boxes nominally 500 × 500 mm in plan fitted together with lugs and plastic fixings. Originally designed as temporary floating platforms for marinas, they are now being used in far more innovative applications as access platforms for bridge and lock inspections, and have even been used as floating bridges to transport cars and tanks. Other similar systems are available, all with the advantage that they can be constructed on land and pushed into the water using a lightweight machine (Figure 18.3).

The perceived weak spot of these pontoon units is the connection lugs. Various independent studies have been undertaken to check the long-term durability of the connections and finite-element analysis has been performed on the stability of the pontoons due to the inherent flexibility of the product.

18.2.3 Jack-up barges

In certain circumstances, the need to provide a fully stable platform in order to carry out works may arise. In this scenario, the use of the jack-up barge may be the most appropriate.

A jack-up barge is, in effect, a barge or pontoon that can be lifted clear of the water. Typically, it will have four spud legs (one in each corner) which embed in the watercourse bed, allowing the pontoon to be raised using jacks (Figure 18.4).

Jack-up barges have been used extensively in offshore works such as wind farms and in the petrochemical industry, but are now also used in tidal rivers and estuaries. They can consist either of a single section or be built up of individual units connected by lugs. This gives the user some flexibility in transporting the barge and the shape of the platform.

Figure 18.3 Modular pontoon used as access platform

Examples of the use of a jack-up barge include the following

- Undertaking ground investigations in tidal conditions where it is necessary for the platform to remain at a constant level. A special unit can be provided for this with a hole (or moon pool) in the centre to allow an investigation rig to be positioned over it.
- To hold the pile gate (or guide) in place during single-pile (circular or H-section) driving, giving more confidence that the pile is being driven in the correct location. This is particularly useful with raking piles.

Figure 18.4 Typical elements of a jack-up barge

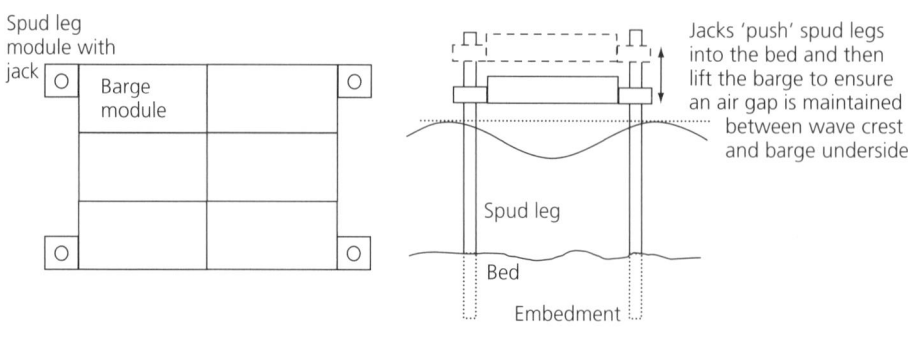

18.2.4 Other

18.2.4.1 Crane barges
These are bespoke vessels designed with an integral crane unit. These vary from the small boats working on inland waterways to the large vessels constructing oil rig platforms.

18.2.4.2 Hopper barges
These are open barges allowing materials such as dredging arisings or aggregates to be transported from one location to another, sometimes with opening or tipping arrangements for self-unloading.

18.3. Design principles
18.3.1 General stability principles
To prove the stability of a barge or pontoon various calculations should be undertaken by the design engineer or naval architect. A number of texts are available, varying from the straightforward, such as Hathrell's *The Bailey and Uniflote Handbook* (1968), to the more complex, such as Tupper's *Introduction to Naval Architecture* (2004).

In basic terms, the stability of any floating plant is a function of its combined centre of gravity and a theoretical height known as its 'metacentric' height. It is easiest to relate distances from the keel. Figure 18.5 depicts the basic principles (see also Webber (1990)).

For the pontoon and its plant to be stable, the theoretical distance KM should be greater than KG (i.e. the metacentric height is higher than the centre of gravity). If not, the craft will rotate until the above is true, which may mean that it capsizes. KB is computed by calculating the centre of buoyancy, which is a function of the water displaced by the weight of the pontoon and the plant thereon. KG is computed by determining the combined centre of gravity for the pontoon and the plant.

Figure 18.5 Key principles for a stability study

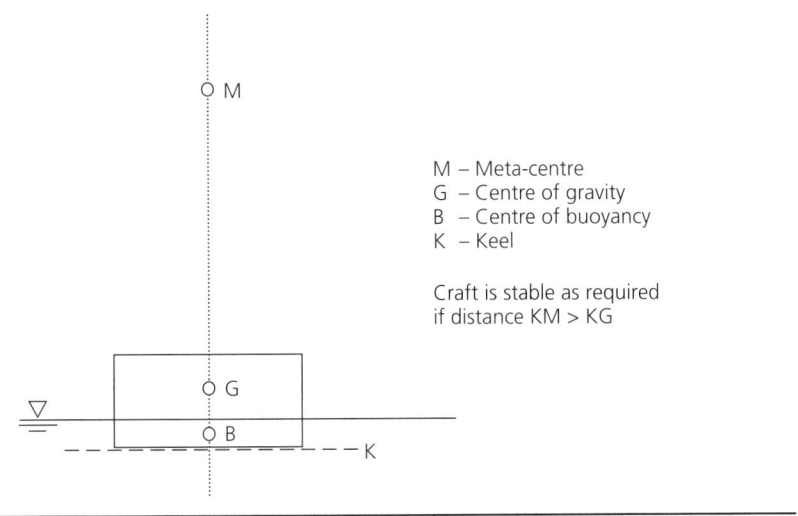

Figure 18.6 Determining BM for a flat pontoon

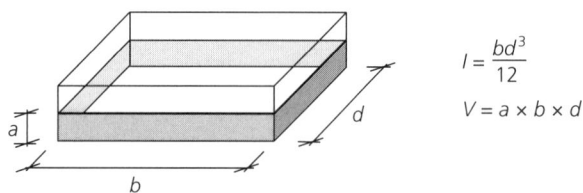

$$I = \frac{bd^3}{12}$$

$$V = a \times b \times d$$

The distance BM is defined as I/V, where I is the second moment of area for the plan of the pontoon and V is the volume displaced by the pontoon. Therefore

KM = KB + BM

For a simple pontoon, see Figure 18.6.

Increasing the size of the pontoon or reducing the volume of water displaced will clearly increase the metacentre height. However, increasing the volume of water displaced will increase KB.

The weight of the pontoon, crane, load and ancillary items depicted in Figure 18.1 are all known, as are their centres of gravity about the pontoon keel; the stability of pontoon can therefore be confirmed. The crane would have been given particular examination and split up into its tracks, body and jib, and the weight and centre of gravity of each item considered separately.

The location of the plant on the deck will also tend to make the pontoon rotate, reducing freeboard on one side and increasing it on the other (Figure 18.7). This should be

Figure 18.7 Effect of eccentric loading on a pontoon or barge

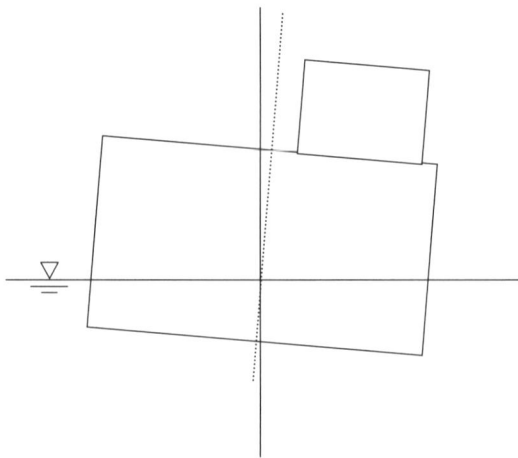

checked because if the roll experienced exceeds 7°, the pontoon can be difficult to work on. In addition, if the freeboard is reduced excessively, there is the possibility water could flood onto the deck.

The crane would also have been closely examined because when it slews around the amount of roll of the pontoon will vary. It may be necessary to down-rate the capacity of the crane while working on the barge. The reason for this is that as the crane lowers its jib to reach out further the pontoon will roll accordingly. This rolling effect will increase the working radius of the crane and could mean it is working outside of its load envelope.

To assist with stability, the barge may have been constructed with a series of internal or baffle walls. These compartments can be individually filled with water to assist with stability by rebalancing the barge should a heavy item of plant be required at one end. The barge itself would need to be designed or checked to ensure it is structurally adequate to support the weight of the plant on the pontoon.

Environmental effects such as wind, water current and tides also need to be considered in the stability of a barge and the plant thereon. The wind load on a lattice jib can be considerable.

The actual types of plant may also give rise to different load patterns. A crawler crane generally operates slowly, both when lifting and slewing, and the loads that are being dealt with are generally fully understood. On the other hand, an excavator undertaking dredging duties will be operating faster as it first excavates the river bed and then drops the arisings into a hopper. It will therefore exert a larger dynamic load on the barge, and this must be taken into account by the engineer checking the pontoon and barge.

A situation where the machine needs to dig into a stiff ground, so that the machine is pulling harder than expected, may occur. The engineer needs to consider the likely conditions the driver will be operating within and ensure that additional loading criteria are checked. This can increase the expected loads on the pontoon by 25–100% of those calculated in theory. Good judgement and experience is just as important as being able to 'crunch' numbers.

18.3.2 Design of jack-up barges

The analysis of a jack-up barge can be quite involved, as many different criteria need to be satisfied. The engineer needs to be proficient in stability, structural engineering, dynamics and geotechnical engineering. These disciplines are covered in depth elsewhere in this book.

The barge will generally be towed to the works location and may for some of its duties act as a floating platform. It therefore needs to be checked for stability in the same manner as a floating pontoon or barge.

When it is in its elevated position, the barge must also be analysed to ensure that it is not overloaded due to the loaded plant. A structural analysis is therefore needed to ensure that the structure can transfer the plant loads onto the spud legs. In a barge this will ensure that the structure is adequate in bending and in shear. In a modular system the connecting lugs between units will also need to be reviewed for shear and tensile capacity. The lugs are a critical item and should be inspected visually on a routine basis; there have been several examples of lugs failing under load.

The loads in the spud-leg jacks should be compared against the jack capacity. The capacity of the leg itself will depend on the length of leg below the barge. The analysis can, in its simplest terms, consider the leg as a pinned column, in which case the capacity is determined by the leg's radius of gyration and the effective length.

However, the critical item with a jack-up barge is the foundation for the spud legs. The legs would typically need to embed in the river bed sufficiently far that when the barge is lifted no further settlement occurs. When planning a jack-up barge operation, topographic and geotechnical investigations should ideally be undertaken. This is not always possible, however, particularly if the jack-up barge is being used to obtain the geotechnical information in question. A successful jack-up relies as much on the skill of the barge master as it does on design and analysis. The barge master will monitor for settlement during barge deployment and will undertake load tests to be satisfied that the set-up is stable. Topography is important to ensure the barge is not being deployed on a slope.

Once jacked up the spud legs will be subjected to forces from the current and wave action. Care has to be taken to ensure the barge is positioned higher than the top of the waves so that it is not lifted and dropped by the resonating wave action. The designer needs to identify the water flows in the location and the expected high-water level. In an estuary or further out at sea the barge legs will be further subjected to wave action. In this situation the designer must identify the critical wave height about the maximum high-water level.

These heights should be related to chart datum, allowing the designer to specify the height of the barge above the bed level so that, in normal working conditions, the water level does not approach the underside of the barge. In an extreme weather event the barge will be jacked up as high as possible so that it is clear of high wave action.

The other main concern for a jack-up barge is when lifting the spud legs so that the barge can be moved. Considerable suction may need to be overcome by the embedded length of tube, and the skill of the barge master is critical for the safe extraction of the spud legs.

The use of a jack-up barge involves considerable risk and the International Jack-up Barge Operators Association has been formed to share best practice. The Health and Safety Executive has produced many documents on the subject, and a best-practice guide, Guidelines for Site-specific Assessment of Mobile Jack-up Units, has also been published by the Society of Naval Architects and Marine Engineers (SNAME) (Bennett *et al.*, 1994).

REFERENCES

Bennett WT, Hoyle MJR and Jones DE (1994) *Guidelines for Site Specific Assessment of Mobile Jack-Up Units.* Society of Naval Architects and Marine Engineers (SNAME), Jersey City, NJ, USA.

Hathrell JAE (1968) *The Bailey and Uniflote Handbook*, 3rd edn. Acrow Press, London, UK.

Tupper EC (2004) *Introduction to Naval Architecture*, 4th edn. Butterworth-Heinemann, Oxford, UK.

Webber NB (1990) *Fluid Mechanics for Civil Engineers*, SI edition. Chapman and Hall, London, UK.

FURTHER READING

Blake LS (1994) *Civil Engineering Reference Book*, 4th edn. Butterworth-Heinemann, Oxford, UK.

Useful web addresses

International Jack-up Barge Operators Association: http://ijuboa.com (accessed 01/08/2018).

Jetfloat: http://jetfloat.com (accessed 01/08/2018).

Pontoonworks: http://www.pontoonworks.co.uk (accessed 01/08/2018).

Society of Naval Architects and Marine Engineers: http://www.sname.org (accessed 01/08/2018).

Temporary Works, Second edition

Pallett, Peter F and Filip, Ray
ISBN 978-0-7277-6338-9
https://doi.org/10.1680/twse.63389.251
ICE Publishing: All rights reserved

Chapter 19
Temporary bridging

Bernard Ingham
Director, BDI Engineering Support Ltd

Temporary bridges combine the performance of a permanent structure with the ability to reuse most or all of the constituent parts. This has set a challenge for military and bridging suppliers to develop solutions that are easily transported, installed and removed but which offer features associated with permanent structures for relatively short periods. A great deal of knowledge has been built up in bridging systems and working methods over many years. The key to success with any temporary bridge scheme is to find the specialist skilled and experienced personnel who can make the best use of this specialist bridging knowledge.

19.1. Introduction

Temporary bridges may be required to carry pedestrians, services, public highways, site access roads, site haul roads, special loads, railways or military vehicles. The obstacles to be crossed can include rivers, canals, footpaths, public roads, railways, site roads, services and construction areas. Bridge spans may also be required to form link spans to jetties or ramps, thus eliminating the use of temporary fill. They can also be used to provide a low-level removable canal crossing, for example, to move construction plant.

The bridge may be required for just a few hours, for example, to carry a special load over a weak bridge or canal, or for several years. Lead times from initial concept to installation can also vary greatly from several months for larger schemes to just 1–2 days for emergency applications such as following flood damage.

Whatever bridge is required, its foundations will also be temporary works and require design. The interface responsibilities between the procurer, temporary bridge supplier and the constructor of the foundations need to be established at an early stage (see Chapter 2 on management). The Temporary Works Coordinator (TWC) has an important role in temporary bridging.

Many different temporary bridge systems and installation solutions are available, from launching the bridge from one side to using heavy craneage to lift the bridge into position. The military solutions have led the development of many of the modern bridge systems.

Detailed design of temporary bridging systems and individual scheme designs can be influenced by the following factors

- the need for easy and rapid transportation to site, installation and removal
- the availability and versatility of components to maximise utilisation
- the need to take economic advantage of the relatively short-term design life
- the use of suppliers' technical knowledge due to lack of Eurocode or UK design codes or standards specifically written for the design of temporary bridges
- the ground conditions for the construction of temporary foundations
- the suppliers' data available from testing of full-scale bridges and their components.

Although the following sections include information about procedures adopted in the UK, many of the principles apply to other sites throughout the world. Information on the temporary works involved in the installation of permanent bridges, for example, 'push-launch' structures, is given in Chapter 27 on bridge installation techniques.

19.2. Temporary bridge types
19.2.1 Historical: the Bailey bridge
Temporary crossings have been used in different forms for hundreds of years. However, the most significant development was that of the Bailey bridge, developed during the Second World War and assembled from distinctive individual panels. The construction industry generic term 'Bailey bridge' is often used to describe all types of temporary bridge.

Named after the civil servant Donald Bailey who designed the radical new steel bridge system, the Bailey bridge was adopted as the standard military bridge in 1941. It was revolutionary in that its light but strong and versatile system could be manually erected without craneage, and proved to be one of the greatest inventions of the War. It played a significant part in the allied forces' victory; by 1947 it had been used to build more than 1500 bridges in north-west Europe.

Other temporary bridge systems have subsequently been developed for both military and civilian use. Figure 19.1 illustrates a modern military bridge, which incorporates many of the principles of Donald Bailey's 1940s design.

19.2.2 Proprietary bridging systems
The many different bridging systems available today range from derivatives of the Bailey bridge to modern all-welded deck and beam systems. Advice and user guidance from specialist suppliers should always be sought. The most common systems are described in the following subsections.

19.2.2.1 Panel bridges
Modern systems copy the original structural benefits of the double diamond truss Bailey panel layout, which provides a combination of strength and lightness yet retains the ability to resist high local roller forces during launching. The manual handling

Temporary bridging

Figure 19.1 Modern day military logistical support bridge. (Courtesy of Mabey Hire Ltd)

limitations in construction mean that most bridges in use in the UK will require craneage for assembly. This has led to the use of larger panels.

A feature of the panel bridge is the use of high-strength pins to connect the panels together. The pins are in shear, and the design of the pin housing has always been a critical area for stress concentrations and welding.

Major improvements have, however, been made in the following areas

- use of higher grades of steel and increases in the size of truss panels
- improvements in welding and detailing for fatigue
- development of numerous deck systems, ranging from footbridge decks to heavy systems designed for up to three lanes of public highway and for heavy construction plant
- parapet and anti-skid surfacing systems.

Current panel bridges such as those produced by Mabey and Leada Acrow remain the most widely used (see Figure 19.2 for a typical example). They have the advantages of

- a large span range, typically 6–80 m
- high efficiency due to the truss proportions and versatility
- are easily transported on standard vehicles on the road and/or in standard containers
- the option available to launch, eliminating the need for large cranes, although craneage is generally required for assembly.

Figure 19.2 Panel bridge with reinforced concrete pads on temporary reinforced fill abutments. (Courtesy of Mabey Hire Ltd)

19.2.2.2 Other truss bridges
Development of other truss bridge systems has taken place for longer spans and semi-permanent situations where the panel bridge trusses become less efficient. Examples are Mabey's Delta Bridge and Unit Construction Bridges, which are designed for spans in the range 50–130 m. Certain types use bolted connections in line with the trusses. This can allow pre-cambering of the structure to cater for deflections during use, a significant benefit when temporary bridge equipment is used as falsework (see Chapter 22 on falsework). Other lightweight truss bridge systems exist for footbridge applications.

19.2.2.3 Plate girder bolted bridges
Advances in girder manufacturing techniques and increases in the sizes of mobile cranes have encouraged the use of plate girders for temporary bridges. Such systems are less versatile than the truss systems and are generally used for relatively long-term use. They are generally site-specifically designed, thus allowing the inclusion of features such as special deck widths and skewed supports.

19.2.2.4 Prefabricated deck sections
The demand for rapid installation has led to the development of all-welded deck systems that can be installed simply and rapidly, such as Mabey's Quickbridge. Multiple units are used side by side to form the deck with a parapet pre-fitted to the outer units prior to delivery. The main longitudinal members are incorporated beneath the deck plate. The span capability (typically 6–20 m) is considerably less than that of the panel bridge systems, but installation speeds and costs are generally more favourable. Figure 19.3 shows a typical application.

Figure 19.3 Deck-type bridge with trestle piers and 3 m high concrete lower piers for impact resistance. (Courtesy of Mabey Hire Ltd)

19.2.3 Special designs

There are a few occasions when a special-purpose design may be the best solution. Examples include

- railway bridges (although temporary systems have been adapted for rail use)
- sites where dimensional constraints preclude the use of any standard system
- simple short-span footbridges where the use of scaffold systems, unit beams and so on is acceptable.

19.2.4 Foundations types and intermediate supports

The foundations for temporary bridging are loaded differently to those of other temporary works, such as falsework or scaffolding. There will be substantial lateral loads from vehicle or plant braking loads, in addition to axle loading. Designs of these foundations are specialised, and guidance should be sought from the suppliers of temporary bridges. Guidance on temporary foundation types is given in other chapters in this book. Many different types of foundation are adopted for supporting temporary bridges, ranging from simple timber mats placed on granular fill to concrete spread footings and various piled solutions. Clause 18.9 of BS 5975:2019 (BSI, 2019) states allowable bearing pressures on compacted fill for use in temporary works. Figure 19.2 shows a reinforced bank seat with end walls supported on temporary fill with gabion walls. Figure 19.4 depicts a typical example of a bank seat arrangement for a temporary site haul road bridge.

Key factors in the choice of type of foundations are economy, speed of installation and method of removal or making good following dismantling of the bridge. Cost and time

Figure 19.4 Typical bank seat detail for temporary panel bridge. (Courtesy of Mabey Hire Ltd)

savings can be made by not adopting specifications of permanent structures, for example

- minimising concrete reinforcement (e.g. anti-crack steel for durability may not be required)
- allowing greater settlement than for a permanent design
- considering reductions in the vertical and horizontal specified loading to the bridge deck (e.g. some load values such as highway longitudinal loads that may be deemed inappropriate for temporary site construction applications)
- the use of pre-cast concrete and timber foundation elements
- the bridge supplier can provide intermediate piers using a proprietary braced trestle system.

19.3. The design process

Considerations of management, safety and other legislation in relation to the provision of temporary works are dealt with in Chapters 1 and 2. BS 5975:2019 (BSI, 2019) sets out the procedures that are to be followed. Some key design and checking aspects specific to temporary bridge design are outlined below.

19.3.1 Programme

Establishing an outline programme from the outset is a vital part of the design process. The requirement for bridging will have been identified in the temporary works register. The likely start date and period of use of the bridge will influence the choice of system due to considerations of economy, performance and availability. Sufficient time should be given to checking the design.

19.3.2 Loading

19.3.2.1 Public use

For public highways and footpaths the relevant highway authority will specify the imposed loading, including any guardrail and/or barrier loading. The authority is most likely to adopt a load rating from a standard specification such as BS EN 1991-2:2003 (BSI, 2003). Where side protection is specified, the solidity of any such barriers can impart significant wind forces onto the bridge.

19.3.2.2 Site bridges

Loading for site access routes will be specified to the bridge procurer in a design brief, usually prepared by the principal contractor after temporary works consultations. The brief can be carried out in a number of ways, and the following aspects should be considered

- *Single-vehicle loading*: adopting one vehicle per span is accepted practice as long as measures are established to control vehicle movements and allow for vehicle breakdown (e.g. towing off by a second vehicle). Emergency vehicles should be considered where lighter vehicles are specified.
- *Bridge assessment standards*: for example, BD 21/01 (Highways Agency, 2001) is a useful document for specifying loading for multiple vehicles where the maximum gross vehicle weight is under 44 t.
- *Vehicle numbers for fatigue*: the approximate total number of vehicle passes should be established.
- *Edge protection requirements*: depending on use of the bridge, guardrails and/or vehicle edge protection barriers may be required, and these can impart high lateral forces.
- *Pedestrian loading for non-public applications*: this can be specified in a number of ways providing measures are established to control the loading on site. It is recommended that service class 1 (0.75 kN/m^2) be the minimum, although each member of any platform should be designed for a minimum of service class 2 loading (1.50 kN/m^2).

19.3.3 Cross-section and span considerations

Finding the optimum width is important for both economy and safety, as excessive width can encourage higher vehicle speeds and possible passing traffic with risks of collision, and incur unnecessary cost.

With regard to spans, the minimum clearance envelopes for vehicles, pedestrians, railways, water-borne traffic, flooding and any other obstacles should be established with the relevant authorities and included in the temporary works design brief. Consider areas that may be required for construction, as this can affect the choice of span arrangement. Establishing the dismantling method is important, as the site layout will often change significantly while the bridge is in use and the method of dismantling may not be the reverse of the assembly procedure.

The TWC should ensure that the deck and foundation design are compatible in order to find the optimum arrangement. Choice of articulation will be a part of this process. An

Figure 19.5 Severe site constraints determined the bridge type and erection method on this site in Scotland. (Courtesy of Mabey Hire Ltd)

example of severe constraints in the design of foundation and bridge is shown in Figure 19.5.

19.3.4 Detailed design and checking

Having established the construction type and layout, design output is generally as follows

- Confirmation of the design brief and submission of any client technical approval documentation (e.g. Form F001 on railway works).
- General arrangement drawings, including foundations.
- Calculations from (a) suppliers and (b) foundation designers.
- Drawings and calculations for any special scheme-specific components.
- Design risk assessments. Particular considerations for temporary bridges are: control of loading on the deck; clearances below for, for example, vehicle impact or flooding; temporary loads and stability during installation and dismantling; ground conditions for installation and dismantling; and site security (bridges are often outside the scope of regular site control measures).
- Design-check certificates.

Note that, although the bridge supplier may issue a design-check certificate to its own check category, there should be a separate design-check certificate for the temporary works as a whole, including the foundations (see Chapter 2 on management).

19.3.5 Design standards and strength data
The specific requirements for temporary bridging are not the same as those required for permanent bridges. However, established bridging manufacturers have the advantages of knowledge gained from experience in developing, testing and using their systems. This is incorporated in published strength tables with specified factors of safety for temporary bridging.

19.3.6 Client technical approval
Typical clients could be Network Rail, Highways England, a local authority or British Nuclear Fuels Ltd, and so procedures to obtain technical approval from clients for temporary bridges can vary greatly. Successfully navigating the more complex client procedures can be confusing, especially as many were developed for permanent works. The keys to success are: (*a*) to identify the procedure to be followed at an early stage, and allocate engineers with suitable specialist experience and qualifications from each organisation involved; and (*b*) to plan the approval process, including the identification of key stages and dates.

19.3.7 Other design considerations
Details that the designer may need to consider include the following

- *Deck surface*: the use of anti-skid surfacing with high levels of skid resistance and durability. Such surfaces can be factory applied to temporary bridge steel deck.
- *Deck drainage*: most temporary bridge decks are formed of a series of relatively lightweight deck units that flex independently under load. Such units are generally designed for water to pass through the joints, where sealing is not a practical option.
- *Parapet*: various proprietary parapet systems are available from the bridge suppliers.
- *Impact protection*: consideration must be given to the risk of vehicle, train or vessel impact on temporary bridge soffits and supports. Common solutions include increasing clearances outside the zone of risk, barrier systems, earth bunds, sacrificial goalposts or beams, and concrete piers (see Figures 19.2 and 19.3).

19.4. Transportation and construction
19.4.1 Transportation
Bridging systems have generally been developed with transportation in mind. Panel bridges and other truss systems can all be transported on standard road vehicles. The main factor in the development of all-welded deck systems is to minimise components and improve the speed of installation; they are therefore manufactured and delivered to the site in fixed span lengths of 20 m or more. The longer spans will often require specialised transportation, but the cost of transportation is offset by the significantly faster erection rates.

19.4.2 Range of installation methods available
The constraints of space, time, foundations, access, safety, environment and cost result in a broad range of installation techniques being adopted. The most common methods

are described below

- *Assembly in place*: this simply involves assembling the span(s) in place, for example, where the bridge is being built on top of an existing bridge or on the ground where excavation beneath will follow installation. Site craneage will be required.
- *Lift into place*: the development and availability of large mobile cranes has permitted the lifting of ever larger and heavier spans. Safety and speed are the main benefits of lifting. Particular attention is required for the design and installation of the temporary works of the crane pad foundations to cater for the outrigger loads.
- *Traditional advancing launch*: this is often the only practicable method for medium- and long-span applications. Standard bridging is used to form a temporary lightweight cantilevered nose connected to the front of the structure. The completed deck is then easily moved forward and controlled using site plant, winches or tirfors. As the bridge is advanced on ground-mounted rollers, the position of the centre of gravity of the bridge must be monitored carefully (particularly where the bridge is advanced and sections of completed bridge are added to the rear end). As the launching nose advances, the tip will deflect and suitable allowances be made. The physical act of rolling a bridge will cause the top and bottom booms to change from tension to compression, and vice versa; in such cases it is even more important that the bridge supplier's instructions are followed.
- *Other launch methods*: for example, the tail launch involves the use of additional bridging at the rear of the structure. Other installation methods include floating on pontoons and cantilever build, as illustrated in Figure 19.6.

Figure 19.6 Side spans of a panel bridge being installed using the cantilever method. (Courtesy of Mabey Hire Ltd)

19.4.3 Site planning and execution

19.4.3.1 Site constraints
It is important to establish access and site constraints at the start of the design process, as the choice of bridge system and span arrangement is very often determined by these details. These should be included in the design brief. Timely site visits and making good use of the bridge supplier's specialist knowledge is the key to success.

19.4.3.2 Construction programme
Time constraints, such as for rail and road closures, will often be the main factor determining the installation and removal methods. Lifting in a pre-assembled deck is favoured if it means that road or rail closure times will be short.

19.4.3.3 Site safety and planning
Careful planning must be carried out and documented. The responsibilities should be established at an early part of the design process. The designer, as defined in the Construction (Design and Management) Regulations 2015, and the site TWC have important roles in controlling the process (see Chapter 2 on management). The overall design must be reviewed to ensure that the structural stability of the temporary bridge is maintained at all phases of construction, use and, often forgotten, during dismantling.

The documentation should include

- a schedule of site facilities, plant and equipment, together with clarification of individual responsibilities
- the construction sequence and drawings, including sufficient detail to ensure health and safety, highlighting any identified design risks
- a risk assessment specific to the scheme, together with measures as required to reduce the risks to acceptable levels (e.g. minimising working at height)
- a detailed and approved method statement and operating manuals for the equipment.

19.4.3.4 Site operations
Although bridging systems are designed for ease and speed of installation there are still many risks involved. Experienced and skilled supervisors and (where appropriate) steel erectors familiar with bridging systems should be used. Good communications between all parties is vital. The use of suppliers' training DVDs and possible site-based workshops are important to familiarise operatives with the specialist equipment in use.

19.4.3.5 Inspections
Inspection requirements should be established before construction commences, and will include short- and long-term inspection regimes. Many factors, such as the bridge type and intensity of loading, will influence the frequency and extent of inspections. Conventional means of access methods are usually employed, such as the use of mobile elevated working platforms. Although temporary bridging will be inspected on a construction site as part of the statutory inspection, the specialised nature of the equipment often requires specific regular checks to be carried out by persons familiar with the equipment who also

have the training to know which part of the structure can be susceptible to fatigue or wear. The advice of the specialist supplier should be sought.

19.5. Other applications of temporary bridging parts

The versatility of many standard bridge parts allows alternative uses in temporary works applications, including the following.

- Panel bridge truss elements used for spanning falsework girders. As the elements are joined by pins in shear, there is no control over the deflection of the assembly, and allowance has to be made either by tapered firring pieces and/or a falsework skeletal system seated on the girders with facilities for height adjustment (see BS 5975:2019, clauses 19.5 and 19.7.1).
- Truss elements in the vertical plane to form supports for bridges and other structures.
- Bridge decking elements used with panel or trestle systems to form site working platforms.
- Deck elements placed on compacted fill and used to spread the applied load and reduce the applied bearing capacity required on the compacted fill.

REFERENCES

BSI (British Standards Institution) (2003) BS EN 1991-2:2003. Eurocode 1. Actions on structures. Traffic loads on bridges. BSI, London, UK.

BSI (2019) BS 5975:2019. Code of practice for temporary works procedures and the permissible stress design of falsework. BSI, London, UK.

Highways Agency (2001) *The Assessment of Highway Bridges and Structures*. HMSO, London, UK, BD 21/01.

FURTHER READING

Harpur J (1991) *A Bridge to Victory*. HMSO, London, UK.

Joiner JH (2001) *One More River to Cross*. Leo Cooper, Barnsley, UK.

Network Rail (2012) *Engineering Assurance of Building and Civil Engineering Works*. Network Rail, London, UK.

Useful web addresses

Health and Safety Executive (HSE) – books: http://books.hse.gov.uk (accessed 01/08/2018).

Temporary Works, Second edition

Pallett, Peter F and Filip, Ray
ISBN 978-0-7277-6338-9
https://doi.org/10.1680/twse.63389.263
ICE Publishing: All rights reserved

Chapter 20
Heavy moves

Martin Haynes
Sales & Marketing Director, Fagioli Ltd

Andrea Massera
Engineering Director, Fagioli SpA

Edited by: **Nick Cook**
Principal Methods Engineer, Balfour Beatty

This chapter deals with the horizontal and vertical movement of heavy loads. It outlines the reasons for choosing this method of construction and discusses the various methods that are available and used within the industry. Although conventional cranes are mentioned, priority is given to alternative techniques, including trailers, hydraulic jacks and various skidding systems.

20.1. Introduction

First, what constitutes a 'heavy move'? For the purposes of this chapter, a heavy move is something that cannot be installed using conventional site plant. It could be as low as 5 t (e.g. retrofitting a vault into an existing bank basement) or as heavy as 30 000 tonnes (lifting of a complete oil platform). The idea of constructing a component somewhere other than its final location must offer a significant advantage over in situ construction for it to be considered. The advantage is usually seen as one or more of the following.

20.1.1 Programme savings

The heavy move technique allows construction activities to take place concurrently. For example, during the construction of Kylesku Bridge in Scotland, in situ construction of the approach spans together with offsite casting of the centre span, and installation using a combination of heavy move techniques allowed lost programme time to be regained (Figure 20.1).

Occasionally, the programme will dictate that a heavy move technique is necessary. This is commonly seen during replacement of rail and motorway bridges. The new bridge is constructed close to the existing bridge, the old bridge is moved out (by skidding or trailers) and the new bridge installed in a similar manner. The on-site programme, and therefore disruption to the public, is kept to a minimum (see Chapter 27 on bridge installation techniques). In addition, off-site prefabrication permits the erection of industrial

Figure 20.1 Kylesku Bridge, Scotland. (Courtesy of PGS Films)

plants, for example, in locations where climatic conditions are adverse and the available site construction period is restricted to a few months a year.

20.1.2 Improved safety
The main improvements in safety are associated with minimising work at height; construction at ground level is inherently safer than working at height. This not only applies to falling of personnel but to the potential injury due to dropped objects.

20.1.3 Improved quality
As with improved safety the improved quality aspect is largely associated with work at height. Workers at ground level are more likely to take care over their work. Likewise, inspections are easier to perform and are likely to be more thorough.

20.1.4 Cost savings
Construction work is generally won by competitive tender. For the heavy move technique to be considered, it must normally offer a significant overall cost saving over in situ construction. The fact that so many techniques and companies offering these services are available is a good indication that the technique works.

20.2. Techniques
20.2.1 Cranes
Two types of crane are widely available for heavy lifts: these are truck-mounted telescopic boom cranes and crawler-mounted lattice boom cranes. Many other types of

crane are available (e.g. tower cranes) but are not considered because of lack of capacity and/or availability.

20.2.1.1 Truck-mounted telescopic boom cranes

Truck-mounted telescopic boom cranes are available in capacities up to 1200 t. These cranes are quick to mobilise and set up. They normally arrive at site with their boom attached; they are ready to lift after placing the outriggers and extending the boom. They can lift, slew (rotate) and boom up or down, but cannot travel under load. Note that this type of crane can only lift its quoted capacity with the boom at minimum extension and radius; capacities reduce quickly as the boom length and/or radius is increased.

Additional capacity can be achieved by adding ballast trays, which are suspended from the rear of the crane. These are commonly known as 'superlift attachments'. The area required for the lift will be considerably larger with this configuration.

The largest cranes will require service cranes to aid assembly and are delivered on numerous vehicles. Consideration will be needed to allow for sufficient assembly space, including space to lay down the jib during assembly and when the crane is winded off.

20.2.1.2 Crawler-mounted lattice boom cranes

Crawler-mounted lattice boom cranes (Figure 20.2) are available in capacities up to 3000 t. Typically they will arrive in several truckloads and require site assembly. They have higher capacities, however, and, unlike truck cranes, can travel with the load.

Figure 20.2 A 750 t crawler crane. (Courtesy of Fagioli Group)

As with the larger mobile cranes, additional capacity can be gained with a superlift arrangement.

Some of the larger capacity crawlers can be mounted on ringers. These are large-diameter ring beams set up on the crane platform. The crane is then mounted on and revolves around the ring beam. Large stacks of counterweights can be carried. One major disadvantage of this configuration is that the crane cannot normally be moved on its tracks and will need to be dismantled to change position.

20.2.1.3 Specialist cranes
There are now available from some specialist heavy lifting companies a range of cranes with capacities in the range 3200–5000 t. These are usually based on ring beam systems. The capacities of these devices are constantly changing, and it is recommended that the specialist companies are contacted for current specifications and capacities.

20.2.1.4 Crane selection
In general, if a truck-mounted telescopic crane can perform the lift then the choice is simple; such a crane will be quicker and cheaper than a crawler crane.

The selection of which crane to use will take into account

- the weight and centre of gravity location of the lifted item (the hook of the crane must be over the centre of gravity or the load will tip)
- the dimensions of the lifted item
- the delivery and final locations of the lifted item
- the space available for the crane, including that required for assembly and dismantling
- the allowable ground bearing, including the point loads if using a crane with outriggers
- any client/contractor/project/site-specific regulations regarding the use of cranes.

It must be remembered that the safe working duty charts do not include the weight of the hook block or any lifting beams, chains or strops. These must all be added to the load to be lifted before choosing the correct crane.

Once these points have been established, the best practice is to consult the crane hire companies, which will select the most suitable crane and prepare the necessary rigging studies. This will allow the selection of the most suitable crane that is available.

An alternative is to carry out the rigging studies oneself. Starting with the information above, the crane location should be set on plan together with the build/delivery and final locations of the lifted item. The distance between the centre point of the crane slew ring and the centre of gravity of the load is the lift radius. The length of the boom is decided next, and it will depend on the lift height required. Once these physical dimensions have been established, crane duties (a series of charts or tables showing lift capacities with various combinations of boom length and lift radii) must be consulted to see which crane and configuration is most suitable. Cross-referencing the lift radius with boom length will give the safe lifting capacity of the crane.

Operationally, crane lifts are relatively quick compared to alternative methods, with most lifts being completed within an hour. When in operation, however, all cranes have a limiting wind speed; in general, the longer the boom, the lower the operational wind speed. The crane duties will specify the limiting wind speed for the crane, which can be as low as 9 m/s.

20.2.2 Trailers

Self-propelled modular transporters (SPMTs) have the ability to move huge loads over long distances, with minimal ground preparation (Figure 20.3). The item to be moved will generally be built on temporary supports, leaving a clear space 1500 mm high underneath to allow the SPMTs to be inserted. The SPMTs have an integral jacking system to lift the item from the temporary supports. Once the load is on the trailer bed it can be moved to location and, if required, the trailer hydraulics can be used to transfer the load onto the permanent supports.

SPMTs come in two basic module sizes of four axles and six axles, with capacities of around 32 t per axle line, which is in excess of even major road capacities. These modules can be connected together to provide the correct capacity and physical configuration necessary for a particular move. Once interconnected they act as a single unit and are

Figure 20.3 Moving a 1250 t bridge using SPMTs. (Courtesy of Fagioli Group)

operated from a single point. Linking SPMTs together allows massive structures to be moved; loads in excess of 10 000 t are not unusual.

If considering the use of SPMTs, consult the specialist transport companies that offer such equipment. They will need to know the following information

- the item weight and the location of the centre of gravity
- the construction location and support details
- the final location and details of the permanent supports (if the SPMTs will be used for load transfer)
- details of the route, including allowable ground loading, potential obstructions and gradients.

Operationally, SPMTs are quicker than alternative systems (e.g. skidding), with operational speeds of up to 5 km/h; speeds in excess of 500 m/h will, however, result in a reduced load-carrying capacity. There is no wind speed rating for using SPMTs as there is with cranes; however, the wind load on the item needs to be considered and may result in a limiting wind speed for any particular operation.

20.2.3 Jacking systems
20.2.3.1 Cylinder jacks
Due to limited stroke, cylinder jacks (and flat jacks, air bags, etc.) are normally used for load transfer only and not for lifting. However, they are normally required in conjunction with skidding systems (see Section 20.2.4) to transfer load.

In cases where a relatively short lift (e.g. less than 2 m) is required, cylinder jacks can be used to jack and pack the load (Figure 20.4). This technique can be used in conjunction with trailers or hydraulic skid shoes to allow the item to be built closer to ground level.

Most cylinder jacks will operate with a full load at around 700 bar of pressure, so expect bearing loads around 70 N/mm^2 above and below the jack. Flat jacks operate at 150 bar and air bags at even less, so there are generally no bearing pressure problems.

20.2.3.2 Jack and pack systems
There are now several versions of jacking and packing systems. These systems rely on fabricated stools that are mounted in hydraulic frames. Hydraulic jacks lift the stools, allowing another section to be introduced below. The frame is then lowered and is clamped around the lower stool. The process is then repeated. This system can also be used for lowering. The whole process is monitored and controlled by computer systems to maintain load stability.

Jack and pack systems can be incorporated in skidding systems (see Section 20.2.4) or SPMTs, allowing the load to be transferred horizontally.

Capacities range between 500 and 2400 t per tower, and are constantly being improved.

Figure 20.4 Jack and pack (12 × 200 t). (Courtesy of Fagioli Group)

20.2.3.3 Gantry lifting systems

These normally consist of pairs of units each of which contains multiple extension hydraulic cylinders (Figure 20.5). Different suppliers have varying arrangements and extensions of the hydraulic cylinders.

The units normally run along beam systems, which must be lined and levelled prior to assembly. Between each unit there is normally a lifting beam from which the load can be suspended.

These systems are ideal for carrying heavy loads into confined buildings where traditional cranes cannot be utilised. For example, the gantry lifting systems and their beam systems can be set up so that the delivery vehicle can reverse between the beam systems. The gantries can then be moved adjacent to the load, which can then be raised and slid into the building where, by retracting the hydraulic cylinders, the load can be positioned in its final location.

Capacities range approximately between 200 and 1200 t, but the manufacturers should be contacted for up-to-date specifications.

20.2.3.4 Strand jacks

Unlike cylinder jacks, strand jacks are not constrained by limitations of jack stroke (see Figure 20.6). They operate by pulling on a strand cable in repeated cycles and are

Figure 20.5 Gantry lifting system. (Courtesy of Fagioli Group)

Figure 20.6 Strand jacks mounted on a purpose-built structure (2000 t lift). (Courtesy of Fagioli Group)

therefore effective lifting, lowering and pulling devices. The strand cables comprise multiple pre-stressing strands: the more strands in a cable the greater the lifting capacity. Strand jacks are available in sizes from 15 t capacity to over 1000 t capacity for a single unit. They can be used individually or in multiples to give the required lifting capacity. Over 150 strand jacks have been used in single lift operations and lift weights of over 20 000 t have been achieved.

Strand jacks are only one part of the overall lifting system; in addition, they will require pumping and control systems. It is important that the systems are correctly matched to ensure a properly controlled and synchronised operation. The control systems available are very precise, which is useful when setting weld gaps or fitting bolts.

Strand jacking arrangements are very flexible and can easily fit with the shape of the item being lifted. They can be used in situations where the use of cranes would be impractical or even impossible (e.g. an aircraft hangar roof where there are multiple lift points with differing loads).

Strand jacks need a support structure (see Figure 20.7). Wherever possible, the supports should utilise the permanent works, with only minor modifications. There are no specific design requirements for strand jack supports; most are steel, which is designed in accordance with normal codes. The speed of operation of strand jacks (m/h rather than m/min) means that dynamic loads are not really applicable.

Figure 20.7 Strand jacks mounted on a modular support structure (1100 t lift). (Courtesy of Fagioli Group)

If it is not possible to utilise the permanent works then it may be necessary to consider a complete temporary support system. This can be purpose-built or configured from a modular support system available to hire. Such support systems are available with very high capacities (500 t on a 100 m unguyed tower, and much more for guyed towers), and the combination of a modular support system and strand jacks can be a viable alternative to heavy cranes.

When specifying a strand jack system, the following should be considered

- number of lift points
- load at each lift point
- jack support detail
- anchor connection detail
- lift speed required.

Limiting wind speeds will depend on the support structure and the wind area of the lifted item. The typical limiting wind speed for strand jack operations is 16 m/s.

20.2.4 Skidding systems

Skidding systems can be fully self-contained units incorporating skid tracks, low-friction skidding interfaces, load-transfer mechanisms and horizontal movement units (typically the hydraulic skid shoes described in Section 20.2.4.2). However, most systems are a combination of items brought together for a particular move. For example, rollers will need skid tracks, a vertical jacking system and a pulling or pushing system.

20.2.4.1 Rollers

Mechanical rollers, often known as 'machinery skates', have been in use for many years. They work by using an endless chain around a central bearing plate to provide a low-friction interface. Typical friction values are quoted at around 5% but could go as low as 2%, so beware, especially if moving downhill. Rollers do not normally require a break-out force (i.e. a greater force to start the movement) and are smooth in operation.

Rollers are available in capacities from very low values up to 1000 t per unit, but this size would be unusual. Multiple units are typically used to give the desired load capacity. The rollers should be suitably sized to suit the load at every support point, and should also have sufficient additional capacity to cope with possible unknown effects (e.g. differential settlement). For loads over 1000 t, the recommendation is that rollers should be used at 50% of capacity.

Some of the rollers will need to be provided with guides to ensure that the item to be moved follows the correct path. Note, however, that providing too many guides can lead to binding during the move. Care is also required at joints in skid tracks to ensure continuity of line and level.

The main drawback with rollers is that they have high localised bearing loads that need to be spread through the skid tracks, possibly using mats or cribbing as an additional

Figure 20.8 Skid shoe (600 t capacity). (Courtesy of Fagioli Group)

layer. This will affect the build height and may also require extensive jacking down on completion of the operation.

20.2.4.2 Hydraulic skid shoes

Hydraulic skid shoes are self-contained movement systems in that they contain load-transfer jacks, a low-friction interface and horizontal movement jacks (Figure 20.8). The item will need to be built on temporary supports, leaving a clear space of at least 1500 mm underneath to allow the shoes to be inserted. The 1500 mm height may need to be increased to allow for the depth of the skid tracks and any load spreading. The low-friction interface is normally stainless steel–PTFE (polytetrafluoroethylene) which needs to be lubricated.

The control systems available with skid shoes allow a high degree of control over the loads and displacements; high tolerances of accuracy can be achieved during both the movement and the final placement.

The limitation of skid shoes is that they work in one direction only. Changes of direction will require a set down of the load onto temporary supports followed by realignment of the track and shoes.

20.2.4.3 Low-friction interfaces

An alternative to mechanical rollers is to use a low-friction skid interface under the moved item. This can be included at the build stage so that no load-transfer system is

Table 20.1 Slide friction coefficients

Layer 1	Layer 2	Dry	Lubricated
Steel	PTFE	0.04	0.04
Steel	Steel	0.80	0.16
Steel	Wood	0.20–0.60	0.20

Data taken from http://www.engineersedge.com/coeffients_of_friction.htm (accessed 01/08/2018).

required. Also, the skid interface can be as large as required on plan so that bearing loads are minimised. The skid interface has a static base layer (the skid track) and a moving upper layer (fixed to the item). The combinations of layers in normal use have the coefficients of friction listed in Table 20.1.

All sources for coefficients of friction advise care be taken when using the values. For the purposes of a heavy move this means that a worst-case scenario should be considered for the movement system and the forces required should be specified conservatively. For example, it would be normal to specify a pulling force of 10% if using PTFE on steel.

With low-friction interfaces there may be a requirement for break-out (a higher pulling force to start movement), particularly if the item has been static for an extended period of time.

20.3. Design
20.3.1 General

The heavy move technique requires accurate preparation. Engineering design is the key element for the success of the operation from the point of view of safety, technical aspects, budget and schedule. Engineering targets are to identify the most safe, robust and economical method and equipment, and to produce complete and clear technical documentation for the safe execution of the operations, avoiding any miscommunication between the engineer and the site.

Input for the engineering activities, which is carefully reviewed and assessed by a competent and experienced engineer, includes the following

- layout, dimensions and drawings of the item
- weight (including an inaccuracy factor) and the location of the centre of gravity (a centre of gravity envelope is recommended to allow for position inaccuracy)
- identification of the handling points of the item (main structures, lifting lugs, trunnions, etc.)
- information about the transport route (site survey) and existing infrastructures (bridges, etc.)
- information about the installation yard (available space, underground services, ground capacity, etc.)

- applicable client documents and national and international codes
- previous experience and lessons learned
- production of lift plans.

The main output for the engineering activities include

- transport and lifting drawings, and installation phases drawings for the item
- installation procedures and operation manuals
- calculation reports for installation hardware
- design, shop drawings for structures, and fabrication and inspection plans
- 3D simulations where appropriate.

Engineering documents form the basis for the analysis of all safety aspects of the operations. HAZID (hazard identification study), HAZOP (hazard and operability study) and safety job analysis are structured review techniques for the identification and assessment of operation hazards. These techniques require sound and detailed engineering documents.

20.3.2 Crane lifting design

Crane lifting engineering consists of

- selection and design of adequate lifting points for the item
- analysis of the structure to be lifted
- selection of crane(s) of adequate capacity
- selection of appropriate lifting hardware (rigging)
- definition of the operational criteria.

The load distribution in a lift is normally calculated as a static load case by applying the dynamic hook load (DHL) at the hook position and distributing the weight and any special load to each element. The DHL is the product of the lifted load (including rigging) and the dynamic amplification factor (DAF). The skew load factor (SKL) allows for the extra loading on slings caused by the effect of sling-length and lift-point manufacturing tolerances, as well as rigging arrangements, which will affect a statically indeterminate lift. For statically indeterminate lifts, such as four slings in a pyramid arrangement, the SKL could vary over a wide range (1.0–2.0), depending on the sling-length tolerance and load shape. A figure of 1.25 is often taken, where the sling length tolerance is within $\pm 0.25\%$ of the sling's nominal length.

The maximum dynamic forces calculated as explained above are the design forces for the lifted item structure and its pad eyes for slings, grommets and lifting hardware (e.g. lifting beams).

Shackles are selected on the basis of the static hook load (SHL), as their safety factor also covers the dynamic effect. Cranes are selected on the basis of the SHL applied at the hook position and by determining the reactions for each crane involved in the lifting operation, considering the centre of gravity envelope as appropriate.

Table 20.2 Dynamic amplification factors

Static hook load (SHL): tonnes	50–100	100–1000
Dynamic amplification factor (DAF) onshore	1.10	1.05

Data taken from DNV (2000).

Recommended DAFs for onshore lifting, according to Det Norske Veritas rules (DNV, 2000; part 2, chapter 5), are shown in Table 20.2. As a general rule, lift operation is studied in such a manner that the maximum tilt of the lifted item is less than 2%, unless the client stipulates a different requirement. The maximum operational wind speed is defined according to the crane manufacturer's instructions, taking into consideration the weight and shape (drag factor) of the lifted item. A normal value for the maximum operational wind speed is $v = 9.0$ m/s (taken at the top of the crane boom). The maximum ratio of exposed surface/weight for the lifted item is 1 m^2/t (otherwise the operational wind speed is reduced).

Lifting a load with two or more cranes requires greater planning because many factors affect the operation: the accuracy of weight and centre of gravity values, the capacities of lifting accessories, the synchronisation of crane motions, crane instrumentation, site conditions and supervision. When such factors cannot be evaluated accurately, good engineering and industry practice dictates that an appropriate down-rating should be applied to all cranes involved (BSI, 1998–2016), as follows

- *single lift*: crane de-rated to 90% of the allowable capacity
- *multiple lift*: all cranes de-rated to 80% of the allowable capacity
- *crane used as tailing crane during lifting with tower system*: crane de-rated to 90% of the allowable capacity.

Higher values for the de-rating factor can be used but should be based on individual cases. The load transferred by the crane to the ground should be evaluated carefully. The pressure under a crawler crane varies over the range 40–60 t/m^2 (crane working at more than 70% of its capacity). Adequate spreader mats or a designed working platform (see Chapter 5) should be placed under the crawler to reduce the soil bearing pressure to an acceptable level.

The crane working area should be adequately designed by a competent person, and constructed and tested for the maximum ground pressure to ensure that

- the maximum inclination of the ground is within $\pm 0.3°$ ($\pm 0.5\%$) in any direction
- the maximum ground settlement under a pressure of 25 t/m^2 for crane lifting is within 30 mm (this will vary with crane model).

Spreader mats should be of good quality, inspected before use and maintained. If on-site changes are required, the designer should be notified and their approval obtained.

Table 20.3 Required clearance for normal voltage in operation near high-voltage power lines

Normal voltage: kV	Minimum required clearance: ft (m)
<50	10 (3.05)
>50–200	15 (4.60)
>200–350	20 (6.10)
>350–500	25 (7.62)

Data taken from ASME (2000).

Where required, adequate installation guides should be provided for placing the lifted items to ensure that the load lands at the correct location. Design load for the guides is a horizontal force of not less than 5% of the weight of the lifted item.

As a general rule, the lifting operation should be studied in such a manner that the load does not get any closer than 1 m to the crane boom or any other structure. This limit can be changed depending on the operational conditions (height, visibility, wind, etc.).

The Electricity at Work Regulations 1989 (UK Government, 1989) state that any work near live overhead power lines must be carefully planned and carried out to avoid danger from accidental contact or close proximity to the lines, and where possible the danger should be eliminated. Clearance from live power lines should be as described in the Health and Safety Executive Guidance Note GS6 (HSE, 2013) and the American Society of Mechanical Engineers document B 30.5-2000 (ASME, 2000) (Table 20.3). Typical output documents of the lifting engineer include

- the lifting plan, showing the crane operating radius
- the crane configuration
- the position with respect to the site and the crane usage factor, with the capacity of each component
- the rigging drawing
- the operational procedure.

20.3.3 Heavy transport using SPMTs

Heavy transport engineering documents include the following

- analysis of the item structure to be transported (inputs for this analysis are the transport drawings and the SPMT's reactions against the item structure, issued by the heavy transport contractor)
- design of the transport beams/stools and of the lashing/stopper system
- selection of an appropriate transport arrangement for the SPMTs
- stability and structural check of the convoy (tipping angle for structural stability and for structural overload within safe limits)

- definition of the operational requirements for the transport and issue of the operational procedure.

For the design transport weight (DTW), the vertical DAF is 1.0. The transported item is subject to horizontal inertia loads in the longitudinal direction (direction of motion) and in the transverse direction due to acceleration and deceleration. It is also subjected to horizontal loads due to the road slope, the SPMT type, the number of driven axles, the convoy arrangement and operational criteria.

For SPMTs (maximum design speed 5 km/h) normal ranges for horizontal accelerations are

- longitudinal acceleration $a_1 = 0.10g$ to $0.30g$
- transverse acceleration $a_2 = 0.05g - 0.15g$
- slope effect (due to the transport path slope, longitudinal or transverse) $a_3 = $ slope (%) $\times g$.

When on a slope the acceleration a_3 is added to a_1 or a_2. Plywood is used between the underside of the transported item and the top steel of the SPMT's frame/stools to achieve a higher friction coefficient. An adequate lashing/stopper system should be provided that has been designed to withstand (with the contribution of the friction, when appropriate) the total horizontal load.

The loads/reactions distribution in the transport system is calculated as a static-load case by distributing the DTW to each temporary transport element and to the SPMT. As a general rule, the SPMT hydraulic circuits are arranged in such a way that a three-point system is achieved. With this system (isostatic), the reactions on the trailers are practically constant and there is no overstress in the transported item or in the SPMT structure or hydraulic system. If tall items are transported with only one SPMT line, then the four-point system is considered to have more stability. SPMTs are also suitable equipment for the execution of load-in/load-out operations, as they can spread heavy loads within the capacity limits of a barge or ship deck.

The design operation conditions that are generally considered are

- DTW of the item and the location of the centre of gravity
- out of verticality due to operator tolerance $\pm 2°$
- maximum transport path slope longitudinal and transverse
- design wind speed 16 m/s (3 s gust at 10 m height) for site transport operations
- design wind speed 12 m/s for load-in/load-out operations
- maximum barge/ship vertical motion due to swell ± 100 mm, minimum period 10 s (for load-in/load-out operation)
- ship maximum trim 1% and maximum heel 1% (for load-in/load-out operations)
- ship in level with the quay tolerance ± 100 mm (for load-in/load-out operations).

In the operational condition, considering the most unfavourable loads combination, the SPMT axle loads should not exceed the allowable loads stated by the manufacturer (which depend on the travelling speed). When several axles are overhanging, the bending capacity of the trailer spine beam should not be exceeded.

The SPMT's tractive power (as per the manufacturer's data sheet) should be at least 20% greater than the theoretical power required to overcome rolling friction and slope.

As a general rule, the transport path should be checked in order to verify there is a minimum lateral and top clearance of 1 m from existing structures. Less clearance can be accepted after checking on a case-by-case basis.

When SPMTs are loaded to full capacity (36 t/axle), the maximum ground gross pressure is 10 t/m^2 (with reference to the SPMT frame projected area). The local contact pressure under the wheels is c. 9.5 kg/cm^2 (tyre inflation pressure 10 bar). The maximum ground pressure for each specific transport shall be transmitted to the client to check the existing bridges, culverts, underground services, retaining walls, quays and so on.

Typical output documents include transport drawings (showing the arrangement of the trailers under the item, hydraulic circuits, maximum operation axle loads, axle capacity and lashing details) and the transport operational procedure (with a definition of responsibilities of key personnel, the operation criteria to be respected and the contingency plan).

REFERENCES

ASME (American Society of Mechanical Engineers) (2000) *Mobile and Locomotive Cranes*. ASME, New York, NY, USA, B 30.5–2000.

BSI (British Standards Institution) (1998–2016) BS 7121. Code of practice for safe use of cranes. BSI, London, UK. (Parts 1–5 and 11–14; see Further Reading below.)

DNV (Det Norske Veritas) (2000) Rules for planning and execution of marine operations. DNV, Høvik, Norway.

HSE (Health and Safety Executive) (2013) *Avoiding Danger from Overhead Power Lines*, 4th edn. HSE, London, UK, Guidance Note GS6.

UK Government (1989) Electricity at Work Regulations 1989. The Stationery Office, London, UK.

FURTHER READING

Bates GE, Hontz RM and Brent G (1998) *Exxon Crane Guide: Lifting Safety Management System*. Specialized Carriers and Rigging Association, Fairfax, VA, USA.

BSI (1998–2016) BS 7121. Code of practice for safe use of cranes. BSI, London, UK.
- BS 7121-1:2016. Code of practice for safe use of cranes. General.
- BS 7121-2:2003. Code of practice for safe use of cranes. Inspection, testing and examination.
- BS 7121-2-1:2012. Code of practice for safe use of cranes. Inspection, testing and thorough examination. General.
- BS 7121-2-3:2012. Code of practice for safe use of cranes. Inspection, testing and thorough examination. Mobile cranes.
- BS 7121-2-4:2013. Code of practice for safe use of cranes. Inspection, testing and thorough examination. Loader cranes.

- BS 7121-2-5:2012. Code of practice for safe use of cranes. Inspection, testing and thorough examination. Tower cranes.
- BS 7121-2-7:2012 + A1:2015. Code of practice for safe use of cranes. Inspection, testing and thorough examination. Overhead travelling cranes, including portal and semi-portal cranes, hoists, and their supporting structures.
- BS 7121-2-9:2013. Code of practice for safe use of cranes. Inspection, testing and thorough examination. Cargo handling and container cranes.
- BS 7121-3:2017. Code of practice for safe use of cranes. Mobile cranes.
- BS 7121-4:2010. Code of practice for safe use of cranes. Lorry loaders.
- BS 7121-5:2006. Code of practice for safe use of cranes. Tower cranes.
- BS 7121-11:1998. Code of practice for safe use of cranes. Offshore cranes.
- BS 7121-12:1999. Code of practice for safe use of cranes. Recovery vehicles and equipment. Code of practice.
- BS 7121-13:2009. Code of practice for safe use of cranes. Hydraulic gantry lifting systems.
- BS 7121-14:2005. Code of practice for safe use of cranes. Side boom pipelayers.

CIRIA (Construction Industry Research and Information Association) (1977) *Lateral Movement of Heavy Loads*. CIRIA, London, UK, R68.

Energy Networks Association (2007) Look Out Look Up! A Guide to the Safe Use of Mechanical Plant in the Vicinity of Electricity Overhead Lines. Energy Networks Association, London, UK.

GL Noble Denton (2010) *Guidelines for Marine Lifting and Lowering Operations*. GL Noble Denton, London, UK, 0027/ND.

Lloyd D (ed.) (2003) *Crane Stability on Site*, 2nd edn. CIRIA, London, UK, C703.

MacDonald JA, Rossnagel WA and Higgins LA (2009) *Handbook of Rigging*, 5th edn. McGraw-Hill Professional, New York, NY, USA.

Shapiro H, Shapiro JP and Shapiro LK (1999) *Cranes and Derricks*, 3rd edn. McGraw-Hill Professional, New York, NY, USA.

UK Government (1998) Lifting Operations and Lifting Equipment Regulations 1998. Statutory Instrument 1998/2307. The Stationery Office, London, UK.

UK Government (1998) Provision and Use of Work Equipment Regulations 1998. Statutory Instrument 1998/2306. The Stationery Office, London, UK.

Useful web addresses

Ainscough – telescopic cranes: http://www.ainscough.co.uk (accessed 01/08/2018).
ALE – specialist cranes: http://www.ale-heavylift.com (accessed 01/08/2018).
CIRIA (Construction Industry Research and Information Association): https://www.ciria.org (accessed 01/08/2018).
Enerpac – cylinder jacks: http://www.enerpac.com (accessed 01/08/2018).
Fagioli – strand jacks and SPMT trailers: http://www.fagioli.com (accessed 01/08/2018).
Flatjack – flat jacks: http://www.flatjack.co.uk (accessed 01/08/2018).
Hilman Rollers – rollers: http://www.hilmanrollers.com (accessed 01/08/2018).
Mammoet – specialist cranes: https://www.mammoet.com (accessed 01/08/2018).
Sarens – specialist cranes: http://www.sarens.com/en.aspx (accessed 01/08/2018).
Weldex – crawler cranes: http://www.weldex.co.uk (accessed 01/08/2018).

Temporary Works, Second edition

Pallett, Peter F and Filip, Ray
ISBN 978-0-7277-6338-9
https://doi.org/10.1680/twse.63389.281
ICE Publishing: All rights reserved

Chapter 21
Access and proprietary scaffolds

Peter F Pallett
Pallett TemporaryWorks Ltd

Ian Nicoll
Independent Consultant

The provision of temporary safe working platforms for the erection, maintenance, construction, repair, access, inspection, and so on, of structures is known as 'scaffolding'. Scaffolding can be formed from individual tubes with fittings or from proprietary components. The design philosophy for the stability of scaffolds and loading limits, together with the information necessary to source the safe height of most UK scaffolds without the need for further calculations, is discussed. The designation and simple rules for inspection are included.

21.1. Introduction

Wherever it is required to provide a safe place of work for the erection, maintenance, repair or demolition of buildings and other structures and to provide the necessary access, a temporary structure known as a 'scaffold' is erected.

In the middle ages, a 'skaffaut' was a mobile tower equipped with battering rams used for assaulting castles. Shakespeare referred to the gallery structure at the Globe Theatre in London as the 'scaffoldage'. The first tubular steel scaffolding was seen in the UK around 1920, and it comprised standard 2 in. diameter water pipes that were available in 21 ft lengths. This type of scaffolding with 2 in. tubes connected together remains a common method, but more recent scaffolding equipment comprises proprietary components connected together. By their nature, such structures are usually temporary and, unlike permanent structures, need to be dismantled after use. This introduces a reuse aspect to scaffold structures, which are almost always erected using previously used material. This generates an industry of supply (generally on hire), erection, use, inspection and maintenance, followed by dismantling, removal and inspection prior to reuse. Managing the scaffold is important in terms of maintaining its capability to provide adequate access.

Falls from height account for over 50% of fatal accidents in construction (see Chapter 1 on safety) and scaffolders are particularly at risk. The recommended procedures for controlling scaffolds and all temporary works are given in BS 5975:2019, 'Code of practice for temporary works procedures and the permissible stress design of falsework' (BSI, 2019). Industry guidance on scaffolding recognised by the Health and Safety Executive (HSE) is provided by the National Access & Scaffolding Confederation (NASC). The NASC has published and continually updates its TG20:13 Operational Guide (*A Comprehensive Guide to Good Practice for Tube and Fitting Scaffolding*) (NASC, 2013a), TG20:13 Design Guide (which provides technical guidance on the use of BS EN 12811-1:2003) (NASC, 2013b) and TG20:13 eGuide (software for establishing safe heights for scaffold) (NASC, 2013c) to provide technical guidance on the use of BS EN 12811-1:2003 (BSI, 2003a) for scaffolding in the UK. The NASC has also launched TG20:13 'toolbox talk' training videos. Guidance on proprietary scaffolding should be sought from the supplier/importer of the particular scaffold. Training courses are run through the Construction Skills Certificate Scheme (CSCS) and by the NASC, and there is also a national certification scheme for scaffolders.

All users of scaffolding on site should be aware of the statutory requirements under the Work at Height Regulations 2005 (UK Government, 2005, 2007) and the Construction (Design and Management) Regulations 2015 (CDM 2015) (UK Government, 2015). They should also be aware that it is their duty to provide a 'safe place of work'. The authoritative guidance is given in SG4:15, *Preventing Falls in Scaffolding Operations* (NASC, 2016a,b). This guidance note features a broader scope to reflect the significant increase in the number of TG20:13 compliant scaffolds, changes to scaffolding good practice and innovation in the industry.

21.2. Managing scaffolding

There can be various different parties involved in procuring and constructing a scaffold. These range from an individual acting as a domestic client, a client (architect, consultant), the principal or main contractor, a trade company (plumber, roofer, decorator), the scaffold subcontractor, proprietary supplier(s) and, possibly, a labour-only organisation carrying out erection services. Designers, checkers, erectors, managers and workers all have stated duties and responsibilities under CDM 2015.

The sequence of events to provide a scaffold structure is similar for all projects

- Planning
  - A person or organisation wants to provide access, support, protection and so on, to facilitate work of some kind.
  - A design brief is produced, including a definition of the scope and use, the overall dimensions and load capacity, the ground conditions, protection, permits, existing hazards and so on.
  - On projects where multiple temporary works are required, a schedule/register of temporary works should be produced.
- Design
  - All scaffolds need to be designed (Work at Height Regulations 2005). The design could be one regularly used for that type of work, essentially a 'standard solution'. Either a partial or a full design may be required. The TG20:13 Design Guide (NASC, 2013b) contains basic compliance sheets for guidance.

- The design responsibilities for various items (e.g. scaffold foundations) need to be established.
- There should be coordination between the Temporary Works and Permanent Works Designers to ensure their designs are compatible.
- Risk assessment
  - Designers need to carry out a design risk assessment and try to eliminate hazards wherever possible. The designer must communicate any residual risks.
- Check
  - Whatever design is used, that design has to be checked. The degree of checking will vary on the complexity and/or location of the scaffold.
- Installation
  - A safe means of erection takes place as defined by the site- and task-specific risk assessments and method statements.
- Inspection
  - Prior to use the scaffold must be inspected for adequacy, and regular mandatory inspections may be required throughout its use.
- Use
  - A means of safe access to the working area is required.
- Modifications and dismantling
  - A safe means of modifying and dismantling takes place as defined by the site- and task-specific risk assessments and method statements.

The procedures for controlling these events will vary between an individual procurer and those required by a major contracting organisation. The format for the procedural control of all temporary works adopted in section 2 of BS 5975:2019 (BSI, 2019) and discussed in Chapter 2 of this book is designed to ensure that relevant responsibilities are highlighted and passed down through organisations. In practice, when you read the roles required the majority of site staff are carrying out those roles already as part of daily site activities – all BS 5975:2019 has done is formally define these roles. The emphasis is that each organisation must establish a procedure to ensure the risks associated with scaffolding are controlled.

21.3. Selection and designation

Scaffold companies will already have procedures for training their operatives and scaffolders using the Construction Industry Scaffolders Record Scheme (CISRS), with regular training under the Site Management Safety Training Scheme (SMSTS) or Site Supervision Safety Training Scheme (SSSTS). Anyone designing, checking, installing, altering or dismantling, or supervising works on a temporary structure that uses scaffolding components must be sufficiently experienced and knowledgeable to carry out their role safely.

21.3.1 Selection

Do I actually need a scaffold? When carrying out the initial design risk assessment to perform a task, the activity should be considered carefully. Ideally, the operation would be carried out at ground level, in which case no access would be required; conversely, painting the outside of a building will require access to the working area. The procurer needs to consider the risk alternatives; these might include rope access, mobile elevated

Figure 21.1 A typical independent tied scaffold in tube and fittings. (Courtesy of Pallett TemporaryWorks Ltd)

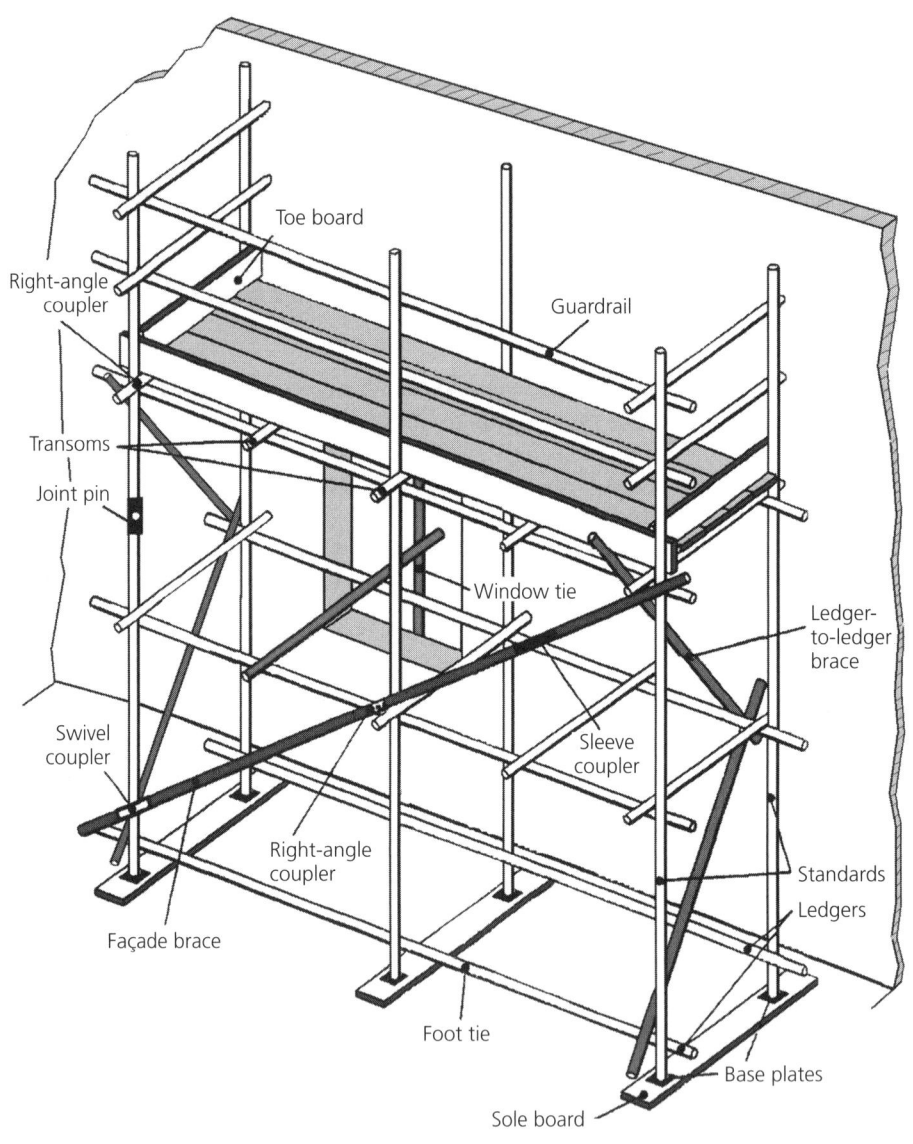

working platforms (MEWPs), mast climbing scaffolds or the provision of accessible working platforms. In modern multi-storey new-build construction the design has often removed the need for external access platforms, and scaffolding can be avoided. There will, however, be many applications where provision of regular access platforms will be required. In fact, these may often need to be considered by the Permanent Works Designer under the CDM 2015 for future access requirements, such as planned long-term maintenance.

The two main types of scaffold are

- free-standing scaffolds (e.g. independent towers)
- independent tied scaffolds (where the scaffold itself is free-standing but it relies on the adjacent structure for its stability).

A typical arrangement of an independent tied scaffold using tube and fittings is shown in Figure 21.1, which also includes the terms regularly used in scaffolding.

Other types of scaffolds include

- birdcage scaffolds – generally on a 2.1 × 2.1 m grid with a single platform
- mast climbing scaffolds – a rigid tower with a bridging unit connected to it that elevates vertically up and down (proprietary items of equipment)
- putlog scaffolds – a standard outside vertical with the inside of the platform supported by the structure (commonly used on small housing projects with brickwork such that the inside support is a flattened tube end built into the bedding planes of the brickwork)
- slung/suspended scaffolds – where the platform is hung from above on either vertical tubes or wire ropes.

With the exception of mast climbing scaffolds, each type of scaffold discussed above is available either as traditional tube and fitting scaffolds or as proprietary prefabricated system scaffolds. The big advantage of traditional scaffolding is its adaptability on complex-shaped structures. Adaptability is often hard to achieve within the dimensional constraints of a proprietary system. Further decisions include whether the scaffold is required to be left unclad, fitted with debris netting or fully sheeted. Obviously there are significant design implications when sheeting or debris netting is considered on a scaffold, due to the increase in the applied wind loading.

Once the type of scaffold has been selected, the procurer needs to consider the activity for which the scaffold is required, the number of platforms, its width, any projections and, most importantly, the imposed loading per platform, which will change depending on the activity to be performed. BS EN 12811-1:2003 (BSI, 2003a) introduced six designated imposed load categories for scaffolds, discussed later in this chapter.

21.3.2 Designation

The NASC introduced a three-number designation for scaffolds which describes the type of scaffold used and assists procurers in the selection of the majority of scaffolds, namely N1-N2-N3

- N1 is the BS EN 12811-1:2003 (BSI, 2003a) load class (1–6).
- N2 is the number of boards between the vertical standards.
- N3 is the number of inside boards (limited to 0, 1 or 2 maximum).

A typical designation would be 3-5-2, meaning load class 3 with five scaffold boards between the uprights and two scaffold boards fitted on the inside adjacent to the structure being scaffolded. The cantilevered inside board is assumed to be lightly loaded.

The scaffold designation has been found to be a very useful way to describe a scaffold requirement, as it suits those initially requesting the scaffold in the design brief, procurers, scaffold suppliers, erectors and contractors. It is essential for all those who regularly check and sign for scaffolds, because they know from the designation what to expect and, importantly, how a scaffold can be loaded.

Many scaffolds in use have boards fitted to the inside of the scaffold, between the inside standard and the structure. These inside boards, either one or two boards wide, are only lightly loaded when bricklaying is in progress, and thus the main scaffold can be of a higher load class, but the inside boards are only designed for the very low load of 0.75 kN/m^2 (service load class 1). The electronic calculations in TG20:13 (NASC, 2013c) and the compliance sheets only permit this light loading on the inside boards. However, there are many instances where the construction activities to be undertaken require the *same* service load class over all the scaffold platform. To suit this loading case earlier editions of TG20 introduced the designation 'F' added to the three-figure designation, so 3-5-2F meant a class 3 scaffold, five boards wide with two inside boards but with the full class 3 loading allowed on the inside boards. The method of calculating the safe height of a scaffold with fully loaded inside boards is to use annex B in TG20:13 (NASC, 2013b) with the additional load given in table 5.5 of the TG20:13 Design Guide (NASC, 2013b).

The importance of the designation is that it defines the size and loading of the scaffold platforms, and it is used by the designer to establish the safe height. The safe height of basic tube and fitting scaffolds for a range of wind conditions in the UK and Ireland may be calculated using the TG20:13 eGuide (NASC, 2013c). This eGuide is designed as user-friendly software that calculates and prints A4 compliance sheets for TG20:13 compliant scaffolding. It allows TG20:13 to incorporate a wide range of scaffolding configurations, and calculates safe heights, tie duties and leg loads. Version 1.2 of the eGuide (2017) provides the following updates

- automatic site reference
- checking a compliance sheet
- reporting the permitted seasons
- company logo
- compliance sheet illustrations
- Build UK support.

21.4. Materials and components
21.4.1 Scaffold tube
21.4.1.1 Steel scaffold tube
The most common material in scaffolding is the scaffold tube, specified in BS EN 39:2001 (BSI, 2001) with two wall thicknesses of scaffold tube (type 3, 3.2 mm; type 4, 4 mm) for the same outside diameter. The general UK practice is to use type 4 tube, the thicker

walled tube of 4 mm. Users should be aware that certain scaffold companies are using a thinner walled tube, but to a higher steel specification. Information and guidance on its use can be found in TG20:13 (NASC, 2013b). In the UK, an allowance for corroded tubes the wall thickness of which has reduced by not more than 10% (known as 'used' tube) is considered. The majority of scaffold tube used in the UK is galvanised, so use of the 'as new' values is recommended in TG20:13 (NASC, 2013b).

The safe load of any strut (member in compression) is limited by its buckling, so its effective length L_E and slenderness ratio L_E/r should be considered. For scaffold tubes in compression, the slenderness ratio should meet the condition $L_E/r < 271$ for struts and braces intended to carry wind and lateral loads (i.e. the lacing and diagonal bracing should be less than 4.25 m long). Safe working loads of scaffold tubes related to the effective length L_E are listed in Table 21.1. Scaffold tube struts that have a free cantilever

Table 21.1 Allowable axial load (kN) in scaffold tube struts[a]

Effective length of scaffold tube: mm	Safe axial load in steel scaffold tubes: kN[b]						
	BS EN 39:2001, type 4		High yield S355, 3.2 mm (BS EN 10219-1:2006)		BS 1139-1.1:1990		
	'As New'	'Used'	'As New'	'Used'	'As New'	'Used'	
500	73.5	65.6	87.0	77.6	66.5	59.2	
1000	58.6	52.0	63.4	56.4	53.9	47.9	
1500	42.1	37.1	41.0	36.3	39.9	35.2	
2000	**29.1**	25.6	**26.7**	23.6	28.2	**24.8**	
2500	20.6	18.1	18.4	16.2	20.1	17.7	
2700	18.2	15.9	16.2	14.2	17.9	15.7	
3000	15.1	13.3	13.3	11.8	14.9	13.1	
3200	13.6	11.9	**11.9**	10.6	13.4	11.7	
3500	11.6	10.1	10.1	8.9	11.4	10.0	
4000	9.1	8.0	7.9	7.0	9.0	7.9	
4250	8.1	7.1	7.0	6.2	8.1	7.1	← Limit as brace or strut
4500	7.3	6.4	6.3	5.6	7.3	6.4	
5000	6.0	5.3	5.2	4.6	6.0	5.2	
5500	5.0	4.4	4.3	3.8	5.0	4.4	
6000	4.3	3.8	3.7	3.2	4.3	3.7	← Limit as tie
7000	3.2	2.8	2.7	2.4	3.2	2.8	
8000	2.4	2.1	2.1	1.8	2.4	2.1	

[a] The values shown shaded are for comparison purposes, and should *not* generally be used for scaffolds, unless the scaffold has been so designed.
[b] Axial loads derived with a quasi-permissible stress approach using the provisions of BS EN 1993-1-1:2005 + A1:2014 (BSI, 2005b) with partial safety factors $\gamma_f = 1.5$ and $\gamma_m = 1.1$.

at the end, such as at the bottom of a free-standing scaffold, are regarded as having an effective length of the adjacent strut length plus twice the length of the free cantilever (see figure 5.1 in TG20:13 (NASC, 2013b)).

The safe axial tensile load for type 4 scaffold tube is 79.1 kN but is usually limited by the safe slip load capacity of one coupler (i.e. 6.1 kN).

21.4.1.2 Aluminium scaffold tube

In certain cases, for example, to reduce weight, aluminium scaffold tube can be used. Some guidance is given in TG20 Design Guide; appendix C (NASC, 2013b). However, because of the different elastic properties, steel and aluminium scaffold tube should never be used together within a structure. Care may be necessary in the selection of suitable scaffold fittings for use on the 'softer' aluminium tube to avoid unnecessary damage to the tube and develop the load capacity.

21.4.2 Scaffold fittings

Also known as couplers, there is a great variety of different types of scaffold fittings, for example, right-angled (doubles), swivels, putlogs, band and plates, and sleeves. Each type can be made as pressed steel, spring steel or drop forged. BS EN 74-1:2005 (BSI, 2005a)

Table 21.2 Safe working loads for individual couplers and fittings

Type of fitting	Class	Type of load	Safe load: kN
Mills right-angle coupler and SGB Mark 3A coupler	n/a	Slip along tube	12.5
Right-angle coupler	BS EN 74-1:2005 class A	Slip along tube	6.1
	BS EN 74-1:2005 class B	Slip along tube	9.1
Mills swivel coupler	n/a	Slip along tube	6.1
Swivel coupler	BS EN 74-1:2005 class A	Slip along tube	6.1
	BS EN 74-1:2005 class B	Slip along tube	9.1
Supplementary fittings	Class A/class A	Slip one way on tube	9.1
	Class B/class B	Slip one way on tube	15.2
Parallel coupler	Class A	Tension	6.1
Sleeve coupler	Class A	Tension	3.6
	Class B	Tension	5.5
Putlog coupler	BS EN 74-2:2008	Force to pull tube axially out of coupler	0.63
Internal joint pin	–	Tension	nil
		Shear strength	21.0
Band and plate coupler	–	Slip along tube	≈4.0

Figure 21.2 Some system scaffold connections. (Courtesy of RMD Kwikform (UK) Ltd)

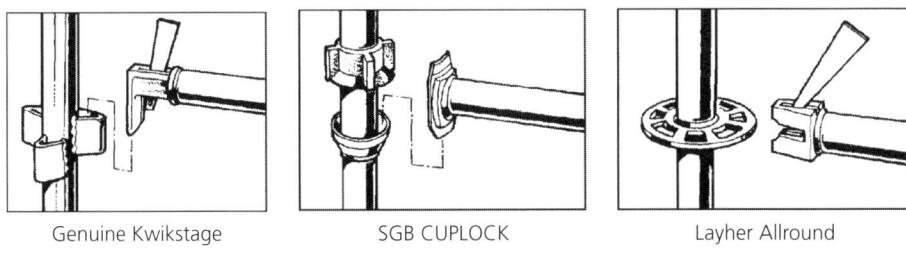

| Genuine Kwikstage | SGB CUPLOCK | Layher Allround |

covers the specification and testing of many couplers for use with steel tubes. The safe working loads summarised in Table 21.2 are from TG20:13 (NASC, 2013b). All couplers should be tightened correctly using the correct spanner.

21.4.3 Proprietary scaffolds
System scaffolds are relatively simple arrangements of scaffold using patented connections to reduce and sometimes eliminate the need for scaffold couplers (Figure 21.2). They essentially comprise separate standards, ledgers and transoms in modular lengths, and are designed for fast assembly and dismantling. Although the standards generally have the same outside diameter as the scaffold tube, some have a thinner-walled tube, which reduces the weight and therefore makes transporting and handling easier. The decking members are rarely interchangeable, often in steel and fitted into special transoms.

21.5. Scaffold design
21.5.1 General
A very simple view is generally taken in the design, and imponderables such as stiffness of couplers is generally ignored. (The use of a complex computer program to analyse scaffolding allowed the NASC to take into account some cruciform stiffness when preparing the TG20:13 eGuide (NASC, 2013c)). The design methods cater for the safety factors for the materials and components. Provided that the scaffold is erected within certain workmanship limits for verticality, joint connections and so on, the couplers are rarely considered as giving any stiffness and merely act as connections. In tube and fitting scaffolds, the couplers will eccentrically load the standards as the tubes are separated by about 60 mm, so, in theory, they apply torsion. This is ignored, provided the diagonals are connected within 300 mm of the standards and the arrangement of bracing to counter the torsion is considered.

When considering the design of any scaffold the fact it can buckle or fall over in any direction should be taken into consideration. The accepted rule is: 'Think vertical, think horizontal, then think horizontal again'. The procurer will have previously decided whether the scaffold will be unclad, sheeted or debris netted.

21.5.2 Permissible stress or limit state design
The design method traditionally used for scaffolding was 'permissible stress design', where the failure load was divided by a factor of safety (usually 2) to derive a safe

working load. BS EN 12811-1:2003 (BSI, 2003a) considers 'limit state analysis', where the structure is designed for the ultimate condition at failure and also at the time of loading (serviceability). Limit state analysis requires knowledge of the partial safety factors and the characteristic strength of the members in order to be able to establish their ability to resist load. This is a very difficult concept for site personnel who are familiar with and understand the 'safe working load' concepts. To enable scaffolding to be understood by the regular users, the NASC commenced the preparation of TG20 in 2001 and adopted a quasi-permissible stress approach to scaffold design. TG20 was first published in 2005, and was updated in 2008 and again in 2013. The work undertaken in these publications provided the data for the axial load in tubes and the capacities for couplers, and accordingly these have been stated as 'safe loads' in Tables 21.1 and 21.2.

21.5.3 Loading on scaffolds

In addition to the self-weight of the scaffold, its boards, guardrails, toe boards and so on, the scaffold has to withstand the imposed allowable load on the working platform(s) and environmental loads such as the wind.

21.5.3.1 Imposed loads

The imposed load depends on the nature of the work, and all loads are assumed to be distributed

- Service class 1: 0.75 kN/m^2; inspection and very light duty access.
- Service class 2: 1.50 kN/m^2; light duty, such as painting and cleaning.
- Service class 3: 2.00 kN/m^2; general building work, brickwork and so on.
- Service class 4: 3.00 kN/m^2; heavy duty, such as masonry work and heavy cladding.

BS EN 12811-1:2003 (BSI, 2003a) states that a scaffold in use shall be loaded with one platform (generally the top one) with the full service class load and an adjacent platform with 50% of that service class load. If more platforms are in use or the scaffold is taller than permitted then the scaffold needs to be designed. Appendix B of TG20:13 Design Guide (NASC, 2013b) provides a method for calculating the safe height of an unclad independent tied scaffold which has more than the two platforms loaded, by considering individual loads in the standards.

BS EN 12811-1:2003 (BSI, 2003a) also requires a notional horizontal load of 2.5% of the vertical imposed load (minimum of 0.3 kN/bay) applied to each bay of the scaffold, and also states specific concentrated and partial area loads on the platforms.

21.5.3.2 Loading towers

When material has to be moved into a building a specific loading tower may have to be considered. Loading towers require a separate design, as the nature of the loading, its type and the safety aspects of loading towers are important considerations. The TG20:13 Design Guide (NASC, 2013b) states that where storage of material in palletised form is envisaged the platform should be designed for a minimum distributed load of 10 kN/m^2 or the actual weight of material to be stored. Proprietary system scaffolds designed as loading towers will generally support the actual weight of material to be stored on more than one lift. The procurer of the scaffold should specify which use the platform is to be put to and the location and levels of such platforms.

21.5.3.3 Wind loading on scaffolds and the wind factor, S_{wind}

The wind is a variable force and gets stronger the higher the position above the ground. It is also dependent on location, being stronger in exposed locations. The wind force acts on the tubes, bracing, couplers, rakers and any debris netting or sheeting attached to the scaffold. The magnitude of the wind forces will alter the required capacity of the ties and may affect the tie frequency.

To cater for the wind on all temporary works a simplified method of calculating the wind forces was developed in BS 5975:2019 (BSI, 2019) by introducing a wind factor, S_{wind}, for any particular site (see also Chapter 22 on falsework). It should be noted that the wind factor is only required to be determined once for each site.

The wind factor S_{wind} needs to be evaluated to check the safe height of a scaffold at a site location

$$S_{wind} = T_{wind} \times v_{b,map}\left(1 + \frac{A}{1000}\right) \qquad (21.1)$$

where $v_{b,map}$ is the fundamental wind velocity at the location (m/s), T_{wind} is a topography factor that takes into account the ground conditions and A is the altitude of the site (m). Values for $v_{b,map}$ and T_{wind} are listed in BS 5975:2019 (BSI, 2019). Analysis of the maximum height of a scaffold within the TG20:13 eGuide (NASC, 2013c) was based on wind loading in accordance with BS EN 1991-1-4:2005 + A1:2010 (BSI, 2005c) by investigating the wind loads according to location. When the TG20:13 eGuide is connected to the Internet the location can be generated more precisely from the Building Research Establishment UK wind map by inserting the relevant postcode.

When using debris netting or sheeting it is recommended that it is placed on the outer side of the outside standards. This will reduce the wind forces on the scaffold, especially those associated with drag when the wind blows parallel to the façade.

21.5.4 Design of tube and fitting scaffolds
21.5.4.1 General
The use, loading class, number of platforms in use, maximum bay length, number of boards and so on are determined using table 2.1 in TG20:13 (NASC, 2013b). This is the starting point of all design work.

The Work at Height Regulations 2005 (UK Government, 2005, 2007) require calculations to be completed for all scaffolds unless there is a pre-existing design (i.e. already completed in the office for a similar job) or the scaffold is erected in accordance with a recognised standard configuration. The latter refers to the use of scaffolds detailed in TG20:13 (NASC, 2013b,c) that are erected correctly and conform to the established bracing, loading and tying patterns. In addition, the latter also refers to technical data contained in the user guide for each system scaffold.

As with all structures, when carrying out a full design the three design requirements to consider, having established the loads, are the strength and stability of the members, followed by the lateral stability of the structure, and finally the overall stability of the structure.

21.5.4.2 Element strength and element stability

Lift heights in scaffolding are generally 2.0 m, although in certain conditions a pavement lift of 2.7 m may be considered. For a pavement lift the first tie should be at the 2.7 m level. The bay length will vary depending on the load class; for example, a general-purpose class 3 scaffold will have a maximum bay length of 2.1 m. The transoms will be spaced to suit the permissible span of the chosen boards; generally, 225 mm wide × 38 mm thick scaffold boards will safely span 1.2 m, but a better quality of board is required to safely span 1.5 m.

Designers will be aware of the effect of effective lengths on the load capacity of the scaffold members. Buckling of struts is prevented by adequate sideways restraint in two directions at right angles, known as 'creating effective node points'. The value of the restraint force is usually only 2.5% of the load in the strut. The amount of fixity of the coupler to the tube is generally neglected; Table 21.1 lists the safe loads for various effective lengths.

The effective length of the vertical standards in an independent tied scaffold is not always obvious. Lift heights were previously considered relevant, but engineering analysis supported by full-scale testing of scaffolds by the NASC (2008) showed conclusively that two parameters control the effective lengths in a tube and fitting scaffold: the vertical spacing between the levels of the tie positions and the bay length. A fuller treatise is given in TG20:13 (NASC, 2013b). The effective lengths of type 4 scaffold tube related to the vertical tie arrangement and the bay length for normal 2.0 m lift scaffolds are listed in Table 21.3.

Table 21.3 Effective lengths, L_E, for fully ledger braced independent tied scaffolds with 2.0 m lifts

Vertical interval between lines of ties	Bay length: m						
	1.2	1.5	1.8	2.0	2.1	2.4	2.7
2 m (every lift)	2.5	2.5	2.5	2.5	2.55	2.75	2.9
2 m with 2.7 m pavement lift	2.7	2.7	2.7	2.7	2.7	2.9	3.0
4 m (alternate lifts)	3.2	3.2	3.2	3.2	3.2	3.2	3.2
6 m (every third lift)	4.0	4.0	4.0	4.0	4.0	4.0	4.0

Taken from TG20:13 NASC (2013b).

A scaffold tied at alternative lifts (4 m) with any bay length therefore has an effective length of 3.2 m, giving a safe axial load of 13.6 kN using type 4 'as new' tube (see Table 21.1).

21.5.4.3 Tying scaffolds to structures

Independent tied scaffolds are restrained by tying the scaffold to the building and using the permanent structure work for stability. The ties will prevent the scaffold from moving into the building, away from the building and from longitudinal movement parallel to the building (left and right). The general main tie layouts for scaffolds in the TG20:13 eGuide (NASC, 2013c) are: (*a*) lines of ties fitted to every lift and (*b*) lines of ties fitted to alternate lifts.

This spacing applies to unclad, debris netted and sheeted scaffolds. TG20:13 (NASC, 2013a) classifies three capacities of tie: light duty (capacity in tension 3.5 kN), standard duty (capacity in tension 6.1 kN for BS EN 74-1:2005 class A couplers and 9.1 kN for class B couplers (BSI, 2005a)) and heavy duty (capacity in tension 12.2 kN).

Four main types of tie are used: the through tie, box tie, reveal tie and anchor tie. Ties should preferably be fitted to the standard or at least within 300 mm of the standard on the ledger.

The selection of tie positions, whether the tie fixing into the building should be tested before use and even the suitability of the fabric of the building to carry ties are all part of the discussions at the early stage of procurement of scaffolds. Structures with minimal provisions for ties will obviously require more detailed design.

21.5.4.4 Lateral bracing to scaffolds

Bracing is required in tube and fitting scaffolds to stiffen the scaffold in two directions

- façade bracing – parallel to the building
- ledger bracing – away from the building.

Although BS EN 12811-1:2003 (BSI, 2003a) states that ledger bracing should be omitted at working lifts to give 'a completely unimpeded area', the opinion of the HSE and the NASC on best-practice is that, for the majority of scaffolds in the UK, all scaffolds should have full-height ledger bracing on alternate bays.

The bracing is fitted as diagonals at an angle of 35–50°, creating stiff triangles. Loads are transmitted down through the scaffold to ties and/or the ground; the brace direction is unimportant (Figure 21.3). If the bays are less than 1.5 m, then ledger bracing is fitted onto every third pair of standards. Façade bracing is fitted to the outside standards at least every six bays, either across two bays, as shown on the right-hand side of Figure 21.3, or continuously from bottom to top. Where façade bracing is across a single bay or not fitted between ledger braced frames, plan bracing is required.

When a client requests the omission of ledger bracing to a scaffold consideration should be given to introducing a unit transom at each lift. Unit transoms are prefabricated

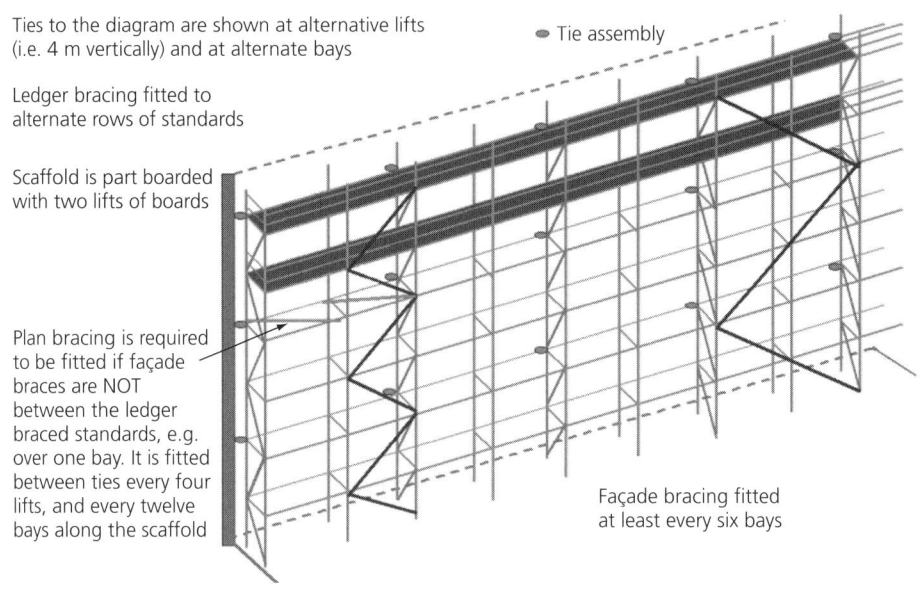

Figure 21.3 Typical part boarded independent tied scaffold. (Courtesy of Pallett TemporaryWorks Ltd)

structural transom units that are inherently stronger and stiffer than traditional scaffolding using transom tubes and putlog couplers. In many cases it is possible to erect a scaffold with unit transoms without the need for ledger bracing. Guidance on the use of unit transoms and the safe height of unclad scaffolds may be derived as described in the following section.

21.5.4.5 TG20 compliant scaffolds

Provided the scaffold is braced correctly, is within the loading limits and is correctly tied to the building it can be regarded as a 'TG20-compliant scaffold'. The safe height in metres for either an unclad, debris netted or fully sheeted scaffold can then be obtained using the TG20:13 eGuide (NASC, 2013c). This is considered a standard configuration of scaffold and no further calculations are required. Compliance of a scaffold compliant with TG20:13 (NASC, 2013a) can be demonstrated through the use of TG20:13 compliance sheets which are produced using the TG20:13 eGuide (NASC, 2013c).

The TG20:13 eGuide (NASC, 2013c) also calculates TG20-compliant scaffolds for those arrangements using high-grade, thin-walled scaffold tube and permits the use of certain prefabricated structural transom units, but only if the transom units conform to the onerous NASC specification for such structural transom units (NASC, 2014b).

As soon as any of the parameters alter, such as a decision to have the full imposed load on more than one platform at a time, the structure is invalidated as a TG20-compliant scaffold and calculations are required by law. A manual method of evaluating the safe heights for such scaffolds, for unclad scaffolds only, is given in appendix B of TG20:13 (NASC, 2013b).

21.5.5 Design of proprietary system scaffolds
21.5.5.1 General
There are many types of proprietary system scaffolds available, and some of the components were discussed in Section 21.4.3. Typical examples are shown in Figures 21.4 and 21.5. Under the Health and Safety at Work etc. Act 1974 (UK Government, 1974) the supplier or importer of articles for use at work has a duty to ensure that the items are suitable, have relevant safety information and are fit for purpose. BS EN 12810-1:2003 and BS EN 12810-2:2003 (BSI, 2003b,c) provide specifications and design information for scaffolds made of prefabricated components.

Suppliers/importers have duties under the Sale and Supply of Goods Act 1994 (UK Government, 1994) to ensure the items are supplied as described. This imposes duties on the users of such scaffolds to ensure that scaffolds are supplied, used and subsequently dismantled following the instructions of the suppliers or importers. This

Figure 21.4 Typical proprietary unclad system scaffold. (Courtesy of SGB Ltd)

Figure 21.5 Typical proprietary unclad system scaffold to large house. (Courtesy of Richard T. Mair, Safe Access (Highland) Ltd)

is usually in the form of a user guide, which provides the loading capability and safe heights of scaffolds, including tying patterns. Proprietary equipment suppliers will often provide erection manuals and safety DVDs, and in certain cases will run training courses and/or workshops to familiarise users with the system. It is important that such scaffolds are used as intended.

Fixed-length transom units with patented fittings to each end (known as 'prefabricated structural transoms') are commonly used, saving erection time and predetermining the scaffold width. These are proprietary scaffold items, not tube and fitting scaffolds, and information on their use and specification should be sought from the manufacturer or supplier. The TG20:13 eGuide (NASC, 2013c) includes certain prefabricated transom units provided they conform to a NASC specification (NASC, 2014b).

21.5.5.2 Design considerations

Although the vertical standards generally have the same outside diameter as the scaffold tube, some have a thinner-walled tube, which reduces the weight and makes transportation and handling easier. The individual member connections have different strengths in each plane and often provide enough stiffness to eliminate ledger bracing in the direction away from the building. Façade bracing is generally still required, but there are alternatives on some proprietary systems.

One significant design consideration is the lack of continuity along the length of the scaffold. The ledgers are not continuous and stop at each standard; each pair of standards generally needs to be tied to the building, and this can increase the amount of tying needed. It is always recommended to refer to the latest supplier's technical data.

The imposed load on proprietary systems is often higher than with the equivalent tube and fitting scaffolds. This can have advantages due to the speed of erection, particularly on large and straightforward façades. The platforms have to fit at the modular connection points of the scaffold, so platforms cannot always be fitted to suit the ideal work location. This is in contrast to tube and fitting scaffolds, which can be fitted to suit the work required.

21.6. Workmanship and inspections

Scaffolds should be erected and dismantled by competent people under supervision and following industry standards. The recognised trade association for tube and fitting scaffolds is the NASC, which provides up-to-date guidance and codes of good practice. The CISRS has a series of 'toolbox talk' training videos providing technical guidance for tube and fitting scaffolding erected to TG20:13 (NASC, 2013a–c). The document SG4:15, *Preventing Falls in Scaffolding Operations* (NASC, 2016a), provides recommendations for safe systems of work and best practice. All scaffolds will require means of access/egress and the document *Access and Egress from Scaffolds via Ladders and Stair Towers etc.* (NASC, 2014a) provides detailed guidance on access to scaffolds. For recommendations on working and safety issues with proprietary equipment, refer to the supplier or importer.

Regulation 12 of the Work at Height Regulations 2005 requires that all working platforms be inspected prior to use after exposure to conditions likely to have caused deterioration (i.e. after high winds or local flooding) and at suitable intervals. The latter refers to scaffolds that have been in position for some considerable time, say more than 3 months, where the fittings may have become loose. It should also be considered that the lifespan of a scaffold should not be longer than 2 years. After this time, more permanent loads (especially wind loading) should be considered, and the scaffold may require additional design.

In addition, the Work at Height Regulations 2005 (UK Government, 2005, 2007) state that all platforms used in construction where a person could fall 2.0 m should be inspected every 7 days. All contractors and subcontractors should be aware of this and should maintain a register of such inspections.

REFERENCES
BSI (British Standards Institution) (1990a) BS 1139-1.1:1990. Metal scaffolding. Tubes. Specification for steel tube. BSI, London, UK.
BSI (1990b) BS 1139-1.2:1990. Metal scaffolding. Tubes. Specification for aluminium tube. BSI, London, UK.
BSI (2001) BS EN 39:2001. Loose steel tubes for tube and coupler scaffolds. Technical delivery conditions. BSI, London, UK.
BSI (2003a) BS EN 12811-1:2003. Temporary works equipment. Scaffolds. Performance requirements and general design. BSI, London, UK.

BSI (2003b) BS EN 12810-1:2003. Facade scaffolds made of prefabricated components. Product specifications. BSI, London, UK.

BSI (2003c) BS EN 12810-2:2003. Facade scaffolds made of prefabricated components. Particular methods of structural design. BSI, London, UK.

BSI (2005a) BS EN 74-1:2005. Couplers, spigot pins and baseplates for use in falsework and scaffolds. Couplers for tubes. Requirements and test procedures. BSI, London, UK.

BSI (2005b) BS EN 1993-1-1:2005 + A1:2014. Eurocode 3. Design of steel structures. General rules and rules for buildings. BSI, London, UK.

BSI (2005c) BS EN 1991-1-4:2005 + A1:2010. Eurocode 1. Actions on structures. General Actions, Wind Actions. BSI, London, UK.

BSI (2006) BS EN 10219-1:2006. Cold formed welded structural hollow sections of non-alloy and fine grain steels. Technical delivery requirements. BSI, London, UK.

BSI (2008) BS EN 74-2:2008. Couplers, spigot pins and baseplates for use in falsework and scaffolds. Special couplers. Requirements and test procedures. BSI, London, UK.

BSI (2019) BS 5975:2019. Code of practice for temporary works procedures and the permissible stress design of falsework. BSI, London, UK.

NASC (National Access & Scaffolding Confederation) (2008) *Guide to Good Practice for Scaffolding with Tube and Fittings*. NASC, London, UK, TG20:08.

NASC (2013a) *TG20:13 Operational Guide. A Comprehensive Guide to Good Practice for Tube and Fitting Scaffolding*. NASC, London, UK.

NASC (2013b) *TG20:13 Design Guide. A Comprehensive Guide to Good Practice for Tube and Fitting Scaffolding*. NASC, London, UK.

NASC (2013c) *TG20:13 eGuide. A Comprehensive Guide to Good Practice for Tube and Fitting Scaffolding*. NASC, London, UK.

NASC (2014a) *Access and Egress from Scaffolds via Ladders and Stair Towers etc*. NASC, London, UK, SG25:14.

NASC (2014b) *Minimum Structural Properties and Test Procedure for TG20 Compliant Prefabricated Structural Transom Units*. NASC, London, UK.

NASC (2016a) *Preventing Falls in Scaffolding Operations*. NASC, London, UK, SG4:15.

NASC (2016b) *User Guide to Preventing Falls in Scaffolding*. NASC, London, UK, SG4:You.

UK Government (1974) Health and Safety at Work etc. Act 1974. Statutory Instrument 1974/37. The Stationery Office, London, UK.

UK Government (1994) Sale and Supply of Goods Act 1994. Statutory Instrument 1994/34. The Stationery Office, London, UK.

UK Government (2005) Work at Height Regulations 2005. Statutory Instrument 2005/735. The Stationery Office, London, UK.

UK Government (2007) Work at Height (Amendment) Regulations 2007. Statutory Instrument 2007/114. The Stationery Office, London, UK.

UK Government (2015) Construction (Design and Management) Regulations 2015. Statutory Instrument 2015/15. The Stationery Office, London, UK.

Useful web addresses

CONSTRUCT (Concrete Structures Group): http://www.construct.org.uk (accessed 01/08/2018).

National Access & Scaffolding Confederation (NASC): http://www.nasc.org.uk (accessed 01/08/2018).

Temporary Works Forum: http://www.twforum.org.uk (accessed 01/08/2018).

Pallett, Peter F and Filip, Ray
ISBN 978-0-7277-6338-9
https://doi.org/10.1680/twse.63389.301
ICE Publishing: All rights reserved

Chapter 22
Falsework

Peter F Pallett
Pallett TemporaryWorks Ltd

Andrew Jones
Chief Engineer, RMD Kwikform Ltd

Falsework comprises the temporary structure used to support other structures, usually permanent, until they can support themselves. Falsework is loaded for a short time, and is often highly stressed, and after loading has to be destressed under load. Constructed generally of reusable equipment, the skeletal nature of such temporary structures makes their stability during erection, use and dismantling a crucial design consideration. In certain cases the structures may rely on parts of the completed permanent works for stability. The design parameters and boundary conditions for falsework are significantly different from permanent works and this necessitates special consideration. Good procedures and workmanship are essential to ensure safety.

22.1. Introduction

Any temporary structure used to support a permanent structure while it is not self-supporting is known as 'falsework'. Although usually considered as the support of in situ concrete for slabs, bridges, and so on, it is also used to support pre-cast units/segments, structural steelwork and timber structures, and is used as backpropping (see Chapter 28). Falsework has different design parameters and boundary conditions to those used in the permanent works: it is loaded for a short time; it regularly involves the use of reusable components; the erection tolerances are greater; it is often destressed under load (so that it can be removed); it is rarely tied down, and therefore relies on its own weight for stability; it is usually stressed to 90% of its safe capacity; and the design period is often short.

The development of the modern British falsework standards can be traced back to a report on falsework published by the Concrete Society and the Institution of Structural Engineers in 1971 (CS/ISE, 1971). It is interesting to note that it proposed a number of classes of falsework. A British Standards Institution (BSI) code committee on falsework was formed shortly afterwards, in 1972. Coincidentally, this was only 2 weeks before the collapse of bridge falsework over the River Loddon near Reading, which generated a government advisory report known as the Bragg Report (Bragg, 1975). BS 5975 was eventually published in 1982, taking into account the recommendations made in the

Bragg Report. BS 5975:1982 contained procedures as well as sufficient technical content to enable a permissible stress design of falsework to be completed without reference to other sources of information. The current version of this standard, BS 5975:2019 (BSI, 2019), makes it clear that the procedures apply to all temporary works, including falsework (see Chapter 2).

A limit state code on falsework, BS EN 12812 (BSI, 2008), was first published in December 2004. It classifies falsework design into three classes (A, B1 and B2), but makes no provision for the safe management of either falsework or temporary works. Note that the existing code BS 5975:2019 exists in parallel with BS EN 12812:2008.

This chapter provides information on the design and use of falsework. It is based on the recommendations of the UK falsework standard, but the principles are generally applicable.

22.1.1 Permissible stress versus limit state

The principal difference between permissible stress and limit state design is how safety factors are applied. The former uses a single factor applied to the capacity of the member to obtain a safe working capacity, often referred to as the 'safe working load'. In the limit state the partial factors are applied to both the load and the member capacity. Limit state design is more complex and can provide a more efficient result. However, the design effort is greater, and for basic falsework design where design time is often limited, limit state is rarely justified.

In limit state design two values of member strength are used: characteristic strength and design resistance. Characteristic strength is the minimum assumed failure load of the item and design resistance is the characteristic strength divided by a suitable partial material factor. Neither of these give the allowable capacity of the item, and it is only when a partial factor on load has been applied that a safe design is produced. This has the potential to cause confusion and possible safety issues for site personnel who work with and understand the 'safe working load' concept. It is therefore important that authoritative guidance that has been prepared by engineers is provided to site personnel, communicated in terms they understand. For example, a crane on site is known to be able to lift, say, 10 t. Its rating says so, and although its original design will have been to the limit state, the end-user information is defined in 'safe working load' terms. This is also discussed in Chapter 21 on access scaffolding.

In essence, limit state design checks that the design value of the loads is less than the design resistance of the structure.

Unfortunately, different partial safety factors are used in various types of temporary works, and often are not similar to those used in permanent works design. To assist the industry, PAS 8812:2016 (BSI, 2016) provides guidance on the application of European design codes for temporary works.

The different methodology of the two design codes means that they cannot be combined.

22.1.2 Choice of standard

Unless specified by the client, the falsework designer is free to choose to use permissible stress to BS 5975:2019 (BSI, 2019) or limit stress to BS EN 12812:2008 (BSI, 2008), discussed in the sections below. If the client does specify BS EN 12812:2008 for the design they should also specify that the procedures in BS 5975:2019 (BSI, 2019) are used.

Similarly, unless specified, the checker is free to choose which standard to use for the check. The results of designs using permissible stress and the limit state to BS 5975:2019 and BS EN 12812:2008 should prove broadly similar. The main difference is likely to be with regard to overall stability, for which the requirements in BS EN 12812:2008 are more stringent.

22.1.3 Use of BS 5975:2019 – permissible stress

Permissible stress using BS 5975:2019 (BSI, 2019) assumes that the applied load is less than the safe working load of the material

$$\text{safe working load} = \frac{\text{failure load}}{\text{factor of safety}} \quad (22.1)$$

BS 5975:2019 for permissible stress design recommends a factor of safety against collapse for temporary works of 2.0. Although not explicitly stated, as the steel design is based on BS 449-1:1970 (now withdrawn) the factor of safety against yield is 1.65.

22.1.4 Use of BS EN 12812:2008 – limit state

The European falsework code BS EN 12812:2008 (BSI, 2008) has three design classes: A, B1 and B2.

Class A is defined as falsework which follows established good practice and which may be deemed to satisfy the design requirements. This is only applicable to defined simple situations where the structure supports a slab less than 300 mm thick, with support to the underside of the permanent works less than 3.5 m and a span less than 6 m – in other words, about 95% of all building construction within Europe. BS EN 12812:2008 gives no structural guidance on designing class A falsework but the UK National Foreword recommends that class A falsework be designed using the philosophy of BS 5975 (BSI, 2019).

Class B1 falsework is one for which a complete structural design is undertaken in accordance with the Eurocodes but modified by specific clauses from BS EN 12812:2008.

Class B2 is similar to class B1 but some simplifications are allowed. This class was introduced when tube and fitting and/or timber bent falsework structures were commonly used in Europe.

Limit state design checks that the design value of the loads is less than the design resistance of the structure. BS EN 12812:2008 introduces a further factor (the class factor) to take into account simplifications in the design process. Applying the partial factors and the additional class factor gives

$$(\text{actual loads}) \times (\text{partial load factor}) \leq \frac{\text{characteristic strength}}{(\text{partial material factor}) \times (\text{class factor})} \quad (22.2)$$

where the partial load factor is 1.35 for self-weight and 1.50 for all other loads; the partial material factor for both steel and aluminium falsework is 1.1. The class factor for design class B1 is 1.0 and for design class B2 is 1.15.

Note that the recommended partial factors used in falsework differ in two significant ways from those used in permanent design

- The 1.35 load factor can only be used for the self-weight of the falsework itself and not for the structure being supported (i.e. the permanent work is regarded as an imposed load in falsework, with a load factor of 1.50).
- The partial material factor for steel and aluminium is increased to 1.1. The UK National Annexes to the Eurocodes state a partial material factor of 1.0 for permanent work.

A more rigorous design is required for class B1 than for class B2. Note that the characteristic strength, or characteristic value, of a member or component is a statistically proven minimum ultimate load (usually stated with a 95% confidence limit, i.e. 5% of components will fail below the value stated).

22.2. Materials and components
22.2.1 Proprietary falsework equipment

There are many different types of falsework system on the market; many are available for hire as well as for purchase (Figure 22.1). Always refer to the recommendations of the supplier for the load capacity and so on. The load capacity is not always obvious, for example, a steel modular system using vertical standards of 48.3 mm outer diameter (generally to suit the scaffold couplers) may look the same as traditional scaffold tube but actually have a different wall thickness and/or be of different grades of steel. The thinner/lighter materials are less expensive, reducing transportation weight and making erection and dismantling easier.

Proprietary aluminium support systems and bearers from suppliers, which are being increasingly used, have high strength/weight ratios. As aluminium is more ductile than steel, it results in greater elastic deflection and the stiffness is more critical in designs. To utilise the higher load capacities, the vertical legs are often spaced further apart. This may not always be economical when you consider other factors such as providing access for operatives to strike out the soffit forms and the permanent work deflection limits as set in the specification. Figure 22.2 depicts a typical example of an aluminium support system.

The majority of aluminium systems in use rely on the soffit formwork being restrained by elements of the permanent works to provide the lateral restraint for the falsework. This is

Figure 22.1 Typical fully braced proprietary steel falsework to a bridge. (Courtesy of Harsco Infrastructure)

Figure 22.2 Typical partially braced proprietary aluminium falsework with ledger frames. (Courtesy of Jim Murray, PERI (UK) Ltd)

known as 'top-restrained' falsework and is discussed in Section 22.4.4. Refer to BS EN 12813:2004 (BSI, 2004) and BS 5975:2019 (BSI, 2019) for the methods of structural design for the load on bearing towers when using prefabricated components.

22.2.2 Scaffold tube and fittings

The use of scaffold tube and fittings, described in more detail in Chapter 21 on access scaffolding, is usually limited in falsework to providing additional components to the system, such as diagonal bracing, extra working platforms within the falsework and connections of falsework areas, and to assist alignment of members.

22.2.3 Adjustable telescopic props

The adjustable steel telescopic prop, generically referred to in the industry as an 'Acrow' (named after the inventor's accountant!), was the traditional method of support for many years. BS EN 1065:1999 (BSI, 1999) introduced 32 types of European adjustable telescopic steel props in five strength classes, and states their load capacity in terms of characteristic strength. BS 5975:2019 (BSI, 2019) states the capacity of these props in terms of safe working load.

Adjustable aluminium props are now used extensively in falsework, mostly for building construction. Their low weight and high capacity make them easily transportable and economical to use on site as falsework and as individual backprops. Many of the systems have 'gates' or 'ledger frames' to connect the legs together. BS 5975:2019 refers to these as 'bracing frames'. The structural capacity of such systems should be obtained from the supplier. Ledger frames are often fitted as aids to erection and, while not designed as such, may contribute to the stability of the falsework. Designers and users should be clear as to their purpose. When joined together, the systems often make tables of soffit formwork. This is discussed in more detail in Chapter 24 on soffit formwork.

As already inferred, care should be taken when using props with very different elastic properties, as the stiffer ones will tend to attract a higher proportion of the loads. The amount a prop shortens under load depends on its cross-sectional area, length and material elastic properties. Although aluminium has lower elastic properties than steel, the cross-sectional area for a given capacity is normally greater than that for the equivalent steel prop. If props are mixed the actual axial shortening under load of the different props should be calculated. Collapses have occurred because of a site mixing steel and aluminium props. The axial shortening of a member can be calculated using (see also Section 22.3.6)

$$d = \frac{FL}{EA} \tag{22.3}$$

where d is the overall shortening of the member (m), F is the applied axial load (kN), L is the length of the member (m), E is the elastic modulus of the material (for steel use 2×10^8 kN/m²; for aluminium use 6.9×10^4 kN/m²) and A is the cross-sectional area (m²).

22.3. Loads on falsework
22.3.1 General
The loads defined for falsework in BS 5975:2019 (BSI, 2019) and BS EN 12812:2008 (BSI, 2008) are aligned and are taken from the Eurocodes. Loads can either be permanent actions (i.e. self-weight) or variable actions (i.e. the supported structure and the weight of operatives).

Loads on structures, referred to as 'actions' in European standards, are given in Eurocode 1. Densities, self-weight and imposed loads are covered in Part 1-1 (BSI, 2002), snow loads in Part 1-3 (BSI, 2003), wind loads in Part 1-4 (BSI, 2005a), and variable actions (live load) and self-weight wet concrete in Part 1-6 (BSI, 2005b).

22.3.2 Permanent loads
The self-weight of the soffit formwork is usually taken as 0.5 kN/m^2. The falsework weight itself is highly critical in the design. Falsework is rarely tied to the ground and its self-weight is often the only load preventing it from overturning under the worst wind conditions. Falsework self-weight is often related to the volume; typical values of self-weight loads are 0.10–0.15 kN/m^3.

22.3.3 Imposed loads
The weight of the structure being built (i.e. the permanent works) is an imposed vertical load. The load can be just vertical (static and impact) or it can have some non-vertical components, such as a lateral load imposed by concrete pressure. In such cases, discontinuities in the soffit may cause lateral forces (see Chapter 24 on soffit formwork). If void formers are used in the concrete their flotation can be significant; similarly, pipes or ducts cast into concrete also exhibit flotation. Remember that the uplift forces will be counteracted by downward pressure, and the design depends entirely on the method of restraint used (see the Concrete Society's guide to formwork (CS, 2012)). Generally there is no resultant upwards force in the supporting falsework.

For design purposes, BS EN 12812:2008 (BSI, 2008) recommends the density of concrete to be taken as 2500 kg/m^3 (i.e. assume 25 kN/m^3). Where smaller building slabs are under construction, the flat slab guide (CONSTRUCT, 2003) recommends that the concrete density for backpropping be 24 kN/m^3 but that the falsework supports be designed for 25 kN/m^3.

22.3.4 Loading from construction operations
- *Working area load.* To allow for the imposed load from construction operations, a distributed load of 0.75 kN/m^2 (i.e. service class 1) is applied over the whole working area.
- *Variable transient in situ loading allowance.* An allowance for heaping of concrete and small plant required to place concrete. When placing in situ concrete a load is applied over a 3×3 m area. The applied load is between 0.75 kN/m^2 for slabs up to 300 mm and a maximum of 1.75 kN/m^2 for solid slabs greater than 700 mm. Intermediate thicknesses are designed on 10% of the slab self-weight (BSI, 2019; clause 17.4.3.1). Figure 22.3 depicts a working area and transient in situ concrete

Figure 22.3 Working area and transient in situ concrete loading

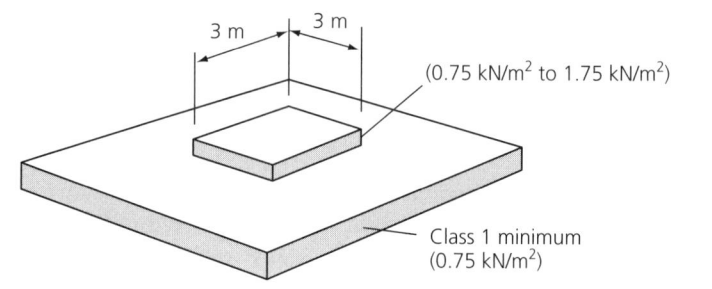

loading. The implication of this transient load when using permanent profiled metal decking formwork, which is typically used on spans up to 3 m, is that all such metal decking should be designed for a minimum imposed load of 1.5 kN/m².
- *Striking formwork load.* To assist the striking and handling of individual items from under a soffit, BS 5975:2019 (BSI, 2019; clause 19.1.1) recommends a working platform be fitted about 2 m underneath the soffit forms. The platform is designed for an imposed class 1 service load of 0.75 kN/m².

22.3.5 Environmental loads

Unless shielded, falsework will be exposed and must be designed to withstand the maximum wind force. The UK wind code BS EN 1991-1-4:2005 + A1:2010 (BSI, 2010) is based on the maximum 10-min mean wind speed likely to occur once in 50 years. BS 5975:2019 (BSI, 2019) introduces the option of using a simplified method to calculate the likely maximum and working wind force on a falsework structure. Because wind codes are written for permanent structures, factors specific to the wind on the soffit and parapet forms are included to give an upper limit to the maximum wind force on the falsework.

Where falsework is exposed to high winds, and there are long slender members, such as bracing, incorporated in the falsework design, the effects of fatigue caused by vortex shedding may need to be considered.

The basic equation, using the simplified method in BS 5975:2019, for the peak wind velocity pressure q_p at a location for a falsework is given by

$$q_p = 0.613 \times c_{prob}^2 \times c_e(z) c_{e,T} \times S_{wind} \tag{22.4}$$

where q_p is the peak velocity pressure (N/m²), c_{prob} is the probability factor (a minimum value of 0.83 is recommended), $c_e(z) c_{e,T}$ is the combined exposure factor (BSI, 2019; clause 17.5.16) and S_{wind} is the wind factor at the location (m/s) (see Chapter 21).

The wind factor S_{wind} incorporates the fundamental wind velocity ($V_{b,map}$) for the site, a topographical factor to allow for the terrain and the altitude (m) of the site (BSI, 2019; clause 17.5.1.33).

The likelihood of the maximum wind occurring during the life of the falsework is catered for by a probability factor. This has conventionally been associated with a 2-year return period for falsework, but the Eurocodes take a more risk-based approach in the decision process.

Generally, placing concrete on falsework will not be undertaken when the wind speed exceeds the operating limits set for the construction plant. This is usually at Beaufort scale force 6, which corresponds to a wind speed of 18 m/s. This is known as the 'working wind', and it generates a working wind pressure of 0.20 kN/m^2.

The procurer will have previously decided whether the falsework will be unclad, sheeted or debris netted. This will have a significant effect on the magnitude of the wind forces that have to be considered.

22.3.6 Indirect loads: settlement and elastic shortening

As loads are applied, the supported structure will move because of the elastic shortening of the actual falsework members and the settlement of the foundations under the load. In certain cases there could be differential settlement between the various members. According to BS 5975:2019 (BSI, 2019; clause 19.3.1), typical values of elastic shortening of falsework are

- for steel falsework, $c.$ 0.5 mm/metre of height (+0.5 mm per joint +1.0 mm per timber joint)
- for aluminium falsework, $c.$ 0.9 mm per metre of height.

These values are based on typical falsework legs and loadings. The actual values should always be calculated (see Equation 22.3).

22.3.7 Minimum horizontal disturbing force

Investigations during the 1970s and in 2005 (Burrows *et al.*, 2005) identified that one of the principal causes of falsework collapse was the absence of stability in members and the structure as a whole (this is also discussed in Chapter 2 on management). For this reason, BS 5975:2019 (BSI, 2019; clause 19.2.9.1) recommends that all falsework be designed for a minimum horizontal disturbing force F_H applied at the top of the falsework. This is defined as the greater of either 2.5% of the vertically applied loads W or all known horizontal loads plus 1% of the vertical load W, as an erection tolerance load to cater for workmanship during erection.

22.3.8 Variable persistent horizontal imposed load

BS EN 12812:2008 (BSI, 2008; clause 8.2.2.2) introduces a horizontal load of 1% of the vertical load, in addition to any sway imperfections, as an externally applied load to allow for minor forces not otherwise identified.

Note this is not a requirement of BS 5975:2019 and is different from the minimum horizontal disturbing force (see Section 22.3.7).

22.4. Falsework design
22.4.1 Method of analysis
The detailed analysis of falsework structures with multiple members pinned together, with joints of various stiffness, is complex and often does not lend itself to elastic analysis software. Traditionally, simplifications have been used to calculate the internal forces, with the effects of vertical and horizontal forces being considered separately and then added together.

One of the differences between BS 5975:2019 (BSI, 2019) and BS EN 12812:2008 (BSI, 2008) is in the approach to the analysis. BS 5975:2019 recommends first-order analysis with a large amplification factor (minimum horizontal disturbing force) to take into account any second-order effects. BS EN 12812:2008 recommends second-order analysis, taking into account appropriate initial sway imperfections and eccentricities. There is therefore no requirement in BS EN 12812:2008 for the minimum horizontal disturbing force. Notwithstanding this, the UK Foreword to BS EN 12812:2008 (BSI, 2008) states (referring to BS 5975:2019) that 'the application of this force has made a significant contribution to the safe use of falsework since its introduction'.

22.4.2 General
When designing falsework, consideration must be given to the fact that it could either buckle or fall over in any direction. A general rule to follow is: 'Think vertical, think horizontal, then think horizontal again'. In practice, this means completing four checks, namely

1. the structural strength of the members and the connections
2. the lateral stability of the falsework structure
3. the overall stability
4. the positional stability (i.e. will it slide?).

The method adopted for the design, whether limit state or permissible stress, will consider these four checks, which are written into BS 5975:2019 (BSI, 2019; clause 19.4.1.1) and are recommended as the starting point of any falsework design. The importance of an accurate and relevant falsework design brief, usually prepared by the Temporary Works Coordinator (see Chapter 2 on management) for use by both the Temporary Works Designer and the design checker, cannot be overstressed. A good brief will provide safe, economical and effective falsework structures.

Falsework needs to be considered at different phases of its construction. BS 5975:2019 (BSI, 2019; clause 19.3.3.1) leaves it open to the designer to identify the critical phases, while BS EN 12812:2008 (BSI, 2008; clause 8.4; table 1) recommends four combinations of load cases that will normally be taken into account. Each of these needs to be considered separately, but there are likely to be two critical phases

1. Phase 1 erected but with only self-weight, and maximum wind (Figure 22.4(a)) (i.e. prior to concreting)
2. Phase 2 on the day of concreting with the full vertical load but with the working wind (Figure 22.4(b)).

Figure 22.4 Typical load combinations for lateral stability. (Courtesy of Pallett TemporaryWorks Ltd)

VERTICAL forces from self-weight of formwork and falsework

LATERAL force
 = 1% (self-weight)
 + maximum wind force

(B2) = 2% (self-weight)
 + maximum wind

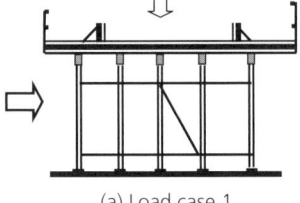

Typical load combinations to check lateral stability

LOAD CASE 1

Erected but
not concreted, with
maximum wind force
(no working on site!)

(a) Load case 1

VERTICAL forces from self-weight of formwork and falsework,
the construction operations load and the permanent works (e.g. concrete)

LATERAL force
 = 1% (vertical loads listed)
 + working wind force

(B2) = 2% (vertical loads)
 + working wind

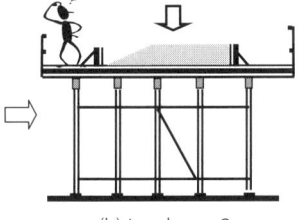

LOAD CASE 2

During placing of
permanent works, with
full vertical load and
reduced wind force

(b) Load case 2

VERTICAL forces from self-weight of formwork and falsework,
the construction operations load and the permanent works (e.g. concrete)

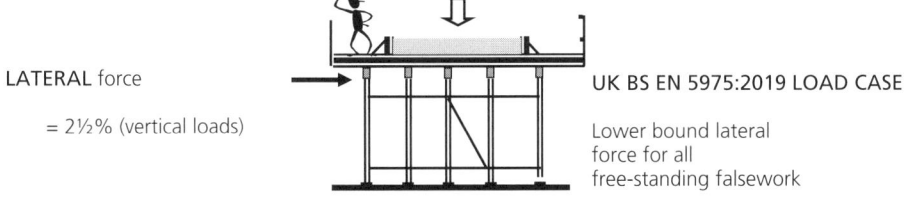

LATERAL force

 = 2½% (vertical loads)

UK BS EN 5975:2019 LOAD CASE

Lower bound lateral
force for all
free-standing falsework

(c) Minimum horizontal disturbing force (BS EN 5975:2019)

Note to (c). The lower bound minimum specified load is considered to act at the point of contact between the vertical load and the falsework. If though the falsework is restrained at its head then a different load case needs to be considered, as outlined in (d) below.

RESTRAINT force
 = 2½% (vertical loads)

(B2) = 2% (self-weight)
 + maximum wind

OR = 2% (vertical loads)
 + working wind

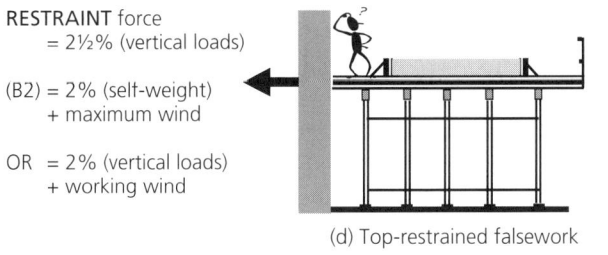

TOP-RESTRAINED LOAD CASE

Where the structure is
restrained by the permanent
works the formwork will
have to transmit the lateral
force to suitable restraints

(d) Top-restrained falsework

22.4.3 Check 1: Structural strength

The designer will check the axial load, bending, shear, deflection, (occasionally) torsion and the local stability of individual members and connections, such as welded or bolted joints. When detailing individual elements, account should be taken of the maximum tolerances for workmanship (unless provisions are taken to minimise the eccentricities, etc.).

The designer will need to consider the bracing required to restrict the effective length of any axially loaded members. BS 5975:2019 (BSI, 2019; clause 19.2.9.2) gives the horizontal restraint force required to provide a fixed node in a strut as 2.5% of the axial load in the strut. Limit state codes recommend 1% per member per joint, so two members intersecting at a node point require a horizontal restraint of 2% – similar in magnitude to that given in BS 5975:2019. This is a notional load and is used to check that the bracing is stiff and strong enough to prevent the struts buckling. Many proprietary falsework systems have patented joints that provide some moment fixity to the standards. These can reduce the effective length factor, thus increasing their safe load capacity. The recommendations of the suppliers should be followed. Particular care should be taken with unrestrained cantilever extensions on falsework at either the head or the base jacks. The effective length of such members in cantilever depends on whether or not the falsework is top restrained or free-standing (see BSI, 2019; clause 19.4.2.4.4).

For structural steel beams carrying concentrated loads, web stiffeners should be provided at all loading transfer points, including supports, unless calculations are provided to show that such stiffeners are not required (BSI, 2019; annex J). The omission of web stiffeners has been the cause of fatal accidents, so their importance must not be underestimated.

22.4.4 Check 2: Lateral stability

The lateral stability of a structure is its ability to remain upright and stable under both horizontal and vertical loads without swaying sideways significantly or falling over. Consideration of the lateral stability of falsework is particularly important and is one of the primary recommendations of the Bragg Report (Bragg, 1975).

BS 5975:2019 (BSI, 2019; clause 19.2.9.1) states that the lateral stability of all falsework should be checked by applying a horizontal disturbing force at the head of the structure, which from Section 22.3.7 will be a minimum of 2.5% of the vertically applied loads W (see also Section 22.4.1). BS EN 12812:2008 (BSI, 2008) is not so specific but does specify a variable persistent horizontal imposed load and second-order analysis (see Sections 22.3.8 and 22.4.1). Although this will provide a broadly similar result it is recommended that the check given in BS 5975:2019 is used to ensure lateral stability. Any bracing required for lateral stability is not cumulative with the bracing for structural strength, and if bracing has been inserted to cater for the horizontal disturbing force then, on a braced structure, the structural strength requirement stated in Section 22.4.3 has been satisfied.

In the design of falsework an important distinction is made between falsework which is top restrained and falsework which is free-standing. Top-restrained falsework does not provide its own lateral stability but relies on the top soffit formwork to provide the restraint; effectively it is the structure's walls and/or columns that provide the stability.

The three load cases depicted in Figures 22.4(a)–22.4(c) illustrate free-standing falsework structures with diagonal bracing providing the restraint for structural strength and lateral stability.

The load case depicted in Figures 22.2 and 22.4(d) illustrate top-restrained falsework. The bracing in Figure 22.4(d) would only be fitted to satisfy the structural strength requirement, although some bracing might be required to aid initial erection. The assumption made by the designer is that the head of the falsework is restrained from movement in both lateral directions. This requires that the permanent works are able to resist these notional stability forces. The Temporary Works Designer should provide any required stability reactions as part of their design output, and the Permanent Works Designer should then confirm that the permanent works can safely resist them. Communication and coordination between the parties can be an issue, and the responsibility for ensuring that this task is adequately completed remains with the Temporary Works Coordinator (see Chapter 2).

All parties should be absolutely clear about whether falsework has been designed on the assumption that it is top restrained and, if so, whether the permanent works have been designed to provide the required restraint (i.e. it is stiff enough in the temporary condition). If not, the falsework should be designed as free-standing. Because of this extra complexity, Table 2.2 in Chapter 2 excludes top-restrained falsework from design check category 1 (simple design). More information on top-restrained falsework is given in BS 5975:2019 (BSI, 2019).

Lateral stability also applies to the webs of steel beams at reaction points and at positions of concentrated loads. Web stiffeners are required at all loading transfer points, including supports, unless calculations are provided to show that such stiffeners are not required (BSI, 2019; annex J). The comment at the end of Section 22.4.2 applies equally here.

All spanning beams should also be considered for lateral torsional buckling. Torsional bracing at the supports and lateral bracing of the compression flanges should be provided where necessary. Permanent works designers are aware of the potential instability of steel beams on composite bridge designs (where relatively small top flanges are designed for shear connection to the concrete), but not necessarily of the forces in the compression flange during erection, which can require additional lateral bracing. See BS 5975:2019 (BSI, 2019; annex K) for the effective length of steel members in axial compression.

22.4.5 Check 3: Overall stability

Tall slender structures and those exposed to high winds will have a tendency to be blown over. As falsework is generally a gravity structure (relying on its own weight for stability), the most onerous condition is often just after the soffit formwork is erected and before reinforcement has been fixed (i.e. with minimal restoring self-weight moment). If the falsework is unstable, then holding-down bolts or kentledge may have to be used. Remember that whenever using kentledge (weights) there is no leeway; you know the point at which it will turn over! BS 5975:2019 (BSI, 2019; clause 19.4.3.1) requires a minimum factor of safety of 1.2 against overturning. BS EN 12812:2008 (BSI, 2008; clause 9.2.2.3) refers to overturning as 'static body equilibrium', and gives different partial load factors for destabilising loads compared to those stabilising the falsework.

When using systems that have cantilever working platforms there is always a risk of overturning. This is particularly the case when narrow aluminium tables are used, where it may be necessary to tie these down to the slab.

22.4.6 Check 4: Positional stability
The lateral disturbing forces on the falsework may be transmitted by static friction or mechanical connection into the foundations, but not both at the same time. This interface is usually through base plates and supporting timber sole plates (see BSI, 2019; clause 19.4.5). The recommended values for the coefficient of static friction μ are listed in BS 5975:2019 (BSI, 2019; table 24), where

$$\mu = \frac{\text{limiting frictional force}}{\text{reaction normal to the surface}} = \frac{F_f}{R} = \frac{W \sin \theta}{W \cos \theta} = \tan \theta \qquad (22.5)$$

where F_f is the frictional restraint force (kN), W is the vertically applied force (kN), R is the reaction normal to the surface, θ is the minimum angle to the horizontal at which sliding will commence (degrees) and P is the force applied to overcome the static friction (kN) (see Figure 22.5).

The friction restraint does not depend on the area of contact. The applied load multiplied by the coefficient of static friction μ gives the value of frictional restraint (the value at which it slides) with no factor of safety.

BS 5975:2019 (BSI, 2019) recommends that the minimum factor of safety against sliding is 2.0, whereas BS EN 12812:2008 (BSI, 2008; table 2) gives different partial safety factors for stabilising loads ($\gamma_i = 0.9$) and destabilising loads ($\gamma_i = 1.35–1.5$).

Figure 22.5 Friction forces on slopes

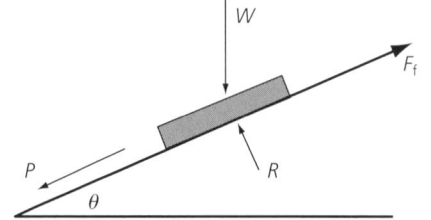

F_f, frictional restraint force (kN)
W, vertically applied force (kN)
R, reaction normal to the surface (kN)
θ, minimum angle (degrees) to the horizontal at which *sliding will commence*
P, force applied to overcome static friction (kN)

22.4.7 Difference between fully and partially braced falsework

In a fully braced tower each falsework leg is considered as an individual strut restrained at regular intervals by lacing and bracing. The legs are designed using simple strut theory and the lacing and bracing is checked to ensure that it provides adequate nodal restraint.

Although at first glance partially braced and fully braced towers look similar, their structural behaviour and design methods are completely different. The discontinuity in bracing means that there is no nodal restraint and, instead of considering each leg as an individual strut, the pair of legs and bracing frames are considered as a composite battened strut with an effective length based on the full height of the legs. They cannot be analysed simplistically and it is necessary to use a second-order analysis, taking into account initial bow imperfections and end conditions. Because of this complexity the analysis is carried out on standard configurations and then published in the form of charts or tables by the suppliers. The jack extension and number and location of frames can have a significant effect on the capacity of partially braced towers.

22.4.8 Erection tolerance

Whenever falsework structures are erected, the design should take into account any inaccuracy in the erection. As most falsework structures are not positively bolted together but rely on clamps and variable fixings, such as scaffold couplers, it is hard to erect them exactly vertical and without any eccentric connections or loads. This gives rise to 'sway imperfections'.

BS 5975:2019 (BSI, 2019; clause 20.3.2) gives maximum allowable erection tolerances, and as long as these are not exceeded a nominal horizontal reaction of 1% of the applied vertical forces is used to take account for them (BSI, 2019; clause 19.2.9.1(b)). Where there is both eccentricity and sway, a more rigorous analysis taking into account second-order effects known as P-Δ may need to be considered (BSI, 2019; clause 19.3.3.3).

Class B1 design to BS EN 12812:2008 (BSI, 2008) requires the designer to assess the structure and make appropriate allowance for eccentricity and sway in the analysis.

Class B2 design advice is given in BS EN 12812:2008 on the amount of sway imperfection allowed in the design (BSI, 2008; clause 9.3.4.2). The sway imperfection φ for structures taller than $h = 10$ m is calculated as

$$\tan \varphi = 0.01 \sqrt{\frac{10}{h}} \qquad (22.6)$$

where h is the height (m) and φ is the angular deviation from the theoretical line.

Where the structure is <10 m high, a lower limit of $\tan \varphi = 0.01$ is considered (i.e. use 1% of applied loads for falsework up to 10 m high).

22.5. Workmanship and inspections

As falsework often involves the use of reusable items (often second-hand), and is generally erected, destressed and then dismantled and removed, the procedures and controls are often more rigorous than for permanent works. Constant emphasis is needed on attention to detail.

Analysis of falsework is increasingly being carried out by computers. The theory is often complicated but, at the end of the day, engineers on site need simple design rules and experience in order to assess and inspect falsework. The '10% continuity rule' for random bearers is an example (BSI, 2019; clause 19.3.3.2). Although booklets like the falsework checklist produced by the Concrete Society (CS, 1999) are useful, they are poor substitutes for experience when checking and designing falsework structures.

The reasons for experience and checks are highlighted in the following examples.

- The incorrect lapping of single primary bearers in forkheads, by forgetting to cross-lap, led to a major collapse.
- 'Installing cross-bracing at every fourth bay' in a large falsework was interpreted on site as 'every four bays sideways' (not in its length). Only one in four rows of falsework were therefore braced, again leading to a major collapse.
- While installing soffit formwork on a 10 m high falsework, the operatives removed the bracing because it got in their way. The bracing was not put back, and this was not noticed during inspection. The 10 m high falsework collapsed during concreting, with seven men falling 10 m, luckily sustaining only minor injuries.
- A proprietary aluminium system using ledger gates for bracing was used on a bridge falsework; each leg carried over 100 kN. One eye bolt connecting a ledger frame to a leg, in one place, was not tightened up, and the falsework leg buckled sideways under the load.
- Very large props were used at each end of a pre-cast concrete beam for stability while the top in situ deck was concreted. Unfortunately, one piece of softwood plywood had been fitted to pack one of the 300 kN leg load props. The plywood crushed from 20 to 5 mm, causing the entire bridge to drop at one end and skew on its bearings, causing a temporary closure of a main line London station!

All the above failures of falsework could have been avoided had correct procedures for checking and inspection by competent persons and trained Temporary Works Coordinators been adopted. The integrity of the whole structure may be impaired by the omission of a bolt or wedge, or even just failing to tighten them up correctly. The designer's and/or supplier's advice should always be sought.

Care must be taken when erecting and dismantling falsework. The size of falsework structures means that there should be some clearly defined stages in the construction. For example, formal checks and/or hold points might need to be established after completing the foundations, when it reaches its support level and, obviously, prior to loading. In addition to the provision of safe working areas for erecting, striking and so on, are the conditions of use as envisaged in the design brief?

A common error on falsework is failing to take account of the elastic settlement under load (see Section 22.3.6). For this reason, experienced erectors will set the top level of the falsework higher than the final required level to allow for shortening under load. For example, a 5.4 m high reservoir roof slab support would be expected to elastically shorten by c. 5–6 mm if steel falsework is used, this shortening increasing to nearer 10 mm if aluminium falsework is used. An example of the consideration of elastic shortening when casting two slabs of different thickness at high level in a concrete silo is given for backpropping in Chapter 28.

Safe falsework requires competent experienced people, with an eye for detail. Failures usually occur because small items are missed or forgotten, or because of a lack of proper process and control.

REFERENCES

Bragg SL (1975) *Final Report of the Advisory Committee on Falsework*. Department of Employment and Department of the Environment. HMSO, London, UK.

BSI (British Standards Institution) (1999) BS EN 1065:1999. Adjustable telescopic steel props. Product specifications, design and assessment by calculation and tests. BSI, London, UK.

BSI (2002) BS EN 1991-1-1:2002. Eurocode 1. Actions on structures. General actions. Densities, self-weight, imposed loads for buildings. BSI, London, UK.

BSI (2003) BS EN 1991-1-3:2003 + A1:2015. Eurocode 1. Actions on structures. General actions. Snow loads. BSI, London, UK.

BSI (2004) BS EN 12813:2004. Temporary works equipment. Load bearing towers of prefabricated components. Particular methods of structural design. BSI, London, UK.

BSI (2005a) BS EN 1991-1-4:2005 + A1:2010. Eurocode 1. Actions on structures. General actions. Wind actions. BSI, London, UK.

BSI (2005b) BS EN 1991-1-6:2005. Eurocode 1. Actions on structures. General actions. Actions during execution. BSI, London, UK.

BSI (2008) BS EN 12812:2008. Falsework. Performance requirements and general design. BSI, London, UK.

BSI (2016) PAS 8812:2016. Temporary works. Application of European Standards in design. Guide. BSI, London, UK.

BSI (2019) BS 5975:2019. Code of practice for temporary works procedures and the permissible stress design of falsework. BSI, London, UK.

Burrows M, Clark L, Pallett P, Ward R and Thomas D (2005) Falsework verticality: leaning towards danger? *Proceedings of Institution of Civil Engineers – Civil Engineering* **158(1)**: 41–48.

CONSTRUCT (Concrete Structures Group) (2003) *Guide to Flat Slab Formwork and Falsework*. Concrete Society, Camberley, UK, CS140.

CS (Concrete Society) (1999) *Checklist for Erecting and Dismantling Falsework*. Concrete Society, Camberley, UK, CS123.

CS (2012) *Formwork: A Guide to Good Practice*, 3rd edn. Concrete Society, Camberley, UK, CS030.

CS/ISE (Concrete Society/Institution of Structural Engineers) (1971) *Falsework: Report of the Joint Committee*. Concrete Society, London, UK, TRCS4.

Useful web addresses

CONSTRUCT (Concrete Structures Group): http://www.construct.org.uk (accessed 01/08/2018).

National Access & Scaffolding Confederation (NASC): http://www.nasc.org.uk (accessed 01/08/2018).

Temporary Works: http://www.temporaryworks.info (accessed 01/08/2018).

Temporary Works Forum: http://www.twforum.org.uk (accessed 01/08/2018).

Temporary Works, Second edition

Pallett, Peter F and Filip, Ray
ISBN 978-0-7277-6338-9
https://doi.org/10.1680/twse.63389.319
ICE Publishing: All rights reserved

Chapter 23
Formwork

Peter F Pallett
Pallett TemporaryWorks Ltd

Laurie York
Consultant

'Formwork' is the term used to describe the fabrications and constructions used to form the shape of concrete structures, acting as the mould. It is normally removed once the concrete has achieved sufficient strength, although it is sometimes left in place as permanent formwork, which may or may not contribute to the structural capacity of the formed concrete. Formwork is either vertical (for walls, columns, beam sides, etc.) or horizontal (for supporting slabs, cantilevers, underside of beams, etc.). Horizontal formwork is described in Chapter 24 on soffit formwork.

23.1. Introduction
Traditional formwork is fabricated using timber-based materials but steel, glass-fibre-reinforced plastics (GRP), glass-fibre-reinforced cement and other materials are also used. Proprietary formwork includes the walings, bearers, soldiers, supports and various panel systems and is available from specialist formwork supply companies. Choice of material is often based on the required number of uses of the formwork and the finish specified for the concrete by the Permanent Works Designer, with more expensive materials often justified by more onerous surface finish and durability requirements for the concrete.

Designing formwork and ensuring its subsequent control on site requires a thorough understanding of the pressures generated during the placing, compacting and setting of concrete. Knowledge of formwork materials and practical experience are both essential to ensure that formwork is economical and easy to fabricate and use. The procedures for and management of formwork are described in Chapter 2.

23.2. Vertical formwork
Formwork is defined as the structure, usually temporary but in some cases wholly or partly permanent, used to contain poured concrete, to mould it to the required dimensions and to support it until it is able to support itself. It consists primarily of the face contact material and the bearers that directly support the face contact material. The authoritative industry guidance for formwork is the Concrete Society book *Formwork:*

Figure 23.1 A typical arrangement of double-faced wall formwork with soldiers. (Courtesy of Pallett TemporaryWorks Ltd)

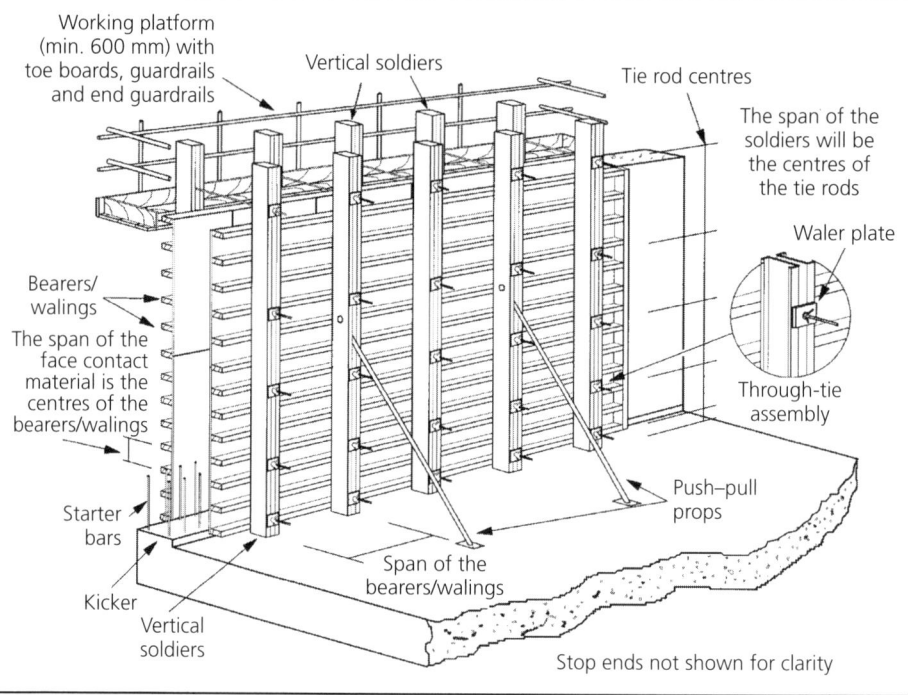

A Guide to Good Practice (CS, 2012). To resist the large pressure of the concrete on the form face, metal rods in tension, known as 'tie rods', are used to connect opposing faces of formwork. This is known as 'double-faced formwork' and a typical example comprising a face material, horizontal members, known as 'walings', and vertical stiff steel soldiers is shown in Figure 23.1. The tie rods balance the forces and the inclined props provide stability and allow alignment of the forms.

Where there is only one face of the form, such as against an existing wall or the ground, the formwork is known as 'single-faced formwork'. The forces can be significant and are discussed in Section 23.8.3.

23.3. Economy

Many people in the process can affect the economy of the construction. These include the Permanent Works Designer, the contractor, the subcontractor and specialist suppliers, as well as the site operatives. The cost of formwork will vary depending on the contract specification and also on the number of uses required of the formwork. On a concrete frame building, the cost of formwork and falsework can be 39% of the cost of the structure, increasing on a 'civils' contract to as much as 55% of the cost of the concrete structure. A better-quality face contact material may be more expensive but may be necessary either to produce the required finish on the concrete or to enable reuse of the formwork without costly refurbishment. Similarly, a more robust form may be more expensive but allow more reuse. A considerable percentage of the costs is associated with the labour involved; formwork design that reduces the labour requirement will generally be more economical. Examples include the reduction in the number of tie rods, the use of proprietary panels to reduce initial make-up costs and use of aluminium bearers to reduce the weight of the formwork panels (permitting larger form areas for the same size of crane).

23.4. Specifications and finishes

The surface finish required for the concrete will generally be described by means of a specified performance, where the standards and tolerances required are stated but the contractor is left to decide how to achieve the finish. The concrete surface, together with the surface zone of the concrete, will be affected by the choice of material used to create the face.

BS EN 13670:2009, 'Execution of concrete structures' (BSI, 2009), defines the generic surface finishes. The two main UK specifications are: *National Structural Concrete Specification for Building Construction* (NSCS) (CONSTRUCT, 2010; section 8.6) and the Highways Agency specification (HA, 2004; volume 1, clause 1708). Table 23.1 compares the two specifications. Other specifications used include the UK Water Industry Research (UKWIR) *Civil Engineering Specification for the Water Industry* (UKWIR, 2011; clause 4.28) and the National Building Specification 'E20: Formwork for in situ concrete', which is published annually online.

The NSCS (CONSTRUCT, 2010) refers to full-size reference panels at seven locations in the UK, which may be viewed by visiting the various locations. Other sources of information include the Concrete Society technical report *Plain Formed Concrete Finishes* (CS, 1999). As well as generic surface finishes, special finishes required under the contract may be defined.

Table 23.1 Comparison of surface finishes

Highways Agency (HA, 2004)	NSCS (CONSTRUCT, 2010)	Examples
Class F1	Basic	For pile caps, etc.
Class F2	Ordinary	Rear, unseen faces of retaining walls (panel systems give F2)
Class F4	Plain	Quality surfaces to walls
Class F3	Plain	Class F4 but no ties; visible parapets (hardest to achieve)
Class F5	Plain	Class F4 embedment of metal parts allowed (intended pre-cast)
–	Special/fair worked	High-class finish (e.g. potable water tanks)

23.5. Tolerances/deviations

Tolerances specified for the formwork are not necessarily the same as those specified for the permanent works. The three sources of deviations in the surface of the finished structure are: (*a*) inherent deviations, such as the elastic movement of the formwork under load; (*b*) induced deviations, such as the lipping between two sheets of plywood (possibly within plywood manufacturing tolerances); and (*c*) errors, such as inaccurate setting-out. Inherent deviations include deflection.

It is generally accepted that appearance and function are satisfied if deflection of individual formwork members is limited to 1/270th of the span. In certain cases (such as where decoration is to be applied directly onto concrete walls in housing) reduced limits may need to be specified, but it should be noted that it is the deflection of the individual formwork members that is limited and not the overall final concrete shape.

The acceptable magnitude of the deviations will depend on the specified quality of the work and also the distance from which the concrete surface will be viewed. Typical deviations in verticality for normal work will be 20 mm from a grid line for a 3 m high wall. More information is available in the Concrete Society formwork guide (CS, 2012; section 2.6).

23.6. Formwork materials
23.6.1 Face contact material

Perhaps the most common sheet material that is used as face contact material is plywood. Plywood is a layered material made up of sheets, which are normally 1.22 × 2.44 m (4 × 8 ft), with properties related to the direction of the face grain of the outer ply layer. It is generally used in 17.5, 19 or 25 mm thicknesses. The Concrete Society formwork guide (CS, 2012) gives details of working properties taken from the previous British Standard and for various plywood and other wood-based panel materials, together with details for other commercially available sheet materials. Plywood has different properties depending on the direction of the face grain; some plywoods have less than 50% capacity if used the wrong way round. Economical design is based on the plywood spanning in its 'strong' direction, and it is important that it is fixed in the correct orientation.

The use of expanded metal steel products (Hy-rib) in construction joints, in place of plywood, almost eliminates the need for scabbling (i.e. removes the risk of problems associated with vibration known as 'white finger') as the metal is left in place. Trials by the British Cement Association have shown that there is a significant reduction in concrete pressure when expanded metal products are used (see CS, 2012).

The use of fabric such as Zemdrain on the face contact material introduces control over the permeability. This is known as 'controlled permeability formwork' (see CIRIA/CS, 2000). By allowing the pressure of the concrete to force the surplus pore water and any air into the fabric and to be drained away, there is an almost complete elimination of blowholes on the concrete surface. The face strength of the concrete is increased by 30% in the critical cover zone of the concrete. A recent development is the use of composite sheets having a foamed polystyrene core with a synthetic thermoplastic material face to both sides or, for wall formwork, with a thin reinforcing sheet of aluminium to both sides. Repair techniques on such sheets are specialised.

23.6.2 Bearers

In wall formwork, bearers are typically aligned horizontally, supporting the face contact material; in such cases they are often referred to as 'walings'. These are often composed of softwood constructional timber (strength classes C16, C24 or C27). Site conditions, such as exposure to wetting, affect the strength of timber. The duration of loading and the facility for load sharing will also affect the apparent strength of the timber, as will the depth of the timber section used. The Concrete Society formwork guide (CS, 2012) provides details of safe working properties for various commonly used timber section sizes.

Alternatively, proprietary beams fabricated from timber, aluminium or steel can be used. Many of the European systems of formwork incorporate proprietary timber bearers fitted vertically. Such beams should be used in accordance with the manufacturer's instructions at all times.

23.6.3 Soldiers

Soldiers can be timber or proprietary beams; in Figure 23.1, the soldiers have central slots for the positioning of tie rods. The soldiers support the bearers, which span horizontally between them. Either design properties given in the Concrete Society formwork guide (CS, 2012) or information from the manufacturer should be used.

Alternatively, proprietary beams or 'soldier-type' members may be installed horizontally to support vertically fitted bearers.

23.6.4 Formwork ties

To resist the bursting effect of the pressure of the concrete, the opposing faces of wall formwork (in double-faced formwork) are connected by tie rods. The most common type of tie rod is the through tie. A typical through-tie system comprising a high-strength threaded bar (generally 15 mm diameter) passed through an expendable plastic tube is depicted in Figure 23.2. Simple plastic cones are fitted at the ends of the tube.

After use, the bar is removed and reused. The safe load taken by the bar can be large (of the order of 110 kN) and waler plates of sufficient size will be required to spread the load into the soldiers or bearers. The Concrete Society formwork guide (CS, 2012) recommends that, when designing with these recoverable tie assemblies, a minimum factor of safety of 2 is adopted based on the minimum guaranteed ultimate strength of the bars used. Alternatively, safe working loads may be supplied by the tie manufacturer.

Other types of tie rods include taper ties, lost ties, wire ties, coil ties and mild steel all-thread ties. The lost-tie system utilises tapered end 'she-bolts' screwed to each end of threaded tie rods. This allows the entire assembly to be placed through the wall form

Figure 23.2 Typical through-tie assembly. (Courtesy of RMD Kwikform (UK) Ltd)

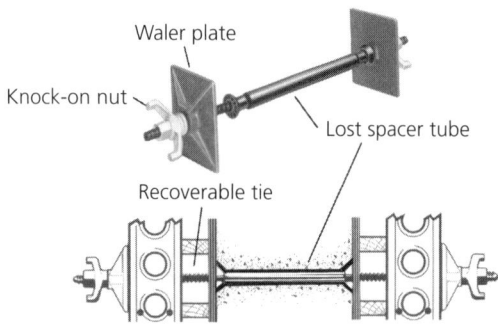

from one side during erection and permits the she-bolts to be subsequently removed, leaving the tie rod embedded in the concrete. The length of the lost tie rod is specified to suit the wall thickness and the cover required to the reinforcement.

It is important to note that suppliers of tie-rod systems will state recommended safe working loads for the ties *used in tension* in wall formwork. Whenever used in other applications, the safe loads will alter as the factors of safety increase, and reference must be made to the tie-rod supplier.

23.6.5 Proprietary panels

Proprietary panel systems are now common, and typically have plywood face contact material fitted within a steel or aluminium frame. A typical example is shown in Figure 23.3.

Formwork comprising proprietary panels gives a greater potential for reuse for the plywood by reducing damage during handling. Panels are either small enough for manual handling (say 1200 × 600 mm), or they may be large crane-handled panels (up to 2400 × 2700 mm). The main advantage is the speed of initial use, with little 'make-up' time. It should be noted that such panel systems, when on hire, will generally give 'ordinary' finish (class F2, suitable for rear, unseen faces of retaining walls), rather than 'plain' finish (class F3, F4, F5, suitable for quality surfaces to walls). They will often have far fewer components and they have the significant benefit of being available on hire.

23.6.6 Release agents

To obtain satisfactory finishes it is very important that the release agent used on the formwork is suitable. A satisfactory finish will not be achieved if an incorrect release agent is used. It should be stored, used and applied in accordance with the recommendations of the supplier. Any subsequent treatment to the concrete surface may be affected and the supplier's recommendations on compatibility should always be sought. Release agents are classified by numbered categories. A fuller treatise is given in section 3.10.2 of the Concrete Society formwork guide (CS, 2012) but the following should be taken into consideration with regard to chemical and vegetable release agents.

Figure 23.3 Typical proprietary panel formwork system. (Courtesy of PERI (UK) Ltd)

23.6.6.1 Chemical release agents
Chemical release agents (category 5) are various blends of chemicals in suspension in light oil solvents. Coverage rates are 35–50 m^2/l. On being sprayed onto the formwork the light oil evaporates, leaving a chemical to react with the cement to form the barrier. They are of a 'drying' nature, and are suitable for plywood, timber, steel, and so on, and most forms. However, over-application resulting in build-up on the forms can lead to dusting of the resultant concrete surface. There are inherent health and safety implications, as the chemicals are dissolved in oils; disposal needs careful consideration.

23.6.6.2 Vegetable oil release agents
Vegetable oil release agents (VERAs; category 7b) are finding increased usage in construction. Being non-toxic, biodegradable vegetable oil (generally based on rapeseed oil), they are suitable for use in confined spaces and for all formwork. Some are supplied in plastic bags, to be dissolved in and diluted with water for use.

23.7. Concrete pressure calculation
23.7.1 General
The pressure exerted by the concrete influences the choice of face contact material, as well as dictating the design of the formwork. Correct calculation of the pressure is very important in designing formwork for walls and columns. The magnitude of the design pressure for a particular formwork arrangement should be communicated to the site team so that those placing the concrete will be able to limit the rate of placing the concrete such that the design pressure is not exceeded. Drawings of formwork should state

Figure 23.4 Fluid pressure on vertical formwork

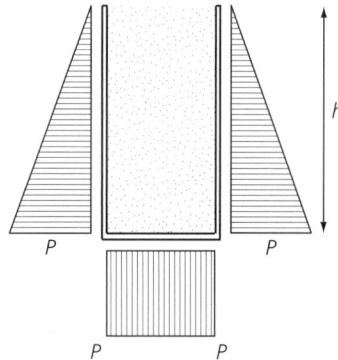

the maximum pressure in units of kilonewtons per square metre (kN/m²), and users of proprietary panels should be aware of the limiting design pressure for their particular arrangement. Changing the tie-rod diameter and/or locations can affect the limiting pressure on the proprietary panel system. Whenever a fluid such as wet concrete is placed in a vessel, the pressure exerted by the fluid at any depth P (Figure 23.4) is defined as

$$P = hD \qquad (23.1)$$

where P is the pressure (kN/m²); h is the depth (m) and D is the density (kN/m³). For concrete, the density is taken as 25 kN/m³. Therefore, under 3 m head of fluid concrete the pressure is 75 kN/m².

The pressure always acts at right angles to the face. The force on the formwork can therefore be calculated as pressure × area acted upon.

23.7.2 Effect of stiffening of the concrete

Concrete does not always act as a fluid because, with time, it starts to stiffen and become a solid. For concrete, pressures may be reduced if stiffening occurs during the placing operations, as shown in Figure 23.5. The full fluid pressure head is not realised on the vertical faces of the lower part of the formwork. It should be noted that the soffit form has applied to it the full vertical weight of concrete without stiffening (i.e. the mass). It is obviously more economical to design the wall/column formwork to the limiting value of concrete pressure, P_{max}. There are many factors which affect the concrete pressure, including the temperature of the concrete and whether or not retarding admixtures, pulverised fuel ash (PFA) or ground-granulated blast furnace slag (GGBS) are incorporated.

Basic research on the pressure imposed by concrete on formwork was published in CIRIA R108 (Harrison and Clear, 1985). Since then there have been changes in the types of cement and in concrete specifications (to suit British Standard EN documents, etc.). Research at the University of Dundee (Dhir *et al.*, 2004) concluded that the methods

Figure 23.5 Effect of concrete stiffening on vertical formwork

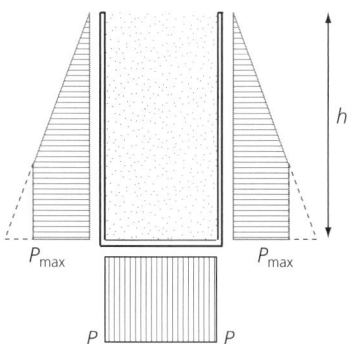

adopted in CIRIA R108 were safe for use, subject to an understanding of the type of concrete used. Concrete mixes using different cement combinations were first classified into three groups for purposes of concrete pressure determination by Pallett (2009). The groups are

- group A – basic concrete
- group B – retarded concrete
- group C – heavily retarded concrete.

23.7.3 The pressure of concrete on formwork

When designing formwork for parallel-sided walls up to 10 m tall or columns up to 15 m tall, tables PAA, PBB and PCC in section 4.4.3 of the Concrete Society formwork guide (CS, 2012) provide concrete pressures linked to rates of rise of the concrete within the form, for various concrete temperatures and for the three concrete groups. The rate of rise is the speed of the rise in the concrete vertically up the forms (in m/h) and, for parallel-sided forms, is assumed reasonably uniform. The concrete temperature used is not the ambient air temperature; for UK work it is likely to be 5°C in winter, 15°C in summer and 10°C in autumn and spring. An example of a table of the maximum concrete pressure, P_{max}, is reproduced in Tables 23.2 and 23.3. Note in particular the significant change in concrete pressure as the temperature reduces.

For the specified concrete, temperature, form height, required rate of rise of the concrete within the form and the maximum concrete pressure, P_{max}, can be determined. On site, the formwork arrangement is known, the limiting concrete pressure, P_{max}, is known (from drawings or from suppliers' data sheets) and the concrete group is known, but the concrete temperature can vary from day to day. To limit the maximum design pressure the rate of rise of the concrete within the form must be determined so that the required rate of delivery of the concrete can be arranged. There are two ways to determine the rate of rise where the value of P_{max} is known: either use the pressure tables or, more accurately, use the rate of rise tables RWA, RCA, RWB, RCB, RWC and RCC in the Concrete Society formwork guide (CS, 2012; section 4.4.3).

Table 23.2 Example of concrete pressures for group A – basic concrete

Concrete temp.: °C	Form height: m	Walls and bases: A wall or base is a section where at least one of the plan dimensions is greater than 2 m							Form height: m	Columns: A column is a section where both plan dimensions are less than 2 m					
		Rate of rise: m/h								Rate of rise: m/h					
		0.5	1.0	1.5	2.0	3.0	5.0	10.0		2.0	4.0	6.0	10.0	15.0
5	2	40	45	50	50	50	50	50	3	75	75	75	75	75
	3	50	55	60	65	70	75	75	4	85	100	100	100	100
	4	60	65	65	70	75	85	100	6	95	115	125	145	150
	6	70	75	80	80	90	100	115	10	115	135	145	170	190
	10	85	90	95	100	105	115	135	15	130	150	165	190	210
10	2	35	40	45	45	50	50	50	3	65	75	75	75	75
	3	40	45	50	55	60	70	75	4	75	90	100	100	100
	4	45	50	55	60	65	75	90	6	80	100	115	130	150
	6	50	55	60	65	75	85	105	10	95	115	130	150	175
	10	60	70	75	80	85	95	115	15	105	125	140	165	190
15	2	30	35	40	45	50	50	50	3	60	75	75	75	75
	3	35	40	45	50	55	65	75	4	65	85	95	100	100
	4	35	45	50	50	60	70	90	6	75	90	105	130	150
	6	40	50	55	60	65	75	95	10	80	100	115	140	165
	10	50	55	60	65	75	85	105	15	90	110	125	150	175

Note: Concrete group A – basic concrete – comprises
(i) concretes without admixture: CEM I, SRPC, CEM IIA with silica fume or metakaolin; (ii) concretes with any admixture except with retarding properties: CEM I, SRPC, CEM IIA with silica fume or with metakaolin (formerly groups 1 and 2).
Data courtesy of the Concrete Society.

Formwork

Table 23.3 Example of concrete pressures for group B – retarded concrete

Walls and bases:
A wall or base is a section where at least one of the plan dimensions is greater than 2 m

Concrete temp.: °C	Form height: m	Rate of rise: m/h						
		0.5	1.0	1.5	2.0	3.0	5.0	10.0
5	2	50	50	50	50	50	50	50
	3	65	70	75	75	75	75	75
	4	75	80	85	90	95	100	100
	6	95	100	105	105	110	110	135
	10	120	125	130	130	140	150	165
10	2	40	45	50	50	50	50	50
	3	50	55	60	65	70	75	75
	4	60	60	65	70	75	85	100
	6	70	75	80	80	90	100	115
	10	85	90	95	100	105	115	135
15	2	35	40	45	45	50	50	50
	3	40	45	50	55	60	70	75
	4	45	50	55	60	65	75	90
	6	50	60	65	65	75	85	105
	10	65	70	75	80	85	100	120

Columns:
A column is a section where both plan dimensions are less than 2 m

Form height: m	Rate of rise: m/h				
	2.0	4.0	6.0	10.0	15.0
3	75	75	75	75	75
4	100	100	100	100	100
6	120	130	140	150	150
10	145	160	175	195	215
15	170	190	205	225	245
3	75	75	75	75	75
4	85	95	100	100	100
6	95	110	125	145	150
10	115	130	145	170	190
15	130	150	165	190	210
3	65	75	75	75	75
4	75	90	100	100	100
6	80	100	115	135	150
10	95	115	130	155	175
15	105	125	140	165	190

Note: Concrete group B – retarded concrete – comprises
(i) concretes without admixture: CEM IIA, CEM IIB, CEM IIIA; (ii) concretes with admixture that retards: CEM I, SRPC, CEM IIA with silica fume or with metakaolin (formerly groups 3, 4 and 5).
Data courtesy of the Concrete Society.

Note that where pour heights are small (such as in building work), if the formwork is designed for full fluid head (see Figure 23.4) there is no limit to the speed of concreting, as it is not possible to generate a pressure larger than fluid head. Designing a 3 m column form for maximum 75 kN/m² pressure would therefore not require control of the rate of rise at any concrete temperature or type of concrete mix.

23.8. Wall formwork design
23.8.1 Double-faced formwork
Having established the design concrete pressure for the formwork (which always acts at right angles to the face), the formwork is designed in five basic steps. Whether the bearers are horizontal (Figure 23.6(a)) or vertical (Figure 23.6(b)) or whether a proprietary formwork system is being used, the basic principle is *follow the load*. Start at the formwork face and follow the forces, designing through to the restraint which, in double-faced forms, is provided by the tie rods. This is described in the five steps below

1 Check the safe span of the face contact material or plywood. This is usually derived from tables plotting pressure against span. This obviously gives the maximum spacing for the bearers fixed horizontally (see Figure 23.6(a)) or vertically (see Figure 23.6(b)). The span of the face contact material is taken as being the spacing of the centres of the bearers/walings.
2 Check the safe span of the bearers. This can also be derived from tables but more economical designs will be produced by analysing the bearers as continuous beams (where appropriate). It is good practice to design the bearers to project past the last support by a maximum of one-third of the adjacent span, as this improves the economy of the design. The span of the bearers when used horizontally as walings is taken as being the spacing of the centres of the soldiers (or of the stiff horizontal members when the bearers are fixed vertically).
3 Check the design of the soldiers. The tie-rod spacing determines the span requirements for the soldiers, but additional checks may be required if, for example, there are joints in the soldiers. As the design pressure reduces near the

Figure 23.6 Follow the load. (Courtesy of Pallett TemporaryWorks Ltd)

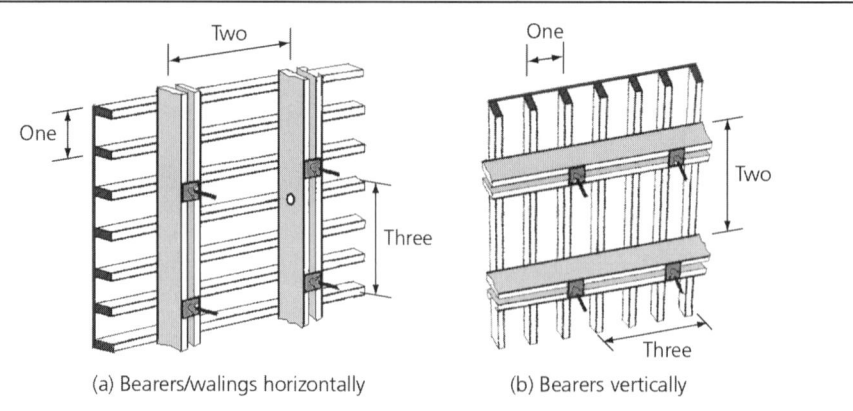

(a) Bearers/walings horizontally (b) Bearers vertically

Figure 23.7 Cross-section of double-faced formwork to wall with propping. (Courtesy of RMD Kwikform (UK) Ltd)

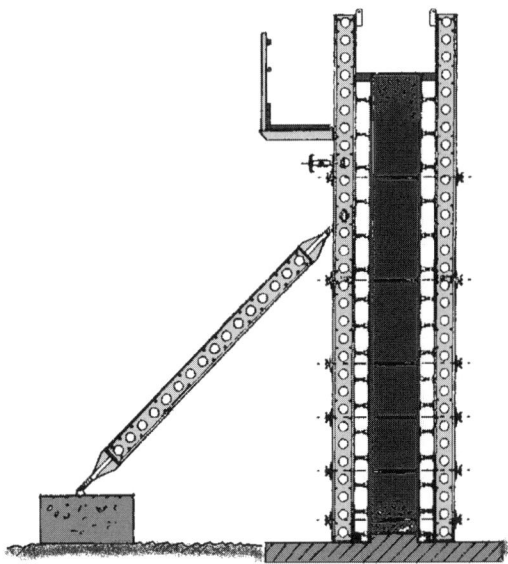

top of the wall, fewer ties are required (Figure 23.7). The layout of ties (and therefore the soldiers) may have been specified by the Permanent Works Designer to suit, for example, a particular surface finish with a regular pattern. The span of the soldiers is taken as being the spacing of the tie rods.
4   Check the capacity of the tie rods for the chosen arrangement of soldiers.
5   Check the deflection of the face of the forms. With such large pressures being imposed on the formwork system, the design should take careful account of deflections with reference to the acceptable tolerances (see Section 23.5). It should be remembered, however, that the deflection limit (normally 1/270th of the span) is applied individually to the face contact material, to the bearers and then to the soldiers; it is not cumulative.

Other considerations to be observed during the design process will include safety issues (working platforms for placing the concrete, Construction (Design and Management) Regulations 2015 (UK Government, 2015) and the Work at Height Regulations 2005 (UK Government, 2005)) and handling of the formwork (weight limit for manual handling and crane or traveller capacity for mechanical handling). See Chapter 1 on safety and the section on mechanical handling of formwork in the Concrete Society formwork guide (CS, 2012; section 5.9).

23.8.2 Stability of formwork
A check will be needed on the stability of the formwork (and any required propping designed for a minimum factor of safety) against overturning, considering the most adverse conditions. The minimum factor of safety on overturning is 1.2. A typical

double-faced wall formwork arrangement with a single inclined push–pull prop, connected to a kentledge block, is shown in Figure 23.7.

Stability checks are carried out for overturning caused by

1. 'maximum wind', plus overturning from a nominal load on any platforms
2. 'working wind' (upper limit at which plant may be operated on site) on the day of concreting, plus overturning from the full load on the platforms
3. 'minimum stability force' (10% of the total self-weight of both faces of the formwork) acting three-quarters of the way up the form plus any overturning from the full load on the platforms.

A simplified method to establish the wind forces (both 'maximum' and 'working') on free-standing wall formwork can be found in the Concrete Society formwork guide (CS, 2012; section 4.5.1).

23.8.3 Single-faced formwork

In certain conditions it is not possible to have two faces of formwork. Examples are walls cast against existing secant pile walls or diaphragm walls. Only one face of formwork can be erected and tie rods cannot be used. These single-faced wall forms will have the same concrete pressures imposed on them as double-faced formwork. The design process is similar, with the exception that the final restraint cannot be provided by tie rods and must, therefore, be provided externally. Single-faced formwork always requires particularly careful attention to detail; it must be remembered that the force generated by the concrete pressure will be horizontal when using vertical formwork.

Figure 23.8 Typical arrangement of single-faced formwork using soldiers. (Courtesy of Pallett TemporaryWorks Ltd)

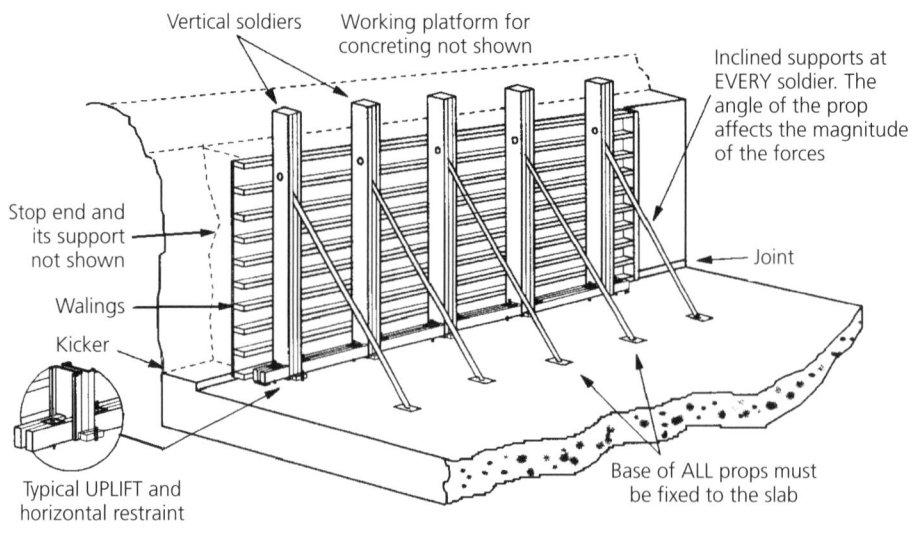

Figure 23.9 Typical arrangement of single-faced formwork using frames. (Courtesy of Pallett TemporaryWorks Ltd)

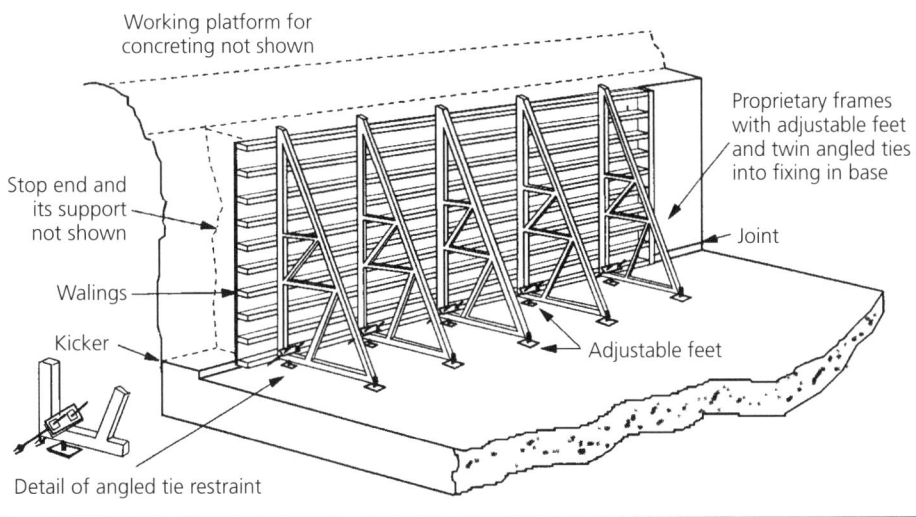

The most common method of support for small single-faced formwork is to use inclined props as shown in Figure 23.8. Because of the inclined props, there will always be a corresponding and significant uplift force at the base of the form. This introduces the need for fixings into the existing slab to provide restraint against the uplift, as well as providing the necessary horizontal restraint.

An alternative method of support for single-faced formwork is the use of proprietary large frames. These incorporate a single connection between the frame and the existing slab, providing both vertical and horizontal restraint to the form; a typical detail is shown in Figure 23.9. It is important to install the ties accurately and at the correct angle to prevent them being incorrectly loaded, which may cause them to fail.

Control of movement of the top of the single-faced formwork is always difficult. Care is required to ensure that the specified tolerances are not exceeded. Loads imposed on the existing slab by the anchors, fixings and props are significant; approval from the Permanent Works Designer should always be sought.

23.9. Column formwork design

Because of their comparatively small cross-section and relatively high rates of rise, column forms are subject to higher concrete pressures than walls. Due to the confined space into which the concrete is placed, tall and inclined columns may require pockets or windows at intervals to facilitate placing and compacting of the concrete (CS, 2012; section 5.5).

Columns may have circular, square, rectangular or other irregular cross-sections. These irregular cross-sections can often be formed by placing inserts or box-outs within standard square or rectangular column moulds. Circular column forms are available made from steel or GRP, and single-use cardboard forms are also available; the latter are particularly useful if only a few columns are required. Steel and GRP column forms can incorporate splayed column heads.

Variations in column shapes and sizes, particularly on the same project, should be kept to a minimum. Ideally, column sizes should be selected in modular increments of 75 mm (e.g. 300, 375, 450, etc.) enabling both the designer and the builder to standardise (and hence economise).

A range of column forms using proprietary panels is available, with some designed to cater for varying rectangular shapes by adjustment on site using 'universal panels' arranged in a 'windmill' pattern. As panel formwork is designed for a specific limiting concrete pressure, it may be beneficial to sequence the construction of the columns with other site concreting work and/or to pour multiple columns simultaneously to reduce the risk of overloading the column formwork. Restricting the pour rate so that the allowable pressure on the forms is not exceeded may not always be a workable solution, and specifically designed formwork may be more appropriate.

23.10. Striking vertical formwork

The time at which striking of vertical formwork can be carried out should be carefully controlled. BS EN 13670:2009 (BSI, 2009) states that 'Formwork shall not be removed until the concrete has gained sufficient strength, and can resist any damage caused by striking'.

A minimum value of in situ concrete cube strength of 2 N/mm^2 is recommended in all cases when striking vertical formwork; see the guidance in *National Structural Concrete Specification for Building Construction* (CONSTRUCT, 2010). The informative annex C.5.7 in BS EN 13670:2009 suggests a concrete strength of 5 MPa (equivalent to 5 N/mm^2) to resist damage when a value is not given in the project specification.

The Concrete Society formwork guide (CS, 2012) recommends a minimum strength of 2 N/mm^2 to reduce the risk of mechanical and frost damage for striking plain and special concrete finishes (classes F3, F4 and F5 (BSI, 2009)).

A lower criterion is generally possible for striking forms with a basic or ordinary concrete finish (classes F1 and F2 (BSI, 2009)), and is usually related to the maturity of the concrete. In practice, provided the mean concrete temperature is above 10°C overnight, the vertical formwork can be struck next morning. If the mean air temperature is above 10°C, it can safely be assumed that the concrete temperature was above 10°C. Where the newly struck concrete may be exposed to frost, a minimum in situ concrete equivalent cube strength of 2 N/mm^2 is recommended.

Note that certain project specifications, for example, the specification for the water industry (UKWIR, 2011), may require higher strengths prior to removal of formwork.

23.11. Workmanship/checking

To produce concrete of a high standard of appearance and precision, a high standard of site workmanship is required; this will apply even if formwork is specially constructed for the job in question. Where the amount of work does not justify the use of specially designed and fabricated formwork, the degree of success in achieving the required result will depend solely on the skill and expertise of the site operatives. In a few cases, the economic solution may be to fabricate a high-quality form for only one or two uses.

There are cases, however, when good-quality work can be achieved with less skilled labour. These will normally be where work is simple, of a highly repetitive nature and where the size of the job justifies the use of sophisticated special-purpose formwork which is designed for simplicity. A considerable amount of the work carried out on site does not demand good appearance or close tolerances, although appreciable skill and experience is still necessary to ensure that the formwork is stable, safe and to the required standard.

All blemishes in the formwork will appear on the surface of the finished concrete. Blemishes will obviously not become apparent until the formwork is struck, by which time the concrete has hardened, making the repairs difficult.

When using proprietary systems, it is important that site management ensures that the operatives construct the systems correctly using the manufacturer's approved components. Once the forms have been constructed the site management must ensure that the concrete placement methods are appropriate for the system being used, and are being used within the pressure limits set by the manufacturer and/or formwork designer (see also Chapter 2).

Checking of formwork should be carried out systematically, as stated in Section 23.8.1. The axiom 'follow the load' also applies to checking of formwork. Two useful guides are *A Guide to the Safe Use of Formwork and Falsework* (CONSTRUCT, 2008) and the Concrete Society's *Checklist for the Assembly, Use and Striking of Formwork* (CS, 2003).

REFERENCES

BSI (British Standards Institution) (2009) BS EN 13670:2009. Execution of concrete structures. BSI, London, UK.

CIRIA/CS (Construction Industry Research and Information Association/Concrete Society) (2000) *Controlled Permeability Formwork*. CIRIA/Concrete Society, London/Camberley, UK, C511.

CONSTRUCT (Concrete Structures Group) (2008) *A Guide to the Safe Use of Formwork and Falsework*. Concrete Society, Camberley, UK, CSG/005.

CONSTRUCT (2010) *National Structural Concrete Specification for Building Construction*, 4th edn. Concrete Centre, Camberley, UK, CCIP-050.

CS (Concrete Society) (1999) *Plain Formed Concrete Finishes: Illustrated Examples*. Concrete Society, Crowthorne, UK, TR52.

CS (2003) *Checklist for the Assembly, Use and Striking of Formwork*. Concrete Society, Crowthorne, UK, CS144.

CS (2012) *Formwork: A Guide to Good Practice*, 3rd edn. Concrete Society, Camberley, UK, CS030.

Dhir RK, McCarthy MJ, Caliskan S and Ashraf MK (2004) *Design Formwork Pressures for the Range of New Cement, Superplasticised and Self-Compacting Concretes*. University of Dundee, Dundee, UK, DTI research contract 39/3/739 (CCC2399), Report CTU/3004.

HA (Highways Agency) (2004) *Manual of Contract Documents for Highways Works*. Vol. 1, *Specification for Highways Works*. HMSO, London, UK.

Harrison TA and Clear C (1985) *Concrete Pressure on Formwork*. Construction Industry Research and Information Association (CIRIA), London, UK, R108.

National Building Specification (annually) E20: Formwork for in situ concrete. National Building Specification, Newcastle-upon-Tyne, UK. (Published online.)

Pallett PF (2009) Concrete groups for formwork pressure determination. *Concrete* **43(2)**: 44–46.

UK Government (2005) Work at Height Regulations 2005. Statutory Instrument 2005/735. The Stationery Office, London, UK.

UK Government (2015) Construction (Design and Management) Regulations 2015. Statutory Instrument 2015/15. The Stationery Office, London, UK.

UKWIR (UK Water Industry Research) (2011) *Civil Engineering Specification for the Water Industry*, 7th edn. WRc plc, Swindon, UK.

FURTHER READING

BRE (Building Research Establishment) (2007) *Formwork for Modern, Efficient Concrete Construction*. HIS BRE Press, Bracknell, UK, BR495.

Useful web addresses

Concrete Society: http://www.concrete.org.uk (accessed 01/08/2018).

CONSTRUCT (Concrete Structures Group): http://www.construct.org.uk (accessed 01/08/2018).

Temporary Works: http://www.temporaryworks.info (accessed 01/08/2018).

Pallett, Peter F and Filip, Ray
ISBN 978-0-7277-6338-9
https://doi.org/10.1680/twse.63389.337
ICE Publishing: All rights reserved

Chapter 24
Soffit formwork

Peter F Pallett
Pallett TemporaryWorks Ltd

The support to the underside of in situ concrete, either a flat or an inclined slab, requires formwork to contain and mould the concrete to the desired shape. Often supported on foundations by means of falsework, the soffit formwork is an important part of temporary works. The supports can only be removed when the concrete has gained sufficient strength, so the assessment of concrete strength and the procedure and sequence for striking needs to be understood and approved before concreting.

24.1. Introduction

The forming of the underside of concrete structures while the concrete gains strength utilises soffit formwork, either inclined or nominally level. Unlike vertical formwork (discussed in Chapter 23), the concrete has to have gained sufficient strength before the structure can support itself, allowing the formwork to be removed. Generally, in situ concreting on soffit formwork will have supporting falsework (see Chapter 22) to transfer the load to suitable foundations. In certain cases, such as permanent formwork or in the case of cantilever soffit formwork, supporting falsework may not be required and the loads are transferred directly to the permanent works.

Soffit formwork is found in civil structures and to the undersides of bridge decks, beams and parapets. In building it is used to form the underside of slabs: ideally flat slabs or, with moulds, slabs with downstands. It is also used to form the underside of beams and in a wide variety of situations. A typical arrangement of soffit formwork to a structure with supporting falsework is depicted in Figure 24.1.

This chapter provides information about the different philosophies of design and use of soffit formwork for both civil and building applications. It includes bridge and slab formwork and cantilever soffits, and details the fast-track method for striking of soffit formwork in building construction in the UK.

24.2. Preamble to soffit form design
24.2.1 General
The main sources of information for soffit formwork and its supporting falsework are the Concrete Society's *Formwork: A Guide to Good Practice* (CS, 2012), CONSTRUCT's *Guide to Flat Slab Formwork and Falsework* (2003) and the British

Figure 24.1 Typical arrangement of soffit formwork and falsework. (Courtesy of Pallett TemporaryWorks Ltd)

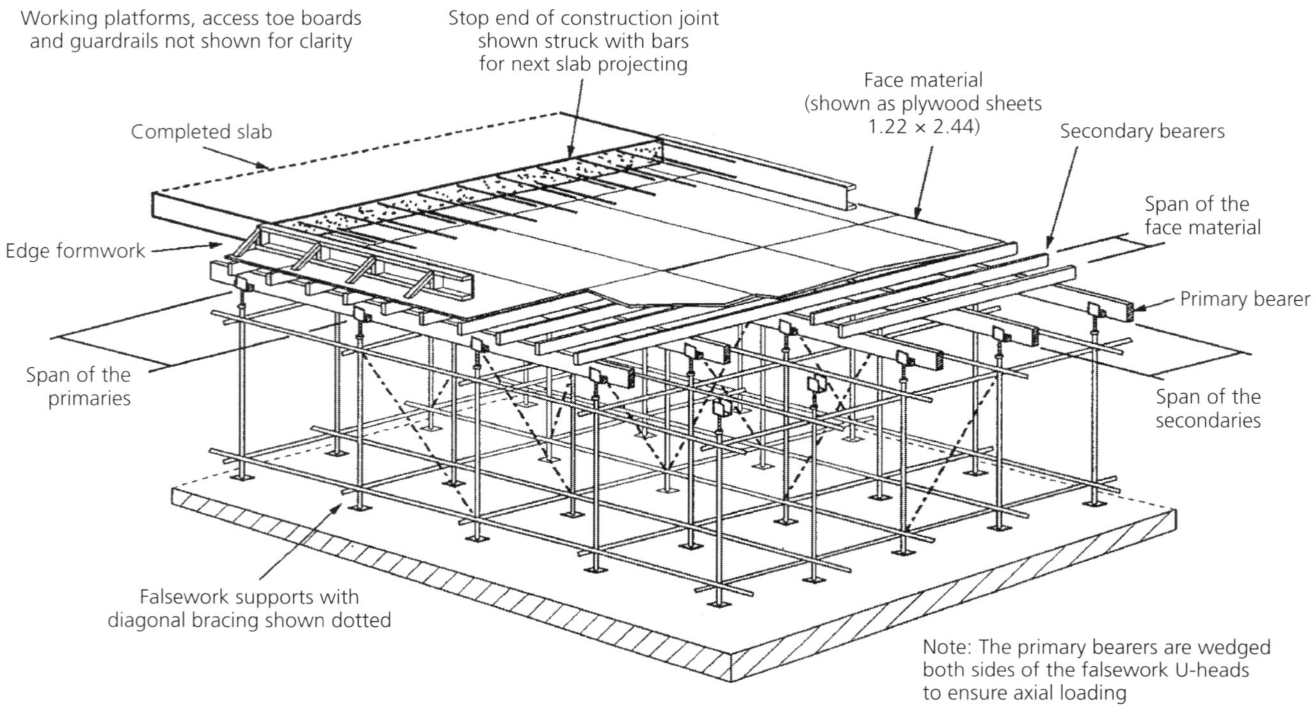

Standards on falsework, namely BS 5975:2019 (BSI, 2019) in permissible stress and BS EN 12812:2008 (BSI, 2008) in limit state terms.

Consider the components of soffit formwork. The face contact material in contact with the concrete could be plywood, wood-based panels (particleboard or oriented strand board), plastic composites, steel or a proprietary panel system. These are usually supported on bearers known as 'secondary bearers'. In turn, the secondary bearers are supported on more substantial bearers, known as 'primary bearers'. The primary bearers fit onto the falsework uprights, usually centralised in adjustable forkheads. The primary bearers may be considered as part of the formwork and designed using the Concrete Society formwork guide (CS, 2012) or as part of the falsework (BSI, 2019); for example, if the primary bearers are constructional softwood the safe load tables are identical in the Concrete Society formwork guide and BS 5975:2019.

24.2.2 Specification and finishes

The contract documents will specify the standard of finish required for the concrete surface of the soffit and this will affect the selection of material used to create the face that will form the visible surface finish.

BS EN 13670:2009, 'Execution of concrete structures' (BSI, 2009), defines the generic surface finishes. The two main UK specifications are the *National Structural Concrete Specification for Building Construction* (CONSTRUCT, 2010; section 8.6) and the Highways Agency specification (HA, 2006; volume 1, clause 1708). Table 23.1 in Chapter 23 provides a comparison of the two specifications. Other specifications used include the UK Water Industry Research (UKWIR) document *Civil Engineering Specification for the Water Industry* (UKWIR, 2011; clause 4.28) and the National Building Specification 'E20: Formwork for in situ concrete', which is published annually online.

Particular care is necessary in specifying the surface finish for bridges, especially for their parapets. The Construction Industry Research and Information Association (CIRIA) report *Bridges: Design for Improved Buildability* (Ray et al., 1996) recommends that the Highways Agency class F3 finish be limited to small vertical areas of the parapet, such as those visible from the highway. Unfortunately, if specified for the entire parapet edge and soffit, the 'no-tie' requirement of class F3 makes restraint of such soffit formwork extremely complex and unnecessarily expensive, and often leads to grout loss at connections with unsightly marks. The solution is to specify class F4 for most of the parapet, leaving the small vertical upstand (visible) face as a class F3 finish. Realistic specifications will produce better results.

24.2.3 Equipment selection

The equipment used to create the soffit formwork for one-off structures, such as in situ bridge decks, will generally be a proprietary falsework support system (steel or aluminium) supporting bearers in two directions (primary and secondary) to support a plywood face contact material. The bearers, traditionally in timber sections, are more likely to be proprietary aluminium beams. The benefit of using aluminium is that, whereas the timber has to be purchased in advance, the aluminium bearers can be from a company's stock or hired from a supplier (see Figure 24.1). Soffits may include voided slabs such as trough or waffle floors or bridge decks incorporating square or circular voids.

Where there are several similar flat slabs to be cast, such as in multi-storey buildings, proprietary components such as formwork panels, proprietary timber and aluminium bearers as well as lightweight aluminium falsework components are frequently used. There are two ways of handling the equipment: in pre-assembled tables, moved as crane-handled sections from use to use; or by 'strip and re-erect' in individual components. There is also a subdivision of the 'strip and re-erect' arrangement in that certain proprietary systems allow 'left in place' props. This permits the more expensive formwork components (often panels) to be struck very early and moved to the next use without disturbing the falsework props supporting the slab. This obviously has implications on the load transfer and sequencing, which is discussed further as backpropping in Chapter 28.

The choice of handling method will have an effect on the speed and economy of the new construction. 'Flying' the tables is considered a high risk activity, mainly because of the possibility of debris collecting on the table and falling off during moving, and 'no-go' zones may need to be established. Thus the location, nearby roads, pedestrian areas and so on can all affect the choice of the method of operation (such items should be included in the design brief – see Chapter 2, Section 2.4.2).

Whereas the 'strip and re-erect' method was popular in the 1980s it was replaced by the use of tables in the 1990s, but the safety implications in today's construction, and in particular the client requirements for full external protection (see Chapter 25), means that 'strip and re-erect' systems are often now used.

Table systems for flat slab construction generally comprise aluminium beams in both directions at the top. A typical example is shown in Figure 24.2. Table systems comprise large-diameter aluminium props (typical diameters are 100–150 mm) with long threaded sections to allow for adjustment, connected together with ledger frames and/or cross-bracing. To reduce the labour costs of dismantling and re-erecting at the next floor, sections of made-up table can be handled between uses either by large hoists between floors or by a crane. The site safety restrictions on the craneage might limit the sizes that can be handled.

The stiffness of the table assembly is derived from long lengths of aluminium primary beams. These tend to limit the lengths of table handled to 12 m with standard components, but longer lengths are possible.

Tables have the benefit that once made up they enable rapid construction, and the time benefits increase with repetitive use. They can be used with cross-wall construction or columns, but are most economical when there is access to opposite faces of the building for direct removal of the tables. Table systems ideally suit flat slab and repetitive

Figure 24.2 Typical aluminium table with ledger frames. (Courtesy of Jim Murray, PERI (UK) Ltd)

construction and become economical at over eight uses. Cycle times as short as 4 days have been achieved with careful planning; see the flat slab guide (CONSTRUCT, 2003) and Section 24.6.3.

Tables need space to be 'flown' out of the building (either to one side or to both sides), with a minimum end allowance of 500 mm for clearance from the building to adjacent structures or objects and to allow for cantilever access platforms. A minimum clearance to columns or walls of 40 mm per side should be allowed to the sides of each table; some infill support is therefore necessary at arises when used with cross-wall construction.

The various methods of handling tables are outside the scope of this book; see the Concrete Society formwork guide (CS, 2012) for detailed information. Consideration must be given to the handling and operation of the system. For example, if handled as individual tables, the lengths of aluminium bearers become critical in striking out after pouring. Physically removing long lengths is extremely difficult and places unnecessary risk on the operatives. Note that BS 5975:2011 (BSI, 2011; clause 19.1.13) recommends that, when handling individual units, a working platform should be fitted about 2 m below the underside of the soffit.

'Strip and re-erect' systems at their most simple are proprietary panels supported on props. An example is shown at Figure 24.3. There are choices of aluminium- or steel-framed panels, matched with either steel or aluminium props. Care must be taken to

Figure 24.3 Typical panel and prop system. (Courtesy of Hünnebeck (a BrandSafway company))

follow the supplier's erection and dismantling procedures to ensure that the systems are stable at all times during erection. The initial use of a bay of ledger frames, as shown on the left-hand side in Figure 24.3, to establish a stable start point for the general erection is often recommended. Such 'strip and re-erect' systems can be manhandled onto pallets for transferring to the next use, possibly using material hoists between floors. Such systems are labour intensive but have little or no effect on safety issues outside the building area.

An adaptation of the simple panel and prop system is the 'drop-head' system, also known as 'quick-strip' systems. The introduction of a proprietary beam connected to a 'drop head' on the prop allows the panels to be supported off the beams. Once the new slab has been cast, the act of striking the 'drop head' lowers the beam and panels by a few centimetres, allowing them to be removed for another use while leaving the slab undisturbed on its supporting props. Once the slab has gained sufficient strength, the supporting props can be removed. Such 'drop-head' systems would typically comprise two or more sets of props for each set of beams and panels. The props, being undisturbed, may become 'left in place' supports for backpropping; this is discussed further in Chapter 28.

24.3. Loading on soffit forms
24.3.1 Vertical
The soffit formwork has to carry not only its self-weight and the weight of the wet concrete being supported (note that wet concrete is regarded as an imposed load, not

a dead load) but also the two construction operations loads (see Chapter 22). These are (*a*) the working area load distributed over the whole working area of 0.75 kN/m^2 (i.e. service class 1) and (*b*) the variable transient in situ loading allowance applied over a 3 × 3 m area, which varies from 0.75 kN/m^2 for slabs up to 300 mm up to a maximum of 1.75 kN/m^2 for solid slabs larger than 700 mm. Intermediate thicknesses are designed on 10% of the slab self-weight (BSI, 2019; clause 17.4.3.1).

24.3.2 Horizontal

In addition to vertical forces, soffit formwork can have applied horizontal forces. These could be from surges in concrete pump lines, impact forces and possibly from arrangements of the stop-ends. All slabs will have stop-ends and/or construction joints. Particular care is necessary on all slab and deck stop-ends greater than 400 mm deep.

When casting a slab against an existing structure or against a previous cast section there will be a lateral force generated from the reaction to the pressure of the concrete (i.e. the fluid head of concrete on the connection) acting on the existing face. For example, a 900 mm slab cast against an existing slab generates a lateral force of about 10 kN per metre run of joint.

A common site error is to ignore discontinuities in the soffit formwork. Where there is a discontinuity in the formwork a lateral force (equivalent to the fluid head of concrete at that point) exists laterally and attempts to move the forms apart. Generally the arrangement of staggered bearers supporting the face material will provide the restraint, but where a section of falsework and formwork is not connected (e.g. between tables of formwork) or where the falsework is staggered to allow for changes in levels, discontinuities can be formed. The solution is usually quite simple: join the sections of falsework together below the soffit formwork. This is discussed in more detail in both the Concrete Society formwork guide (CS, 2012) and BS 5975:2019 (BSI, 2019).

24.3.3 Notional force

One of the topics discussed in Chapter 23 was the use of the soffit formwork to provide the stability of the falsework, known as 'top-restrained falsework'. Where this occurs, the soffit forms will have to transmit the relevant lateral notional forces to a suitable restraint, such as the columns or walls, with the soffit formwork acting as a plate. This is discussed in more detail in BS 5975:2019 (BSI, 2019; clause 19.3.2.4). Designers need to consider the forces in all directions so that both tension and compression may be applied to the soffit formwork.

24.4. Design

The design of soffit formwork is different from that for walls, as discussed in Chapter 23; the loads are usually less. For example, a 900 mm thick in situ concrete bridge deck imparts an imposed load of 22.5 kN/m^2 on the soffit, while a 250 mm slab imparts only 6.25 kN/m^2, compared to the pressures on the face of wall and column forms of the order of 60–130 kN/m^2. As a result, the supporting members will safely span greater distances, which will lead to larger deflections in the bearers. Often it is the deflection criteria that will govern the allowable span of the members.

Figure 24.4 Beam reactions for one, two or three spans with distributed load

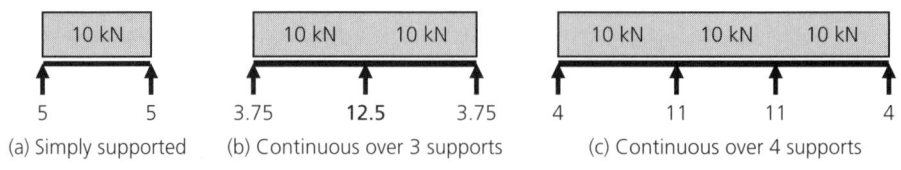

(a) Simply supported   (b) Continuous over 3 supports   (c) Continuous over 4 supports

The basic design principle is that the loads are transferred from the face to the supporting falsework. The distribution of load into the vertical members from the soffit formwork bearers will often be random. The face contact material, the secondary bearers and often the primary members in the forkheads will be continuous over several supports, giving rise to increased reactions at internal supports from their elastic reactions. The support reactions of beams change when they are continuous over more than two supports, and this applies to all types of members used as bearers.

Figure 24.4 shows the support reactions caused by a distributed load on each equal span of 10 kN/span for one, two and three spans. The worst case is a single beam continuous over two spans (i.e. with three supports, see Figure 24.4(b)), giving a central reaction of the static load times 1.25 for continuity – a staggering 25% increase in load. The '10% continuity rule' in BS 5975:2019 (BSI, 2019; clause 19.3.3.2) accepts that, in the case of falsework comprising random bearers with the formwork and falsework all in various lengths, the vertical load is calculated on the area supported by the standard plus 10% to allow for continuity. (Note that the 10% is added only once, not for each level of bearers.) In certain cases, for example, over two spans, a more precise calculation may be justified. Certain proprietary systems incorporate simply supported beams and this 'rule' may not apply.

Whatever system of falsework support is used, under load it will shorten elastically. This is the reason why operatives will invariably set the soffit formwork higher than the required finished level. Aluminium vertical support members will elastically shorten more than equivalent steel members for the same configuration. This is discussed further in Chapter 28. Particular care is necessary for tall soffit heights, such as in building construction where the ground to first floor dimension can be larger to allow for foyers and so on; the additional elastic shortening will not only affect the design of the soffit supports but any fixings and/or connections will need to accommodate the shortening under load.

24.5. Cantilevered soffits

A common method for casting the concrete parapets of bridges that are constructed with pre-cast concrete or steel beams is to project beams out either below or above the parapet to be supported; typical arrangements are depicted in Figure 24.5. Known as 'cantilevered soffits', they have to be designed for the actual loads applied, including any side forces from edge forms. Allowance also has to be made for deviations caused by the method adopted, such as deflections caused by the elastic extension of the tie rods and

Figure 24.5 Typical cantilever soffit arrangements: (a) projecting beam below; (b) connected to permanent beam. (Courtesy of Pallett TemporaryWorks Ltd)

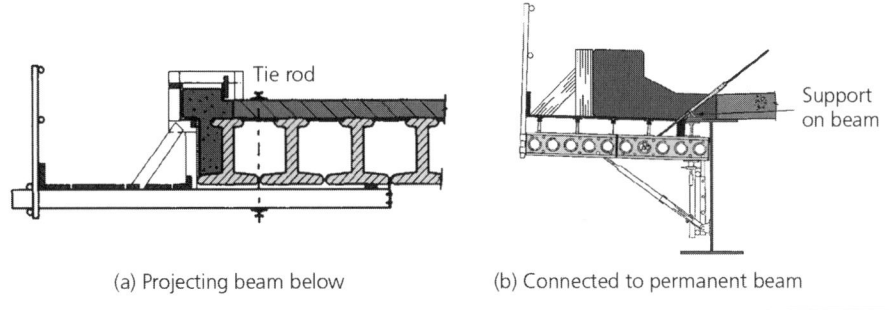

(a) Projecting beam below   (b) Connected to permanent beam

so on. The effective length of cantilevered beams is given in BS 5975:2019 (BSI, 2019; table K.3) where, for example, when the tip of the cantilever is free to rotate with the applied load the effective length is 7.5 times the actual length. Design is not easy, and consideration of differential deflections and subsequent movements can become complicated. Further guidance is given in the Concrete Society formwork guide (CS, 2012).

24.6. Striking soffit formwork
24.6.1 General

The time at which soffit forms can be removed and the exact procedure for striking the forms requires detailed consideration. Without such consideration, there is a real risk of damaging the permanent works and possibly initiating an accident. The specification should define any requirements for striking, taking into account considerations of frost and mechanical damage, reduction in thermal shock and limiting excessive deflections. Detailed guidance on use of fast-track equipment for early striking procedures is given in the flat slab guide (CONSTRUCT, 2003) and the criteria are listed in Table 24.1.

Unlike wall formwork, in order to remove soffit formwork the structure has to be capable of carrying its own weight plus any imposed load at the time of striking. The total service load on the member at the time of striking will depend on the sequence of construction and, in multi-storey construction, whether there is any backpropping.

Table 24.1 Striking soffit forms (slabs and beams, etc.)

Finish description	Criteria for striking soffit formwork
Basic, ordinary, plain or special finish	Use specifications, codes of practice or tables (e.g. CIRIA R136 (Harrison, 1995))
Highways Agency, for all classes	Assess the concrete strength at time of striking (knowing the maturity and concrete mix, etc.) using either a method based on the pro rata strengths or, for flat slabs, an assessment based on crack width

The load can include the following

- self-weight of the member: it is usual to assume a concrete density of 24 kN/m^3 in building and a higher value of 25 kN/m^3 in civil engineering structures
- formwork self-weight from any further construction: usually about 0.50 kN/m^2
- falsework self-weight from any further construction: usually 0.10–0.15 kN/m^3
- working area load for access: minimum 0.75 kN/m^2 (service class 1 loading)
- stored materials: should be avoided on newly struck slabs.

Particular consideration should be given to the striking of suspended slabs which are designed for light imposed loads; examples are roof slabs and reservoir roofs. In these cases, the self-weight may be the predominant load and the construction operations load will represent a load very similar to the Permanent Works Designer's imposed load. For such slabs with very low design imposed loads it may not be possible to strike until the concrete approaches its characteristic strength. Instances of early cracking and excessive deflections on reservoir roofs have been attributed to erroneous early striking.

To assist readers a flowchart of the likely steps to be considered in the early age striking of soffit formwork for flat slabs in building construction is shown in Figure 24.6.

Some proprietary systems actually have quick strip arrangements, allowing the panels and beams (formwork face) to be struck early (normally at a minimum concrete strength of 5 N/mm^2 to avoid damage) while leaving the main slab supported until approval to strike is received. The management and control of such soffit schemes is discussed in Chapter 2.

24.6.2 Striking bridge soffits

The Highways Agency specification (HA, 2006; volume 1, clause 1710.4) limits striking to either 10 N/mm^2 concrete strength or three times the stress to which the member is subjected. This is discussed in detail in the Concrete Society formwork guide (CS, 2012; section 5.3.6.1). The criteria for striking should be agreed with the Permanent Works Designer.

24.6.3 Striking slabs up to 350 mm thick

The Concrete Society formwork guide (CS, 2012; appendix A) will have identified the method of strength assessment for striking, the design service loads on the structure and the characteristic strength of the concrete. In building construction the two methods adopted for considering the required strength at early age for striking soffit formwork are either based on the ratio of loads, or the faster method, based on determination of maximum crack width. The latter method is discussed in more detail in Section 24.6.3.2.

24.6.3.1 Ratio of loads method

Methods of striking soffit and beam formwork have traditionally been based on considering the ratio of the loads on the slab at the time of striking and the designer's known service load. This method assumes that the slab is elastic (i.e. if you double the total load

Soffit formwork

Figure 24.6 Flowchart for striking flat slabs in buildings

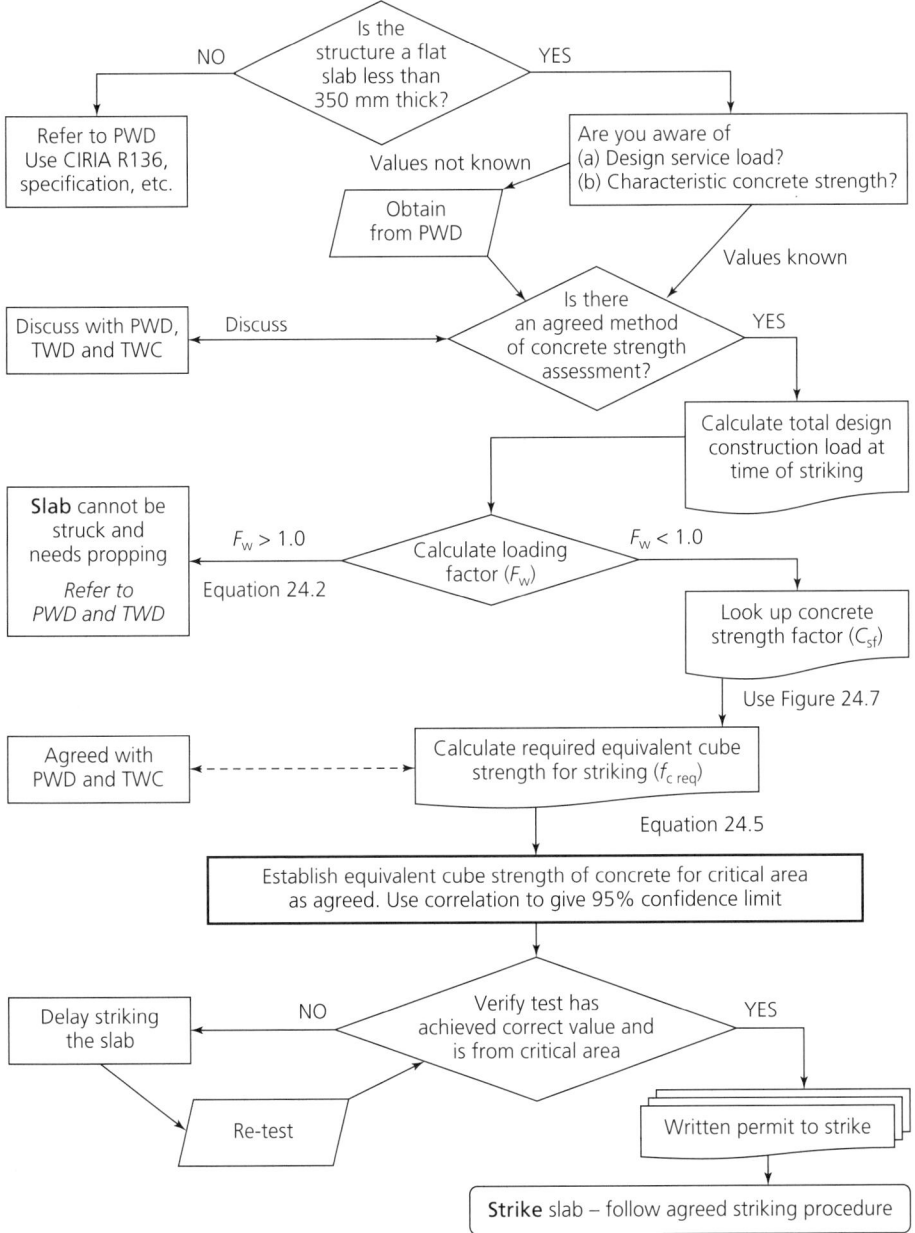

Notes:
1. It is assumed that the safety requirements for striking have been considered. These would include adequate working platforms, methods of material handling, etc.
2. PWD, Permanent Works Designer; TWC, Temporary Works Coordinator; TWD, Temporary Works Designer.

then the bending moment, deflection, etc., are also doubled). Striking may therefore commence if the ratio of the loading on the slab at the time of striking to the design service load is similar to the ratio of concrete strength at the time of striking to the characteristic concrete strength, as shown in

$$f_{c,req} = \left(\frac{w_{con}}{w_{ser}}\right) f_{cu} \qquad (24.1)$$

where $f_{c,req}$ is the required equivalent cube strength of the concrete at the critical area at the time considered (N/mm^2), w_{con} is the total unfactored design construction load on the slab considered (kN/m^2), w_{ser} is the total unfactored design service load (kN/m^2) and f_{cu} is the characteristic strength of the concrete (N/mm^2), generally at 28 days.

24.6.3.2 Crack width method

The research work, first published in 2000 by the Building Research Establishment (Beeby, 2000) showed that faster and safe construction methods are possible using crack width and not the ratio of loads as the criterion. This was confirmed in the European Concrete Building Project, where full striking of soffit formwork was regularly carried out at only 19 h after the floor slab had been concreted, without the need for temporary propping. The method is described below, but is given in more detail in the Concrete Society formwork guide (CS, 2012; Section 5.3.7.2).

The assumption for loading a concrete slab is that the crack width is proportional to the stress in the steel reinforcement which, in turn, is proportional to the load. Hence if load is removed or added there will be a proportional reduction or increase in crack width. Although the slab is designed for the ultimate limit state, the actual maximum load on the slab at the time considered will be the sum of the unfactored loads, because the consideration of crack width is at the serviceability limit state, not the ultimate limit state. The load applied to a slab during any stage of construction should not be greater than the designer's unfactored design service load. In other words, the crack width experienced by the concrete at early striking will not be larger than that intended by the Permanent Works Designer if the full load were applied in service. Obviously, if the concrete slab is struck earlier than intended the structure may be permanently damaged.

The method relies on accurately evaluating the actual concrete strength of the new concrete (see Section 24.7).

Generally the permanent work slab or beam will be allowed to take up its instantaneous deflected shape *before* further floors are built. If props or supports are then replaced to carry additional loads it is known as 'repropping'.

The method described below is taken from the *Guide to Flat Slab Formwork and Falsework* (CONSTRUCT, 2003) and can be applied to in situ reinforced concrete flat slabs with the following parameters

- up to a maximum thickness of concrete of 350 mm

- not post-tensioned or cast on thick pre-cast planks
- are to be struck and become self-supporting before any additional loads are placed on top
- where used, there are no more than two levels of backpropping.

Given the above, the two main criteria for ensuring that a concrete slab that is being struck is not overloaded are the loading factor (F_w) and the cracking factor (F_{cr})

$$F_w = \frac{\text{total design construction load on slab}}{\text{total design service load on slab}} = \frac{w_{con}}{w_{ser}} \leq 1.0 \quad (24.2)$$

$$F_{cr} = \left(\frac{w_{cr}}{w_{ser}}\right)\left(\frac{f_{cu}}{f_c}\right)^{0.6} \leq 1.0 \quad (24.3)$$

where w_{con} is the total unfactored design construction load (kN/m²), w_{ser} is the total unfactored design service load (kN/m²), w_{cr} is the total unfactored construction load on the slab causing cracking (kN/m²), f_c is the equivalent cube strength of the concrete at the critical area at the time considered (N/mm²) and f_{cu} is the characteristic strength of the concrete (N/mm²), generally at 28 days.

The construction site requires the equivalent cube strength to enable a particular total construction load w_{con} to be applied to a given slab. The required equivalent cube strength can be calculated by rearranging Equation 24.3

$$f_{c,req} = f_{cu}\left(\frac{w_{con}}{w_{ser}}\right)^{1.67} = f_{cu} C_{sf} \quad (24.4)$$

where $f_{c,req}$ is the required equivalent cube strength of the concrete at the critical area at the time of striking (N/mm²), w_{con} is the total unfactored design construction load on the slab considered (kN/m²) and C_{sf} is the concrete strength factor.

Hence the method to establish the required equivalent concrete cube strength of a given reinforced concrete flat slab, not exceeding 350 mm in depth, during any particular construction operation is first to calculate the loading factor (satisfying Equation 24.1) and then to establish the concrete strength factor from Figure 24.7.

The required equivalent concrete cube strength, $f_{c,req}$ (N/mm²), of the concrete at the critical area at the time of striking is given by

$$f_{c,req} = f_{cu} C_{sf} \quad (24.5)$$

where f_{cu} is the characteristic strength of the concrete (N/mm²) and C_{sf} is the concrete strength factor from Figure 24.7.

Where there are several other levels of construction, there may also be a requirement to support additional loads. There will be occasions in multi-storey construction when the

Figure 24.7 Graph of loading factor against concrete strength factor

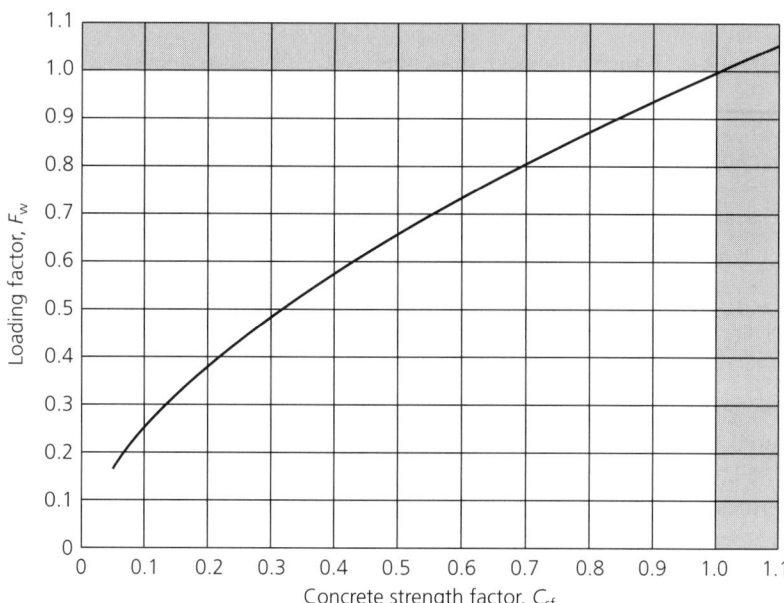

Notes:
1. Values of ratio of loading above 1.0 should only be used with care. The approval of the Permanent Works Designer and Temporary Works Coordinator should be sought in such cases.
2. The graph here is the graphical representation of Equation 24.4.

slab immediately below the level to be supported has not achieved full maturity, and construction loads will need to be supported through several levels. This is known as 'backpropping' (see Chapter 28).

The method shown relates to reinforced concrete slabs. The use of this method for post-tensioned flat slabs, has, to date, not been verified by research, and engineering judgement is required.

24.6.4 Sequence of striking

The sequence for striking the supports should be agreed before any striking commences. On complex structures this may be specified by the Permanent Works Designer. The following procedures should generally be adopted

- *Slabs spanning between walls*: commence striking at the middle, working towards the walls.
- *Slabs supported on beams*: commence striking the slabs at the middle, working towards the beams. Once the slabs have been fully struck, strike the beams, commencing at the middle of the beam's span and working towards the columns.
- *Cantilevers*: commence striking at the tip and work towards the support.

24.7. Assessment of concrete strength

Accurate knowledge of the *actual* strength of the concrete slab at the time of striking is crucial to safe construction. The compressive strength of concrete is determined by many factors: temperature, location, water/cement ratio, and type of cement used. The use of cement replacement materials, called 'additions', such as pulverised fuel ash (PFA) and ground granulated blast-furnace slag (GGBS), to produce 'blended cement' is now common in the concrete industry. Although useful for thick concrete structures to reduce generated internal heat, they are predominantly used to reduce cost; unfortunately, such blends have a side effect – the rate of gain in strength is appreciably slower compared to a plain CEM I concrete. A typical comparison of the strength development for a designated concrete class C 30/37 is shown in Figure 24.8. It can be seen that it is impossible to achieve the full 28-day strength at early age for the concrete using blended cements. This has significant implications for designers and contractors when building multi-storey structures, as the backpropping requirements (see Chapter 28) often need the supporting slab to have achieved nearly the full 28-day strength before casting the next level slab.

Both Permanent Works Designers and Temporary Works Designers need to be aware of the trade-off between achieving a low cost structure using less expensive concrete and the disadvantage of low early age strength and the subsequent longer construction time. To achieve fast-track multi-storey construction the commercial decision to revert to CEM I concrete is often more cost-effective.

The use of concrete cubes to assess concrete at early age is not recommended. The cubes give a lower bound result and are rarely representative of the actual strength of the in situ

Figure 24.8 Typical strength development for a designated C 30/37 concrete

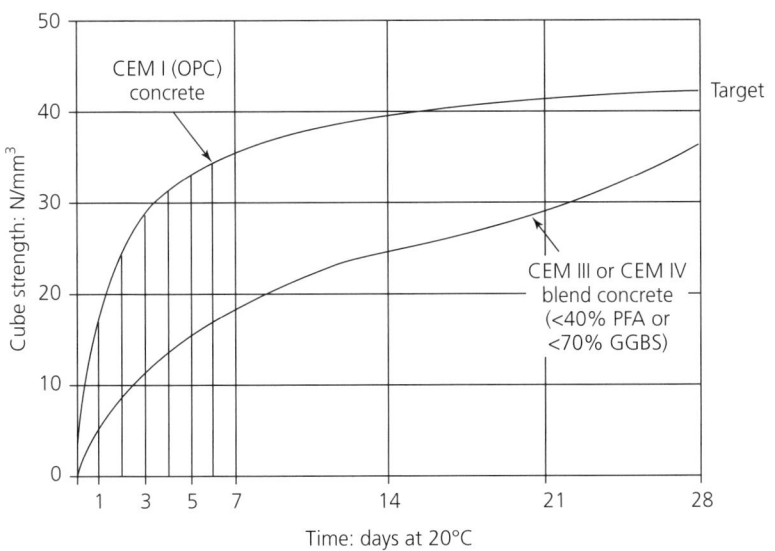

concrete in the building. Several methods of strength evaluation at early age are given in the guides published by CONSTRUCT (2003) and the British Cement Association (BCA, 2000). The method most commonly used today for early-age strength assessment of flat slabs is the maturity measurement method, based on the time–strength curve for the concrete. The favoured method from the European Concrete Building Project is the LOK test based on the average of four readings. This method was used in the construction of the Storebælt Bridge linking East and West Denmark.

The maturity measurement system is mix specific and requires casting temperature thermocouples into the slab. Reading the values electronically and then comparing the reading with the strength–temperature curve for that concrete allows the determination of the actual strength at that time for the concrete. This method requires laboratory pre-testing in order to generate the strength curves for the particular concrete used. It is therefore generally more suitable for larger construction sites where multiple slabs are cast.

The LOK test also requires pre-planning, as inserts are placed into the top surface of the slab. However, it has a distinct benefit over other methods in that, because it is carried out on the actual concrete, it is *not* mix specific. The top surface is used as the test area for a pull-out test, and minor making good is necessary. This test was accepted by the British Standards Institution (BSI) as a test method in 1992. Suppliers give correlation charts to convert site readings directly to 'equivalent cube strength'. This has the advantage that, unlike cube and maturity measurements, the concrete strength is obtained almost instantly from the LOK test hand-held equipment. It therefore suits both large and small sites.

Reference should be made to the Permanent Works Designer before using such methods.

24.8. Checking and inspection

Useful guides to aid checking and inspection of soffit formwork are *A Guide to the Safe Use of Formwork and Falsework* (CONSTRUCT, 2008) and *Checklist for the Assembly, Use and Striking of Formwork* (CS, 2014).

REFERENCES

Beeby AW (2000) *A Radical Redesign of the In-situ Concrete Frame Process. Task 4: Early Striking of Formwork and Forces in Backprops*. Building Research Establishment, London, UK, BR394.

BCA (British Cement Association) (2000) *Best Practice Guide – Early Age Strength Assessment of Concrete on Site*. BCA, Crowthorne, UK, 97.503.

BSI (British Standards Institution) (2008) BS EN 12812:2008. Falsework. Performance requirements and general design. BSI, London, UK.

BSI (2009) BS EN 13670:2009. Execution of concrete structures. BSI, London, UK.

BSI (2011) BS 5975:2008 + A1:2011. Code of practice for temporary works procedures and the permissible stress design of falsework. BSI, London, UK.

BSI (2019) BS 5975:2019. Code of practice for temporary works procedures and the permissible stress design of falsework. BSI, London, UK.

CONSTRUCT (Concrete Structures Group) (2003) *Guide to Flat Slab Formwork and Falsework*. Concrete Society, Camberley, UK, CS140.

CONSTRUCT (2008) *A Guide to the Safe Use of Formwork and Falsework*. Concrete Society, Camberley, UK, CSG/005.

CONSTRUCT (2010) *National Structural Concrete Specification for Building Construction*, 4th edn. Concrete Centre, Camberley, UK, CCIP-050.

CS (Concrete Society) (2012) *Formwork: A Guide to Good Practice*, 3rd edn. Concrete Society, Camberley, UK, CS030.

CS (2014) *Checklist for the Assembly, Use and Striking of Formwork*, 3rd edn. Concrete Society, Crowthorne, UK, CS144.

HA (Highways Agency) (2006) *Specification for Highway Works. Manual of Contract Documents for Highway Works*. Highways Agency, London, UK.

Harrison TA (1995) *Formwork Striking Times: Criteria, Prediction and Methods of Assessment*. CIRIA, London, UK, R136.

National Building Specification (annually) E20: Formwork for in situ concrete. National Building Specification, Newcastle-upon-Tyne, UK. (Published online.)

Ray SS, Barr J and Clark L (1996) *Bridges: Design for Improved Buildability*. CIRIA, London, UK, R155.

UKWIR (UK Water Industry Research) (2011) *Civil Engineering Specification for the Water Industry*, 7th edn. WRc, Swindon, UK.

Useful web addresses

Concrete Society: http://www.concrete.org.uk (accessed 01/08/2018).

CONSTRUCT (Concrete Structures Group): http://www.construct.org.uk (accessed 01/08/2018).

National Access & Scaffolding Confederation (NASC): http://www.nasc.org.uk (accessed 01/08/2018).

Temporary Works: http://www.temporaryworks.info (accessed 01/08/2018).

Temporary Works Forum: http://www.twforum.org.uk (accessed 01/08/2018).

Temporary Works, Second edition

Pallett, Peter F and Filip, Ray
ISBN 978-0-7277-6338-9
https://doi.org/10.1680/twse.63389.355
ICE Publishing: All rights reserved

Chapter 25
Climbing and slip forms

Charlie McKillop
Engineering Director, Hünnebeck

Climbing and slip forms are systems for the construction of in situ reinforced concrete vertical wall elements several lifts in advance, and therefore do not rely on support or access from other parts of the permanent works. A combined formwork assembly and access platform is supported on anchors or tracks bolted to the previous section of wall or, for continuous pouring, supported on climbing rods cast into the concrete. Protection screen systems, either integrated in climbing formwork or as separate entities on building perimeters, have similar design considerations. The type of structure and its suitability for use on repetitive wall elements affects the economy of these systems.

25.1. Introduction

The operation of climbing formwork, often referred to as 'jump-form' formwork, involves individual pours with the formwork being moved ('jumped') between the pours, whereas the operation of slip form requires the concrete to be poured continuously with the forms being raised to suit the speed of concrete placement.

Climbing, or jump-form, formwork (see Section 25.3) typically comprises the formwork and safe working platforms for cleaning or fixing the formwork, and access for fixing reinforcement and concreting works. Unlike conventional formwork (see Chapter 23) it supports itself on the previously cast concrete and does not therefore rely on support or access from other parts of the structure or permanent works.

Climbing formwork is suitable for vertical elements in high-rise structures such as bridge piers and columns, dams, and stair and lift shafts, core walls and shear walls in buildings. These are constructed in a staged process. It is a highly productive system designed to increase safety, speed and efficiency while minimising labour and crane time. Systems are normally modular and can be joined together to form long lengths to suit varying construction geometries. Further detailed guidance on the use of climbing formwork is given in the Concrete Society's *Formwork: A Guide to Good Practice* (CS, 2012; section 6.2).

Slip-forming (see Section 25.4) is a system whereby the entire formwork system, including its safe access platforms, is incrementally jacked upwards as the reinforcement and concrete are placed. Slip-forming is used for constructing chimneys, silos, water tanks and shaft linings, as well as towers, lift shafts and bridge piers. Continuous pouring with no construction joints is common, which is particularly important for leak-free structures. Fast rates of production can be achieved, but this requires expert planning, monitoring and logistics. In addition, more initial site preparation is required, and this is generally carried out by specialist subcontractors. (See also *Good Concrete Guide 6: Slipforming of Vertical Structures* (CS, 2008).)

25.2. Climbing and slip-form viability assessment
25.2.1 General

Jump-form formwork, where the formwork assemblies are moved from position to position with the aid of a crane, is widely used in high-rise buildings, typically those having in excess of five storeys. As the height of the structures increases so does the complexity of the system, with fully self-climbing systems, independent of craneage, becoming viable in excess of 20 storeys. However, a combination of crane-handled guided rail platforms and self-climbing platforms can be viable on lower structures.

Self-climbing formwork offers the advantage of considerably reducing the requirement for crane time, thus allowing the crane to be used for other construction work. Another advantage is that climbing formwork can be designed, where required, to operate in high winds, when the use of a crane may be limited, thus lessening the risk of delays in the construction programme.

Slip-form formwork has a longer initial set-up time than climbing or jump-form formwork and is more expensive per square metre of formwork, but the resulting equipment cost per square metre of slip-formed surface will be much lower. As the process is often a 24-hour process, in shifts, the labour costs can be higher, but production rates will be faster. Slip-forming will usually suit tall structures that are not less than 20–25 m tall and which do not have frequent plan changes. Slip-forming may be economical for structures as low as 12 m high, or even less if several identical structures can be constructed in sequence.

Whether it is a crane dependent fully self-climbing or slip-form system that is selected, the general principle is the same. The formwork is independently supported, relying on the concrete cast earlier, so that core walls, for example, can be completed ahead of the rest of the main building structure (Figure 25.1).

The construction in advance of such core/lift shaft walls can provide stability to the main structure during its construction, and can have the beneficial effect of taking the core off the project critical path. This then permits different phases of work to be carried out concurrently without interference.

25.2.2 Economy of construction
Economy and speed of construction are key factors in choosing between climbing formwork and slip-form formwork. For bridge piers and columns and similar structures the question is not so much whether to climb, but how to climb. For high-rise buildings the decision-making process is not so straightforward and can depend on many factors. With

Figure 25.1 Typical climbing formwork system to a stair/lift core. (Courtesy of Hünnebeck (a BrandSafway company))

an ever-growing emphasis being placed on improvements in safety and productivity, and with the increasing complexity of structures and limited building footprints on congested city sites, more often than not climbing is nowadays the preferred solution. The vertical elements can be progressed quickly in advance of the floors, thus taking a significant part of the overall structure off the critical path.

The benefits of reduced demand in terms of the crane use and the efficiency of construction due to the repetitive nature of the construction method lead to fast cycling times. Where sites have sufficient ground-level space and a well-considered construction programme, linked with good site management and experienced operatives, the use of more traditional techniques can be just as effective.

25.2.3 Assessment/suitability of structure

The first point to consider when investigating the feasibility of using climbing formwork or slip-form techniques is the structure itself. All systems require that the permanent wall under construction be sufficiently strong and have sufficient stability in the temporary condition to support it.

The Permanent Works Designer (PWD) is the only person who has sufficient knowledge of the structure to assess whether it can withstand the loading induced, particularly that induced by the climbing formwork system. The principal designer has a duty to ensure that the PWD and Temporary Works Designer (TWD) cooperate, and that the permanent structure can withstand the loads applied to it by the temporary works (see

Chapter 2 on management). The Temporary Works Coordinator (TWC) has an important role at this stage. It is highly probable that the use of climbing formwork as a method of construction was not considered during the structural design phase. The PWD should verify whether the vertical elements can support climbing formwork, albeit within defined constraints (e.g. under extreme wind conditions). However it is the responsibility of the PWD to review their design and liaise with the contractor's temporary works design team.

The PWD also has a role to play when the contractor is considering the climbing cycle or rate of slip. The concrete mix design and the early-age strength of the concreted section both have a significant impact on whether a short cycle time/climbing rate is achievable. The cycle time can be a critical factor in the effectiveness of the system. For example, if the climbing system cannot be climbed after 4 days it may not be a viable solution in relation to the costs involved, particularly when large systems utilising hydraulics are proposed; and, when slip-forming, can materials such as concrete, reinforcement, hoist extensions and so on suit the rate of construction envisaged? Changes to the design, such as increased reinforcement or improved concrete mix for higher early strength, may be viable to produce the output required for a given site.

Openings, features or cast-in components in the permanent structure can significantly affect slip-forming and in certain cases make it unfeasible. The resulting surface finish is always an important aspect when considering the use of slip-form techniques. It should be pointed out that the final tolerances of slip-formed structures will be similar to those in other construction methods.

The proprietary supplier will provide details of the loads being imposed on the structure by the system proposed and of how the brackets then transmit the loads into the structure. The ability of the structure to resist the applied forces is equally as important as the concrete strength local to the anchor. To check these conditions requires a detailed knowledge of the concrete structure and associated site activities. The TWC has a duty to ensure that the PWD is satisfied in the structure's ability to withstand the loads being imposed and remain in accordance with the design specification. Working together with the contractor's temporary works design team the TWC can then consider an appropriate lifting cycle.

25.2.4 Design

The design of climbing formwork and slip-form operations is generally carried out by the design offices of the proprietary supplier or specialist subcontractor, as they have the product knowledge, competency and expertise. With the reduction in contractors' temporary works departments, contracts increasingly rely on proprietary suppliers to provide relevant designs together with the equipment. Those contractors that still have a design capability should be able to understand and produce designs for climbing (jump-form) formwork systems (see Section 25.3.2). However, it is recommended that design responsibility for the specialised design of rail-guided climbing formwork systems and slip-form systems should remain with the proprietary suppliers and/or specialist subcontractor due to their in-depth knowledge of and familiarity with these systems.

25.3. Climbing formwork
25.3.1 System selection

In their most basic form climbing formwork systems can be used simply as working platforms for personnel. In many cases, such as sites in city centres where available space within the site compound is limited, the platforms can be preassembled offsite, with integral guardrails, toe boards and lifting points. The platforms are generally used to allow steel fixers access to the reinforcement, the finished concrete face of a structure, fix soffit edge formwork, and so on. They can be designed to provide a continuous deck around outside corners and can be easily adapted to suit buildings having geometrically complicated shapes.

Climbing formwork is compatible with a variety of formwork systems and allows incremental vertical segments of structure to be poured, commonly to a storey height of 3–4 m. The three types of system in common use are as follows

- *Climbing (jump-form) formwork* – units are individually lifted off the structure and relocated at the next construction level using a crane (see Figure 25.1).
- *Rail-guided climbing formwork* – units are connected to a rail/anchor configuration attached to the structure, which offers greater safety and control during lifting operations. The proprietary systems available differ in that some have a fixed sectional rail attached to the structure which the units climb up, recycling the rail as it climbs, while others have an 'anchor and shoe' assembly attached to the structure which the rail climbs through, recycling the 'shoe' as it climbs.
- *Rail-guided self-climbing formwork* – hydraulic jacks are used to climb the units up the structure, obviating the need to use a crane. With portable hydraulics the platforms are raised sequentially; alternatively, hydraulics are positioned at every rail and the platforms lifted in unison. The latter option is more expensive, only becoming viable on structures of significant height or of a complex plan layout.

25.3.2 Climbing/concreting cycle

Most formwork systems are supported from anchor assemblies cast into the preceding section of wall. A typical sequence for a jump form is shown in Figure 25.2.

Anchors are attached to the formwork (see Figure 25.2(a)) and cast-in with the concrete pour, but the anchor cone and the mounting device are recoverable. Note that a wind brace in tension is fitted to stabilise the assembly from clockwise rotation from the wind. The anchor assembly is mounted on a threaded stud secured to the wall formwork to ensure it is held in the correct position during concrete pour. Once the section of wall has been cast (see Figure 25.2(a)) and an adequate concrete strength has been verified, the formwork can be struck off the wall. The system anchor assembly for the next cycle can then be fitted (see Figure 25.2(b)) and the formwork cleaned and release agent fitted in readiness for the next use. Shortly before moving, the wind brace is disconnected and the whole assembly lifted off its anchors (see Figure 25.2(c)) by a crane and refitted onto the anchors in the next position, usually by moving vertically upwards. The wind brace is replaced, and reinforcement fixing can now commence. The cycle is then repeated.

Figure 25.2 Typical jump form detached and lifted by crane. (Courtesy of Hünnebeck (a BrandSafway company))

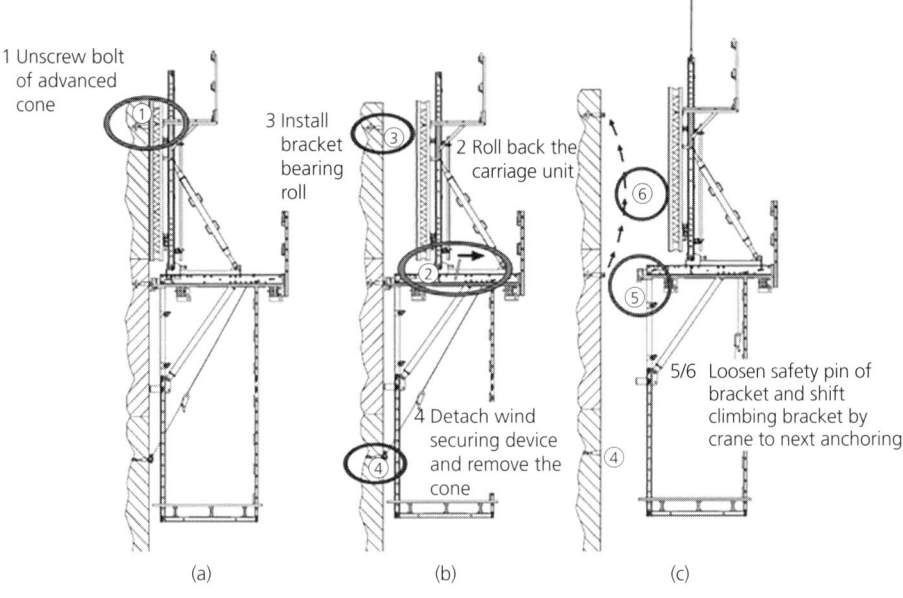

The operation must be carefully planned and coordinated. Access to the working area during lifting should be restricted to essential personnel because when platforms are being moved unprotected edges are created between the individual platforms, which present a significant safety risk and must be cordoned off.

The process for rail-mounted systems is similar (Figure 25.3), although the system is fixed to a rail and therefore does not leave the wall once attached until construction is complete. Often such systems incorporate devices to strike the formwork laterally on the working platform to facilitate cleaning, fixing anchors and so on.

25.3.3 Design considerations

To achieve a viable and cost-effective climbing formwork solution it is necessary to strike a balance between several governing factors, primarily maximising platform size while maintaining the ability to climb the system with minimal limitations (e.g. during periods of high winds).

Some systems offer the option of varying the platform widths. Wider platforms provide the space to allow the associated wall formwork to be fully retracted when struck, thus allowing clear access for cleaning and fixing of reinforcement. Narrower platforms, however, generally incorporate a tilting mechanism for access to the formwork face. Self-climbing systems can be designed to incorporate additional platform levels above the main construction level. This enables multiple operations to be performed

Figure 25.3 Self-climbing form using rails and hydraulics. (Courtesy of Hünnebeck (a BrandSafway company))

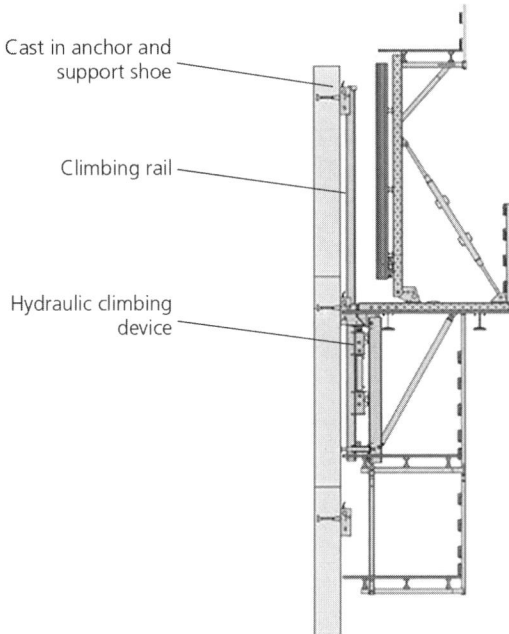

simultaneously (e.g. fixing reinforcement for a wall prior to casting concrete at a lower section), offering the potential for realising shorter construction cycle times.

The achievable platform length is governed by the spacing of anchors and the associated wind parameters to be considered. The longer the platform, the higher the load on the anchors, which then has an impact on the height and width of the associated wall formwork. All these factors will dictate the allowable operational wind speed at which the climbing system can be safely worked on and moved. If this is misjudged then part of the reason for selecting to use a climbing system from the outset may be compromised.

25.4. Slip forms
25.4.1 General
Slip-forming is the process of continuous concrete construction. It is normally carried out vertically with a constant cross-section, and the forms are raised and the concrete placed at such a rate that the concrete achieves sufficient strength before the forms expose it. Rates of rise of 80–300 mm/h can be achieved, and construction usually continues around the clock. Day-shift only working (10-h shift), involving slip-forming for approximately 6 h, with 2 h preparation before and 2 h of clean up after slip-forming, is common practice. With a good level of control and close liaison with the concrete supplier a consistent surface finish can be achieved. The process has the advantage of not

Figure 25.4 Typical slip-form construction to building cores. (Courtesy of Carey Civil Engineering)

requiring night working and it thus eliminates supply of small amounts of concrete throughout the night from commercial batching plant.

A typical example of slip-form formwork is shown at Figure 25.4. Vertical slip-forming can involve large volumes of concrete supplied in small amounts continuously. The concrete is generally of high strength and low water/cement ratio, and contains significant amounts of admixtures to achieve the consistency necessary for placing. The speed of the upward form movement can be adjusted to a significant degree by varying the admixture dosage. The concrete surface finish achieved is peculiar to the process.

The layout is normally simple. It is possible to vary the wall thickness and layout over the height, although each complication will increase the overall cost. The ideal structure should have significant dimensions in both major axes to ensure stability. Variants of the process can operate horizontally, and on sloping structures.

Slip-forming is a highly specialised process and advice should be sought from people with the relevant experience. A useful guide is the *Good Concrete Guide 6: Slipforming of Vertical Structures* (CS, 2008).

Figure 25.5 Typical slip form. (Courtesy of Carey Civil Engineering)

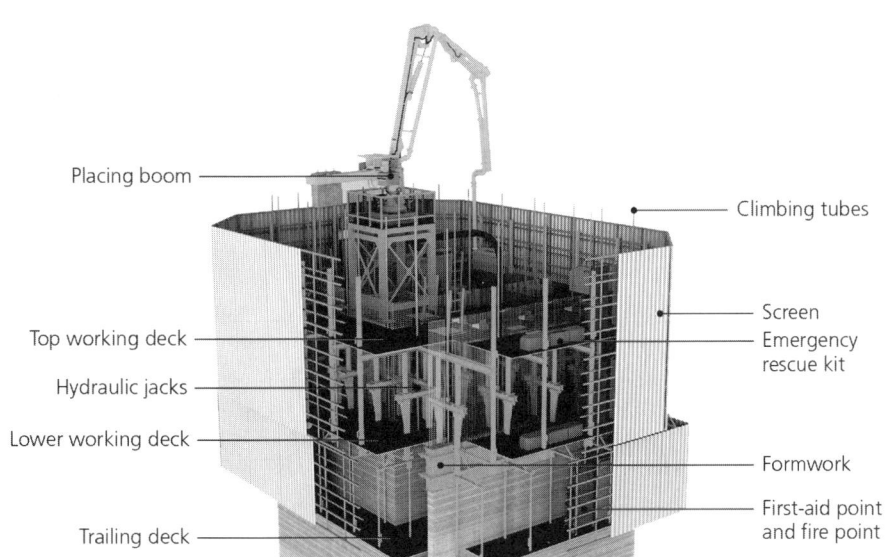

25.4.2 Design considerations

The typical arrangement of slip forms is two panels held in position by a yoke with two legs and a cross-member over the top, as shown in Figure 25.5. A jack is fixed to the cross-member and the jack climbs a rod the base of which is in the structure. The jacking rods are either left in place or subsequently withdrawn from the concrete by reversing the jacks. The various walls are braced together, and working areas and access scaffolds are added. The PWD has an important role in ensuring that reinforcement is appropriately detailed to suit the method.

The forms are generally only 1.2 m high and are set up with a slight taper to prevent the form gripping the concrete and lifting it at the trailing edge. The form is set up with the correct wall thickness at approximately mid-height. The face contact material will most likely be steel panels, although where vertical striations are needed film-faced plywoods with hardwood features have been used. Trailing platforms can be used to carry out any remedial or finishing work on the concrete.

Openings can be formed by using inserts narrower than the wall, thus avoiding vertical displacement of the former as sliding proceeds. An allowance of 18 mm less than the nominal wall thickness will clear any build-up of hardened concrete on the upper edge of the form.

The provision of sufficient ancillary plant (standby equipment, etc.) is recommended. Detailed consideration of access to the continuously changing slip-form location has to be planned, and any material/passenger hoists and concrete pumping boom extensions must keep pace with construction. Detailed planning is important as stoppages on slip-forming should be avoided. Clearly a plan should be in place for stopping the system, but this should be a last resort.

25.5. Climbing protection screens
25.5.1 General
Protection screens create a safe working enclosure, providing protection from wind and other adverse weather conditions, as well as from falling objects, while working at height. They often incorporate access platforms. A typical example is shown in Figure 25.6.

The inherent benefits these systems offer has seen their use become more widespread within the construction industry. Often assembled offsite in units 12–15 m high (four storeys), they are delivered in a programmed just-in-time sequence. The benefits offered by full-height protection with climbing formwork systems have seen protection screen systems deployed on structures in excess of ten storeys high. For shorter structures the systems can be configured in a more economical two-storey height. Further guidance on protection screens is included in the Concrete Society's formwork guide (CS, 2012; section 7.6.3).

The intended use of protection screens should be stated in the temporary works design brief so that the TWD can take full account of the effects of wind on the system.

25.5.2 Types of screens
The basic principles are similar to those of climbing formwork systems. The screens cantilever one or two storeys above the completed floors, enclosing the structure to provide protection to site operatives constructing the upper levels (Figure 25.7).

Figure 25.6 Typical protection system screen: (a) exterior view and (b) inside view. (Courtesy of A. J. Morrisroe Ltd)

(a) (b)

Figure 25.7 Typical protection screen sections. (Courtesy of Hünnebeck (a BrandSafway company))

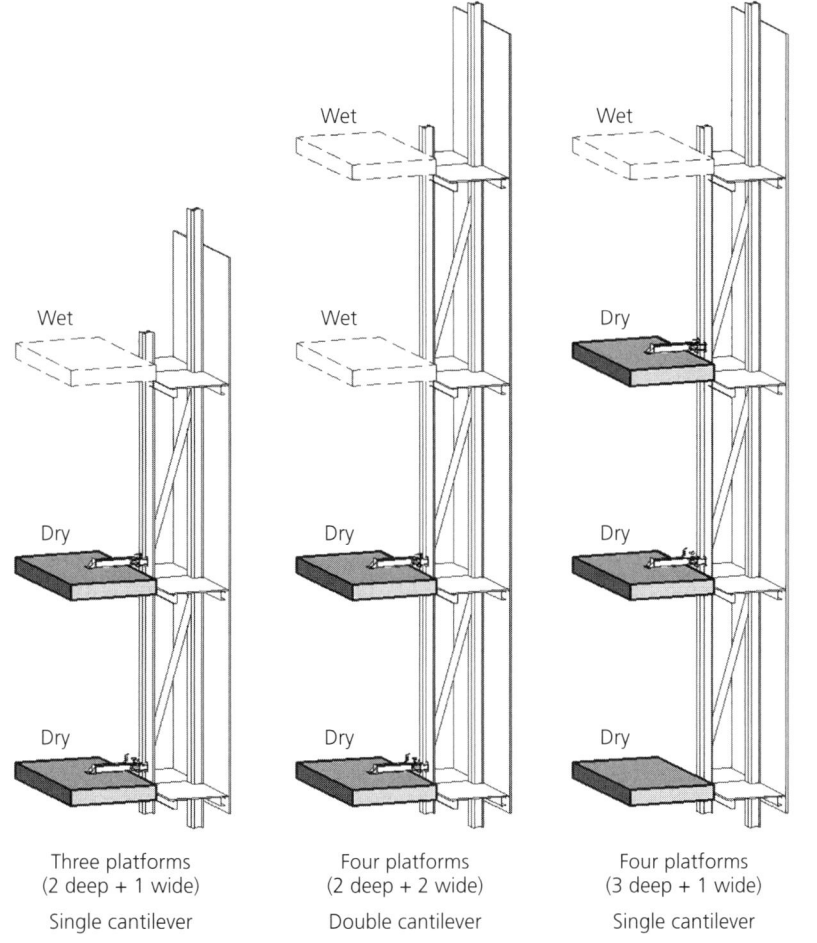

Screens can be clad in either solid panels (e.g. plywood) or perforated mesh or netting, and are supported on rails attached to the already cast slabs of the permanent structure. Commonly, perforated panels or mesh is chosen as it provides a balance between reducing the imposed wind load, while protecting those below who may be struck by falling debris, but still permitting natural light to enter each working level. Screens can incorporate working platforms for essential work such as post-tensioning operations or to provide space for following trades undertaking work on the façade. Some screens are simple barriers without platforms, but do have folding covers to prevent debris falling to lower levels.

25.5.3 Design considerations

Similar criteria to those examined when determining the suitability of a climbing formwork system are applicable when deciding on the suitability of a protection screen system. The permanent structure's capacity to withstand the imposed loads induced, in particular from wind forces, access requirements, method of climbing, and so on, must all be evaluated in the same manner. Again, the principal designer must ensure that the PWD and TWD cooperate and ensure that the permanent structure can withstand the loads applied to it by the temporary works.

The wind effect is often the governing factor to be considered when designing protection screens, as they often travel up to two levels in advance of the construction works, thus generating significant cantilever forces, especially if solid cladding panels are used. Anchor spacing is directly affected by wind load, which in turn determines the spacing of slab rail supports. Additional propping may even be required, in particular at corner sections, which induce even higher loads as described in the Concrete Society's formwork guide (CS, 2012; section 4), the British Standard on wind actions (BSI, 2005) and the associated UK National Guidance (BSI, 2015).

25.6. Checking and inspection

Suppliers and importers have a duty to provide information and user guidance about their products (e.g. project drawings, parts lists, safety instructions) (see CS, 2012; section 3). The TWC should gather such information and ensure that all those working with the proprietary system are familiar with the content of the relevant instructions and safety information for their particular application. All operatives should be trained in the use of the system, and be aware of the precise procedure of using the system. Toolbox talks and similar on-site familiarisation with the equipment are an essential prerequisite for safe operation of these systems. Due care and attention should be taken to assess that no additional guidance is required as a result of any site-specific conditions encountered. Precise guidance will be available from the manufacturer for all working operations, climbing operations and non-operational conditions. Periods of high wind or the application of sheeting to handrails, for example, may result in design loads being exceeded and necessitate a review of the working parameters provided.

Slip-forming requires a continuously controlled inspection regime and needs a small but highly skilled workforce on site; this requires team effort and there is no place for restrictive working practices on slip-form operations. For this reason, specialist subcontractors are often employed to carry out the task, the various responsibilities having been established before slip-forming commences.

Each stage of a climbing sequence, such as a climbing sequence for jump forms, as described in Section 25.3.2, will result in load being transferred onto the supporting permanent structure. The TWD must pass on all relevant information to ensure that stability is considered at every stage and that permissible imposed loads are not being exceeded.

REFERENCES

BSI (British Standards Institution) (2005) BS EN 1991-1-4:2005 + A1:2010. Eurocode 1. Actions on structures. General actions. Wind actions. BSI, London, UK.

BSI (2015) PD 6688-1-4:2015. *Background information to the National Annex to BS EN 1991-1-4 and additional information*. BSI, London, UK.
CS (Concrete Society) (2008) *Good Concrete Guide 6: Slipforming of Vertical Structures*. Concrete Society, Camberley, UK, CS162.
CS (2012) *Formwork: A Guide to Good Practice*, 3rd edn. Concrete Society, Camberley, UK, CS030.

FURTHER READING

BRE (2007) *Formwork for Modern, Efficient Concrete Construction*. HIS BRE Press, Bracknell, UK, BR495.
Concrete Society (2014) *Checklist for the Assembly, Use and Striking of Formwork*, 3rd edn. Concrete Society, Crowthorne, UK, CS144.
CONSTRUCT (2010) *National Structural Concrete Specification for Building Construction*, 4th edn. Concrete Centre, Camberley, UK, CCIP-050.

Useful web addresses

Concrete Society: http://www.concrete.org.uk (accessed 01/08/2018).
CONSTRUCT (Concrete Structures Group): http://www.construct.org.uk (accessed 01/08/2018).
Temporary Works: http://www.temporaryworks.info (accessed 01/08/2018).

Temporary Works, Second edition

Pallett, Peter F and Filip, Ray
ISBN 978-0-7277-6338-9
https://doi.org/10.1680/twse.63389.369
ICE Publishing: All rights reserved

Chapter 26
Temporary façade retention

Ray Filip
Temporary Works Consultant and Training Provider, RKF Consult Ltd

Stuart Marchand
Managing Director, Wentworth House Partnership

The work involved in supporting existing façades or party walls for renovation, during rebuilding or after damage is a specific type of temporary works with different risks, durations and types of loads than the temporary structures of falsework and scaffolding. Although similar equipment is used, the philosophy, procedures and relationship to the permanent works are different. Unlike formwork and falsework, the structure to be supported already exists; it may be old, in poor condition or not vertical. Procedures will need to reflect the status, stressing the importance of initial surveys and possible extensive monitoring during use to reduce the risks. Permanent works designers have an important role in the management and control of façade retention schemes.

26.1. Introduction
The elevations of a building are generally known as 'façades' and the front façade is often highly decorative. Building façades are retained because they are considered to be of architectural, historical or visual importance. Behind the façade, the interiors may have deteriorated or simply may not meet the requirements of modern usage. To meet with planning restrictions and conservation orders (listed buildings), the interiors are rebuilt behind the retained façade using modern methods and materials such as steel frame or reinforced concrete. A shoring (façade retention) scheme is generally required to support the façade and protect workers (and the general public) while the work is carried out. The shoring scheme must be designed and installed with care. The façade will eventually be connected and supported by the new internal structure. Occasionally, other external walls (not just the front elevation), internal walls, floors and roof structure will be retained and restored. Emergency unplanned retention may also be required after fire or explosion to make the remaining structure safe for rebuilding, or safe demolition if deemed beyond repair.

The authoritative guidance document is the Construction Industry Research and Information Association (CIRIA) report C579, *Retention of Masonry Façades – Best Practice Guide* (Lazarus *et al.*, 2003), which gives information for all the parties involved in the planning, design and construction of façade retention schemes. For specific site guidance, CIRIA C589, *Retention of Masonry Façades: Best Practice Site Handbook* (Bussell *et al.*, 2003), is recommended. Both documents highlight the importance of appointing a competent Temporary Works Coordinator, as recommended in BS 5975:2019 (BSI, 2019); see also Chapter 2 on management.

This chapter gives information about the types of façades and the important differences in design principles relative to other temporary works.

26.2. Philosophy of façade retention
26.2.1 Major alternative
The first consideration is: does the façade actually have to be supported? Dismantling the façade as a whole, restoring the stonework, masonry or bricks and rebuilding using traditional methods might be more appropriate.

26.2.2 General principles
Many of the building façades (and other parts of the building) being retained would not conform to modern building techniques and codes of practice, hence significant engineering judgement and experience is required of those involved.

When considering façade retention schemes, the following philosophy and sequence should be adopted.

1. *Plan.* A desk study should be undertaken to establish the age of the building, neighbouring properties and their owners, history and previous usage of the building and listing issues. In addition, a visual inspection should be made to identify potential hazards (including hazardous materials), existing services, working restrictions, other site constraints, and so on. Principal designers (appointed by the client) have a duty to be involved at this stage so that a conceptual design for the temporary works and new permanent works can demonstrate that the works can be built safely. Relevant permissions may be needed and other statutory obligations may have to be met.
2. *Secure the site.* Public protection, protection of neighbouring properties, safe means of access and so on should be provided.
3. *Understand how the building works.* Never underestimate the importance of this. A full structural investigation and dimensional survey (which should include ground conditions, groundwater levels and foundations) should be carried out. Digital surveys can prove very useful and can be used by designers and contractors to identify potential clashes (see Chapter 31). How stability is achieved and the position of load-bearing elements should be established. CIRIA R111, *Structural Renovation of Traditional Buildings* (CIRIA, 1986), is a useful guide on how old buildings may have been constructed. Particular attention should be paid to areas of obvious distress, such as cracks, bowing or ingress of water (signs of poor maintenance), and to any area where repairs or alterations have previously been carried out. The façade may not necessarily be of solid construction; it may be stone faced on brickwork backing or rubble in-filled stone walls. The surroundings should be investigated, including pavement vaults, cellars, services and the

condition of nearby pavements and roads. Preliminary shoring and propping may be necessary. Surface finishes may need to be removed to expose brickwork and embedded steel or timber and their condition assessed. Floorboards may also have to be removed to assess floor capacity. The need for temporary works should be evaluated and competent designers and competent on-site supervision identified.

4 *Make unstable areas safe.* Temporary supports should be installed to allow repairs to be carried out or to remove hazardous items (e.g. asbestos). Chimney stacks are prone to weathering and can be removed. If laid with lime mortar, the bricks can be reused for rebuilding (or the chimneys will require shoring). Existing foundations may need to be underpinned and existing services diverted or made safe. Delicate items that are to be retained may need to be protected from accidental or weather damage, or they may be dismantled for safe storage during the works and then reinstated.

5 *Design the retention scheme.* Designers should carry out risk assessments and any residual risks identified should be clearly communicated. A sequence of works should be provided and provisions made for unexpected items. Structural surveys cannot consider every part of the structure, as many structural items are unseen until demolition commences. The retention scheme should allow the new structure to be constructed as easily as possible. Although the design may be carried out by the client's engineer, the principal contractor or a specialist equipment supplier or specialist subcontractor, the design interfaces should be established along with the design check category (BSI, 2019). Possible modes of failure should be considered and adequate support provided for each eventuality, or the risks eliminated (e.g. the risk of swinging loads impacting the façade retention scheme may be reduced by specifying lifting and craneage zones). The initial design proposals should be reviewed by the site team for practicality and agreed before detailed design commences.

6 *Install the retention scheme.* Installation should be carried out carefully by competent persons to prevent damage to the existing structure. Some partial demolition or minor shoring may be necessary to allow access, so that the main retention scheme can be installed and connected to the façade it is supporting.

7 *Carry out the main demolition of the internal structure to an agreed sequence.* Demolition of the internal structure should be completed carefully and in a systematic manner, avoiding damage to the retained structure and retention scheme. Checks should be made on the façade during demolition and after it has been completed (monitoring is discussed later). Weather protection may be needed to any exposed party walls and to the rear of the façade where water could cause deterioration (i.e. to lime mortar, embedded steel or timber).

8 *Carry out the new works to an agreed sequence.* Monitoring, inspection and maintenance of the retained structure, retention scheme and surroundings should be carried out throughout the works. For longer terms works a detailed maintenance schedule should be agreed and followed.

9 *Load transfer.* The new works are designed to restrain and sometimes support the existing façade. Where the weight of the façade is supported on the new structure, consideration must be given to the means of load transfer from the temporary support to the permanent structure. Unless the façade is to continue to be load

bearing, the connections should allow for differential movement between the façade and the new construction.
10 *Remove the temporary works*. Removal may be progressive as the permanent structure is completed. A detailed removal sequence should be agreed and followed.

It is important to realise that the building has achieved a state of equilibrium over many years and, by changing the support conditions, this equilibrium can be jeopardised; the designer must consider this. Deflection is one of the main criteria for the design, and the support system should be sufficiently stiff to limit the deflection.

Some of the above points may be given lower priority or even be bypassed in the case of emergency retention schemes. Emergency schemes will be selected on the basis of the availability of materials and the ease and speed of assembly.

26.2.3 Party walls
Walls that form part of a building but stand partly on land having different owners is known as a 'party wall'. Temporary retention systems can be used to support party walls. The legislation is complex and beyond the scope of this book. Party walls can generally only fall in the direction of the previously demolished building, but stability of any structures isolated by demolition must be considered. However, if it can be shown beyond doubt that the party wall is adequately tied into the structure beyond, then support will not be required. This may be the case in a row of terraced properties when demolishing a newer property back to a previous end-of-terrace property. If there are any concerns about the adequacy of the tied connection then additional fixings may be considered as an alternative to a retention scheme. For a party wall the effects of deflection need careful consideration, as limits provided in CIRIA C579 (Lazarus *et al.*, 2003) may be inappropriate. Always seek professional advice when dealing with party wall support and its legislation.

26.2.4 Surveying the existing building
The importance of an early thorough investigation of the existing façade fabric is critical to the successful completion of all temporary works façade support systems. When as built drawings are not available then dimensional surveys will be required and are often carried out using digital surveying techniques, see Chapter 31, some of the major issues that will need to be addressed are discussed in detail in CIRIA C579 (Lazarus *et al.*, 2003) and CIRIA R111 (CIRIA, 1986).

Are there any hazards (e.g. asbestos)? How is vertical and lateral stability achieved? Are neighbouring properties affected by the new works? What is the form of construction? Particular attention should be paid to cracks and signs of damage. Chimneys should be thoroughly investigated (e.g. position of flues), as should the verticality of the façade (out of plumb by more than 10% of the wall thickness is a matter of concern).

Are there any basements or pavement vaults? What is the position and nature of existing services? Is the building listed or part listed, or is it in a conservation area? Are there any tree preservation orders? Soil conditions, groundwater level and existing foundations must all be considered. As part of the survey, some materials testing may be necessary to establish strength and so on.

26.3. Types of temporary façade retention schemes
26.3.1 Timber shoring
Timber shoring was previously commonplace, but nowadays its use is rare apart from in minor retention schemes. Timber shoring can be considered to be a fire risk, and timber will shrink and swell and deteriorate over a period of time.

26.3.2 Scaffolding
Scaffolding in tube and fitting (see Chapter 21 on scaffolding) is suitable for relatively low-level façades (up to 3–4 storeys with sufficient space around the base of the façade for installation; the approximate width required will be 50–100% of the height (Lazarus et al., 2003)). Above this height the quantity of scaffolding involved and labour costs may make installation, inspection, rebuilding and removal challenging. A typical example is depicted in Figure 26.1.

26.3.3 Proprietary equipment
Proprietary equipment with formwork soldiers and ties may be suitable for higher façades as the number of components is reduced (compared to scaffolding) and each component is stronger and stiffer (see Figure 26.1). Proprietary equipment has the advantage that components can be easily joined together with push–pull props and tie rods, and easily adjusted to remove 'slack' from the system.

26.3.4 Fabricated steelwork
Fabricated steelwork may prove an economical solution for long-term retention schemes where the cost of hiring proprietary equipment may be prohibitive; however, manufacturing and installation costs may be higher. Combinations of equipment are often used.

26.3.5 Vertical towers
Vertical towers (Figure 26.2) comprising proprietary equipment (soldiers, push–pull props, tie rods) or fabricated steelwork with tie rods clamping the façade are a common solution (approximate width required is 16–20% of the height (Lazarus et al., 2003)). Substantial kentledge blocks (or compression/tension piles) are required to resist overturning and sliding forces; they can also act as foundations to spread the vertical loads. The self-weight of the façade can be utilised as part of the kentledge required.

To minimise disruption to the internal rebuilding, vertical towers are installed to the external face of the façade (where practicable). The frame acts as a vertical cantilever, with wind loads transferred to the foundations through bracing. Deflection of the frame is usually the critical design consideration to prevent damage to the façade (through cracking). Push–pull props or tie rods are used for bracing. Horizontally spanning waling beams, which are tied together through window openings (to minimise drilling though the façade), are placed internally and externally, 'sandwiching' the façade. The towers are connected to the waling beams.

Figure 26.1 Comparison of three-storey temporary façade retention schemes. (Courtesy of RKF Consult Ltd)

(a) Typical using scaffolding

(b) Typical using proprietary equipment

Figure 26.2 Typical vertical tower arrangements. (Courtesy of Pallett TemporaryWorks Ltd)

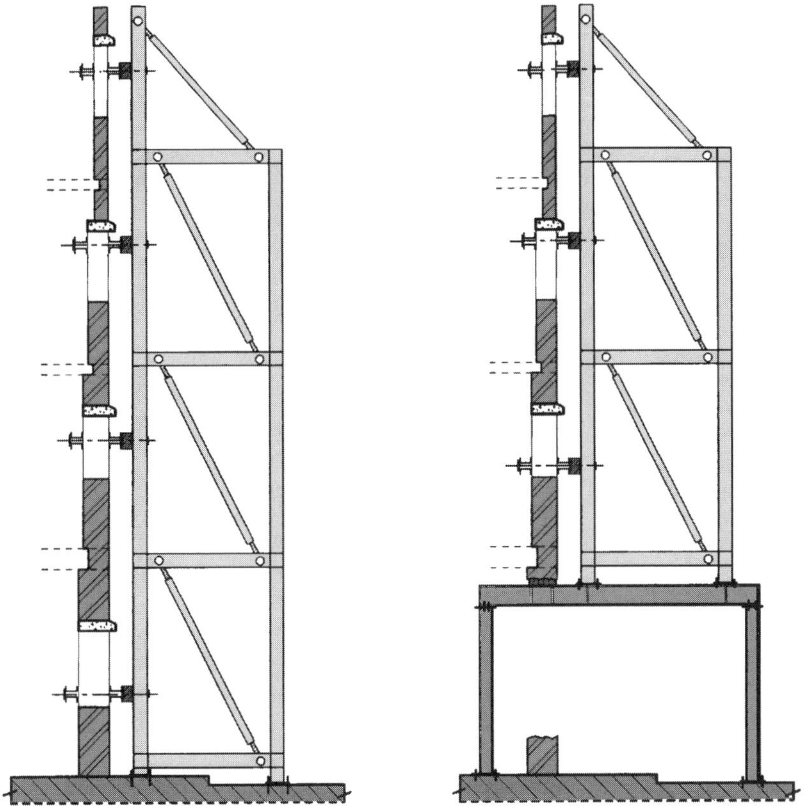

(a) Façade connected to retention system

(b) Façade supported and connected to retention system

It is important to consider the combined vertical and horizontal loads on the foundations as the uplift (due to the overturning moment) will reduce the horizontal capacity. The greater the width of the frame (i.e. greater lever arm), the smaller the uplift will be. An allowance may need to be made for pedestrian access. Portal frames are used for this purpose, as shown in Figure 26.2. Particular attention is required in the design of such schemes, as the eccentric loading from the self-weight onto the portal structure induces lateral sway deflection in the portal. Vertical towers can also be used where party walls are to be retained.

26.3.6 Horizontal frame arrangements

Where vertical towers are impracticable and adequate buttress walls exist, horizontally spanning trusses can be installed. These will have diagonal wind braces, waling beams as described above and vertical column supports or supporting braced towers, as shown in Figure 26.3. When using internal frames (see Figure 26.3) the temporary frames must be positioned in order to avoid clashing with the new floor construction. The frames are

Figure 26.3 Typical horizontal frame arrangements with diagonal bracing

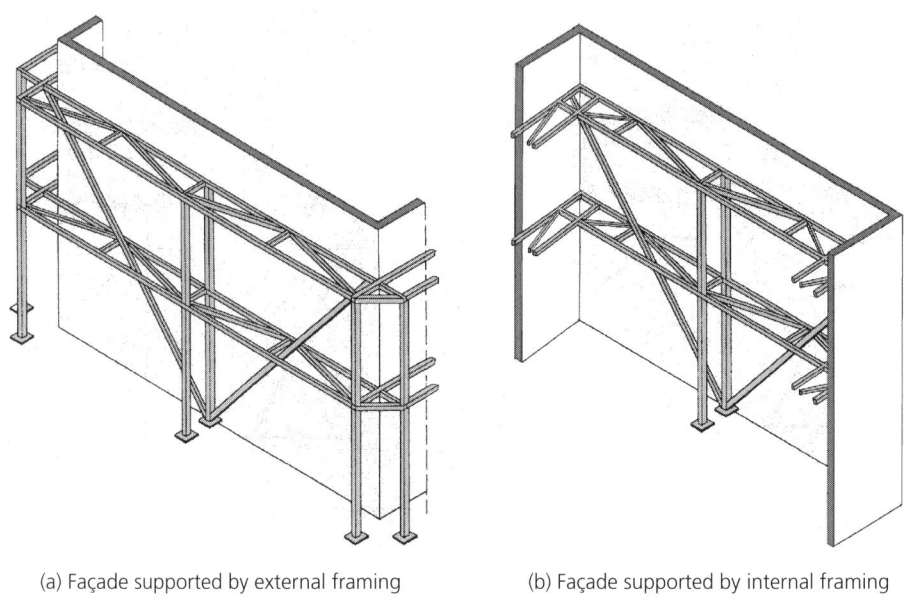

(a) Façade supported by external framing   (b) Façade supported by internal framing

positioned so they can be installed by operatives standing on the existing floors prior to their demolition. As the demolition of floors progresses downwards, subsequent lower-level internal supporting frames can then be installed.

With all support systems, timber or proprietary frames may be used to brace up major openings in the façade, to maintain the shear capacity and prevent lateral displacement.

26.4. Loads to be considered
26.4.1 General
The Temporary Works Designer for a retained façade should consider all potential modes of failure. Temporary works offices and designers have traditionally used permissible stress methods, and equipment manufacturers quote 'safe working loads' in their literature. The actual deflections need to be calculated in the service condition. When the design uses 'limit state', the relevant partial safety factors will need to be included. Particular care is necessary when using Eurocodes, as future suppliers' literature may quote 'characteristic values' for equipment; note, however, that a characteristic value is not a safe working load. The following loads and practical considerations typically need to be addressed in the design process, from concept to installation, monitoring and use.

26.4.2 Vertical loads
Vertical dead loads include the self-weight of the façade and the self-weight of access or storage platforms, hoists, sheeting and site offices (space is often at a premium). If vertical supports are removed (columns or piers replaced due to deterioration or strengthening), the self-weight of the façade must be supported by means of needling

schemes or vertical dead shores. Elastic shortening of the support system and settlement of any temporary foundations can be partially overcome by pre-loading (pre-deflecting the supports by jacking load into the system). Settlement, and possibly undermining, may also be caused by adjacent works such as trenching or excavating for a basement. Differential settlement (between the existing structure and the temporary works) is likely to be most problematic and can lead to the greatest degree of damage to the retained structure. When considering the vertical dead load, 'favourable' and 'unfavourable' conditions exist. The self-weight of the façade alone is the only true dead load when considering resistance to overturning, as cabins, storage, working platforms, and so on may be moved as the work progresses. Further guidance can be found in CIRIA C579 (Lazarus et al., 2003; section 8.3).

26.4.3 Construction operation loads

Imposed vertical loads during construction will include working platforms, stored materials or snow. The retention scheme should be easily inspected. CIRIA C579 (Lazarus et al., 2003) recommends that working platforms be normally designed for service class 2 loading (1.5 kN/m^2) as a minimum for minor repair work, or a higher service class if significant rebuilding is necessary. Material storage areas, loading bays or site cabins need to be considered separately. Further guidance can be found in CIRIA C579 (Lazarus et al., 2003; section 8.4).

26.4.4 Impact loads

Accidental impact loads occur from falling debris, vehicles or a swinging load suspended from a crane. These should be assessed by means of an appropriate risk assessment (generally measures should be in place to prevent impact wherever possible). Minor damage can be repaired; however, disproportionate collapse should be considered, and protective measures may be provided as necessary. Ideally, the design risk assessment should be used to reduce or eliminate any impact on temporary façade structures. Impact loads are listed in Table 26.1 and should be considered to act in any direction on critical members. Vehicle impact is considered in the bottom 1 m of the structure, but a lower impact loading of 10 kN is considered at higher levels to represent a swinging crane load. CIRIA C579 (Lazarus et al., 2003; section 8.5) gives further guidance. BS 5975:2011 (BSI, 2011b; clause 17.4.3.3) recommends that designers should 'refer to contract specification' when considering the design of vehicle crash barriers.

26.4.5 Wind loading

Wind loading is generally the main loading to be considered, and is calculated to the withdrawn BS 6399-2:1997 in CIRIA C579 (Lazarus et al., 2003). The current wind code is BS EN 1991-1-4:2005 + A1:2010 (BSI, 2005), and a simplified method, based on the Eurocode for use in all temporary façade retention structures, recommends that when used in the UK both the seasonal factor c_{season} (formerly referred to as S_s) and the probability factor c_{prob} (formerly S_p) are taken as 1.0. Unlike falsework, scaffolding and formwork, there is no reduction for the short-term loading (less than 2 years) for façade temporary works. The basic equation for the peak wind velocity pressure (q_p) is given in Section 22.3.5 in Chapter 22 on falsework. The façade is generally considered as a solid impermeable face (due to sheeting and boarding up of openings). If the wind can be

Table 26.1 Lateral load combination

Load case	Weight of actual façade	Self-weight of retention structure	Wind load		Impact load	Out of plumb	Other vertical loads
			Maximum	Working			
Overturning greater of either	1.5% of total	1.5%	Full	n/a	No	No	1.5%
or	No	1.5%	Full	n/a	No	Yes	1.5%
Connections to façade as greater of	2.5% at level	No	Local at level	n/a	No	No	No
or	No	No	Local at level	No	No	Full at level	No
Impact loads	No	No	No	Yes	10 kN in any direction	No	No
Vehicle impact	No	No	No	Yes	25 kN horizontal up to 1m	No	No

Data taken from CIRIA C579 (Lazarus et al., 2003).

'trapped' in a corner then the wind force may be increased. Positive wind pressures as well as wind suction are considered. For further guidance see CIRIA C579 (Lazarus et al., 2003; section 8.6).

26.4.6 Notional lateral forces

In a similar way to the minimum disturbing force allowed for in falsework (see Section 22.3.7 in Chapter 22 on falsework), the temporary façade retention system has to resist notional lateral forces. It is recommended in CIRIA C579 (Lazarus et al., 2003) that the supports are designed for a minimum lateral load of the greater of the following loads

- 1.5% of the vertical load at that point acting horizontally which includes the self-weight of façade, retention structure, other dead loads and imposed loads affecting the retention structure (e.g. storage and site cabins), but the actual lateral load should be calculated (the lateral load could be as much as 10% of the vertical load)
- the actual lateral load from eccentricity of the façade (due to out-of-plumb corbels, balconies, etc.) plus 1.5% of the total vertical load on the retention structure, which will include its self-weight plus other dead and imposed loads (the self-weight of the façade is not included).

This lateral load is considered as acting as a uniformly distributed load over the façade surface and is applied to the retention scheme at the connection points. It is added to the

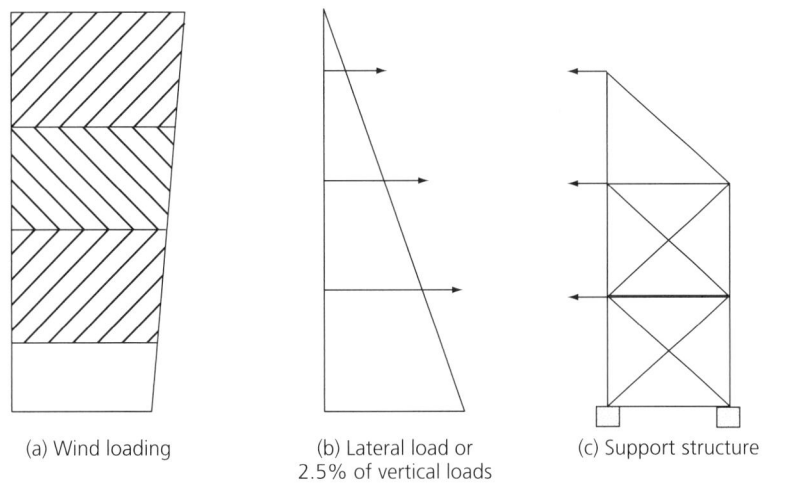

Figure 26.4 Wind loading plus lateral load applied to support points. (Courtesy of RKF Consult Ltd)

(a) Wind loading

(b) Lateral load or 2.5% of vertical loads

(c) Support structure

wind loading as shown in Figure 26.4. The wind loading will generally be significantly higher than the self-weight component of lateral load. A clear load path to the ground should be identified. For further guidance see CIRIA C579 (Lazarus et al., 2003; section 8.8).

26.4.7 Other loads

Existing basement walls will be subject to earth, water and surcharge pressures (traffic, plant, rubble or kentledge blocks). For further guidance see CIRIA C579 (Lazarus et al., 2003; section 8.7).

CIRIA C579 (Lazarus et al., 2003; section 8.9) provides advice on other loads to be considered. Dynamic loads (such as traffic) generating vibration are generally ignored, and any loosening of fixings can be addressed by using lock nuts or regular inspection. Fatigue is generally not taken into account in temporary works. However, there have been cases whereby a retention scheme has had to support a façade for many years due to planning or financial issues, and fatigue and corrosion can become issues. Seismic activity is not taken into account in the UK. Thermal movement, due to expansion and contraction of the materials of the façade and support system, may occasionally need to be considered when specifying 'acceptable' levels of movement in the façade.

Older façades tend to be of substantial stone or masonry thickness and have greater thermal mass, and hence respond slowly to daily temperature variations. Excessive thermal movement can lead to cracking of the façade. CIRIA TN107, *Design for Movement in Buildings* (Alexander and Lawson, 1981), and Building Research Establishment (BRE) Digest 251 (BRE, 1995) give further advice. If a retained façade is within a potential flood-risk area or next to water then these effects should be considered; ice formation is not deemed to be a major issue.

26.5. Design considerations
26.5.1 General
The support scheme must be positioned to allow the demolition and rebuilding to be carried out. This normally involves the temporary works being positioned externally (assuming sufficient available space and local authority permission), as internal temporary works (see Figure 26.3) make demolition and rebuilding more challenging.

External schemes need to consider interfaces with the general public, and appropriate safety measures should be taken. Security is a significant consideration, and adequate protection necessary.

Provision may be required to allow for the scheme to be altered due to variations in wall locations discovered during demolition and/or any unexpected items discovered on site. Large openings in the façade may require bracing with timbers or steel frames to maintain overall shear capability. The effects of fire are occasionally considered, and an appropriate risk assessment should be carried out. Lightning protection with adequate earthing should be provided to steel support frames.

26.5.2 Overall stability
Overall stability must be maintained in all directions, as wind loading or other loads (impact) could act in any direction. The retention scheme must resist the overturning moment as well as moments generated by eccentric dead loads (out of plumb, corbels, balconies, etc.) and imposed loads to provide overall stability. The structure may also tilt during demolition or reconstruction work as support conditions change due to re-supporting the vertical weight of the façade, settlement or heave. The new support positions should be provided as close as possible to the original condition in the existing structure. It must, however, be remembered that wind loading on the façade may change after the existing building has been demolished. The points of support to the façade should be at relatively close centres to prevent excessive deflection.

CIRIA C579 (Lazarus *et al.*, 2003; section 8.14) recommends a factor of safety of 1.5 against overturning; this is larger than the factor used for falsework in BS 5975:2011 (BSI, 2011b). Any kentledge required is then calculated as

$$\frac{\text{maximum overturning moment} \times 1.5}{\text{lever arm to the kentledge}} \qquad (26.1)$$

and potential settlement of the kentledge should be considered.

26.5.3 Deflection criteria
The support system must be stiff enough to prevent excessive movement, which may cause cracking to the fabric of the façade and should ideally replicate that of the existing structure. When retaining a party wall some degree of pre-loading (pre-deflecting) of the support system may be possible to reduce potential movement. The positioning of the jacking points should coincide with strong points in the retained building. The degree of permitted movement of any façade being retained should be discussed carefully with

the structural engineer and client. The nature and state of repair of the façade will determine this, and should be monitored during the course of demolition and rebuilding. See, for example, the monitored movement shown in figure 6.4 of CIRIA C589 (Bussell et al., 2003). CIRIA C579 (Lazarus et al., 2003; clause 8.10.1) recommends that, in the absence of more onerous limits set by the Permanent Works Designer, the following limit be adopted

$$\text{lateral deflection } \theta_H \leq \frac{H}{750} \tag{26.2}$$

where H is the height to the restraint level under consideration above the point at which the façade is considered fixed and θ_H is the lateral deflection. The limit was first quoted by Goodchild and Kaminski (1989).

CIRIA C579 (Lazarus et al., 2003) further states that: 'there appears to be no evidence that working with this limit has resulted in distress to façades. Equally there is no evidence to warrant recommending a more liberal limit.' In addition, a maximum floor-to-floor deflection of 5 mm over 3 m is often quoted. Designers of temporary works façades should be aware that the movement of the ground and foundation scheme to the façade retention scheme, in response to partial demolition and subsequent reconstruction are inherent in the scheme and should be considered by the client's consultant. The lateral deflection limit of the system defined above will be used by the temporary façade designer, unless tighter limits are set. This limit could be unacceptable for a party wall, creating large gaps between it and the return walls/floors within the adjacent building. BRE Digest 251 (BRE, 1995) discusses crack widths and repairs. Damage category 2 is generally considered appropriate, with cracks of < 5 mm requiring only decorative repair, but agreement should be reached with the party wall surveyors. Owners of supporting party walls of occupied buildings may be less tolerant of movement causing cracking, and deflection limits may need to be reviewed. Excessive vibration during demolition should be avoided.

26.5.4 Connections and restraint

Designing the connections to the façade is important. When making connections into old brickwork, expanding anchors should be avoided and bonded (resin or similar) type or thin wall sleeve expansion anchors should be used. Old bricks may have soft cores or may be loose (mortar may have weathered) and may not provide a positive fixing; alternatively, the anchors may work loose under cyclic loading. The number of drilled connections into a listed façade is usually limited by planning and architectural constraints. Fixings into brickwork should be tested to establish capacity for design purposes.

Local restraint is provided by the strength and stiffness of the support scheme prior to demolition of the existing supporting structure. When the existing structure is removed, the effective length of the façade dramatically increases and the supporting connections must provide sufficient restraint to the façade. CIRIA C579 (Lazarus et al., 2003; section 8.11) recommends that connection and walers should be designed for the greater of the following loads (to be effective in resisting lateral buckling of the wall)

- 2.5% of the total gravity load on the façade at the level of the connection being considered, plus the wind force on the area of the façade restrained by the connection
- the lateral load arising from offsets and out of plumb of the façade at the level of the connection, as a uniformly distributed load along the length of a linear element (waling).

26.5.5 Lateral restraint

The load combinations in Table 26.1 should be considered in the design process. The combination of the wind load and the connection lateral load is shown diagrammatically in Figure 26.4. Effectively, the façade itself has to span vertically between the connection points. Because the dead weight of the structure increases near the bottom of the façade, its ability to withstand lateral wind forces increases. It is, therefore, not uncommon to notice that, contrary to expectations, the vertical spacings between connections are closer together near the top.

Where kentledge blocks are used to cater for the lateral loads in sliding, sufficient friction must be generated. If the blocks are partially or fully buried then passive resistance can be considered. The kentledge blocks can also be designed to spread vertical load and limit settlement. If the kentledge is to be cast in situ, polythene must not be used to protect the pavement, as it will act as a slip membrane. CIRIA C579 (Lazarus *et al.*, 2003; section 8.14) recommends a factor of safety of 2.0 against sliding. When considering sliding, only permanent loads should be used to provide beneficial restoring moment or sliding resistance. Uplift and sliding need to be considered in combination as the uplift will reduce sliding resistance.

26.6. Demolition, monitoring and inspection

In order to install a retention scheme, some additional temporary works items may be necessary (e.g. bracing in window openings) and some localised partial demolition (generally using hand-held tools) may be required. Demolition should be a careful and systematic process following an agreed sequence of works in order to maintain stability and prevent damage. Demolition should be carried out by competent persons in accordance with BS 6187:2011 (BSI, 2011a). Further shoring of main structural members or 'protection/containment platforms' may be necessary to facilitate a safe demolition sequence. To prevent overloading, debris should not be allowed to accumulate on existing floors and chutes or conveyors should be used to remove rubble. Particular attention should be given to avoiding or protecting existing services. A common safety measure is to provide vertical scaffold boards or netting around cut edges to prevent loose masonry falling from the façade during the works.

A monitoring system is usually established prior to demolition. The purpose is to check for movement of the façade during the work. Readings are taken on a regular basis (at least weekly during demolition and following high winds). The readings should be assessed to identify potential problems, with the following trigger levels

- green (low movement) – no problems, continue taking readings
- amber (generally three-quarters of the maximum designed deflection) – increased movement, increase frequency of readings
- red (maximum designed deflection) – significant movement, undertake a thorough investigation of the causes and possible remedial measures that have been taken.

Inspection should be carried out on a regular basis, checking: that ties, wedges and fixings are tight (they can work loose in cyclic wind loads); that the structure and its state of repair is as expected by the design; and for any signs of damage due to impact during demolition or construction.

REFERENCES

Alexander SJ and Lawson RM (1981) *Design for Movement in Buildings*. CIRIA, London, UK, TN107.

BRE (Building Research Establishment) (1995) *Assessment of Damage in Low-rise Buildings*. BRE, London, UK, Digest 251.

BSI (British Standards Institution) (1997) BS 6399-2:1997. Loading for buildings. Code of practice for wind loads. BSI, London, UK. (Withdrawn. Replaced by BS EN 1991-1-4:2005 + A1:2010.)

BSI (2005) BS EN 1991-1-4:2005 + A1:2010. Eurocode 1. Actions on structures. General actions: Wind actions. BSI, London, UK.

BSI (2011a) BS 6187:2011. Code of practice for full and partial demolition. BSI, London, UK.

BSI (2011b) BS 5975:2008 + A1:2011. Code of practice for temporary works procedures and the permissible stress design of falsework. BSI, London, UK.

BSI (2019) BS 5975:2019. Code of practice for temporary works procedures and the permissible stress design of falsework. BSI, London, UK.

Bussell M, Lazarus D and Ross P (2003) *Retention of Masonry Façades – Best Practice Site Handbook*. CIRIA, London, UK, C589.

CIRIA (Construction Industry Research and Information Association) (1986) *Structural Renovation of Traditional Buildings*. CIRIA, London, UK, R111.

Goodchild SL and Kaminski MP (1989) Retention of major façades. *The Structural Engineer* **67(8)**: 131–138.

Lazarus D, Bussell M and Ross P (2003) *Retention of Masonry Façades – Best Practice Guide*. CIRIA, London, UK, C579.

FURTHER READING

BRE (Building Research Establishment) (1991–1992) Good Building Guides. BRE, London.
- GBG1: Repairing or Replacing Lintels (1992)
- GBG10: Temporary Support for Openings in External Walls: Assessing Load (1991)
- GBG15: Providing Temporary Support during Work on Openings in External Walls (1992)

BSI (British Standards Institution) (1981) BS 5977-1: Lintels. Method for assessment of load. BSI, London, UK.
BSI (1993) BS 5080-1:1993. Structural fixings into concrete and masonry. Method of test for tensile loading. BSI, London, UK.
BSI (2015) BS 5930:2015. Code of practice for ground investigations. BSI, London, UK.
Doran D, Douglas J and Pratley R (2009) *Refurbishment and Repair in Construction*. CIOB (Chartered Institute of Building), Ascot, UK.
Gilbertson A (2017) *Structural Stability of Buildings during Refurbishment*. CIRIA, London, UK, C740.
Highfield D (1991) *The Construction of New Buildings Behind Historic Façades*. Spon, London, UK.
Historic England (2016) *Stopping the Rot. A Guide to Enforcement Action to Save Historic Buildings*, 3rd edn. Historic England, Swindon, UK.
HSE (Health and Safety Executive) (1984) *Health and Safety in Demolition Work*. HSE, London, UK, GS29, Parts 1–4.
HSE (1985) *Safe Erection of Structures*. HSE, London, UK, GS28, Parts 1–4.
HSE (2006) *Avoiding Structural Collapses in Refurbishment*. HSE, London, UK, research report 463.
Knight LR (1984) The façade can be a nightmare. *Civil Engineering Magazine* March: 29–31.
Lamsden BS (1988) *Remedying Defects in Older Buildings*. CIOB (Chartered Institute of Building), Ascot, UK, Technical Information Service 89.
NSWC (New South Wales Construction) (1992) *Façade Retention. Code of Practice*. NSWC, New South Wales, Australia.
Perry JG (1994) *A Guide to the Management of Building Refurbishment*. CIRIA, London, UK, R133.
Thorburn S and Littlejohn GS (1992) *Underpinning and Retention*, 2nd edn. Taylor & Francis, London, UK.
UK Government (1990) Planning (Listed Buildings and Conservation Areas) Act 1990. Statutory Instrument 1990/9. The Stationery Office, London, UK.
UK Government (2015) Construction (Design and Management) Regulations 2015. Statutory Instrument 2015/15. The Stationery Office, London, UK.

Useful web addresses
CIRIA (Construction Industry Research and Information Association): https://www.ciria.org (accessed 01/08/2018).
Historic England: http://www.historicengland.org.uk (accessed 01/08/2018).
National Access & Scaffolding Confederation (NASC): http://www.nasc.org.uk (accessed 01/08/2018).
National Federation of Demolition Contractors (NFDC): http://www.demolition-nfdc.com (accessed 01/08/2018).
reFURB – journal of repair, replacement and maintenance: http://www.refurbprojects.com (accessed 01/08/2018).

Pallett, Peter F and Filip, Ray
ISBN 978-0-7277-6338-9
https://doi.org/10.1680/twse.63389.385
ICE Publishing: All rights reserved

Chapter 27
Bridge installation techniques

Keith Broughton
Retired

John Gill
HOCHTIEF (UK) Construction

Bridge installation techniques range from straightforward construction methods to highly complex and unusual methods developed for a specific project. Construction techniques covered in earlier chapters are adequate to describe normal bridge construction. In this chapter, more unusual and complex techniques are described which use the bridge structure as part of the temporary works, depending on the bridge design concept and the specific project constraints. With more unusual techniques, the installation technique will often drive the final bridge design, and certainly the detailing. Techniques considered here include partial deck erection schemes, deck erection as a single unit, erection by tunnelling and mining, and segmental deck erection. Factors to consider in the design of the temporary works for these bridge installation techniques include significant stability issues (global and local), robustness, foundations and supports, plant and equipment (mechanical and motive systems design), contingency planning, control and monitoring systems, and independent review and validation.

27.1. Introduction

Installing bridges is one of the more high-profile aspects of civil engineering, and generates significant media and public interest. It takes a brave project team to invite public scrutiny, however, as the risks in these projects can be high. The rewards for engineers working in this field are great.

Most people will have heard of the Millau Viaduct in southern France, watched a local bridge construction project or seen one of the documentaries in the *Megastructures* TV series (Channel 5 in the UK). This chapter will help readers appreciate the effort needed to make the techniques look simple, and enthuse them to investigate further.

The bridge installation techniques discussed here use the structure itself to provide all or most of its support during erection. The required design is covered in this chapter, but only in concept.

Some of the concepts are subject to patents and intellectual property rights. Many companies have particular specialist knowledge which goes much deeper than described in this text. Current best practice is generally project specific and many details do not necessarily translate to another project.

The rate of development is rapid and techniques quickly evolve such that some of the details here will be obsolete within a few years. As with permanent design, future developments in material technology and mechanical plant will change our view of the possible. However, engineers will always return to the basic principles and concepts such as those involved in the erection schemes described here.

27.2. Preparation and selection of installation technique

There has been a continuous drive by UK clients and contractors to achieve cost benefits and improve risk management. This has in turn led to a greater consideration of the erection technique at the design stage. Contractors intuitively look to remove expensive temporary works schemes, particularly the 'bridge to build a bridge' solutions. Within densely populated and industrial areas there is a greater need to devise erection schemes that minimise erection periods, maintain traffic flows, reduce the need for service diversions, or incorporate the efficient demolition and replacement of a worn-out structure.

Recent trends that could affect the bridge installation technique include the move towards build-offsite methods. Such principles need to be enshrined within the design concept stage. The benefits derived from large-scale offsite prefabrication can deliver a reduced programme and costs, improved worker safety and higher product quality. Further improvements may be seen from the inclusion of life-long stress and deflection monitoring gauges within the structure to provide real-time measurement, not just during construction but also through the asset-operation period. Such products have the ability to improve the client's asset maintenance programme.

Selecting a bridge installation technique is a major decision that requires careful and thorough preparation in order to have all the required information to hand. The range of information includes

- health and safety hazards
- installation constraints
- bridge design principles
- experience and current practice
- cost and time
- detailed site information.

A useful debate at an early stage of a project is whether the client should prepare a fully developed scheme, or leave the design and installation techniques entirely to the tendering contractors. Experience suggests that the optimum is part way between these extremes. Early contractor involvement is successful when developing a scheme prior to a full tender. A negotiated contract can also be successful, where tenderers develop and submit their technical solutions and obtain client feedback during a tender process leading to a final pricing stage, but this is expensive for both parties. For such schemes, clients should carefully consider and review the allocation of responsibility of the major risk events.

A large number of solutions are available for bridge installation, and most can be grouped into the headings used for the following sections: partial deck erection schemes, deck erection as a single unit, erection by tunnelling and mining, and segmental bridge construction.

27.3. Partial deck erection schemes
27.3.1 Large crane erection to single and multi-span bridges
27.3.1.1 General description of erection technique
This technique is used for bridge designs incorporating composite steel and concrete decks and pre-cast concrete bridge beams. The composite design requires a reinforced concrete deck, bonded with welded shear studs to an upper steel flange. Bolted or air-welded joints would be located between 1/3 and 1/4 points of the span. Pre-cast beam solutions also require an in situ stitch arrangement. Large-piece erection uses either tracked or telescopic cranes.

27.3.1.2 Base requirements
These include a good access for the crane and materials, stable crane platforms with the ability to carry large outrigger leg loads (see Chapter 5 on site roads and working platforms) and no overhead restrictions. For multiple-span structures, provision may be required for temporary supports to form bolted or welded joints. Large fabrication or storage lay-down areas are also required. Where the bridge is erected on a critical timescale, such as a rail or motorway possession, a trial erection is essential to ensure all components are completed on time and fit properly. For information relating to the crane selection and the lifting design see Chapter 20 on heavy moves.

27.3.1.3 Risks and opportunities
For the sake of safety, working at height should be avoided if possible. This includes using permanent formwork systems for the soffit of in situ decks, installation of parapet formwork and working access onto the outer steel beams prior to lifting. The growth in hydraulic crane capacities now offers the opportunity for far larger deck sections. The capability for a single large deck panel lift over the railway is illustrated in Figure 27.1 in the construction of Bridge 12 at Stratford Olympic Park. There are, however, risks to be considered, such as high winds, when depending on a single lift during a disruptive Network Rail possession.

Stability and robustness. Considering the stability of the beams during the lift, it is preferable to erect the beams as braced pairs; otherwise, temporary bracing is required. The main beam design also needs to consider the construction load cases, including placing of deck concrete. Bracing should be designed to ensure that any permanent formwork cannot slip off its bearing on the steel flanges, particularly in an area over road, rail or occupied sites. Pay particular attention to the lateral stability of edge beams as lifted and landed.

Foundations and support. Temporary trestles will require foundations designed to spread short-term loads into the ground while supporting beams until splices have been completed (and sometimes until the deck is complete).

Figure 27.1 Single large lift at Stratford Olympic Park. (Courtesy of HOCHTIEF (UK))

Plant and equipment. Erection cranes will typically be 250–800 t mobile cranes, but the largest mobile (1000 t) or gantry (1200 t) cranes may be required for high load and large radius lifts. Cranes utilising super-lift require a clear working area of up to 30 m diameter. A limitation on crane set-up near railways will be a significant constraint, and the congestion of city-centre areas often prevents super-lift work.

Temporary stresses. A critical load case during concreting of an in situ composite deck, where additional temporary bracing will be required, occurs when concrete cast on the beams creates locked-in stresses as they deflect under wet concrete load.

Independent review and validation. For any installation over railways, a full design and independent check of the temporary supports, stability and method statements will be required.

Current and future developments. Greater use of pre-cast concrete includes large panel deck and parapet sections, where attention to detail regarding the connections between main beams and the deck and between the deck panels is critical for immediate construction safety and long-term durability. Fibre-reinforced polymer and glued bonding technology will further reduce the on-site work and cost of this method. Increasing use of integral pier/deck design and techniques to permit longer integral viaducts will enable more efficient use of pre-cast concrete and composite decks, as described by Barnes and Gill (2018) for the design of Pont Briwet Viaduct.

27.3.2 Pre-cast concrete arch

This is a variation on segmental construction where a pre-cast concrete arch is formed in segments. There are generally two types of arch

- A single-piece arch structure, spanning springing points. Erection moves progressively along the bridge alignment using a single crane or a tandem lift.
- A three-pinned arch, comprising two pieces joined at the crown. The erection method requires the two initial pieces to be erected together, one held on the crane (or propped) until the other is positioned, offset by half a segment width. Once this is secure, both cranes release their support. The erection of the following segments progresses alternately, using the previously placed arch segment for support.

The large pre-cast concrete sections are lifted onto a pre-constructed edge foundation that acts as a springing point for the arch. The most important aspect of the erection concerns the designer's assumptions of the shear capacity of the foundation and placement of the structural backfill. The temporary loads on the foundation generally govern design. The structural fill must be placed equally on both sides of the arch, taking care not to create an out-of-balance loading.

27.3.2.1 Base requirements
Consideration should be given to the following points

- Early input from specialist suppliers is recommended to inform the arch configuration and the interaction with adjacent structures or earthworks.
- The delivery, handling and storage of pre-cast sections must be considered. Lifting points should be designed to be cast into the back of each section such that the section hangs on the hook in the configuration required for setting down. This avoids the need for pulling, kentledge or coarse adjustment.
- Control of material delivery, placement and compaction must be ensured in the method statement.

27.3.2.2 Risks and opportunities
The key benefit of a pre-cast concrete arch is its speed of erection and the ability to prepare the foundation beams outside of the influence of the traffic. Consequently, such arches are often appropriate during possessions over railways and highways. A typical concrete arch structure constructed over the railway of the Greater Bargoed Community Regeneration Scheme in South Wales is shown in Figure 27.2. An arch profile is sometimes adapted to the gauging of trains and vehicles. Quality control is obviously better with factory production. The placement of fill over the structure often provides the benefit of landscaping and planting. Edge walls can be created with reinforced earth structures or simple pre-cast concrete spandrel panels.

27.3.2.3 Design considerations
Stability and robustness. A fundamental aspect of the design is the changing load distribution from first installation through backfilling and any nearby construction. The

Figure 27.2 Lifting in a pre-cast arch at the Greater Bargoed Community Regeneration Scheme. (Courtesy of HOCHTIEF (UK))

backfill sequence assumed in design must be followed and monitored on site as the arch is susceptible to small changes in load, particularly unequal loading. The case history on the rail tunnel failure at Gerrards Cross is useful for any Temporary Works Designer to reflect on and learn from (NCE, 2005).

Stability of the initial arch is critical and may require temporary propping, kentledge or anchors to prevent toppling or spreading. The subsequent arch units are generally stable once the first units have been fixed in position. Units are generally robust during lifting and handling, provided attention is paid to positioning lifting points and the layout of the working areas. Two-piece arches are not robust when placed until all joint details have been completed; for example, a stitch joint may be required at the crown and grouting may be required at the springing points. They are best used only where the load of a one-piece arch would be beyond the limit of the crane.

Foundations and support. Temporary foundations should not be required, although temporary props may be needed to support units in storage or during final positioning. It is critical that the lifting arrangement is designed carefully to ensure that the final positioning is achieved with the piece in the correct orientation, which requires a calculation of the centre of gravity of the piece.

Contingency measures. A temporary trestle may be required to support the first section if two cranes cannot be used. Rapid-strengthening concrete could be used to complete joints or stitch details to prevent arch spread or to connect units (e.g. Ductal produced by Lafarge).

Control and monitoring. Control measures should cover the final position, alignment and inspection of bearing areas to ensure no point loads are introduced. Monitoring of arch units during and after installation is important to ensure that deflections are within expected limits.

Independent review and validation. It is crucial that the interface between the specialists' design of the arch system and the global design into which it fits is considered and that all temporary load cases are developed. A number of failures during construction have occurred, and it is recommended that the designer review literature on these while developing the design.

Current and future developments. A FlexiArch^{TM} system developed by Macrete allows the arch to be delivered 'flat' and take on its arch shape on erection (see Macrete under Useful Web Addresses at the end of this chapter).

27.4. Deck erection as a single unit
27.4.1 Launching
27.4.1.1 General description of erection technique

This is an installation technique in which the deck is fabricated in part or in whole behind one abutment, and pulled or pushed into position over rollers or skids located at the abutment and the intermediate supports. A lightweight launching nose is usually required and ballast may be required at the tail. Due to the deflection of the nose there may be a vertical jacking requirement at the intermediate supports. The technique is complex and costly and is chosen only for multiple spans or after all other options have been discarded. The effect of the rolling load on the lower soffit or flange means that the temporary loads to deck and substructure are often the most onerous loading conditions and define the design of the structure. A clean straight launch path is preferred, although a circular path is possible. Fully varying soffits or 'corkscrew' decks have been launched, but are extremely complex and require multiple vertical jacking points to prevent overloading of individual elements.

A variation less frequently used is to launch the bridge sideways from a temporary construction position alongside the final alignment. This requires temporary abutments and piers with a slide track to the permanent abutments and piers. A benefit is the shorter timescale for the final installation, which usually requires an existing road deck to be closed for a very short time.

Such methods were employed during the construction of the A38 Marsh Mills Viaduct project near Plymouth (Figure 27.3), where traffic disruption was minimised using a side launch technique (permitting the two viaducts to be replaced with only two weekend road closures). Figure 27.3 shows the viaduct deck on temporary supports, carrying live traffic, before the demolition of the existing pre-cast concrete deck and the eventual slide onto the newly constructed piers.

Figure 27.3 Sideways slide at Marsh Mills Viaduct. (Courtesy of HOCHTIEF (UK))

27.4.1.2 Base requirements
The client should consider a contract that allows for the combination of temporary and permanent design both to the deck structure and the loads applied to the substructure. The length of preparation during the pre-construction/design period should not be underestimated. A specialist jacking subcontractor is required. If this solution is being applied to a project with multiple stakeholders, then it can be very beneficial to bring them in closely to the project at the earliest stages.

27.4.1.3 Risks and opportunities
Launching systems are favoured where the underlying ground is inaccessible (e.g. because of a river or an operating road or railway). A large area behind one abutment is required to erect the deck launch. For side launching, a linear site is required for the full length of the bridge. The design process will go through several iterations due to the effects of the temporary loads on the permanent design. There are advantages in the speed of erection and the avoidance of over-sailing craneage.

The minimisation of friction at the bearing positions will reduce the jacking loads. However, consider restraints for sideward sliding and the requirement for a braking system.

If technically feasible, it is normally beneficial to install the bridge in as close to a complete state as possible (i.e. track and ballast in place, or surfacing and white lining complete). On the Paddington Bridge Project (Figure 27.4), it was possible to incorporate the

Figure 27.4 Deck launch at the Paddington Bridge Project. (Courtesy of HOCHTIEF (UK))

foundations of the temporary loading towers in the permanent substructure. On completion of the deck launch the new deck was also used to transport away the old truss structure.

27.4.1.4 Design considerations

Stability and robustness. Many serious failures have occurred during this form of construction, where temporary stability has played a significant part in the failure. A thorough appreciation of many factors affecting stability is required before designing a launched bridge system, covering the global and local stability of the bridge and its components. Some factors to consider include the following

- loads on and resistances of substructures and foundations
- stability following out-of-tolerance movement
- flexibility of the whole system at all stages
- effects of environmental loads, including wind, ice and temperature
- site controls and technical supervision
- monitoring and feedback control systems
- lower chord buckling at temporary supports.

Foundations and support. Normally the permanent substructures and foundations are used to provide support during the deck launch. This is often supplemented by temporary supports to reduce bending moments, shear forces and deflections. Careful attention is important to the interface between the sliding deck and the support to ensure adequate guidance is provided and to limit the resistance loads on the support.

The pathway on the support can be provided by sliding or rolling bearings. The bearing design accommodates static friction (when the launch commences) and lesser dynamic friction (as the launch proceeds), as well as lateral forces (imposed in restraining any lateral movement). The choice of rolling or sliding bearing will depend on the requirements for tolerances, friction resistance, intensity of load at bearings and robustness. Final fixity of the deck is completed after the launch; permanent bearings will sometimes be carried in during the launch, but isolated out of use to prevent abnormal stresses.

Plant and equipment. The design of a deck launch should only be undertaken with the knowledge and support of specialist plant and equipment suppliers. Indeed, the design erection scheme will be partly defined by the size, capacity and applicability of the suppliers' hydraulic jacks, strand jacks and the associated power pack. The motive power required to move the structure can be calculated from the frictional values estimated in the slide tracks, the roller bearings or at the structure–soil interface. Guidance to estimated frictional coefficients is given in Chapter 20 (Section 20.2.4.3) on heavy moves and in the online Engineer's Handbook (see Useful Web Addresses at the end of this chapter).

Contingency measures. As with all high-risk schemes, contingency measures should be planned at critical stages of the erection. This may include the provision of additional jacking capacity, the means of reducing friction or simply a means to halt progress for a period for repair.

Control and monitoring. It is essential to monitor both movement and stress to the structure during erection. The structure should be modelled through all the temporary design load conditions and allowance made for the effects of temperature. There may be a requirement for a braking restraint. The onsite teams should be highly integrated, with well-understood lines of communication and clearly identified decision-makers. This can involve specific specialists taking control of the site operations for the critical jacking operations, who then hand over to the overall site supervision team upon completion.

Independent review and validation. An independent design check must consider all aspects of the method of erection on the design. The checking engineer should be selected from consultants with the experience and ability to undertake similar complex schemes.

Current and future developments. Lightweight materials including fibre-reinforced polymer make the nose and forward cantilever lighter to reduce the high bending forces during the launch. Increased design and supervision have permitted long spans to be launched over live railway such as the Tennison Road Bridge at Norwood Junction in Croydon.

27.4.2 Self-propelled modular transporter (SPMT)

Note that relevant information relating to the capacity and arrangement of SPMT equipment is detailed in Chapter 20 (on heavy moves).

27.4.2.1 General description of erection technique
The application of SPMTs is a relatively recent development, in which multi-wheeled trailers incorporating hydraulic jacks lift and drive pre-constructed bridge forms into position. The jacks and wheels can be moved by remote control with very fine tolerances, while lifting substantial loads.

27.4.2.2 Base requirements
The most important consideration is the preparation of the roadway formation on which the trailers must run. Clearly, the road formation must be strong enough to withstand the running and skewing of the multiple axles. In addition, the overall settlement of the loaded trailers must be within determinable limits. For this reason, steel plates, proprietary track and even laying an asphalt surface are used for the erection roadway. Distance travelled within the possessions should be limited by the close location of the construction yard or by parking the structure within a short distance of the bridge site. Because of limited jacking heights, SPMTs do not have the ability to go over humpbacks with total vertical differentials of over 750 mm.

27.4.2.3 Risks and opportunities
The main advantage of the SPMT technique is the speed at which the deck or pre-cast form can be moved into position. SMPTs are therefore popular for bridge erections requiring railway possessions in which the tracks are lifted, an embankment is excavated, the bridge is driven into position and backfilled, and the track is replaced, all within a long weekend. They are similarly popular for bridge replacements to over-bridge decks on the motorway system. If ground conditions are very good, consideration might be given to carrying in the bridge foundations with the remainder of the portal. Such a scheme was successful on the Channel Tunnel Rail Link Contract (CTRL) 342, where the portal structure and foundations were installed in one possession (Figure 27.5).

27.4.2.4 Design considerations
Stability and robustness. A critical calculation is to find the centre of gravity of the structure to be transported and to ensure the stiffness in the temporary transport configuration is accurately defined in the permanent structure modelling.

Foundations and support. While the transporter will spread the load well, the competence of the surface layers must be ensured such that the transporter does not settle under load. The risk of transporter failure due to settlement of the wheels is significant in determining whether the work can be completed within a road or rail possession.

Plant and equipment. Appropriate transporters are readily available from a number of companies, including Mammoet UK Ltd, Fagioli Ltd and ALE Ltd.

Temporary stresses. The arrangement of the support system located on top of the transport bogeys will determine the stresses during the lifting process. If possible, it is preferable to incorporate the temporary load case in the permanent design. Beware of the danger of stresses created by distortions generated by the transport path.

Figure 27.5 Installation of a pre-cast portal box at CTRL 342 using SPMT units. (Courtesy of HOCHTIEF (UK))

Independent review and validation. An independent review is needed to ensure that all load cases have been considered within the final design and method, and incorporated either in the permanent design or within temporary works.

Current and future developments. Integration of design, control and monitoring will allow more efficient design and use for more complex bridges. For example, Wakefield Eastern Relief Road rail bridge, including the cill beam, bearings and deck, was carried by SPMT.

27.5. Bridge and deck erection by tunnelling and mining
27.5.1 General description of erection technique

Advances in the manufacture of hydraulic jacks and the development of slide tracks with very low values of friction have led to the jacking of ever-larger structures under existing road and rail embankments. The mining technique involves the pushing of a structure through an embankment, while material is excavated from within a mining shield at the front. This is done either by jacking a single box structure or by jacking the abutment and roof as separate elements. A key element in the design is the means by which the embankment above the roof is restrained in its original position, often known as the anti-drag system (ADS). Solutions include the use of thin steel sheets or steel ropes to minimise forward movement of the ground above the structure as it is jacked into position.

Whichever method is chosen, provision of a large anchorage restraining mechanism is required at the entrance portal. The design of the mining shield is also critical and must fully consider the ground conditions, proposed methods of excavation and shoring. The safety of the miners and operatives is imperative.

An alternative method is to construct foundations and slide tracks in advance through the embankment, for example, under a railway. Once the slide tracks are complete, the section of railway track and embankment above are removed and the pre-constructed bridge is jacked into position along the slide track onto the new foundations. The choice of method is determined by the construction programme and possession time available, the volume of material to be excavated and an assessment of risk by the contractor.

27.5.2 Base requirements
This method does not suit poor ground or high water tables. Mining within chalk and cohesive soils above the water table give the best results. The minimising of surface settlements is a major requirement. This can generally be mitigated by minimising the free area being mined at the face shield; the greater the potential for face loss, the greater the surface deflections. The limiting factor on the design will usually be the amount of jacking force required to move the structure. Large structures can be moved in sections by the introduction of intermediate jacking stations. A structure is initially jacked at the interface between the front sections and then progressively backwards, resulting in a caterpillar-like motion.

27.5.3 Risks and opportunities
In general, tunnelling and mining techniques should be considered high risk and only undertaken when all other options have been exhausted. The temporary works are expensive and require a high level of engineering expertise. The advantage of such a technique is in minimising the effect of construction on a key transport link. Contingency planning is essential in this type of scheme and fall-back alternatives should be considered at every step, whether this involves the provision of more jacking power or the injection of bentonite to improve the frictional values.

The calculation of the frictional resistance to jacking is critical and should only be undertaken by an experienced designer using reputable data sources. The final solution will be obtained by numerous design reiterations in which the estimated costs and risks are repeatedly measured and balanced. On the A23 Coulsdon Town Improvement Scheme (Figure 27.6), a horizontal line of 600 mm diameter steel tubes was installed at high level beneath the rail ballast to act as the ADS. The tubes were filled with reinforced concrete, and locked into a concrete slab over the jacking pit to provide horizontal restraint to the rail track.

27.5.4 Design considerations
27.5.4.1 Stability and robustness
One of the most important design requirements is the ability of the structure to withstand the substantial local jacking loads imparted both at the back of the structure and at any intermediate stations. Typical horizontal jacking arrangements may consist of several

Figure 27.6 Jacking a full road box on the A23, Coulsdon. (Courtesy of HOCHTIEF (UK))

500 t capacity jacks, each with a stroke of up to 2.0 m. Upon achieving full extension, the jacks are retracted, spacer blocks are inserted and the cycle repeated.

The safety of the miners, the loads from their machinery and the requirement to support the open face of the excavation determines the robustness of the design of the front mining shield. The shield is typically made either from heavy steel sections or from reinforced concrete with a steel cutting edge. Guidance is available in the preparation of the specification for jacked box tunnelling, including the requirements for the working environment, predicting and monitoring ground conditions and grouting (BTS/ICE, 2010).

27.5.4.2 Foundations and support

The provision of foundations at an early stage (e.g. within tunnels) has the advantage that the slide path for the roof and walls provides the guidance of a straight jacking path. Box jacks that have no such restraints or guidance are more likely to incur errors in their final alignment. The largest frictional jacking forces are generated at the underside of box portal structures. There is a requirement for a casting bed or a pre-construction area at the box invert level within approach retaining walls, and consideration must be given to the means of resisting the substantial jacking forces. Often the casting slab and the retaining walls can be arranged to provide the jacking resistance. A head beam must be constructed at roof level of the entrance portal, to anchor the ADS mechanism that prevents the soil block over the roof from moving with the box (see Figure 27.6). The loads to be resisted here require a substantial beam and appropriate supports.

Various means are available to the designer to reduce the jacking loads. These might involve low friction materials such as PTFE (polytetrafluoroethylene) on the slide paths, tunnelling muds such as bentonite injected into the over-excavated void around the structure, and drag shields to the roof incorporating steel sheets or steel ropes.

27.5.4.3 Plant and equipment
Specialist suppliers provide the hydraulic equipment. Strand jacks could be employed to pull a bridge structure along a slide path. For further details regarding the design and capacity of such equipment, see Chapter 20 on heavy moves.

27.5.4.4 Independent review and validation
The assistance of an experienced checking engineer in the team will provide the assurance required by the contractor and client that all loading assumptions are reasonable and within an acceptable range.

Current and future development. This technique is very expensive and the design is specific to each situation. However, innovation can influence the health and safety of the workforce in underground construction, confined spaces and manual handling. Robotics and 3D printer technology will be particularly influential.

27.6. Segmental bridge construction

It is not possible for this chapter to cover the design and erection techniques of large bridge structures, such as cable-supported bridges and balanced cantilever construction. These structures incorporate many of the temporary works described in the above and in earlier chapters. A project such as the Queensferry Crossing bridge near Edinburgh, employed offsite factory construction of the deck sections, the movement of large deck sections using SPMTs, bridge deck launching and utilisation of the balanced cantilever method, to name but a few. As an example, a particular specialist process requiring high levels of skill for this segmental bridge structure involved the accurate manufacture of the steel box deck elements in long prefabricated sheds overseas, and their repeated erection in factory conditions to match cast the upper reinforced concrete decks in the adjacent shipyard.

The balanced cantilever construction technique for the Queensferry Crossing is shown in Figure 27.7. The individual deck sections are transported by floating barge, with the deck sections incorporating many of their final fittings, such as drainage, internal lighting and monitoring gauges. The sections are erected alternately to maintain balance by the specially designed mobile lifting frames located at the end of each deck span.

A more detailed description of the construction methodology used for the Queensferry Crossing is given by Carney (2015).

Figure 27.7 Deck erection method on the Queensferry Crossing. (Courtesy of Forth Crossing Bridge Constructors)

Current and future development. With the Queensferry Crossing and the new Mersey Gateway records such as tower height, length of cantilever and volume of concrete poured demonstrate the potential for further development. Use of building information modelling (BIM) (digital construction, collaboration and integrated data), point cloud survey and 'big data' analysis, robotics and 3D printer manufacturing, new materials and even drone flying will all change what is possible in long-span bridge construction (Parag *et al.*, 1999; Oliveira and Reis, 2016).

REFERENCES

Barnes JN and Gill JC (2018) The design and construction of the Pont Briwet Viaduct, UK. *Proceedings of the Institution of Civil Engineers – Bridge Engineering* **171(2)**: 80–90.

BTS/ICE (British Tunnelling Society/Institution of Civil Engineers) (2010) *Specification for Tunnelling*, 3rd edn. ICE Publishing, London, UK.

Carney CT (2015) How to build a road deck. *Forth Replacement Crossing Project Update*, September: 4–5. See https://www.transport.gov.scot/media/35842/frc-project-update-september-2015.pdf (accessed 01/08/2018).

NCE (New Civil Engineer) (2005) Backfilling thought to be culprit in Gerrards Cross rail tunnel collapse. *New Civil Engineer*, 1 August 2005.

Oliveira PJJ and Reis AJ (2016) Composite cable-stayed bridges: state of the art. *Proceedings of the Institution of Civil Engineers – Bridge Engineering* **169(1)**: 13–38.

Parag CD, Frangopol DM, Nowak AS (1999) *Current and Future Trends in Bridge Design, Construction and Maintenance*. Thomas Telford, London, UK.

FURTHER READING

Allenby D and Ropkins JWT (2015) The use of jacked-box tunnelling under a live motorway. *Proceedings of the Institution of Civil Engineers – Geotechnical Engineering* **157(4)**: 229–238.

Rosignoli M (2002) *Bridge Launching*. Thomas Telford, London, UK.

Troyano LF (2003) *Bridge Engineering: A Global Perspective*. Thomas Telford, London, UK.

Watt D (2017) *The Queensferry Crossing Vision to Reality*. Lily Publications, PO Box 33, Ramsey, Isle of Man.

Useful web addresses

ALE – examples of heavy lifting: http://www.ale-heavylift.com/case-studies (accessed 01/08/2018).

Britain's Greatest Bridges – Channel 5: http://www.channel5.com/show/britains-greatest-bridges (accessed 01/08/2018).

Ductal – rapid strengthening concrete for jointing pre-cast segments: http://www.ductal-lafarge.com (accessed 01/08/2018).

Engineer's Handbook: http://www.engineershandbook.com (accessed 01/08/2018).

Macrete – video of the construction of a FlexiArch bridge system: http://www.macrete.com/flexiarch-flexiarch-projects (accessed 01/08/2018).

Mammoet – examples of heavy lifting: http://www.mammoet.com/cases (accessed 01/08/2018).

Queensferry Crossing – YouTube, bridge construction videos: https://www.youtube.com/channel/UC2GKRiYa1SylyQI9QJVXX8A (accessed 01/08/2018).

Tennison Road Bridge – ICE London Awards 2016 (video): https://www.youtube.com/watch?v+5vwBWI8Yzbs&feature=youtu.be (accessed 01/08/2018).

Pallett, Peter F and Filip, Ray
ISBN 978-0-7277-6338-9
https://doi.org/10.1680/twse.63389.403
ICE Publishing: All rights reserved

Chapter 28
Backpropping

Peter F Pallett
Pallett TemporaryWorks Ltd

David Szembek
Temporary Works Manager, Byrne Bros (Formwork) Ltd

Backpropping is the method by which the loads imposed during construction can be transferred to lower parts of the structure so that the structure is not overloaded. Backpropping is usually required when newly constructed parts of the works have either not gained sufficient early-age strength or have been designed with superimposed loads less than those imposed during construction.

Typical examples of structures that require backpropping are found in multi-storey building construction, where floor slabs are designed for light imposed domestic loads that are less than the self-weight of the new floor slab. An example, for heavier civil engineering structures, is a thick transfer slab at high level required to be cast above a thinner slab not designed to accommodate the full self-weight of the high-level slab as an imposed load.

28.1. Introduction

Backpropping of concrete slabs during construction is a subject often misunderstood in the industry. Regrettably, many Permanent Works Designers (PWD) still consider the subject as 'not relevant and being a contractor's issue'. This chapter explains the theory and background to backpropping. It gives advice on the methods recommended to be adopted for backpropping calculations for both building and civil engineering applications.

The safe transfer of the loads through the structure with backpropping has legal implications for the client, PWD and contractor(s). Legal Guidance (L153) (HSE, 2015) on the Construction (Design and Management) Regulations 2015 (CDM 2015) (UK Government, 2015) has specific requirements for designers to control temporary works. The law also states that 'Any temporary structure must be of such design and installed so as to withstand any foreseeable loads which may be imposed on it' (regulation 19 (2a)). Backpropping during construction creates a totally foreseeable load on the structure. Hence both constructors and designers have to consider backpropping and understand the mechanics of load transfer during construction of the designed structure.

A senior and respected engineer talking about whether concrete slabs get overstressed during construction has said: 'It's not a question of whether they crack, but by how much they crack!'

In building, the issue is really quite simple; nearly all multi-storey buildings are designed for imposed loads that represent only a small proportion of the total design load. Many commercial buildings have a ratio of imposed load to total design load of 0.43 and for apartment buildings this ratio is often less at 0.35. Hence the self-weight of the next slab to be constructed cannot be taken on the recently completed slab, and the construction load needs to be distributed to lower, already completed, floor slabs. Where the floor slabs are of similar construction, the likely effects on the backpropping can be estimated by making assumptions about the stiffness of the supporting slabs and of the propping used.

In civil engineering, where single thick slabs/decks (e.g. high-level heavy plant rooms) are required to be cast over lower slabs not designed for the appropriate imposed construction loads, the varying stiffness of the slabs and the stiffness of the temporary works can require intriguing engineering solutions in order to construct the works without detriment to the permanent works.

28.2. The theory

The mechanics of how loads transfer through slabs is basic physics: within elastic limits, the deflection of a slab is proportional to the total applied load on the slab, to carry load it needs to deflect. So, if there are two identical floor slabs separated by rigid (non-elastic) props, applying a load to the upper slab would cause both slabs to deflect by the same amount (Figure 28.1(a)). Hence as load is proportional to deflection, each slab would effectively be taking 50% of the applied load. This is theoretically not correct, because the props will themselves be elastic members and need to shorten in order to take load. So if you have the same two identical floor slabs, but now separated by elastic props (see Figure 28.1(b)), as the load is applied to the top slab the props physically shorten in order to transfer load to the lower slab. The upper slab *must* now deflect more than the lower slab, as the distance between the slabs is reducing and the upper slab therefore carries more load.

The amount of load transfer obviously depends on the relative stiffness of the system. When the elastic props are pre-loaded (see Figure 28.1(c)) the upper slab is pushed upwards and a corresponding force increases the load on the lower slab, so that when the additional load is added to the top slab the distribution of load will change. This might also be affected by whether or not the slabs have been post-tensioned. However,

Figure 28.1 Slab deflections under applied load with rigid and elastic props

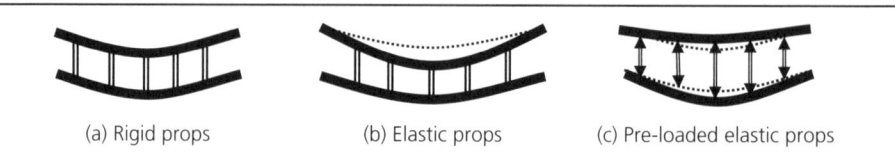

(a) Rigid props (b) Elastic props (c) Pre-loaded elastic props

in general, backpropping considerations are made with the slabs acting elastically, and whether post-tensioned or just reinforced the slabs will act similarly, although the post-tensioned slabs are likely to be thinner and are likely to give greater deflections in service conditions.

The three main factors that affect the relative stiffness of the backpropping system are

- the number of backprops
- the span/depth ratio of the slabs being constructed
- whether the backprops are left in place or whether the new slabs are allowed to take up their instantaneous self-weight deflection.

Having fewer backprops than the number of supports forming the falsework to the new slab results in the load per backprop being increased, and hence the actual backprop shortening will increase. This chapter considers both 'one-for-one' backpropping (an identical layout on each floor) and '50% backprops' (the number of backprops is no more than half the number of falsework supports to the new slab). This is discussed further in Section 28.5.4.

As the span increases, the PWD's limit of allowable maximum deflection will increase, often designed on a ratio of 1/250 (i.e. a 10 m bay gives 40 mm deflection). The storey heights of floors are often similar, about 3.0 m, so the elastic shortening of the backprops will be of similar magnitude (usually of the order of 1–1.5 mm). Hence on longer spans the deflection of the supporting slab becomes significantly more than the elastic shortening of the props, and the overall arrangement behaves in a manner approaching that for rigid props.

With regard to whether left-in-place backprops are used, certain proprietary systems allow the face contact material to be released early, leaving the newly cast slab supported by undisturbed props (see Chapter 24). Where the progress is fast, the next new floor may be cast before the supporting slab has taken up its deflected shape. This procedure not only alters the load distribution between floors but also means that the falsework will require designing for higher loads than a single floor. For example, when starting from the foundation level without striking the falsework/backprops, by the time the third floor is cast the lower supports will be subjected to more than three times the self-weight of one floor (see Section 28.5.2).

The nomenclature usually used for backpropping in multi-storey buildings is shown in Figure 28.2.

Another aspect of load transfer is whether or not the backprops have been inserted with some pre-loading, which would push the floor above upwards, decreasing its load, while at the same time increasing the load into the lower floor (see Section 28.4.3).

When the three-dimensional deflected shape of a slab is considered, plus the variations in the deflected shape of internal, corner or edge slabs, the movement of the various members and their method of support becomes complex.

Figure 28.2 Backpropping in multi-storey buildings – nomenclature

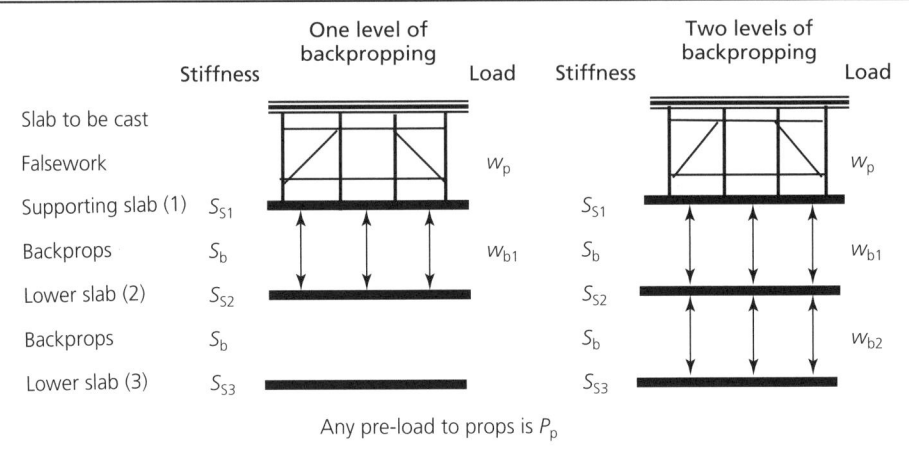

The simple assumptions above take no account of the different physical stiffnesses of the completed floor slabs, older floors being stiffer than newly constructed ones. Hence they have different structural properties, further adding to the complexity.

28.3. Loads (actions) to be considered

The actions (loads) considered in backpropping calculations differ from those used in the permanent works design.

The self-weight of the completed concrete slabs is generally assumed in backpropping calculations to be based on a density of $24\,\text{kN/m}^3$ for building multi-storey flat slab structures. When placing the new slab, the falsework and the backpropping should be designed for the new concrete, being 'wet', using a density of $25\,\text{kN/m}^2$. The density generally used on civil engineering structures is $25\,\text{kN/m}^3$, unless specified otherwise.

The self-weight of the formwork and falsework for most applications, up to (say) 4 m high, may be considered as $0.5\,\text{kN/m}^2$ based on the plan area. Depending on the sequence of installing backpropping, the self-weight may or may not be included in the backpropping calculations (see Section 28.8).

The construction operations loads imposed during backpropping will not be the same as those considered during physical construction. The minimum construction operation live load allowance of $0.75\,\text{kN/m}^2$ over the entire plan area of each floor under construction, allowed for in design, has never been observed in research. It is therefore recommended that for backpropping calculations no live load allowance is considered on the completed floors (i.e. the supporting slab and lower slabs). It should be noted that where sites decide to store or load out slabs in advance of other works (e.g. with pallets of bricks), this additional loading must be taken into consideration.

To allow for the imposed load during construction operations (e.g. placing reinforcement) the minimum construction operation load of $0.75\,\text{kN/m}^2$ (service class 1) is considered to act on the new slab's formwork. It is recommended that this 'working area' load is considered in the backpropping calculations. The additional 'variable transient in situ concrete load' (BSI, 2019; clause 17.4.3.1) applied over any 3×3 m area is considered in the falsework design but is *not* considered in the backpropping calculations.

In backpropping calculations, it is also recommended that the applied loads from the formwork or falsework, including the self-weight and construction loads, are considered as a distributed load applied to the supporting slab.

28.4. Research

28.4.1 European Concrete Building Project (ECBP)

Academic concerns in the 1990s were formulated into a research project, culminating in full-scale trials at the European Concrete Building Project (ECBP), which was completed in 1998. The research demonstrated that it is the supporting slab below the falsework that takes the majority of the load when backpropping (BRE, 2000). It further confirmed that backpropping through more than two levels is unnecessary, as the load does not get distributed to the lowest level.

The research was written up, for industry use, in CS140, *Guide to Flat Slab Formwork and Falsework* (CONSTRUCT, 2003).

The rationale of the research was to promote fast-track construction, economise on equipment and minimise the labour content. This meant also that the slabs were conventional reinforced concrete with ready-mixed concrete of a strength that was readily available at the time (i.e. a C30/37 mix).

It was also realised that using aluminium backprops with 100 kN capacity was not an economical use of material when these are installed on a one-for-one basis, in which case less than 60% of the load is passing between floors. Hence the research used individually installed backprops and fitted fewer than 50% of the members as backprops. On the 7.5×7.5 m bays of slab, the ECBP had nine falsework legs but only four backprops per bay. This halved the amount of equipment and was deemed much faster to install.

Early striking of soffit formwork and falsework was a key issue, and the research introduced a new method of considering early striking of slabs. The project also published research on a more accurate method of predicting the concrete strength required. The method is based on determination of crack width as opposed a simple ratio of loads as in earlier methods. This is discussed in more detail in Chapter 24.

28.4.2 Latest research

The ECBP research was carried out on reinforced concrete slabs, not post-tensioned slabs. The tensioning of a cast slab will obviously have an effect on the relative stiffness of the various slabs. Furthermore, when the slab is tensioned the distribution of load will change if propping is already in place. Although research into backpropping post-tensioned slabs was first identified as a subject by the Concrete Society in 2012 (CS, 2012a), at time of writing (2018) no research has been commissioned. Backpropping of post-tensioned slabs requires engineering judgement based on an understanding of structures and the basics of backpropping.

More recently, industry has studied the effects on current (2018) structure types and found that the ECBP research published as original Method One is not valid for the more flexible, larger span/depth ratios seen in modern construction. This chapter has been informed by these latest findings, and gives the current viewpoints on design as Method One – Revised (see Section 28.6.1).

The unanimous view of the engineers consulted is that Method Four from the ECBP, which is a three-dimensional approach using an Excel spreadsheet, still gives satisfactory results. It is certainly significantly simpler to use than some of the three-dimensional linear frame software, and it gives results in line with those checked against sophisticated frame analysis software.

28.4.3 Pre-loading backprops

Further investigations by Alexander (2004) and Vollum (2008) into pre-loading of backprops showed that pre-loads vary significantly in the range 7–14 kN. This research and site experience show that methods to control pre-load such as the use of strain gauges, load adjusting washers, x turns of the prop, and so on, are difficult to evaluate and control in a site environment and are not practical in construction. Faced with 10–15 backprops, how can a uniform pre-load actually be practically achieved with one or two operatives?

The industry now accepts that operatives will rarely, if ever, install a backprop 'finger-tight' with zero pre-load (one of the assumptions in the ECBP research). Operatives know from experience that it needs one wallop with their hammer. This overcomes any slackness in joints, and is estimated to give a pre-load of 2–3 kN in the prop. If, however, the operative gives it two hits with the hammer, a load of about 5 kN is realistic. Some proprietary systems have 'jack spanners' about 500 mm long, and these can impart a pre-load of 10–15 kN.

It is also realistic to assume that an operative will install an individual prop with greater pre-load to avoid it falling over rather than one attached to a framing system for stability. In other words, the pre-load is likely to be greater than 5 kN and near 10 kN.

Although the theory may consider differing loads in backprops when installed, the practical reality is that engineering judgement justifies the operative using two hits as the norm, which gives a 5 kN pre-load. There is a precedence for the operative defining the loads – in scaffolding, although the scaffold fittings are tested with a defined torque in the bolts, the reality is that the length of scaffold podger provides the correct torque when an average scaffolder tightens the fitting.

It is also realistic to assume that the operator(s) will progressively install backprops and, because of the flexible nature of the slabs, the effect will be to reduce the pre-load on previous props as the tightening progresses because the floors will move. For this reason it is considered good practice to provide framing or bracing to the props to prevent them falling over if they become loose.

This chapter makes the assumption that, in general, backprops will be progressively inserted to give an equivalent overall pre-load of about $0.50\,\text{kN/m}^2$ (note that this is similar to the $0.3\,\text{kN/m}^2$ recorded during the ECBP research).

28.5. Methodology – multi-storey construction

28.5.1 General

The complexity of modern flat slab buildings can broadly be categorised as 'regular' or 'irregular'. Examples of regular flat slab buildings are commercial developments on a basic grid, about 10×10 m, with the flat slab supported on columns. The ECBP research described in Section 28.4.1 is considered a regular structure. In contrast, there are many high-rise irregular shapes of differing spans, often with short stub walls for support; many apartment buildings have this arrangement. The advice given in this chapter caters for both types, and engineering judgement should be sought when backpropping irregular shapes. The irregular shapes often have shorter spans, and different criteria may apply.

Although this chapter introduces a revision to the ECBP simple Method One analysis for backpropping, users should be aware that use of the Method Four spreadsheet will provide more accurate answers, but requires more input data from the user.

As discussed previously, there are two arrangements to consider, left-in-place or struck-and-moved, and these are discussed in detail in the following sections. There are also two ways in which backprops can be placed relative to the falsework legs: on a one-for-one basis or, as per the ECBP, with at least 50% fewer backprops (see Section 28.5.4).

28.5.2 Left-in-place prop-and-panel systems

The typical arrangement in multi-storey construction for strip and re-erect systems (see Section 24.2.3 in Chapter 24) is the 'drop-head' system, whereby the props and/or beams remain undisturbed while the formwork, either conventional plywood or panels, is struck. This means that the most recently cast slab is not allowed to take up its full deflected self-weight profile. When a second set of props and/or beams is erected on the top of the slab and the new slab cast, the load in the undisturbed props will now be carrying a percentage of the newly cast slab as well. The support system is effectively left in place.

Left-in-place prop-and-panel systems are operationally efficient because there is typically no backpropping to install and remove. The falsework props remain in place to become the backpropping until they can be removed and recycled. The formwork and soffit panels can be struck early by virtue of the drop head, to allow recycling.

The distribution of load between floors depends on how stiff the floors actually are. If they have span/depth ratio less than 40 and are stiff compared to the propping, then the load transfer between the various floors will not be evenly distributed. In such cases, as the elastic backprops shorten the higher level concrete slabs will attract a greater percentage of the load from the newly cast slab, as the floors will deflect different amounts.

The left-in-place method, when used for flexible slab construction where the span/depth ratio exceeds 40, gives the closest approximation to the rigid-prop scenario (see Figure 28.1(a)), as the props are in intimate contact with the concrete and are already pre-loaded.

However, careful consideration must be given to the sequencing to ensure that cumulative loading is identified and designed for. In particular this applies to the first two or three slabs above the ground-bearing foundation slab. Similarly, on the typical cycle it is important to remove the lowest props at the correct point in the sequence to allow relaxation of the supporting slabs above, thus avoiding overloading both the props and the newly cast slabs.

The Concrete Society (CS, 2012b; worked example 7) highlights the limits of such a technique and illustrates the significant role of the PWD in accepting that loads greater than designed are regularly being applied during construction. The control of the sequence as planned is paramount (see Chapter 2 on management), as the slightest change in the order of removal and replacing of props or backprops will have a significant effect on the cast slab.

28.5.3 Struck-and-moved props

Whereas the use of tables for soffit formwork (see Chapter 24) was popular in the 1990s, the safety implications in today's construction, and in particular client requirements for full external protection (see Chapter 25), limits the use of handling formwork in large tables. The corollary is that prop-and-panel systems (left in place) or erect-and-strip systems are more common. For erect-and-strip systems the formwork and support are fully removed once the slab has gained sufficient strength to support itself and any construction activities. This may slow the cycle time compared to left-in-place systems. The additional operation of installing backprops to the new slab can then begin.

In terms of backpropping calculations there is more certainty regarding the load path when a slab is struck and allowed to deflect under its own weight prior to becoming a supporting slab for the floor construction above. The struck slab self-weight will be transferred directly to the permanent supports of the columns, walls and so on. Hence any loads transferred through this floor from the construction of higher floors will all be 'additional loading' to that already on the slab.

Consider the general arrangement of the construction of a concrete floor slab, with its soffit formwork and grid of supporting falsework legs standing on the previously cast floor. A typical building under construction with two levels of backpropping is shown

Figure 28.3 Backpropping through two levels: (a) with 50% fewer backprops; (b) with one-for-one backprops. (Courtesy of (a) Building Research Establishment and Byrne Group and (b) Byrne Bros (Formwork))

(a) (b)

in Figure 28.3. When the fresh concrete is placed, does the load distributed into the supporting slab act as a distributed load or as individual point loads from each of the falsework legs?

28.5.4 One-for-one or 50% fewer backprops

The grid of backprops below the supporting slab, transferring load to lower floors, will either be at the same centres as the falsework legs (i.e. one for one), or at much greater centres to use the props more efficiently (i.e. with at least 50% fewer backprops than falsework legs).

There are benefits to using a one-for-one system because the arrangement to stabilise the falsework legs, such as ledger frames, can also be used for the backpropping. This has the benefit that the load passes straight through the recently cast supporting slab into the backprops, and does not impart bending in the supporting slab. It is also much safer during erection, and overcomes the issue of pre-loaded individual props destressing adjacent backprops, which are then likely to fall over as they become loose. A typical layout would have supports laid out in an approximately 1.8×2.4 m grid, so that the backprop pre-load of about 5 kN (see Section 28.4.3) gives a theoretical equivalent slab pre-load of about 1.10 kN/m^2. As already discussed, as adjacent props are tightened there will be a relaxation in the pre-loaded props, so the likely average slab pre-load approaches the recommended value of 0.50 kN/m^2.

Whereas the 50% fewer backprops arrangement utilises the backprops better, the concrete supporting slab will now have an influence on load transfer as it is acting as a beam transferring load from the falsework legs into the backprops. In addition, the higher prop loads resulting from 50% fewer backprops and the bending induced in the slab will 'soften' the rigidity of the backpropping (see Figures 28.1(b) and 28.1(c)). In using this method the designer will need to check that the slab is capable of accepting the induced hogging moments and punching shear. This may be a particular problem for post-tensioned slabs, where there is commonly no top reinforcement in the slab at midspan.

A typical layout would have backprops laid out in an approximately 3 × 3 m grid, so that the backprop pre-load of about 10 kN (see Section 28.4.3) gives a theoretical equivalent slab pre-load of 1.10 kN/m², reducing to 0.55 kN/m² if the props are pre-loaded to 5 kN.

There is a safety consideration because with fewer props and increased spacing the methods for stabilising individual props become more difficult. As already stated, an operator pre-loading one individual prop can easily cause the slab above to go up, causing an adjacent prop to become loose and fall. For this reason, contractors should consider using a one-for-one arrangement, where the props can be easily braced, and it is for this reason that this chapter recommends an average pre-load of 0.5 kN/m² is adopted.

28.5.5 Concrete slab strength
The methods of calculating backpropping loads all highlight the fact that it is the strength of the supporting slab ((1) in Figure 28.2) at early age that dictates the speed of construction. The analysis methods often require concrete strength values close to the full 28 day strength at the time of casting the new slab. While this may be possible with concrete without additions, the low rate of gain of strength of concretes comprising blended cements can impose time delays on construction; this is rarely considered by specifiers wanting to use less expensive concrete. The implications of this low rate of gain of strength and methods of assessing concrete strength at early age are discussed in more detail in Section 24.7 in Chapter 24.

28.5.6 Slab stiffness
The theory (see Section 28.2) and the latest research (see Section 28.4.2) show that the amount of load transferred through slabs is related to the deflection of the slabs and the stiffness of the backprops. Slabs of different thickness and support conditions will behave in different ways. Although the PWD may have given limits on deflection after construction (span/500 is common), the calculation of slab deflection during backpropping along with the likely shortening of backprops is outside the scope of the Temporary Works Designer (TWD). The use of the spreadsheet in Method Four (see Section 28.6.4) takes these imponderables into account by automatically introducing deflection factors into the arrangement.

A simpler method for most sites is to consider the span/slab depth ratio – both items are known to the TWD. This has been quantified by using the ECBP spreadsheet (Method Four) on different spans, thicknesses and prop pre-loads to give a reasonable representation of load transfers, hence the recommendation to use Method One – Revised (see Section 28.6.1) and the span/depth ratio as the qualifier.

28.6. Calculation methods – flat slabs
There are four methods by which designers can complete backpropping calculations. These are detailed in the following sections.

28.6.1 Method One – Revised

Based on the ECBP research and updated to allow for slab stiffness (see Section 28.4.2) this method uses a simple assumption about the percentage of load transferred through the supporting slab(s) related to the span/depth ratio. Values for either one-to-one backpropping (solid line) or 50% fewer backprops (dotted line) than falsework legs are shown in Figure 28.4.

This method is generally conservative, and recommendations on percentages for either one or two levels of backpropping are given. This is the method most likely to be used in calculations to assess the amount of backpropping necessary.

Worked examples of backpropping calculations using Method One for span/depth ratios less than 30, including 'What if?' scenarios, are published in a booklet that accompanies the Concrete Society's formwork guide (CS, 2012b).

The percentages of load transmitted through the lower supports for a falsework system with one level backpropped, and then with two levels of backpropping is shown in Figure 28.4. It assumes elastic backprops that are inserted with a pre-load equivalent to about 0.50 kN/m², and, where there are two levels of backpropping, they are identical (i.e. the backprops are inserted exactly above each other between the supporting slab (1) and the lower slab (3)).

It is important to state that the distributed load applied on the existing floor slabs is *additional* to the load already being supported by the floor (self-weight, imposed load, storage, etc.). On the stiffer slabs this can result in loads in the backpropping less than that assumed for rigid backprops. The corollary being that more load is required to be carried by the supporting slab (1).

28.6.2 Method Two

This method uses the equations established in the University of Leeds research (BRE, 2000) to predict the load transfer knowing the stiffness of the slabs and the stiffness of the backpropping. It considers deflection of the system in two dimensions only.

For more detailed information on this method refer to *Guide to Flat Slab Formwork and Falsework* (CONSTRUCT, 2003).

28.6.3 Method Three

This method, using simplified equations, is given in detail in CS140 (CONSTRUCT, 2003) and the Concrete Society's formwork guide (CS, 2012a; section 5.4.2.5).

The equation for two levels of inserted backprops is reproduced below; it assumes that the slabs have been struck individually, and have taken up their deflected shape, prior to installation of the backpropping. The analysis assumes that the structure is in two dimensions only and that, to calculate the loads in backpropping, the slabs will be at least twice the stiffness of any backpropping introduced. Method Three is therefore only suitable for slabs that are very stiff, and have low span/depth ratios: $S_{S1}/S_b = 2$ and $S_{S2}/S_b = 2$ (see Figure 28.2).

Figure 28.4 Percentage of load transfer for flat slabs less than 350 mm thick. (Courtesy of Pallett TemporaryWorks Ltd)

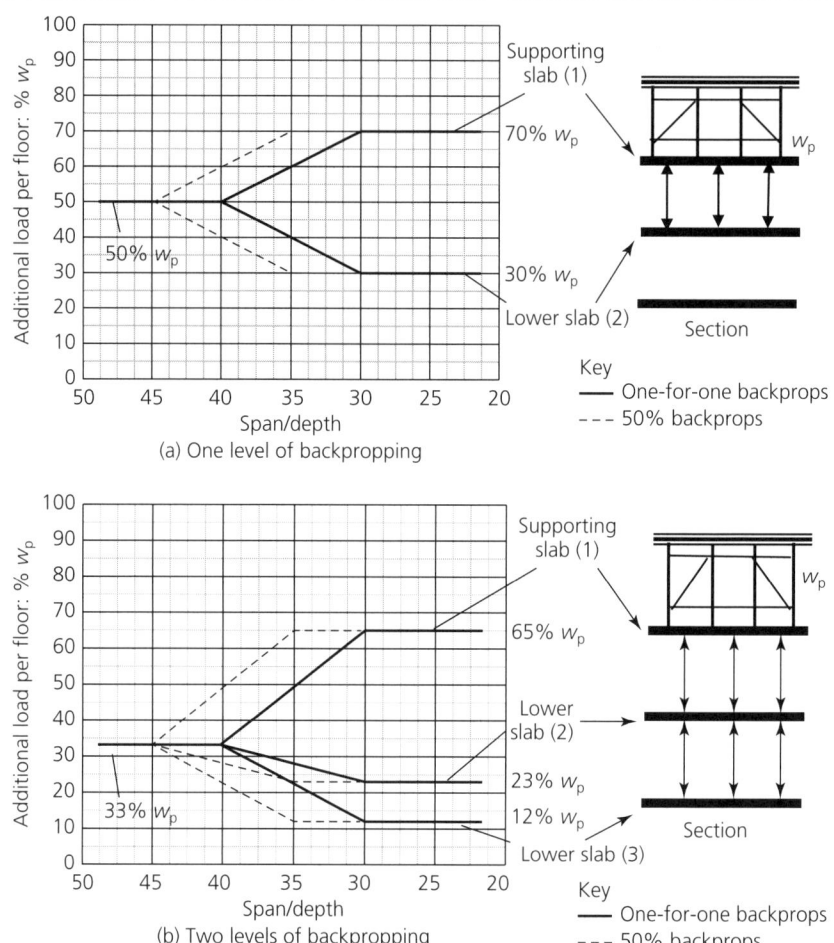

Notes:
1. The load w_p is the load applied to the supporting slab from construction of the new slab.
2. Assumes all the floors are of similar construction, are suspended floors, are less than 350 mm thick and have similar stiffness at the time considered.
3. Assumes the backprops have been inserted with a pre-load approximating 0.50 kN/m², and where there are two levels of backprops they are fitted in identically similar locations above each other.
4. The values represented by the dashed line indicate where there are less than 50% of the number of backprops compared to the total number of supports used in the falsework for the slab.
5. Assumes the lower and supporting slabs have been struck and have taken up their deflected shape, and are carrying their own weight.
6. The distribution is that percentage of the applied load on the supporting slab. Each floor slab will also have to carry its own self-weight and any imposed construction loads already on the floor.
7. Determination of the characteristic strength of the slabs to carry the applied loads is not considered.
8. The load in the backprops at each level is the sum of the additional load applied to the slabs below the backprop level considered.

For *two levels of backprops*, as shown on the right-hand side of Figure 28.2, the load in the top backprops is

$$w_{b1} = \frac{w_p}{[3 + (S_{S1}/S_{S2})] - \{(S_{S1}/S_{S2})/[3 + (S_{S2}/S_{S3})]\}} \quad (28.1)$$

and the load in lower backprops is

$$w_{b2} = \frac{w_{b1}}{3 + \left(\dfrac{S_{S2}}{S_{S3}}\right)} \quad (28.2)$$

28.6.4 Method Four

This method is a more accurate determination of backpropping loads using a three-dimensional representation of the equations used in Method Two. It introduces deflection coefficients and allows for the location of the slab and its deflected shape. Edge panels will behave differently from internal panels of the slab and so on. The calculation is presented as an Excel spreadsheet on a CD Rom accompanying the *Guide to Flat Slab Formwork and Falsework* (CONSTRUCT, 2003).

The spreadsheet allows selection of interior panels, edge panels, corner panels or panels supported on four sides by walls or beams. The stiffness of the concrete slabs and back-propping can be varied, and props can be pre-loaded. The output gives a loading factor, a cracking factor and an effective deflection factor. If all these factors are less than unity then the limits are safe for striking. If any factor is greater than unity then reference must be made to the PWD. The philosophy of loading a slab above its design service load is discussed extensively in the CONSTRUCT flat slab guide (CONSTRUCT, 2003; annex E).

28.7. With one level of backpropping

The previously cast floor slab is now the supporting slab for the next level of construction, as shown on the left-hand side of Figure 28.2.

The Temporary Works Coordinator (TWC) will need to establish whether the supporting slab has sufficient capacity at its very early age to support the self-weight of the temporary works and, possibly, some imposed construction operations load at the time considered. As the supporting slab matures, its capacity should increase up to its design service load capacity. Note that the supporting slab should *always* be considered to take the weight of the formwork and falsework for the next slab. This removes the onerous requirement to place the backprops in position *before* the formwork can be moved vertically up the building. The intention should be to install the backpropping at the earliest available opportunity following removal of the falsework.

The load in the backprops is the same as the load transferred to the lower slab as estimated using Method One – Revised (see Figure 28.4(a)). Alternatively, if it is a very stiff slab arrangement, the load can be calculated using the simplified Method Three equation. The additional load imposed on the supporting slab will often be the critical condition and govern the speed of construction. The TWC must ensure that both the supporting slab and the lower slab have gained sufficient strength before the new slab is cast.

The most accurate method for predicting the loads, once the arrangement of the falsework and the backpropping is known, is Method Four (the Excel spreadsheet in the CONSTRUCT flat slab guide (CONSTRUCT, 2003)).

28.8. With two levels of backpropping

Three previously cast floor slabs are now the supports for the new slab, with the most recently cast slab being the critical supporting slab, as shown on the right-hand side of Figure 28.2.

Obviously, the TWC will need to first establish whether this supporting slab has sufficient 'spare capacity' at its very early age to support the self-weight of the temporary works and some imposed construction operations load. As the supporting slab matures, its capacity should increase up to its design service load capacity. As in the case of one level of backpropping, the supporting slab should *always* be considered to take the weight of the formwork and falsework for the next slab. This overcomes the onerous requirement to place the backprops in position *before* the formwork can be moved vertically up the building.

In the backpropping calculations for construction of the new slab, the TWD will need to establish the total load during construction (w_p). This will include the self-weight of the new slab, but with *no* superimposed construction load on the already cast slabs. The self-weight of the falsework and formwork may not necessarily be carried through to the backprops, because if erection has commenced before installing the backprops the supporting slab will already be supporting this construction load. The working area imposed load (0.75 kN/m^2) during the placing of reinforcement and during concreting of the new slab (see Section 28.3) will be included in the backpropping calculations.

The additional load (w_p) applied on the three floors can be estimated using Method One – Revised (see Figure 28.4(b)). Alternatively, if it is a very stiff slab the additional load is estimated using equations 1 and 2 in Method Three, which requires advance knowledge of the relative stiffnesses. The most accurate method for predicting the loads is Method Four (the Excel spreadsheet in the CONSTRUCT flat slab guide (CONSTRUCT, 2003)). The load in each level of backprops is the algebraic sum of the additional load transferred to the floors below the backprop in question.

The TWC must ensure that both the supporting slab and the lower slabs each have gained sufficient strength before the new slab is cast. It should be noted that where sites decide to store or load out slabs in advance of other works (e.g. with pallets of bricks) this additional loading must also be taken into consideration.

28.9. Worked examples – multi-storey construction

The complexity of the calculations of the load transfer through slabs in multi-storey construction should never be underestimated. It is not an easy or quick task; it requires a knowledge of the sequence to be adopted, the nature and location of the equipment, as well as full information from the PWD about the design parameters. Detailed worked examples are given in both the Concrete Society formwork guide worked examples (CS, 2012b) and in Pallett (2017). These examples, based on site experience on typical structures, also include 'What if?' scenarios, where operatives accidentally remove back-propping and/or falsework in the wrong sequence, emphasising the need for site management (see Chapter 2).

28.10. Methodology – heavy construction

Where thick in situ slabs are required to be cast above thinner lower suspended slabs, such as casting a high-level heavy plant room slab or transfer slabs for higher level superstructure, the effects on the lower suspended slab must be considered. It is most unlikely that the thinner lower slab will have been designed to take the full self-weight of the slab to be cast on top of it.

One obvious solution is to cast the heavier upper slab first and then, once the slab can support itself, dismantle the formwork and falsework down to the underside of the thinner lower slab, fix the formwork and reinforcement and concrete the lower slab. This solution is rarely practicable because there is no crane access for materials, and placing concrete becomes very difficult under an existing slab. Furthermore, this construction sequence can only be considered when the PWD has made provision for it.

The concern in casting the heavier upper slab on top of the lower slab is not that the lower falsework cannot be designed to take the weight of both slabs. Rather it is with the effect of elastic shortening of the lower falsework when the heavier slab is cast. The lower slab will be physically restrained from moving downwards by its permanent supports and, when the backpropping shortens under loading from the new slab, it could be seriously distressed at connections to the permanent works.

The following example of a silo construction (Figure 28.5) illustrates both the problem and the solution adopted to allow for differential elastic settlements and to avoid inducing unplanned stress into the permanent works.

The tall silo walls, about 25 m high and 6 m in diameter, shown in Figure 28.5 were first slip formed in one continuous operation with arrangements at each floor level for 'pull-out' shear reinforcement. This gave a flush surface for slip-forming. The tall falsework (A) was then erected (see Figure 28.5(a)), with the soffit formwork deliberately set high to allow for the elastic shortening for the self-weight of slab 1 only. The thin slab (slab 1) was poured (see Figure 28.5(b)) but a gap was left around the circumference so that there was no physical connection between slab 1 and the walls. The soffit of slab 1 now moved to the correct final level.

Construction of the thicker slab (slab 2) was started by erecting falsework B (see Figure 28.5(b)). This soffit formwork had been set deliberately high but this time to allow for shortening of falsework B and the additional shortening of falsework A under the thick slab's self-weight load. At this time falsework A has already shortened with the load from the thin slab.

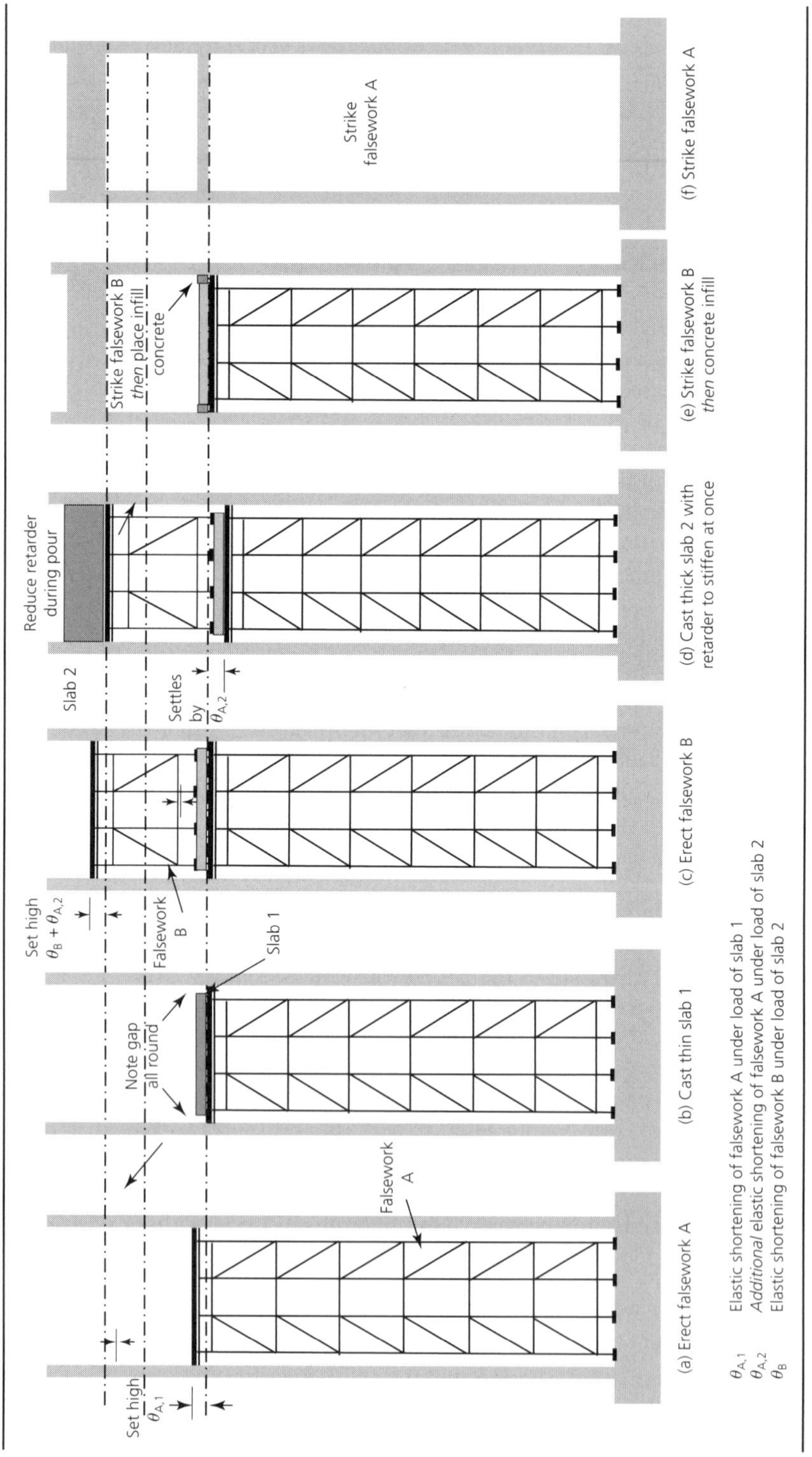
Set high
$\theta_{A,1}$
Falsework
A
Note gap
all round
Slab 1
Set high
$\theta_B + \theta_{A,2}$
Falsework
B
Slab 2
Settles
by
$\theta_{A,2}$
Reduce retarder
during pour
Strike falsework B
then place infill
concrete
Strike
falsework A
(a) Erect falsework A
(b) Cast thin slab 1
(c) Erect falsework B
(d) Cast thick slab 2 with
retarder to stiffen at once
(e) Strike falsework B
then concrete infill
(f) Strike falsework A
$\theta_{A,1}$ Elastic shortening of falsework A under load of slab 1
$\theta_{A,2}$ *Additional* elastic shortening of falsework A under load of slab 2
θ_B Elastic shortening of falsework B under load of slab 2

Figure 28.5 Silo construction sequence of two slabs, allowing for elastic shortening of the falsework

As slab 2 was then poured the entire falsework structure (A and B) moved downwards (see Figure 28.5(d)), including the freestanding slab 1, as a result of the elastic shortening. The consequence of the movement of slab 2's supports is that the reinforcement to slab 2 is also moving downwards, increasing in movement as the concrete is placed. To allow this movement the reinforcement has to be fixed so that it does not clash with projecting wall shear reinforcement; in addition, the entire slab 2 concrete has to undergo initial set at around the same time. This requires planning the pour and decreasing the amount of retarder added as the pour progresses (see Figure 28.5(d)). Note that the effect is to move slab 1 below its final level.

The next stage is crucial: once slab 2 has reached the agreed strength to be self-supporting falsework B can be removed. This has the effect of releasing the majority of the pre-compression in falsework A, which now relaxes, and slab 1 moves upwards back to its correct position. Only now can the infill concrete be placed (see Figure 28.5(e)). Once the infill concrete has sufficient strength falsework A can be removed (see Figure 28.5(f)).

The corollary of the movement vertically, up and down, of the formwork/falsework system is that all fixings providing stability to the temporary works have to be hinged or designed such that they allow for the elastic movement of the system under load.

It is worth noting that if, in error, the site completed the infill concrete before removing any falsework, whether they started with falsework A or B, in both cases slab 1 would become cracked and seriously overloaded. If falsework A is removed first the thin slab is known not to be able to take the self-weight of the thick upper slab, and if falsework B were removed first, the release of pre-compression in falsework A would push slab 1 upwards!

The example above highlights the likely pitfalls of transferring loads and the consequences of not understanding backpropping and its effect on structures.

28.11. Conclusion

The conclusion from this chapter is that the whole construction team, both contractors and PWDs (as competent designers), need to be aware of the implications of specifying design loads, and must understand the effects on the slabs during construction of load transfers between floors and slabs through backprops. The arrangement of backprops as one-for-one, whether left in place or not, and the building shape, its stiffness (span/depth ratio) and floor layouts can all affect the transfer of load. Even slight changes in sequencing can impart significant unplanned loads into new slabs. In particular, the use of blended cements with low rates of gain of early strength, exacerbated by low temperatures, can have a crucial effect on the speed of construction, and may introduce the need to mitigate damage to the existing concrete slabs caused by backpropping loads.

This chapter has highlighted the coordination needed between the PWD, the TWD and the TWC to ensure that the sequence of construction assumed at the design stage is carried out in practice, thus ensuring a safe method of work without detriment to the final structure.

REFERENCES

Alexander R (2004) Propping and loading of in-situ floors. *Concrete* **38**: 33–35.

BRE (Building Research Establishment) (2000) *A Radical Redesign of the In Situ Concrete Frame Process. Task 4: Early Striking of Formwork and Forces in Backprops*. BRE, Watford, UK, BR394.

BSI (2019) BS 5975:2019, Code of practice for temporary works procedures and the permissible stress design of falsework. BSI, London, UK.

CONSTRUCT (2003) *Guide to Flat Slab Formwork and Falsework*. Concrete Society, Camberley, UK, CS140.

CS (Concrete Society) (2012a) *Formwork: A Guide to Good Practice*, 3rd edn. Concrete Society, Camberley, UK, CS030.

CS (2012b) *Formwork – A Guide to Good Practice. Worked Examples*, 3rd edn. Concrete Society, Camberley, UK, CS169.

HSE (2015) *Managing Health and Safety in Construction. Construction (Design and Management) Regulations 2015. Guidance on Regulations*. HSE Books, Sudbury, UK, Publication L153.

Pallett PF (2017) Temporary Works Toolkit. Part 6: Backpropping of flat slabs – design issues and worked examples. *The Structural Engineer* **95(1)**: 30–32.

UK Government (2015) Construction (Design and Management) Regulations 2015. Statutory Instrument 2015/15. The Stationery Office, London, UK.

Vollum R (2008) *Investigation into Preloads Induced into Props during their Installation*. Imperial College, London, UK.

FURTHER READING

BCA (British Cement Association) (2000) *Best Practice Guide – Early Age Strength Assessment of Concrete on Site*. BCA, Crowthorne, UK, 97.503.

BCA (2001) *Best Practice Guide – Early Striking and Improved Backpropping for Efficient Flat Slab Construction*. BCA, Crowthorne, UK, 97.505.

Beeby AW (2001) Criteria for the loading of slabs during construction. *Proceedings of the Institution of Civil Engineers – Structures and Buildings* **146(2)**: 195–202.

Useful web addresses

Concrete Society: http://www.concrete.org.uk (accessed 01/08/2018).

CONSTRUCT (Concrete Structures Group): http://www.construct.org.uk (accessed 01/08/2018).

Temporary Works: http://www.temporaryworks.info (accessed 01/08/2018).

Pallett, Peter F and Filip, Ray
ISBN 978-0-7277-6338-9
https://doi.org/10.1680/twse.63389.421
ICE Publishing: All rights reserved

Chapter 29
Pressure testing of pipelines

David Cooper
Principal Designer's CDM Advisor, Mott MacDonald Bentley Ltd

Stewart Carolan-Evans
Sector Chief Engineer, Costain Ltd

29.1. Introduction

The purpose of pressure testing a pipeline system is to ensure that the system can be operated safely. The pressure test gives confidence that the system as constructed/installed is suitable for its operational requirements. A system pressure test also demonstrates the integrity of all the components, including pipes, joints and fittings. Pipes and fittings should have been manufactured such that they can withstand the system test pressures. However, they may be defective, have been damaged in transit or during installation, could be incorrectly installed, have inherent material defects or defects caused during manufacturing, and so on. Other components of the pipeline system (e.g. flanges, gaskets, welds, bedding and pipe support, thrust restraint) will have been constructed and installed on site and the pressure test provides assurance that the system as provided is safe and correctly constructed (Figure 29.1).

The traditional pressure test developed over a period when pressure pipes were made from cast iron with run-lead or hand-caulked lead wool joints. Over the decades other materials were introduced, such as asbestos cement, welded steel, unplasticised polyvinyl chloride (uPVC) and polyethylene (PE). Test methods for these pipelines have remained the same and have, in most cases, performed satisfactorily.

For butt-welded steel pipelines carrying oil and gas, much higher pressures are used, sometimes with the pipelines being tested to yield. However, for continuously welded steel pipes, no significant leakage is allowed. Cyclic testing has also been introduced for some pipelines. This chapter does not cover such areas as the high-pressure tests required for pipelines at chemical process works and refineries, gas supply pipes and so on where pressures can reach in excess of 80 bar. Problems can occur with pipes that absorb significant quantities of water when they are filled (concrete pipes, mortar linings) or when they are pressurised. These problems increased when PE pipes were introduced as the material creep made the pressure test rather meaningless. Glass-reinforced plastic (GRP) pipelines have also been problematic, as leakage at permissible rates is masked by the elasticity of the pipeline.

Figure 29.1 Digital gauge used to monitor pressure at downstream end of pipeline during pressure loss method test. This was connected to an independently verified data logger. (Courtesy of Mott MacDonald Bentley Ltd)

The design of joints has also developed over the years with the introduction of flexible joints either with spigot and socket joints or flexible couplings. Simultaneously with the increasing quality of joints, designs have improved and the pressure-test criteria have become tighter. Recently, additional test criteria have been introduced for joints to sewer pipes, which must now also withstand negative pressure to avoid water ingress.

Pressure pipelines are subject to internal forces which tend to cause joint separation at any changes of direction, blank ends and tapers. There is a wide variety of pipe jointing options, ranging from traditional push-fit spigot and socket joints through to restrained and mechanically tied joints. If the pipe is wholly above ground it is likely that the joints will be either bolted flanged joints or tied joints. Flanged joints resist tension forces through the bolted connection. Due to the bending moments induced across the rigid joint it is not recommended to use flanged joints below ground.

29.2. Gravity sewer pipelines

The testing by water and air of gravity sewers is detailed in *Sewers for Adoption* (WRc, 2012; section 5.7) and *Civil Engineering Specification for the Water Industry* (UKWIR, 2011; sections 7.4–7.8), with further detail provided in BS EN 1610:2015 (BSI, 2015a). Both specifications detail the available tests, including an infiltration test.

29.3. Pressure pipelines

Depending on the pipe material there are ways of determining the test pressures. Table 29.1 outlines the methods specified. The test pressure would normally be specified

Pressure testing of pipelines

Table 29.1 Methods for determining test pressures

Reference	Section	Quote
BS 8010-1:1989 (BSI, 1989; withdrawn, replaced by BS EN 14161:2011 + A1:2015 (BSI, 2011)) and PD 8010-1:2015 + A1:2016 (BSI, 2015b)	13.4	'The site hydrostatic test pressures for ductile iron pipes and fittings and flanged joints in accordance with BS 4772 should be not less than: (a) the working pressure + 5 bar; (b) the maximum pressure under surge conditions.' BS 4772:1988 is now withdrawn and has been replaced by BS EN 598:2007 + A1:2009, BS EN 969:2009 and BS EN 545:2010 (BSI, 2007, 2009, 2010). In the replacement document (PD 8010-1:2015 + A1:2016 (BSI, 2015b)) this has changed to 'not less than 1.1 times the MAOP of the pipeline', where MAOP (the maximum allowable operating pressure) is the sum of the operating pressure, the static head pressure, the pressure required to overcome friction losses and any necessary backpressure. Note: PD 8010-1:2015 + A1:2016 is not to be regarded as a British Standard.
BS EN 805:2000 (BSI, 2000)	11.3.2	'For all pipelines the System Test Pressure (STP) shall be calculated from the Maximum Design Pressure (MDP) as follows: surge calculated: STP = MDP_c + 100 kPa surge non-calculated: STP = MDP_a × 1.5 or STP = MDP_a + 500 kPa (whichever is the least) The fixed allowance for surge pressure included in MDP_a shall be not less than 200 kPa.'
IGN 4-01-03 (Water UK, 2015)	3.0	'The BS EN 805:2000 method for choosing the test pressure is that STP should be the lowest of: 1.5 × PN or PN + 5 bar.'
Water suppliers (Southern Water, 2018)		'(f) The PN rating of the lowest rated component in the system should be used. (g) The value of STP should apply at the lowest elevation of the pipeline and should therefore include the initial maximum static head applied (P_o). (h) The test pressure at the highest elevation should be at least the maximum operating pressure. If this is not possible due to the elevations involved then the line should be split prior to testing.' Note: Some companies may prefer to use an STP value of 1.5 × design continuous maximum.

by the Permanent Works Designer. BS EN 805:2000 (BSI, 2000) gives advice on different test methods that may be used to assess pipelines for leakage. These methods are not mandatory; it is left to the Permanent Works Designer or client to choose the appropriate procedure.

The pressures given in BS EN 805:2000 apply notionally to water supply pipes. However, these are generally also specified for application to sewerage pressure mains due to their similarities to water supply applications. *Sewers for Adoption* (WRc, 2012) specifies testing to BS EN 805:2000 or Water UK's Industry Information and Guidance Note IGN 4-01-03 (Water UK, 2015), although it specifies a simplified test pressure for rising mains of 1.5 times the maximum operating pressure at the lowest part of the main (WRc, 2012; section 5.7.9). It also specifies the testing of PE pressure mains to IGN 4-01-03. A test pressure according to this method should also be applied to surcharged sections of gravity systems.

29.4. Design

Pressure testing involves applying stored energy to an assembly of pipes, valves, bends, manifolds and so on in order to verify its strength, integrity and/or functionality. The Health and Safety Executive (HSE) has issued Guidance Note GS4, *Safety Requirements for Pressure Testing* (HSE, 2012). The guidance is aimed at all employers, supervisors and managers responsible for pressure testing. The safety regulation that covers pressure testing is the Provision and Use of Work Equipment Regulations 1998 (PUWER) (UK Government, 1998). Pressure testing is a high-risk activity. When applying stored energy to a pipe system or an assembly, especially for the first time, there is potential for an unintended or premature pressure release while people are in the danger zone. An exclusion zone should be set up around the test.

Release of stored energy under pressure can cause

- test assembly rupture, creating flying debris
- component or connector failure
- test hose failure, including detachment, with consequential hose whip
- sudden release of the test medium, causing injury, such as burns, eye damage or pressure injection into body tissue, and potentially pollution of the ground or watercourses.

Common issues include the following

- *Incorrect or insufficient propping.* There is a common misunderstanding, particularly on site, that low pressure means low force. Pressure does not equal force – a large-diameter low-pressure pipeline could produce much higher forces than a small-diameter high-pressure one. It is imperative that all pressure tests are designed for, and calculations produced detailing the forces produced in the test and how they are dealt with. If a thrust block is used it could also be undersized or designed for incorrect ground conditions. Large factors of safety are usually applied to thrust blocks to prevent movement (see the Construction Industry

Research and Information Association (CIRIA) report R128 (Thorley and Atkinson, 1994)).
- *Presence of groundwater*. If there is water in the excavation containing the restraint (e.g. a temporary thrust block) the temporary works must be designed to cope with its presence. The presence of water can significantly reduce the friction between a concrete block and the soil. If the design assumes dry conditions the restraint could fail in sliding once the test pressure is applied.
- *Presence of air in the test fluid*. The presence of air in a pipeline will have a number of effects. It will markedly increase the pressure rise time and will distort the interpretation of pressure decay results, and may result in a considerable increase in risk as compressed air (and gases in general) can result in explosive failures of pipelines under test. Attempts must be made to purge air from the main during and after charging with water and before the start of the pressure test. Plastic materials, particularly PE, creep under stress, and therefore the analysis of pressure test data is more complicated than for other materials.

29.4.1 Preventing injury and damage
Assume the pipe will fail and take appropriate precautions, including

- have a proper test procedure
- train those carrying out the test
- ensure test equipment is fit for purpose
- monitor the pressure and have a relief device
- limit the stored energy
  - a water test preferred over an air test
  - venting is important in hydraulic tests
- segregate the test area
- minimise the number of people close to the equipment
- always reduce the pressure before approaching the equipment.

29.5. Can restraint be provided without temporary works?
Some pipeline systems are designed as 'tied' or 'anchored' systems whereby the pipe lengths are welded together (or connected with other pressure or thrust-bearing joints – what BS EN 805:2000 (BSI, 2000) calls 'restrained joints'). The advent of anchor gaskets has led to many systems being classed as self-restraining. In these cases, the whole pipeline (or at least significant lengths that are tied together) can be treated as one 'component'. Examples are

- PE continuously welded pipelines
- pipelines with thrust-bearing couplings
- steel continuously welded pipelines
- 'tied' ductile iron pipelines with anchor gaskets
- flanged pipeline systems.

For these 'tied' systems there are no thrust blocks or similar components that would (for other pipe systems) be designed based on working pressures. Therefore, the governing pressure for the system will be the pipe nominal rated pressure (PN) of the pipe, assuming air valves and similar fittings are rated at least as high as the PN. If the tied/anchor joints or flanges have a lower PN than the pipe, it is the PN of the tied/anchor joints that governs. Thrust blocks may exist at bends and junctions for resisting thrusts from fluid moving through the pipe when in service. These are not required for the test but they may be present.

The advantage of this approach is that it means the pipeline has been proven based on its PN, which means that if the duty changes in the future the pipeline will be safe to operate at any pressure up to the PN. Furthermore, this PN-based test approach demonstrates that the pipeline is properly restrained, and that the welds have been properly made and are watertight, a key requirement for welded PE pipe. In other words, applying the definitions from IGN 4-01-03 (Water UK, 2015), the field test pressure for tied pipe systems (such as continuously welded PE pipelines) should be generally 1.5 times the PN of the pipe (or anchor/tied gaskets, if lower), rather than some factor of the working or occasional pressure.

In some applications there will be a case for using the BS EN 805:2000 method. For example, the test pressure for a PE pipeline may be higher than the working pressure where the pipe standard dimensional ratio (SDR) – the ratio of the nominal outside diameter to the nominal wall thickness – has been sized based on structural design considerations. Likewise, for tied ductile iron pipelines, consideration should be given as to whether testing based on the PN is appropriate if the intended working pressure is far lower than the rated pressure. If joints are welded or bolted, restraint will be provided by the buried pipe. The Ductile Iron Pipe Research Association (DIPRA) method can be used to determine the restrained length required to resist the forces produced, and therefore how much of the pipe needs to be backfilled prior to the test (DIPRA, 2016).

29.6. If temporary works are required, what are the options?
29.6.1 Propping of existing structures
If the test location can be located adjacent to an existing structure such as a tank, pump well or inlet works and if the pipework is self-restraining, the only element of the test requiring design may be the supporting of the blank flange back to the structure. For a fully flange-jointed pipe system the test flange is normally simply bolted to the last flange on the pipe run, thereby eliminating the need for any additional support. In this instance, the temporary works comprises checking that the blank flange and connecting bolts are sufficient. Checks on whether the existing structure is sufficient to withstand the applied loads and to transfer the loads into the surrounding ground should also be made, as follows

- punching shear checks on existing walls, and so on
- sliding and overturning checks.

29.6.2 Constructing new structures
The provision of new structures (blocks, walls etc.) should be avoided, as this will be expensive. Considerable efficiencies may be made by consideration of temporary pressure

test loading during the design of the permanent works. The Permanent Works Designer and the Temporary Works Designer should work together to understand whether any new permanent structures can be configured to allow the testing of the pipeline.

One option could be to use a thrust block designed to act in the permanent works, perhaps at a bend to act in a temporary case as a foundation for props. This might mean changing the profile of the block to allow the temporary propping to be seated safely and securely. This could be achieved at minimal cost to the permanent works but with a significant saving during testing.

29.6.3 Design of steel supports

Temporary propping can be used, from proprietary props to large-scale box sections, to transfer loads from end plates onto new or existing structures. The design of the props should take into account eccentric loading, which could be produced due to erection tolerances. A 25 mm eccentricity is often taken as the minimum to determine the loads in the props. The design of props should also take account of other risks, such as accidental impact from plant. All loads arising from the imposed load, erection tolerance and other incidental loads should be considered to act in combination.

29.7. Design of thrust blocks

For normal or traditional pipeline construction, each pipe length is not tied to the adjacent pipe lengths (what BS EN 805:2000 calls 'non-restrained joints' (BSI, 2000)). Thrust is accommodated by thrust blocks or similar (Figure 29.2), which are generally designed based on working pressures. It is also more usual to select a pipe with a PN far in excess of the working pressure, and so it is likely there will be other components (flanges, line valves, air valves and similar fittings) with a lower PN. Examples are

- push-fit ductile iron pipelines
- concrete pipelines (generally)
- many GRP pipelines
- push-fit PE pipelines
- pipelines with compression fittings that are not self-anchoring
- flange adaptors.

In these cases, the test pressure for the system should be calculated from the working pressures using the BS EN 805:2000 method. This generally results in a lower test pressure than for a tied system, and means that, for example, the thrust blocks may not need to be designed to a test pressure far in excess of what they would ever see in service. However, it does mean that the pipe duty cannot be increased in future.

Thrust blocks can be massive for large-diameter pipelines and may need to incorporate piles. Designs are normally carried out to CIRIA R128 (Thorley and Atkinson, 1994). These designs will need to undergo a check by the Temporary Works Designer to check that thrust blocks designed for the permanent case can be used in the testing phase. Thrust blocks will be required at all changes in direction, tees, valves, blank ends and reducers, whether vertical or horizontal. It is also necessary to ensure that the thrust

Figure 29.2 Temporary concrete thrust block and props to transfer blank end forces during pressure testing of upper section of pipeline. The thrust block was later broken out and a spool section of pipe laid to connect the upper and lower sections following successful testing. (Courtesy of Mott MacDonald Bentley Ltd)

block will not move, so where practicable the blocks should be cast against undisturbed ground. In poorer soils requiring support this may not be practicable for the vertical faces of the block. In this instance careful consideration will need to be given to the staged withdrawal of ground support as the thrust block is constructed, or the use of structural backfilling between the completed block and ground support.

At minor changes in direction the backfill over and alongside the pipe may provide sufficient restraint, provided this has been properly assessed. It is unlikely that the weight of the pipe and its contents will be sufficient to provide this restraint, so a pipe should never be tested without backfill unless the joints are restrained.

29.8. Internal (puddle) flanges

In some instances, integral flanges are cast into concrete chambers, which are specifically designed to take the thrusts during normal operation and testing. It is necessary to check with the designer of the pipeline that this is in fact so, because a lot of pipelines have valve chambers that are not designed to withstand the thrust and contain movement joints to allow the valves to be replaced. These chambers require extensive temporary works and checking to ensure that the flanges do not burst out of the wall and that the chamber does not move when generating sufficient resistance to overcome the thrust.

29.9. On-site safety considerations

The main hazard during pressure testing is the unintentional release of stored energy. As water is virtually incompressible, the pressure falls as soon as there is any movement due to a leak or component failure. If the same failure occurs using air compressed to the same pressure the energy release is more than 200 times greater, resulting in large displacements and flying debris. Therefore, it is essential that hydraulic pressure tests are carefully planned to avoid the presence of trapped air within the system (Figure 29.3). Pneumatic testing should only be carried out where either the use of water (or some other fluid) is either corrosive to or contaminates the system being tested, or in instances where the pipeline system supports and foundations cannot support the weight of the test fluid.

29.9.1 Testing procedures
29.9.1.1 Pre-operations activities
- Agree test pressures with the designer. The pipework designer has an obligation under the Construction (Design and Management) Regulations 2015 (UK Government, 2015) to consider how the pipeline is to be built and tested during their design. However, this does not mean that the details of construction have been fully considered.
- Determine and acquire a water source and agree a method of water disposal after the test.

Figure 29.3 Failure of poorly designed pressure test restraint system. (Courtesy of Temporary Works Forum)

- Identify the scope of any temporary works required and issue the design brief.
- Check that all instrumentation, gauges, manifolds, bolts, hoses, connectors, restraints, temporary works and blank flanges are in place and tightened in sequence to the required torque.
- Ensure all test equipment has been recently calibrated, and check that the operating instructions for the testing equipment are available, have been read, understood and briefed out to the testing team. If testing equipment has been refurbished or repaired after a period of service a separate inspection may be necessary to ensure that the existing components meet the design test pressure.
- Prepare the risk assessment to determine the extent and content of the safe system of work that needs to be in place to carry out the test. Review and challenge the test procedure and the risk assessment, agree who is to undertake the test and the communications protocols. Agree with the client or other stakeholders who should be present at the test.

29.9.1.2 Site preparation
- Install thrust blocks, tied flanges and any other designed temporary components required to perform the test. Establish an exclusion zone around the pipe system being tested.
- Set up the pressure testing equipment outside the exclusion zone, but ensure it is also cordoned off from the workforce and general public.
- Check that accesses into the area are clear and trip hazard free.
- Make sure that emergency arrangements including equipment are in place and have been briefed out to the testing team.
- Check the ground conditions against the temporary works design. For example, is there a requirement for groundwater to be at a certain level? If different, seek advice.
- Check that any live plant or pipework that requires protection has it in place.
- Install safety/relief valves and pressure gauges to ensure the design test pressure is not exceeded.

29.9.1.3 During the test
The Temporary Works Coordinator should issue a permit to load prior to commencing the test. The system should then be filled slowly with water, from the lowest point if practicable. Air should be vented from the pipeline high points to prevent air pockets from forming.

The pipe should be left full for up to 24 h to allow the air in solution to come out and for the water to soak into the pipe lining. The pressure should be gradually applied, or in steps of 10% increments, until the design test pressure is reached. The system being tested should not be subjected to close inspections while under the test pressure until a reasonable period of time has elapsed. For the safety of the testing personnel remote viewing methods may be more suitable. Prior to the completion of the test, isolate the system from the pressure source and vent/depressurise, checking for any instances of re-pressurisation or failure to vent prior to declaring the test complete and removing the safety exclusion zone.

REFERENCES

BSI (British Standards Institution) (1988) BS 4772:1988. Specification for ductile iron pipes and fittings. BSI, London, UK. (Withdrawn. Replaced by BS EN 598:2007 + A1:2009, BS EN 969:2009 and BS EN 545:2010.)

BSI (1989) BS 8010-1:1989. Code of practice for pipelines. Pipelines on land: general. BSI, London, UK. (Withdrawn. Replaced by BS EN 14161:2011 + A1:2015.)

BSI (2000) BS EN 805:2000. Water supply. Requirements for systems and components outside buildings. BSI, London, UK.

BSI (2007) BS EN 598:2007 + A1:2009. Ductile iron pipes, fittings, accessories and their joints for sewerage applications. Requirements and test methods. BSI, London, UK.

BSI (2009) BS EN 969:2009. Ductile iron pipes, fittings, accessories and their joints for gas pipelines. Requirements and test methods. BSI, London, UK.

BSI (2010) BS EN 545:2010. Ductile iron pipes, fittings, accessories and their joints for water pipelines. Requirements and test methods. BSI, London, UK.

BSI (2011) BS EN 14161:2011 + A1:2015. Petroleum and natural gas industries. Pipeline transportation systems. BSI, London, UK.

BSI (2015a) BS EN 1610:2015. Construction and testing of drains and sewers. BSI, London, UK.

BSI (2015b) PD 8010-1:2015 + A1:2016. Pipeline systems, Steel pipelines on land. Code of practice. BSI, London, UK.

DIPRA (Ductile Iron Pipe Research Association) (2016) *Thrust Restraint Design for Ductile Iron Pipe*, 7th edn. DIPRA, Birmingham, AL, USA.

HSE (Health and Safety Executive) (2012) *Safety Requirements for Pressure Testing*. HSE, London, UK, GN4.

Southern Water (2018) *Self-Lay Policy. Appendix One – Commissioning and Handover*. Southern Water, Worthing, UK, Version 1.

Thorley ARD and Atkinson JH (1994) *Guide to the Design of Thrust Blocks for Buried Pressure Pipelines*. CIRIA, London, UK, R128.

UK Government (1998) Provision and Use of Work Equipment Regulations 1998. The Stationery Office, London, UK.

UK Government (2015) Construction (Design and Management) Regulations 2015. Statutory Instrument 2015/15. The Stationery Office, London, UK.

UKWIR (2011) *Civil Engineering Specification for the Water Industry*, 7th edn. WRc, Swindon, UK.

Water UK (2015) *Guide to Pressure Testing of Pressure Pipes and Fittings for use by Public Water Suppliers*. Water UK, Snodland, UK, IGN 4-01-03.

WRc (2012) *Sewers for Adoption*, 7th edn. WRc, Swindon, UK.

FURTHER READING

BSI (British Standards Institution) (1997) BS EN 1295-1-1997. Structural design of buried pipelines under various conditions of loading. General requirements. BSI, London, UK.

BSI (1999) BS EN 1508:1999. Water supply. Requirements for systems and components for the storage of water. BSI, London, UK.

BSI (2010) BS EN 545:2010. Ductile iron pipes, fittings, accessories and their joints for water pipelines. Requirements and test methods. BSI, London, UK.

UK Government (1999) Water Industry Act 1999. Statutory Instrument 1999/9. The Stationery Office, London, UK.
UK Government (2010) Flood and Water Management Act 2010. Statutory Instrument 2010/29. The Stationery Office, London, UK.

Useful web addresses

Ductile Iron Pipe Research Association (DIPRA), USA: https://www.dipra.org (accessed 01/08/2018).

Temporary Works Forum: http://www.twforum.org.uk (accessed 01/08/2018).

Pallett, Peter F and Filip, Ray
ISBN 978-0-7277-6338-9
https://doi.org/10.1680/twse.63389.433
ICE Publishing: All rights reserved

Chapter 30
Basement construction

Ray Filip
Temporary Works Consultant and Training Provider, RKF Consult Ltd

Chris Robinson
Design Manager, Cementation Skanska Limited

30.1. Introduction

In many cities, gaining additional space for a home or commercial property by building outwards may not be possible due to lack of space or existing neighbouring properties, and extending a property upwards may not be permitted by local planning regulations. A popular solution is to extend downwards by constructing a basement. However, restrictions and understanding the construction techniques and associated risks make this work challenging. Constructing a basement beneath a new domestic or commercial property, which could be in an open space, should not be considered as a straightforward operation, because the construction techniques also have to be understood and challenges overcome. Constructing a new basement beneath or adjacent to an existing building can be, and often is, significantly more challenging.

Planning approval, building regulations and building control approval may be required (the UK Government's Planning Portal and the Construction Industry Research and Information Association (CIRIA) publication C740 (Gilbertson, 2017) provide guidance).

In recent years there has been a significant increase in the number of new basements being constructed beneath existing inner-city domestic properties. This type of work has involved a significant number of collapses and fatalities, and as a result the Health and Safety Executive (HSE) often focuses its attention on these types of projects. These collapses and fatalities may be attributed to some or all of the following

- Unrealistic client expectations about what can be achieved for their budget. Domestic clients often are not aware of their duties under Construction (Design and Management) Regulations 2015 (CDM 2015) (UK Government, 2015).
- Work is often carried out by inexperienced builders who do not appreciate the technical challenges and risks (such as ground collapse, undermining, structural stability, damaging existing buried services and falls from height). Often, very little forward planning is undertaken and there is very little experienced site supervision.

- Little or no ground investigation or investigation of any existing structures, prior to construction work commencing.
- Use of inappropriate plant and construction techniques.
- Professional Temporary Works Designers are often not consulted and work is carried out using 'custom and practice'. Even if a temporary works design is commissioned, often there are no temporary works procedures on site (and no temporary works coordination, see BS 5975:2019 (BSI, 2019)) so the design details are not implemented on site and the designer is not consulted about changes made on site.

CIRIA C740 (Gilbertson, 2017) contains guidance for those involved and the potential risks.

30.2. General planning considerations prior to work commencing

For any basement construction the designer and contractor should have an understanding of the following

- Planning constraints and restrictions that may be imposed on the design, and the construction methodology.
- A site investigation should be carried out to identify the types of soil (with engineering properties) and soil boundaries to a depth to suit any temporary or permanent piles that may be required. Indication of any contamination in the ground should also be included.
- Groundwater levels need to be identified, including any seasonal variations in levels and artesian pressures (flotation may become an issue).
- Potential construction tolerance and construction sequences. That is, can the basement be constructed in 'open cut' or will a temporary retaining wall be required to allow construction? If a retaining wall is required, how will it be constructed to suit site constraints (soil types, noise, vibration, space constraints, etc.)? Access for machinery on to site must also be considered.
- How will the spoil be removed? Bulk excavation and loading into transport for immediate removal is an option if space allows. However, hand excavation or using 'mini excavators' and conveyors is often the option in an urban environment where working space will be severely limited (Figure 30.1).
- If the soil is contaminated there will be additional health and safety considerations during excavation. Where will the spoil be removed to?
- How will the new construction materials be delivered, offloaded, stored and placed in their final position (lifting considerations)?
- Existing services that may require diversion or protection.
- Foundation levels for any adjacent buildings or structures which may be undermined by the new works.
- Magnitude of surcharges that the temporary and/or permanent retaining wall will need to withstand.
- Grade of basement required (waterproofing requirements are dependent on the end use, e.g. car park or domestic – see below) and the expected lifetime of the basement (this could also include corrosion protection, fire protection, etc.).

Figure 30.1 Well organised site for construction of a domestic basement: contiguous pile retaining wall (installed using mini piling rig) and steelwork framing, and conveyor being used for spoil removal. (Courtesy of RKF Consult Ltd)

- How will access be gained into the excavation? How will falls from height be prevented (edge protection, etc.)? Considerations regarding confined spaces.

BS 8102:2009, 'Code of practice for the protection of below ground structures against water from the ground' (BSI, 2009), defines four grades of basement

- Grade 1: basic utility (car parking, plant rooms (excluding electrical equipment), workshops).
- Grade 2: better utility (workshops and plant rooms requiring drier environments than grade 1).
- Grade 3: habitable (ventilated residential and commercial areas).
- Grade 4: special (archives, requiring controlled environments).

It also defines three types of water-resistant construction

- Type A: barrier (membrane) protection.
- Type B: structurally integral protection.
- Type C: drained protection.

30.3. Constructing a basement in open cut

For shallow basements (generally single or double depth), if space and groundwater conditions allow and there are no adjacent structures, roads and services that may be affected, the simplest way to construct a basement is in 'open cut'. A risk assessment should be carried out to determine if this form of construction is possible. The angle of battered or stepped slope should be designed (so as not to collapse) to suit the soil type, depth of excavation, surcharges at the top of the slope and the length of time for which the slope will be required. The slope will need to be inspected on a regular basis as per the requirements of CDM 2015 (UK Government, 2015). Weathering of the slope is an issue, and the slope could be protected by using polythene sheeting or 'sprayed concrete', but this would make inspection difficult.

If the groundwater table is well below the excavated depth (and sub-artesian water pressure is not a concern) the excavation can be relatively straightforward. If the groundwater table is near the existing ground level and the soil is relatively permeable, simple groundwater control may prove to be impracticable or even impossible, and a scheme which cuts off the water or a dewatering scheme will be required. If dewatering is required, the process of lowering the water table can have an effect on structures some distance away and settlement is a possibility. In addition, consideration must be given to where the water will be discharged (see Chapter 6). Even if groundwater control is not required, rainwater surface runoff needs to be controlled. Often this can be achieved by a shallow ditch near the base of the slope and sump pumping.

If constructing a basement in open cut, backfill will be required behind the newly constructed basement wall. The backfill will be compacted and the compaction plant and process will exert additional forces on the basement wall. If this is considered an issue a cementitious backfill could be used (placed in layers to reduce the pressure).

Soil nails with steel mesh and a sprayed concrete surface could be used where a temporary retaining wall is not required but there is not sufficient space to cut the ground to a safe angle. Reinforced earth techniques could also be used in a similar manner; both these techniques have been used to reduce active pressures on permanent retaining walls, thereby reducing the required strength of the wall (hence reducing cost).

30.4. Constructing a basement in a supported excavation

For deeper basements or where there is insufficient space to construct the basement in open cut a temporary ground support scheme will be required. A temporary retaining wall (which could subsequently form part of the permanent retaining wall) will be required and the wall may also require temporary support (e.g. propping, anchorage).

The design of the wall and supports should be as follows

- Be stiff enough to support the earth, surcharges and any water pressures without excessive deflection. The designer should allow for the possibility of accidental over-dig, softening of the formation and accidental impact on the supports. Softening of the formation will affect the stability of the retaining wall. This can be controlled by excavating to just above the required formation level and then excavating to the final level just before the basement slab is constructed, or, alternatively, the formation can be protected by blinding concrete.

- Be positioned to allow the basement to be constructed. The temporary retaining wall could be sacrificial and could potentially be removed (if possible) after constructing the permanent wall or may be incorporated in the permanent works (any temporary supports should not clash with or adversely impact on the permanent works).
- Prevent excessive amounts of groundwater from entering the excavation. Some degree of water ingress through a temporary retaining wall is acceptable, as this can be relatively easily controlled by sump pumping.
- Be installed (and removed if required) in a manner to suit the site restrictions (e.g. ground conditions, noise, vibration, access, working space). Clients generally demand as large a basement as possible (to maximise internal space) so the temporary works should be as slender as possible and installed to acceptable tolerances.

30.5. Constructing a basement beneath an existing building or next to adjacent buildings

This is potentially one of the most technically difficult and risky types of projects, and there are numerous examples of poor practice leading to collapse of the ground or adjacent structure. The principal risks are the undermining of foundations, which could cause settlement and possibly collapse of structures, collapse of the basement excavation, and damage to existing services and nearby roads.

Existing foundations could be underpinned or supported (see below) prior to excavating the basement. Underpinning should be installed in a 'hit-and-miss' sequence by specialists. A specific issue with underpinning for a basement construction is that in many cases the underpins have to suit the basement excavation depth. An excavation for a single-storey basement could be in excess of 3 m and temporary stability of the ground during underpinning needs to be considered (and groundwater). For deep underpinning (to suit the basement excavation depth) the earth pressures behind the underpins could cause instability until the basement slab has been cast (unless temporary propping or anchorage is provided). Underpinning pier or column foundations can be particularly difficult as the vertical loads are large, and often piling alternatives are adopted. Concrete piles can be an option when forming a large basement beneath a large existing building (Figure 30.2). The piles can be connected to existing foundations, or combined with steel or concrete beams to support existing foundations. Once the existing foundations are supported the basement can be excavated beneath the existing building.

When working beneath or adjacent to existing buildings the working space is often an issue and careful thought must be given to selection of machinery and construction techniques. Mini piling rigs, mini excavators and conveyors for spoil removal are often used (see Figure 30.1).

30.6. Temporary retaining wall

Permanent basement walls will be designed using the long-term soil properties. However, there may be circumstances whereby the short-term soil properties may be appropriate for temporary retaining walls or the permanent wall acting in the temporary condition.

Figure 30.2 Existing building supported on piles and beams to allow a new basement to be constructed beneath. (Courtesy of RKF Consult Ltd)

The temporary retaining wall should be designed to support earth and water pressures and surcharge loadings. It may also be required to carry horizontal loads from edge protection systems and vertical loads from cranes or hanging access staircases. The installation of the wall is a significant consideration and the chosen technique should suit the site constraints, ground conditions and access for plant.

A temporary retaining wall should be incorporated in the permanent wall and the design should identify the critical loading conditions and protection requirements. Wall design methodologies and failure modes are highlighted in Chapters 11, 13 and 14.

The retaining wall could be one of the following

- *Sheet piled wall (or a combination wall, with tubes or I-sections with sheet piles).* The clutches in a temporary sheet pile retaining wall can be sealed with bitumen or hydrophilic sealants. Sheet piles can be installed by using piling hammers, vibrators or hydraulic jacking (commonly known as 'silent piling'). In urban environments hydraulic jacking equipment is preferred if the underlying soils are cohesive (see Chapter 11 for further details). Modern hydraulic jacking equipment allow sheet piles to be installed extremely close to adjacent buildings in order to maximise the size of the new basement. The sheet pile designer should also consider 'driveability' (see the ArcelorMittal *Piling Handbook* (2016) for further guidance).

- *Post-and-lagging wall*. In this technique holes are augered (holes are generally 450–750 mm in diameter using machinery to suit the site restrictions) around the perimeter of the proposed basement at around 2–3 m centres to a designed depth below the basement slab (similar to fence posts). Steel posts (typically universal column sections) are then placed into the holes and the base filled with concrete to secure the posts. Above the concrete the augered hole can be filled with sand or 'lean mix concrete' to prevent the holes collapsing. The site is then progressively excavated (around 1 m depth at a time) and panels are formed between the flanges of the steel posts. The panels could be timber sleepers, pre-cast concrete, in situ concrete, steel plates (Figure 30.3) and so on. The panels are installed progressively beneath previous panels as the excavation progresses. The panels are designed to

Figure 30.3 Post-and-lagging retaining wall with steelwork support to construct a domestic basement. Steel posts were installed into augered holes and steel plates span between the flanges of the steel posts. The wall is sealed by welding and becomes the permanent wall. The basement slab is cast against the wall (puddle flange can be seen). A blockwork wall with a drained and insulated cavity is built in front of the post-and-lagging wall. (Courtesy of RKF Consult Ltd)

span between posts, and the posts could be designed as vertical cantilevers or to be propped near the top. This technique has the advantage that it is quiet, with relatively low levels of ground disturbance, and the equipment for augering the holes can be small and practicable for restricted sites (restrictions on space, headroom, etc.). The disadvantages of this technique are that it can be extremely difficult to exclude groundwater ingress and it is relatively slow. Therefore it would generally not be suitable for very large or very deep basements.

- *Contiguous or secant piling piles.* These techniques, which are covered in detail in Chapter 14, are suitable for large and deep basements. Secant walling has the advantage of excluding groundwater. For both techniques large piling equipment is often required (although suitable plant is available for cases where there is restricted access), and this will require significant working space, large access provision (on and off the site), support cranes, excavators for spoil removal, lay-down areas for equipment and reinforcement cages, temporary concrete guide walls and a substantial working platform. The pile sizes vary from around 300 to 1200 mm diameter and the depths would suit the basement depth. These walls have significant stiffness and hence could be designed as cantilevers, but if the excavation depth is significant or the deflection criteria onerous, temporary supports will be required. Generally a facing will be required in front of the retaining wall, although this may depend on the end use of the basement.
- *Diaphragm walls.* This technique, which is covered in detail in Chapter 13, is suitable for very deep basements. The piling equipment and support used required are very substantial and will require significant access and working space; the attendances will be even larger than for contiguous or secant piling. Generally a facing will be required in front of the retaining wall.

30.7. Support scheme to retaining wall

If the retaining wall is unable to stand as a cantilever (because of the excavation depth, onerous deflection limits, large surcharges, or poor or weak ground to a significant depth) then a support scheme is required. The positioning and spacing of supports is important. If the spacing is too close then the excavation and subsequent construction becomes difficult. However, if the spacing is large the span (and hence the deflection) of the wall and waling beams can become excessive unless extremely large sections are used.

The supports to a basement retaining wall could be one of the following

- *Horizontal frames (walers and props).* These frames can be fabricated steelwork or proprietary hydraulic systems (Figure 30.4). A reinforced capping beam (to piles) can also be used as a waler beam. For deep basements the retaining wall spans between the frames, the walers span between the props. 'Knee braces' can be used in the corners, but the designer must consider the horizontal reaction which causes axial loading into the waler beams, in which case shear stops can be provided. Where the props meet the waling beams there will be a concentration of loading, and steel waler beams may require web stiffeners. If the basement excavation is wide or long the struts may become very large and may sag

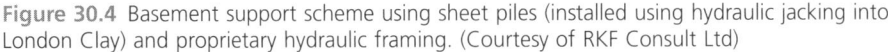

Figure 30.4 Basement support scheme using sheet piles (installed using hydraulic jacking into London Clay) and proprietary hydraulic framing. (Courtesy of RKF Consult Ltd)

significantly under their own self-weight; king posts may be used to support such long props. If the basement slab is cast up against the retaining wall it can act as a prop to allow higher levels of propping to be removed (however, the retaining wall will now have to span further – possibly as a cantilever). For large schemes due consideration should be given to potential thermal effects on the loads and the performance of the propping system.

- *Inclined frames (walers and props)*. When using raking props a common construction sequence is to leave a temporary soil berm against the retaining wall until the waling beam and props have been installed. Often the base of the props can be connected to the partially constructed basement slab. Once the props are installed the berm can be excavated and the remaining section of basement slab constructed (Figure 30.5). Inclined props onto horizontal walers cause torsion in the waler beam. If the walers are horizontal then shear stops will be required to the retaining wall; alternatively, the waler can be inclined to suit the prop angle, but shear stops will still be required. The upward vertical reaction from the inclined prop will ultimately be taken into the retaining wall. As for horizontal frames, due consideration should be given to potential thermal effects on the loads and the performance of the propping system.
- *Ground anchors and steel waler beam*. Multiple levels of inclined anchors may be required, and these will be installed progressively as the excavation progresses. Unless a passive anchorage system is adopted the anchors will need to be tensioned. The anchors could provide permanent support and will need to be

Figure 30.5 Basement support scheme showing the soil berm and inclined props to the sheet pile retaining wall installed using hydraulic jacking. (Courtesy of RKF Consult Ltd)

protected against corrosion. If they are to be removed they have to be destressed. Anchors have the advantage that they do not obstruct the excavation or subsequent basement construction. However, this option might not be possible as permission will be required from owners of the surrounding area. In addition, obstructions such as services, neighbouring basements, tunnels and so on may preclude their use.

The excavation sequence is broadly as follows. First, the retaining wall is constructed and then the excavation progresses to the underside of the uppermost support frame level. The support frame is installed (support brackets or chains may be required to support the frame), and the excavation can then progress to the underside of the next level, and so on. The support frames are progressively removed as each level of basement slab is removed. Each support frame should be positioned so as not to clash with the permanent slab it is to be replaced by.

To reduce elastic shortening of long props, and hence wall movement, pre-loading by using hydraulic jacks is commonly used. Consideration should also be given to how the load is to be released so that the props can be removed. Hydraulic jacks, steel wedges, sand boxes and concrete packs can be used.

The thermal expansion and contraction of props may also need to be considered for large and long props. Axial loads can vary significantly for large props. Props may be painted white to reflect sunlight, and this also has the benefit of making them more visible to crane operators (reducing the possibility of impact).

Designers should also give consideration to the possibility of accidental over-dig, accidental impact loading and robustness, in order to prevent progressive collapse.

30.8. Top-down construction

Top-down construction is a technique that is often used for large and deep basements beneath new multi-storey buildings. The sequence involves first constructing the basement retaining wall (permanent wall) and internal steel columns. Internal bored foundation piles are installed from a piling mat, which is constructed at a level near to the existing ground level, often with 'open bore' from the basement formation level to ground level. Before the concrete at the base of the internal piles has set, the steel columns are 'plunged' into the wet concrete. Proprietary frames are used by the piling contractor to position and plumb the steel columns. The site level is then reduced to the underside of the ground-floor slab, connections are made to the steel columns and the concrete slab is cast. Excavation holes (known as 'mole' holes) are left in the slab through which excavating machinery will work and spoil will be removed. Forced ventilation systems may be necessary due to confined-space considerations. As the excavation progresses downwards, subsequent slabs can be cast. These slabs will again be connected to the steel columns. Once the excavation has reached the formation level, the basement slab is cast. The holes left in the slabs above can now be filled using a falsework system. The steel columns will require permanent protection from fire and corrosion. If the retaining wall is steel it will also require fire and corrosion protection, while a concrete wall (contiguous, secant or diaphragm wall) may require a concrete finish.

This technique has programme advantages, as it allows construction of the multi-storey building above the basement to commence while the basement is still under construction (i.e. work can be carried out simultaneously downwards and upwards). It also has the advantage that temporary supports to the retaining wall are not required because the support will be provided by constructing the permanent slabs as excavation progresses downwards. The basement slab can also be used for storage on very congested sites. However, this technique does make excavating and constructing the basement significantly more difficult.

30.9. Other design considerations

The design methodologies and potential failure mechanisms covered by Eurocode 7 are discussed in Chapter 11, Section 11.4.

The construction and removal sequence for temporary works should be included in the design analysis, with the different construction stages being analysed to find the critical conditions. Often, critical conditions might not occur during the installation sequence but during removal (i.e. when a level of framing is removed the bending moments in the retaining wall are likely to increase, as are the loads on any other levels of framing). Seasonal variations in groundwater levels should be considered.

When a basement is excavated in over-consolidated clays there is a potential for significant base heave as built-up stresses in the ground are released. 'Clay board' (which is designed to crush) can be provided under the basement slab to accommodate this upward movement. However, it has to be remembered that the basement slab is now effectively suspended and may not be able to accommodate significant temporary loads (from plant, material storage, falsework props from constructing the ground floor slab, etc.).

When basements are constructed below groundwater level, buoyancy needs to be considered. In the permanent condition the structure above the basement may have sufficient weight to resist the upward forces. However, in the temporary condition dewatering (see Chapter 6) may be necessary, or a cut-off may be provided by designing the temporary retaining wall to be sufficiently deep so as to penetrate into an underlying impermeable soil layer, or the basement slab could be anchored down to prevent uplift.

REFERENCES

ArcelorMittal (2016) *Piling Handbook*, 9th edn. ArcelorMittal, Luxembourg.

BSI (British Standards Institution) (2009) BS 8102:2009. Code of practice for protection of below ground structures against water from the ground. BSI, London, UK.

BSI (2019) BS 5975:2019. Code of practice for temporary works procedures and the permissible stress design of falsework. BSI, London, UK.

Gilbertson A (2017) *Structural Stability of Buildings during Refurbishment*. CIRIA, London, UK, C740.

UK Government (2015) Construction (Design and Management) Regulations 2015. Statutory Instrument 2015/15. The Stationery Office, London, UK.

FURTHER READING

Admiral H and Corano A (2018) *Underground Spaces Unveiled: Planning and Creating the Cities of the Future*. ICE Publishing, London, UK.

ArcelorMittal (2004) *Steel Sheet Piles. Installation*. ArcelorMittal, Luxembourg.

ArcelorMittal (2014) *Impervious Steel Sheet Pile Walls. Design & Practical Approach*. ArcelorMittal, Luxembourg.

ArcelorMittal (2016) *Steel Foundations Solutions. General Catalogue*. ArcelorMittal, Luxembourg.

BSI (British Standards Institution) (2004) BS EN 1992:2004. Eurocode 2: Design of concrete structures. BSI, London, UK. (Parts 1–4.)

BSI (2004) BS EN 1993:2004. Eurocode 3: Design of steel structures. BSI, London, UK. (Parts 1–12.)

BSI (2004) BS EN 1997-1:2004 + A1:2013. Eurocode 7: Geotechnical design. General rules. BSI, London, UK.

BSI (2009) BS 8102:2009. Code of practice for protection of below ground structures against water from the ground. BSI, London, UK.

BSI (2010) BS EN 1536:2010 + A1:2015. Execution of special geotechnical work. Bored piles. BSI, London, UK.

BSI (2015) BS 8002:2015. Code of practice for earth retaining structures. BSI, London, UK.

Concrete Centre (2012) *Concrete Basements*. MPA/The Concrete Centre, London, UK.

EFFC/DFI (European Federation of Foundation Contractors/Deep Foundations Institute) (2016) *Best Practice Guide to Tremie Concrete for Deep Foundations*, 1st edn. EFFC/DFI, Bromley, UK/Hawthorne, NJ, USA.

Filip RK (2006) Recent advances in quiet and vibrationless steel pile installation. *Proceedings of the 10th DFI Conference on Piling and Deep Foundations, Amsterdam, the Netherlands*, pp. 442–450.

Gaba A, Hardy S, Doughty L, Powrie W and Selemetas D (2017) *Guidance on Embedded Retaining Wall Design*. CIRIA, London, UK, C760.

HSE (Health and Safety Executive) (2012) *Domestic Basement Construction*. HSE, London, UK, CIS66.

ICE (Institution of Civil Engineers) (2009) *Reducing the Risk of Leaking Substructure. A Clients' Guide*. ICE, London, UK.

ICE (2016) *Specification for Piling and Embedded Retaining Walls*, 3rd edn. ICE Publishing, London, UK.

LABC (2014) *Technical Guide: Guidance on the Design and Construction of Basements*. LABC, London, UK.

NHBC (2011) Basements and waterproofing. *Technical Extra* **2**: 2–5.

Nicholson D, Tse CM and Penny C (1999) *The Observational Method in Ground Engineering: Principles and Applications*. CIRIA, London, UK, R185.

Powrie W and Batten M (2000) *Prop Loads in Large Braced Excavations*. CIRIA, London, UK, R77.

Puller MJ (2003) *Deep Excavations: A Practical Manual*, 2nd edn. Thomas Telford, London, UK.

Twine D and Roscoe H (1999) *Temporary Propping of Deep Excavations – Guidance on Design*. CIRIA, London, UK, C517.

UK Government (1990) Planning (Listed Buildings and Conservation Areas) Act 1990. Statutory Instrument 1990/9. The Stationery Office, London, UK.

UK Government (1990) Town and Country Planning Act 1990. Statutory Instrument 1990/8. The Stationery Office, London, UK.

UK Government (1996) Party Wall etc. Act 1996. Statutory Instrument 1996/40. The Stationery Office, London, UK.

Useful web addresses

Association of Specialist Underpinning Contractors (ASUC): http://www.asuc.org.uk (accessed 01/08/2018).

Basement Information Centre: https://www.basements.org.uk (accessed 01/08/2018).

Planning Portal (UK): https://www.planningportal.co.uk (accessed 01/08/2018).

Temporary Works Forum: http://www.twforum.org.uk (accessed 01/08/2018).

Temporary Works, Second edition

Pallett, Peter F and Filip, Ray
ISBN 978-0-7277-6338-9
https://doi.org/10.1680/twse.63389.447
ICE Publishing: All rights reserved

Chapter 31
Digital project delivery – visual planning and BIM

Nick Boyle
Technical Innovation Director, Balfour Beatty Major Projects

The tools being developed to communicate design intent are becoming easy to learn and easy to use. This section covers some of the aspects of virtual design and construction, and how users can integrate this into temporary works used in construction processes and methods. The chapter introduces the tools used to identify the optimum construction methodology and how best to engage all parties in the review process to obtain the best solution.

31.1. Introduction

The way to improve safety, productivity and quality is to optimise the interaction between the permanent and temporary works design and methodology, so that the solutions eliminate the hazards. Key to this is identifying the optimal solutions, which can be constructed efficiently in the right sequence. Doing so means eliminating the hazard and reducing the time required for construction, all while achieving the design intent for the final user. The integration of engineering and design is also key to success. It is critical to learn from our successes and failures, what works and what does not, and to communicate this information in a digestible way so that we can find the best solution.

It is important to visualise the temporary works, what needs to be constructed and how it can be sequenced. This has always been difficult to share while at the same time ensuring that something is not misinterpreted or misunderstood. With the use of the latest digital processes and technologies we can now develop various solutions and test them prior to implementation. Visualisation has the added benefit that it provides information that is not language, dialect or pronunciation dependent: it is a 'common language' and it is universally understandable by all levels of the project team.

(The subject and the technology are changing fast, so this chapter can only be an appraisal based on systems and information current in 2018.)

31.2. Basics of building information modelling

Building information modelling (BIM) is the process of creating a 3D information model containing both graphical and non-graphical information in a common data environment

(CDE). A CDE is a shared repository for digital project information. The information is created to different levels of detail as the project progresses, with the relevant sections of the dataset then handed to a client at completion to use during the operations and maintenance phase, and ultimately into a decommissioning phase.

Typically the model has been built up from different models, each representing a different discipline, which are then combined on the same x,y,z coordinates. This is currently the most commonly used level of complexity in the industry. The discipline models that make up the whole are known as 'federated' models – multiple pieces, brought into a 'federation', which then describes the whole project.

This covers the three spatial dimensions, but a fourth can be added: time. Linking the model to the construction programme/scheduling is known as '4D'. This brings the traditional bar 'Gantt' chart alive, and integrates the design and methodology with both permanent and temporary works sequencing. (A Gantt chart is one in which a series of horizontal lines shows the amount of work done or production completed in certain periods of time in relation to the amount of work planned for those periods.)

And then we can add a further fifth dimension to the model: cost and quantities. This is then known as '5D'.

These dimensions describe the type of content you can expect to find in any given model. BIM also has different 'levels of maturity' in how we manage information

- *Level 0* – 2D CAD (computer-aided design) drafting shared via paper or PDF type formats.
- *Level 1* – Mixture of 2D/3D CAD and standards are managed to BS 1192:2007 + A2:2016 (BSI, 2007). Data are shared through a CDE.
- *Level 2* – 3D CAD models for each discipline – federated models on the same x,y,z coordinates. Design discipline information (data) is shared through a common file format enabling clash detection and multi-disciplinary views.
- *Level 3* – To be developed. A single shared project model to which all parties have access and can modify the model.

31.3. BIM and communication

BIM is all about communicating in a common language. The secret of communicating is to understand what is to be constructed, how the works are planned, how they are integrated in the permanent works design and the temporary works design, the methodology, and the role of the supply chain, in order that we all speak in a common language between the disciplines. The adage 'a picture is worth a thousand words' says it all – visualisation is the common communication tool.

An old riddle that demonstrates the value of visualisation goes something like this. To demonstrate communication, consider in your mind, 'Drive 1 mile south, 1 mile east, 1 mile north. Are you back where you started?' The initial answer is 'No, of course not', as we live in a 2D world. However if this is illustrated using a 3D model (Figure 31.1)

Figure 31.1 3D representation of the path travelled from the North Pole. (Courtesy of Balfour Beatty)

it depends entirely on where you started, and by a simple trick we instantly understand and there is no confusion.

There are lots of different ways of visualising what we plan to do, with different levels of complexity, but the premise is the same. How do we communicate and engage all the relevant parties?

Communicating information, using the latest digital techniques, is to communicate the design intent and methodology to others. We have to demystify the information so that people can start the learning process and remove the fear factor around technologies. Demonstrating the simplicity of what we want to achieve enables us to take the small steps on a bigger journey. Planning the design and construction is an iterative process, and we need to

- engage the people
- have the right processes
- use the right technology.

With the digital tools now available we can share what we plan to do, from using simple methods to using more complex models.

It is commonly known that under stress there is a 1 in 2 chance we can make the wrong decision (Smith, 2017): we must therefore improve the odds. We need to make things simpler to understand, to review and to come up with the optimal solution.

31.4. Key issues

The Health and Safety Executive (HSE) report RR834 (HSE, 2011) identified eight key issues the industry needs to address to avoid catastrophic events during the lifecycle of a project. Those that relate directly to design and construction are reproduced in italics below.

Issue 4 – Communication and interfaces management should be improved.

The research emphasised the need for effective communication about hazards and particularly the importance of effective management of risk at interfaces between and within organisations.

BS 5975:2019, the standard for managing temporary works states: 'Communication tends to be one of the major problem areas of temporary works because of the multiplicity of actions normally required when temporary works are being constructed and put into service' (BSI, 2019).

Issue 6 – Management of temporary works is crucial to success.

It was apparent from many case studies that insufficient consideration was being given to the management of temporary works in its widest sense. This work must be taken seriously and include all temporary works aspects, including issues relating to cranes and scaffolding. The potential impact of failures of temporary works needs to be considered carefully to reduce the likelihood of a catastrophic event occurring and the industry needs to seek to improve performance in this vital area. All stakeholders should be consulted on how to achieve this improvement.

Issue 7 – Independent reviews should be employed.

Evidence was found that the use of independent review, from an early stage and ongoing, would have reduced the risk of a catastrophic event. Evidence was also found of projects where there was inadequate independent review of what was happening on site and there was concern in the industry that levels of effective supervision had been stripped away over recent decades. These issues need to be explored further and encouragement given for clients to seek independent authoritative advice.

The creation of 3D/4D models will enable rapid communications of any issues and will aid understanding and development of the optimum solution. These models also support more effective independent reviews, by making potential issues clearer and intent explicit.

Issue 8 – The industry should learn from experience.

Learning from experiences was not found to be well-rooted in the industry. There was lack of confidence that:

- *Learning was well-shared rapidly.*

The use of 3D/4D models enables rapid learning and sharing of complex issues and the understanding of the causes of failures in the management and installation and dismantling of temporary works.

31.5. Methods and techniques

There are various digital processes and technologies that should be used to inform and add to the understanding of the temporary works brief throughout the project lifecycle.

- The true as-is – required so that we know the actual existing situation and can design the temporary works to fit.

- Integrated 3D and 4D models.
- The use of virtual reality or mixed reality so you can see the context of the permanent and temporary works on site – this area is developing rapidly.
- Completing the cycle – capturing the true as-is so the as-built can be held on record for future temporary works.

31.5.1 Initial baseline information

The start is the identification of the true as-is state at the time considered. Prior to carrying out the design of the permanent works, temporary works and methodology, it is important that the actual, existing situation is known. On a number of projects we have relied on the as-built drawings, which should reflect the existing situation, only to find out belatedly that this was not the case, and we had not validated and verified what was actually there. The latest digital survey techniques can capture the true as-is state to ensure construction is planned on the basis of what is there and, on completion, can capture the final state and replace the old as-built drawings. Techniques include

- static and mobile laser scanning – light imaging, detection, and ranging (LIDAR) utilises millions of laser points to create 3D point clouds for measurement
- photogrammetry – the use of pixels in photos to create 3D models for measurement
- other techniques to capture the above, including drones (i.e. unmanned aerial vehicles (UAVs)).

These techniques enable basic survey data to be captured without putting anyone at risk. Increasingly more uses are being identified for these different techniques, making it possible to survey areas that previously had poor or limited access. The methods can be cost-effective and enable accurate temporary designs to be completed. The level of accuracy required will determine the level of survey required. Static and mobile laser scanning can produce results accurate to within millimetres if used in conjunction with surveyed ground control points. Key to the use of these techniques is to understand the accuracy levels required and therefore the techniques to be employed (e.g. ground control points, density of information required).

The use of these techniques means that we can

- achieve significant cost reductions through early problem identification, resulting in less error and re-work
- drive down risk by minimising workforce exposure to high-risk environments such as live traffic or access to restricted areas, as well as by reducing time spent on site
- use high-definition scanning data to create dimensionally accurate 3D and 4D models, aiding conceptual and detailed design of temporary and permanent works and replacing as-built drawings.

Examples of these techniques are shown in Figure 31.2 for an existing building due for refurbishment, Figure 31.3 for the survey of a bridge without the need for a railway possession and Figure 31.4 for the examination of collision damage to a structure.

Figure 31.2 Existing building refurbishment requiring extensive temporary works – fully dimensioned as-is. (Courtesy of Balfour Beatty)

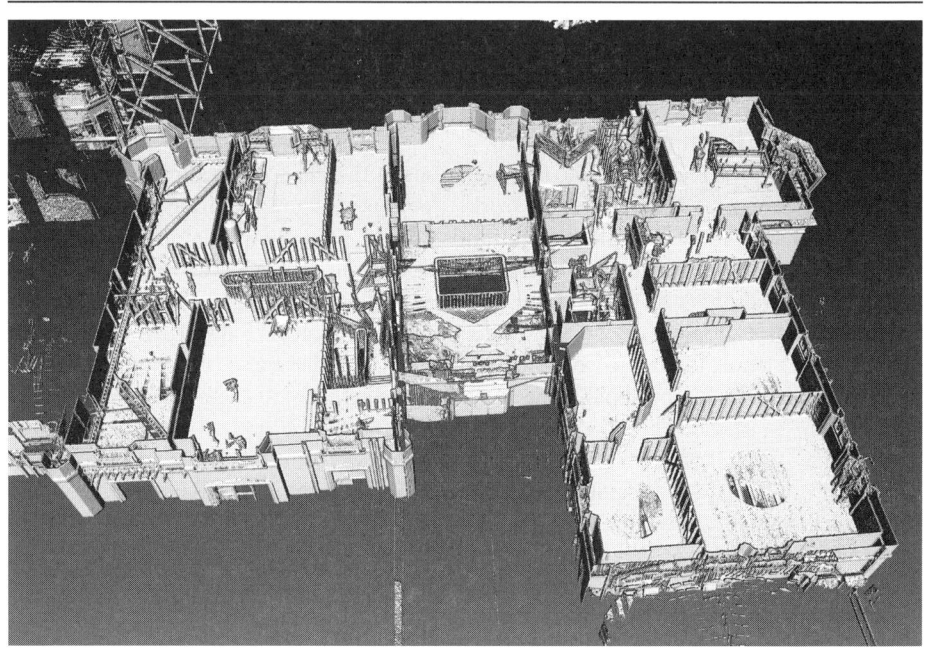

Figure 31.3 Railway overbridge surveyed without a railway possession. (Courtesy of Balfour Beatty)

Figure 31.4 Collision-damaged structure – permanent design and temporary works modelling enabled with full 3D image of structure. (Courtesy of Balfour Beatty)

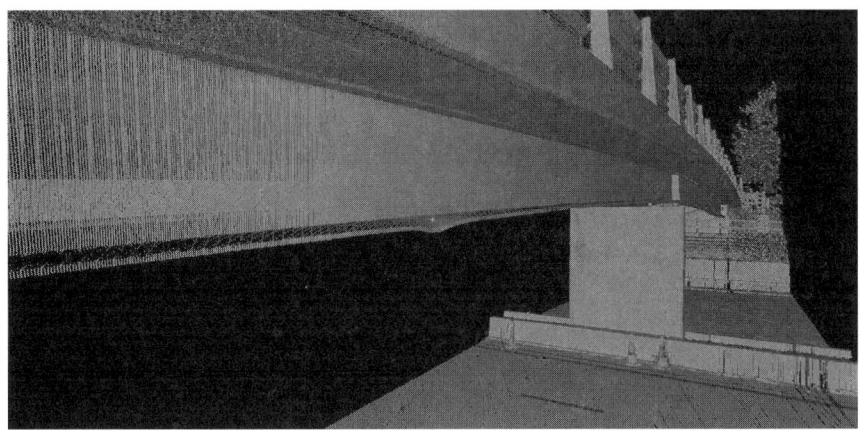

31.5.2 3D/4D modelling

There are numerous techniques and technologies that can be used within the industry, from the simple to learn and easy to use, through to more complex techniques with full multi-disciplinary combined (federated) models. The benefits include

- build twice – once on the screen, once in reality
- clash detection
- interdisciplinary checks
- engagement of all parties at all levels.

The use of simple 3D modelling software such as SketchUp (Trimble), as demonstrated in Figure 31.5, ensures that all personnel can understand and review what is intended and how it can be improved.

31.5.3 Virtual reality, augmented reality and mixed reality

As the power of the available digital tools grows we are able to use further techniques to communicate the temporary works to be installed, checked and approved.

There is rapid development in the latest technologies which literally immerse the viewer into future scenarios.

- In *virtual reality* the viewer experiences the full 3D space in a virtual world. This technique can be used with devices such as the Vive (HTC) and Oculus Rift (Oculus) and can bring the site into the office. The use of multiple wraparound screens in a BIM cave can give a fully immersive experience.
- In *augmented reality* a piece of plant or temporary works equipment can be designed and shared with the project team on site to ensure it works virtually, before it works in reality. For example, Balfour Beatty Ground Engineering designed a piling casing extractor that was shared on site through augmented reality (Figure 31.6), critically reviewed and then implemented.

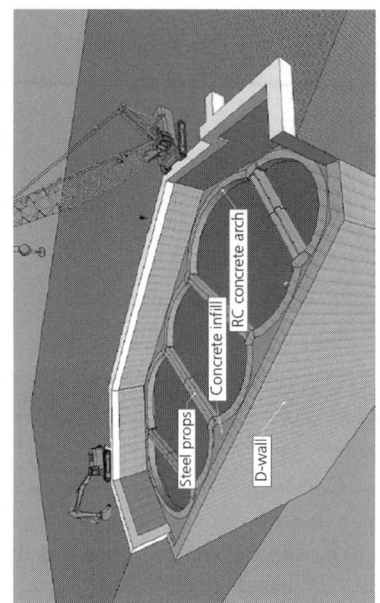
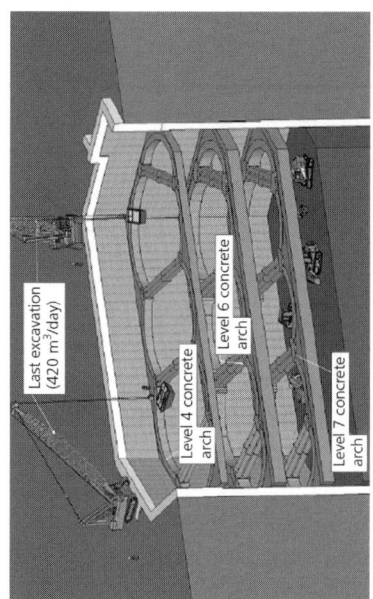
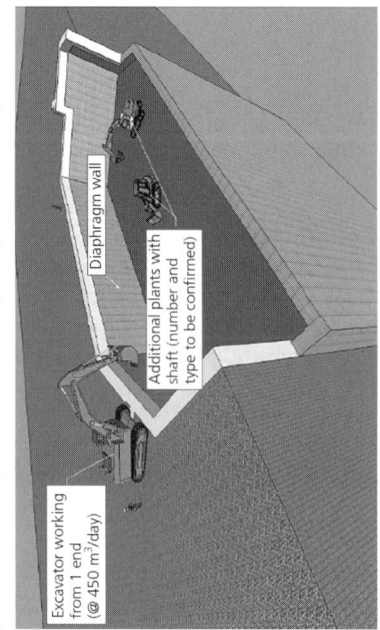

Figure 31.5 SketchUp visualisation output of Crossrail C512 Durward Street shaft. (Courtesy of Balfour Beatty Morgan Sindall Vinci JV)

Figure 31.6 Piling casing extractor. (Courtesy of Balfour Beatty Ground Engineering)

(a) Augmented reality

(b) In use

Figure 31.7 Mixed reality – virtual reality in the real world. (Courtesy of Balfour Beatty)

(a) Virtual beam lift

(b) Virtual footbridge

- In *mixed reality*, the experience of the actual site is combined with the virtual construction, and by using survey points the temporary works on a real site is shown in context, with the actual building site. This uses devices such as HoloLens (Microsoft) – mixed-reality smartglasses that know where they are geospatially (i.e. in x,y,z in space). An example of mixed reality is shown in Figure 31.7.

31.6. Managing and minimising risk

The London Crossrail project adopted many good examples of engineering expertise supported by the use of digital technologies to enable the engagement of all parties and stakeholders during construction, thus controlling the risks.

An example was at Whitechapel station, where a large trapezoidal shaft was to be constructed at Durward Street. The permanent works design had been carried out, complete with an illustrative temporary works design that had significant resilience and redundancy due to the requirement to design for a catastrophic event involving removal of any prop on a scheme incorporating three levels of propping, with loads on the walings of *c.* 3000 kN/m.

An alternative propping arrangement was developed that reduced the number of props and at the same time mitigated the risk of the elimination of a prop. The visualisation concept was completed in SketchUp (see Figure 31.5), which improved all the design and site staff's understanding of the complexity and, importantly, enabled early agreement to the principles. Any changes to the model would instantly reflect through into all the layout drawings and sequencing at the click of a mouse.

31.7. Example – temporary substation

On the Thames Tideway project in London, a temporary substation was required to provide power for the construction plant and equipment. In order to make best use of space on the constrained site an elevated steel deck was required above the transformers to support two storage containers and one backup generator (Figure 31.8).

During construction the site layout was still being refined, which resulted in numerous changes to the layout of the substation as the design progressed. By producing a single 3D model for the structure in SketchUp any layout changes could easily be incorporated (Figure 31.9). From this model, concept sketches and all 24 of the design drawings were produced in Layout (Trimble), SketchUp's sister application (Figure 31.10).

Images and 3D PDFs of the model were emailed to site as the design progressed, which gave the construction team the chance to review the buildability of all elements. As all

Figure 31.8 Image from 3D model of temporary substation. (Courtesy of Balfour Beatty)

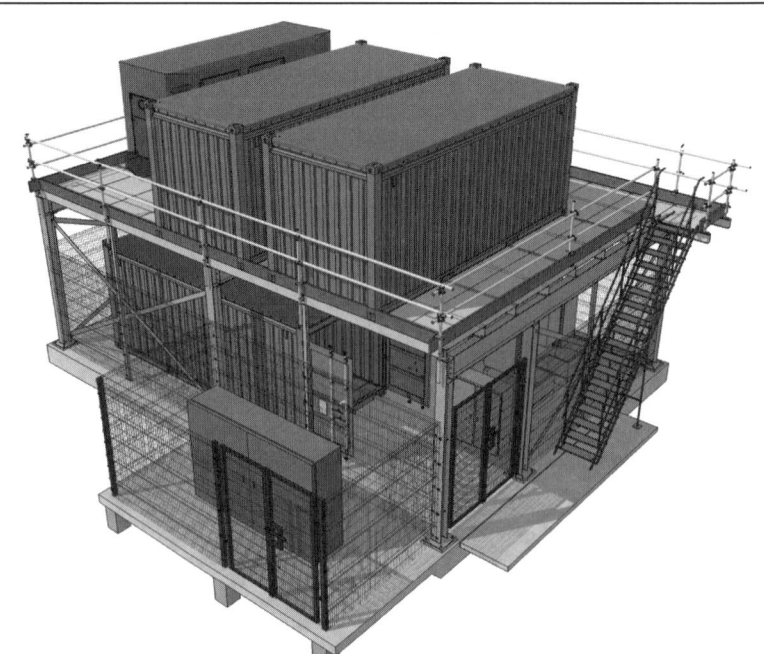

Digital project delivery – visual planning and BIM

Figure 31.9 Screenshot from a 3D PDF shared with the site team as the design progressed. (Courtesy of Balfour Beatty)

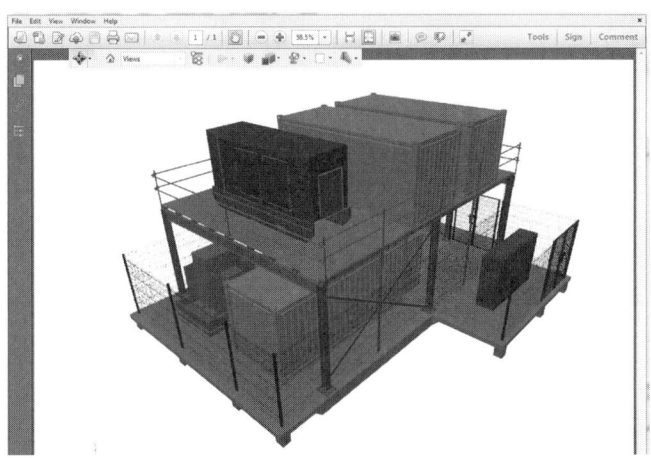

Figure 31.10 Image from fabrication drawings produced in Trimble Layout. (Courtesy of Balfour Beatty)

7.663 m long 406 × 178 × 74 UB steel grade S275 with 10 mm thick S275 endplates

Bracing 100 × 8 mm S275 Steel Flats

2 no. 12.2 m long 180 × 90 × 26 PFC steel grade S275

7.663 m long 406 × 178 × 74 UB steel grade S355 with 10 mm thick S275 endplates

All PFCs connected to primary beams: 3.171 m long 100 × 50 × 10 kg S275 PFC with 10 mm thick S275 end plates

Bracing 100 × 8 mm S275 Steel Flats

3.141 m long 203 × 203 × 46 kg S275 UC with 10 mm thick S275 end plate at top and baseplate at bottom

All PFCs connected to secondary SHSs: 3.524 m long 100 × 50 × 10 kg S275 PFC with 10 mm thick S275 end plates

7 no. 11 m long 180 × 90 × 26 PFCs with alternate flanges facing each other. Steel grade S275

All columns at ends of primaries: 3.494 m long 203 × 203 × 46 S275 UC with 10 mm thick end plate at top and baseplate at bottom

7.663 m long 406 × 178 × 74 UB steel grade S275 with 10 mm thick S275 endplates

Isometric View of Steelwork
Scale 1:50

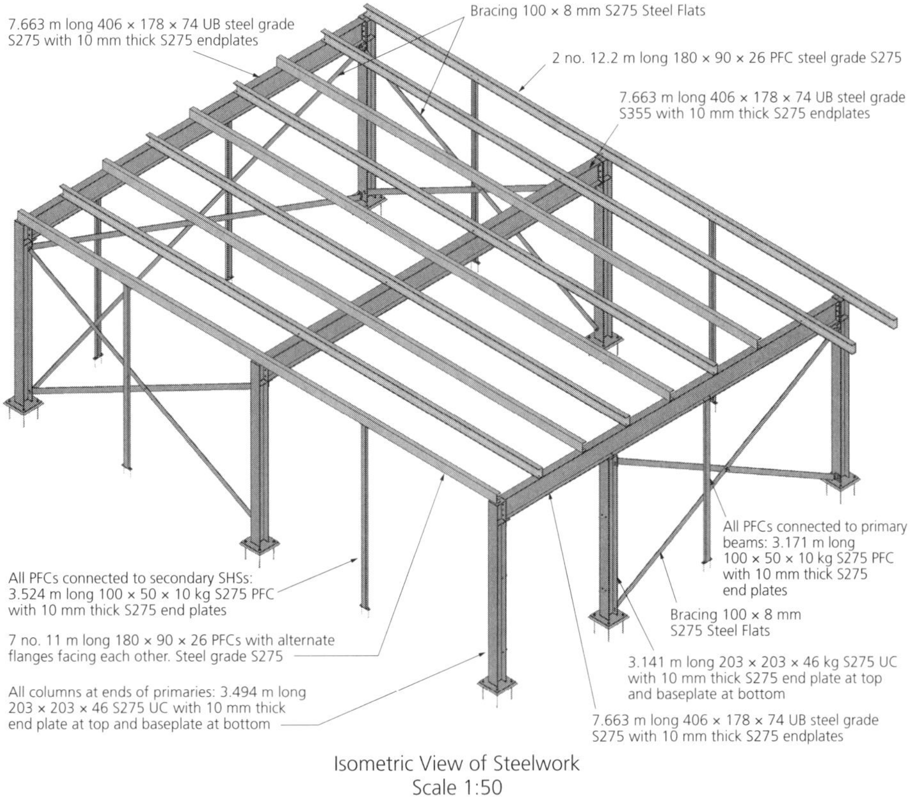

Figure 31.11 Photo of the partially constructed temporary substation. (Courtesy of Balfour Beatty)

the drawings referenced a single model, all changes requested by the site team could be incorporated quickly, right up until the final stages of design.

Further iterations were not required, as the site team had already passed comment on the design. The substation was successfully fabricated and erected from the issued drawings and the 3D model was incorporated in the combined model for the site (Figure 31.11).

31.8. High-quality animations

On a smart motorway scheme a bridge demolition was required and a visual method statement used. The methodology was developed and a 3D model created, which was used for the temporary works and methodology development, and peer review and project and site briefings. High-quality animations (Figure 31.12) enabled a comprehensive peer review that challenged the methods being proposed and enabled high-quality briefings to be undertaken.

An example of the use of visualisation to assist the presentation of system panel formwork is shown in Figure 31.13. A similar screen visualisation for the falsework and soffit formwork to a bridge and the corresponding 2D AutoCAD drawing on which it is based are shown in Figure 31.14. The benefits of using these animations include

- accurate constructible modelling for any structure
- 3D isometric viewpoints from any position to enable engagement of all parties
- automation of the take-off and the bill of quantities
- build twice – once on the screen and once in reality.

Digital project delivery – visual planning and BIM

Figure 31.12 High-quality animation for bridge deck: (a) animation of bridge to be demolished; (b) animation of methodology/temporary works; (c) extent of crash deck. (Courtesy of Balfour Beatty)

Figure 31.13 Sample proprietary supplier intelligent digital model with attributes. (Courtesy of Trimble)

(a) General arrangement of tank formwork

(b) Inside view

(c) Outside view

Figure 31.14 Falsework drawing and visualisation – Medway Bridge. (Courtesy of RMD Kwikform Ltd)

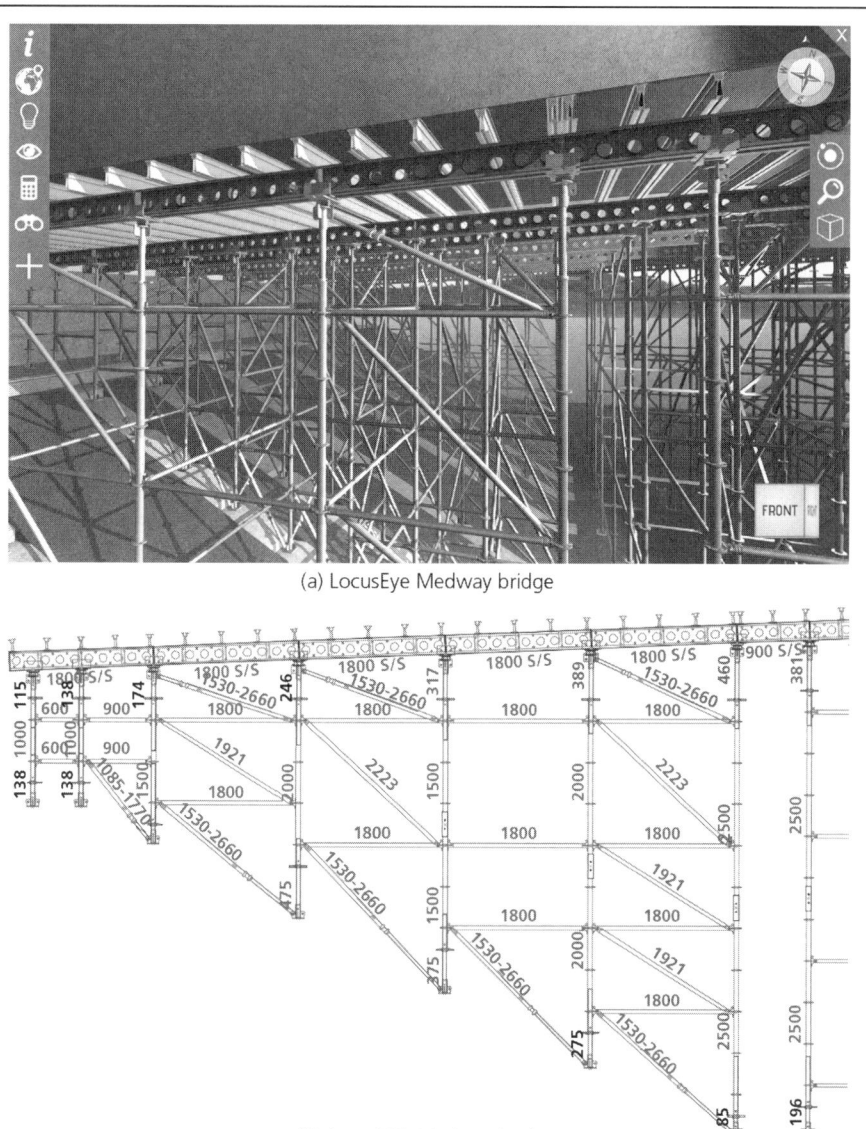

(a) LocusEye Medway bridge

(b) AutoCAD Medway bridge

31.9. Operations and future maintenance

The use of 3D modelling and 4D techniques enables review at the design stage of the future operations and maintenance to ensure that any temporary works required to maintain or replace items can be thought through and, where necessary, incorporated in the permanent works. Schemes can therefore look at future maintenance issues and the methodology required to maintain the permanent works. Examples include the replacement of pumps at pumping stations, escalator replacements and the installation of power plant. These examples could demonstrate where lifting and fixing points or hardstandings could be incorporated in the permanent works to facilitate future temporary works involvement in all parts of the construction, design and management considerations.

31.10. Communication and engagement

Using digital technologies throughout the processes will help explain technically complex activities and projects. By engaging with all the people involved various steps in the design and engineering process can be eliminated. It is possible to

- detect clashes to eliminate errors before they become costly
- integrate the methodology into the permanent and temporary works design, eliminating errors that would be inevitable.

There are many authoritative sources of information on the different technologies. The key is to integrate these technologies in the processes such that the design can be iterated many times, but the structure built only once, in the knowledge that nearly all residual risks have been eliminated.

31.11. Training and support

All the information required is available via the Internet and there are different levels of complexity to the tools, enabling 3D to be delivered at a simple level to more complex projects. New technologies to support processes are continually being developed and new ones upgraded. Identifying the right tools to enable a process is difficult due to the plethora of solutions. However, there is plenty of authoritative guidance available and plenty of opportunities for self-learning through the various platforms available.

REFERENCES

BSI (British Standards Institution) (2007) BS 1192:2007 + A2:2016. Collaborative production of architectural, engineering and construction information. Code of practice. BSI, London, UK.

BSI (2019) BS 5975:2019. Code of practice for temporary works procedures and the permissible stress design of falsework. BSI, London, UK.

HSE (Health and Safety Executive) (2011) *Preventing Catastrophic Events in Construction*. HSE, London, UK, RR834.

Smith DJ (2017) *Reliability, Maintainability and Risk – Practical Methods for Engineers*, 9th edn. Elsevier, London, UK.

FURTHER READING

BSI (British Standards Institution) (2013) PAS 1192-2:2013. Specification for information management for the capital/delivery phase of construction projects using building information modelling. BSI, London, UK.

BSI (2014) PAS 1192-3:2013. Specification for information management for the operational phase of assets using building information modelling. BSI, London, UK.

Useful web addresses

HTC – Vive: https://www.vive.com/uk (accessed 01/08/2018).
Oculus – Oculus Rift: https://www.oculus.com (accessed 01/08/2018).
Trimble – SketchUp and Layout: https://www.trimble.com (accessed 01/08/2018).

Pallett, Peter F and Filip, Ray
ISBN 978-0-7277-6338-9
https://doi.org/10.1680/twse.63389.465
ICE Publishing: All rights reserved

Chapter 32
Rebar stability

Ray Filip
Temporary Works Consultant and Training Provider, RKF Consult Ltd

Mark Tyler
Principal Methods Engineer, Balfour Beatty Major Projects

There have been a significant number of accidents involving the collapse of in situ fixed reinforcement cages and prefabricated reinforcement assemblies during their temporary works life cycle, which have led to fatalities and serious injuries (Figures 32.1 and 32.2). Currently there is limited guidance on this subject. Advancements in the design, detailing and assembly methods of reinforcement have also influenced cage stability. Traditionally the design and detailing of reinforcement was the preserve of the PWD, however it is now frequently the responsibility of the principal contractor or a specialist subcontractor and detailer. The design methodology and analysis have been refined and the amount of reinforcement has been rationalised. This can often lead to smaller diameter, slender reinforcement, hence reducing cage stability. Those responsible for design optimisation might not have sufficient practical site experience to identify fixing risks, including temporary stability of reinforcement.

Site custom and practice has also changed. Reinforcement used to be fixed in situ, tied back to one side of propped formwork system, with scaffolding used for access (which could also be used to provide stability). More recently, offsite prefabrication has become commonplace, with completed cages lifted into a freestanding position. Many sites now use mobile elevating working platforms (MEWPs) instead of access scaffolding.

32.1. Introduction

All reinforcement cages are temporary structures until encapsulated in hardened concrete, and as such must be managed in accordance with the temporary works design, checking and management procedures described in BS 5975:2019 (BSI, 2019). Large free-standing reinforcement cages and those that require lifting (e.g. diaphragm walls and piles cages) will have a higher risk management classification.

Responsibility for considering the stability of in situ reinforcement and lifting prefabricated cages should start at the design stage. The Construction (Design and Management) Regulations 2015 (UK Government, 2015) require designers to carry out a risk assessment and take into account the general principles of prevention to eliminate, insofar

Figure 32.1 Instability of an in situ cage. (Courtesy of Temporary Works Forum)

Figure 32.2 Consequences of failure of an unstable reinforcement cage. (Courtesy of Temporary Works Forum)

Table 32.1 Bar sizes (in the vertical direction) and the corresponding stable height of cage

Bar diameter: mm	Stable height of cage: m
12	2.4
16	3
20	3.5
25	4.5
32 and 40	5

as is reasonably practicable, foreseeable risks to health and safety. The principal designer should ensure these issues are considered. The Permanent Works Designer (PWD) who has designed and detailed the permanent reinforcement has an initial duty to ensure safety risks are reduced through robust design, or if this is not possible, through the communication of these risks to the contractor. Efforts should be made to establish the lifting issues and eliminate specific instability risks at design stage so that these risks can be managed effectively by the contractor. The PWD should advise on the risks of lifting cages and the overall stability of free-standing reinforcement cages.

Good steel fixing practice and adequate temporary works consideration are prerequisites for temporary works stability. Experience, custom and practice of steel fixing is needed to understand the likely benefits and capacities of various tying arrangements. There is limited design information on the strength of different types of ties, and variations in steel fixing tying arrangements can affect stability until the element is poured.

The various parties involved (designers, Temporary Works Coordinator, specialist sub-contractors) should liaise and collaborate as early as possible to identify risks and determine practical solutions. Various responsibilities need to be clearly identified based on competencies. A clear distinction should be made between temporary works design requirements for stability and buildability advice by fixing specialists. The Temporary Works Designer (TWD) might specify an impractical stability solution and a fixing specialist might offer stability advice based on anecdotal opinion, where in fact factual calculation is required.

Table 32.1 gives bar sizes (in the vertical direction) and the corresponding stable height of cage. These values can generally be adopted when considering if a stability check is required.

32.2. Potential problems and modes of failure

Good design and detailing along with good sequencing and good fixing practice will reduce risks, which include the following

- Work at height issues during fixing of in situ cages.

- In plane sway/racking – the cage can effectively act like a scaffolding with no bracing, and each connection point acts as a pin with minimal moment capacity. The cage fails by collapsing in the long face by racking.
- Bending failure – the cage bends over out of plane. This could be caused by the wind, accidental impact, heavy elements at the top of the cage (e.g. slab starter bars) or poor workmanship (cage being built out of plumb). As the cage progressively continues to lean it will exacerbate the problem.
- Buckling failure – this can occur when the self-weight of the cage is large and the individual bars are long. This mode of failure can cause secondary modes to occur. Most cages are slender and lack rigidity, even though they appear rigid.
- Discontinuity failure – this occurs when ties holding the vertical reinforcement to the starter bars fail. Often the whole cage will fall over and become separated from the starter bars. This can occur when the starter bars are too short or too thin, there are insufficient ties connecting the starters and the cage, the tying wire is too thin or not strong enough, or the bars are poorly tied together.
- Lap failure – a discontinuity failure can also occur within the cage itself at lap positions. Designers will often consider manual handling (without considering overall stability) and reduce the length of bars, hence introducing intermediate splices which weaken the structure. Splices weaken the overall cage; they should be staggered with their positioning and number carefully considered.
- Slab/foundation failure – this occurs due to insufficient support to the top mat of reinforcement when loaded by dynamic or static loading from people, materials or plant. Sufficient robust chair spacers should be provided and should be positioned and tied correctly. A particular safety issue is the possibility of collapse of deep cages where access for operatives inside the cage is required.
- Combination failure – it is common for a number of the above modes of failure to act together. For example, as a cage starts to bend greater stress will be applied to the ties at the base connecting it to the starter bars. In this case, although the cage will start failing in bending due to wind or self-weight, it may finally fail as a discontinuity failure when the ties break.
- If a reinforcement cage deforms then the structural performance of the reinforced concrete structure may be compromised and reinforcement cover may be inadequate (which can affect fire protection and corrosion protection).
- Individual bars can be welded together (approved factory conditions required) or clamped together to provide greater rigidity at node points, but the quality of the welding should be checked. Small tack welds are vulnerable to moment and torsion forces at bar connections.
- Temporary frames/jigs can be used to allow reinforcement to be fixed in the same alignment as it will be placed to avoid having to rotate or tilt the cage during lifting.
- Using one side of formwork to support the reinforcement during fixing (when in situ) or placement (when lifting prefabricated cages) – the formwork should be stabilised by using adjustable push–pull props which are anchored to a slab. The formwork, props and anchors should be designed for wind loading, self-weight (or formwork, working platforms for concreting, any box-outs and reinforcement), any live loading on the working platforms and a nominal

Figure 32.3 Sacrificial reinforcement support trusses. (Courtesy of Ward and Burke Ltd)

allowance for out-of-plumb. If access is required to the reinforcement at height then a MEWP or scaffold can be used. Rather than using formwork, a purpose-built tower (from scaffolding or structural steelwork) can be used for access and stability. The tower could be anchored to a slab or kentledge could be used to prevent overturning and sliding.
- Adjustable push–pull props could be tied directly to the reinforcement to provide stability to prefabricated cages. The design of the connection of the prop to the cage and the dispersion of concentrated load on the cage needs to be addressed in the design. Guide ropes can also be used, but the reaction from inclined ropes will cause additional compression in the bars (leading to buckling) and the ropes need to be anchored. The process of tensioning and releasing the ropes can pull the cage over if not done carefully to a designed and specified sequence.
- Sacrificial steel support posts, at regular spacing, can be positioned prior to the erection of in situ cages. These posts can be substituted by columns or trusses formed from reinforcement (see Figure 32.3).
- To prevent sway/racking failure, longitudinal bracing bars can be added to the cage.
- Welding node points between reinforcing bars or using 'bulldog grips' to create a greater amount of moment resistance than tie wire is unable to achieve.
- The Temporary Works Forum has produced a guidance note (*Stability of Reinforcement Cages Prior to Concreting*) which contains guidance on the ultimate capacity of chair elements (TWf, 2013; tables 2 and 3).

The designer should consult with the contractor so they can better understand how the site team will carry out the work and to help identify the potential modes of failure. Multiple failures occurring simultaneously should be considered. Valuable buildability advice regarding the practicalities of steel fixing can be provided by steel-fixing operatives. Typically, this advice will cover favoured bar arrangements to improve safety and ease the assembly process. However, the practical advice must be supplemented by the technical judgement of an appropriately experienced TWD.

- *Failure of lifting points* – welded or tied lifting points could become overloaded and fail. Inclined chains/strops will cause a horizontal reaction in the cage. Poor quality welding can commonly be an issue.
- *Failure of lifting equipment* – self-weight could be incorrectly calculated (in winter ice can add to the overall self-weight) and the lifting attachment or lifting equipment is under sized. Dynamic effects should be allowed for (typically 10–25% of the dead load, depending on the type of crane being used (see TWf, 2013)).
- *Movement of cage during lifting* – if the centre of gravity of the cage is not directly beneath the lifting hook then the cage will 'swing'. A cage may also rotate if the centre of gravity is above the lifting points. If 'L' starter bars are present at the top of the cage then the cage will be eccentric and likely to move during lifting. If winds speeds are excessive during lifting the cage will sway and could deform or could hit adjacent structures or plant. If the cage is allowed to slide along the ground during the lifting it could deform.
- *Falling items* – if reinforcement is not tied correctly (ties could be loose or laps are too short), some bars may fall out under their own self-weight during lifting operations. The quality of the tie wire and quality of tying could be inadequate. Loose items left within the cage during fabrication (tools, spacing blocks, etc.) could also fall out. If 'box-outs' are present they need to be secure, otherwise they will move or fall out during lifting.
- *Cage folding during lifting* – large vertical cages will be prefabricated in the horizontal plane and will need to be lifted from horizontal to vertical. If the cage is not stiff enough it is likely to 'fold' during this operation. Large horizontal cages may fold under their own self-weight and curved cages may fold in the plane of the curve.
- *Stacking and vibration* – due to lack of space on many inner city sites it may be necessary to prefabricate cages off site and then transport them. Cages could deform when stacked during storage. Vibration during transportation could also cause individual bars to become dislodged or whole cages to deform. This can also occur with relatively small cages, which are often lifted and transported on site by excavators travelling on rough ground.

32.3. Common solutions

- A lift plan should be provided, lifting operations should be carried out by competent persons, and any lifting points should be designed, tested and certified as per BS 7121 (BSI, 1999–2017), the Lifting Operations and Lifting Equipment Regulations 1998 (UK Government, 1998a) and the Provision and Use of Work Equipment Regulations 1998 (UK Government, 1998b). Inclined lifting chains

should be avoided for wide prefabricated cages, as the cage may deform due to the horizontal reaction. A designed, tested and certified lifting beam should be provided so that multiple vertical lifting points can be used. The weight of the cage should be evenly distributed among the lifting points.
- TWDs should consider the self-weight of individual reinforcing bars and ensure sufficient ties are provided to prevent bars falling out during lifting. Site supervision and inspection should ensure this, and checks should be carried out to remove any loose items before lifting occurs.
- Tag lines should be used for large loads during lifting to prevent them swinging in the wind.
- Sacrificial stiffeners can be fixed into a prefabricated cage to allow it to be raised from the horizontal to vertical plane without deforming. The weight of these items should be included in any lifting operations.

32.4. Design

It is not necessarily obvious where the responsibility lies for calculating self-weight of cages, determining fixing sequences and designing stiffening or lifting points. However, these responsibilities need to be allocated, and structured collaboration between the various specialist parties involved will be necessary. On site the Temporary Works Coordinator should ensure these issues are considered and addressed.

Designers of reinforcement have a responsibility to ensure they reduce risks wherever possible (any residual risks should be communicated) and that cages (prefabricated or fixed in situ) can be fixed and lifted in a safe manner.

Consideration should be given to the positioning and type of lifting points, stiffeners and bracing. If these items will be removed, access should be provided. However, in order to eliminate issues with access provision it may be easier to consider the items to be sacrificial. If this is the case, the positioning of these items may be critical so that they do not clash with other structural items and they do not protrude into concrete cover zones.

32.5. Design rules

- If the height of a slab chair exceeds 1 m a design check should be completed to ensure the slab chair does not bend excessively under the weight of the top mat and any additional loading during concreting operations.
- The likelihood of slab cage instability rises in proportion to its depth, top mat weight and incline. Slab reinforcement placed on any incline whatsoever is prone to racking instability. A designed means of eliminating this failure mode must be incorporated in the details or called up as a specific residual risk to be addressed in the safe system of work. Traditional chairs (shape code 98 in BS 8666:2005 (BSI, 2005)) provide vertical support of top mat reinforcement and should not be relied on for lateral resistance to slab racking. A specific design check is required on the combined effect of vertical plus lateral sway load. On slabs without incline, a nominal 2.5% of the total vertical load is on the chair.
- Chair centres should not exceed 50 times the supported mat bar diameter in order to prevent excessive deflection of the top mat.

- Chairs should be staggered to ensure a row of chairs does not pick up a single bar.
- The chair feet (bottom horizontal) should be long enough to span three bars, with allowance for the bend radius to the chair.
- Chairs must be orientated to provide a robust load path through the supported bars, the chair's lower mat and the cover blocks. For bars and chairs in compression under load, do not rely on the tying wire in tension to provide any part of the principal load-bearing solution.
- The TWD should consider the concentrated weight of bundled reinforcement onto a partially completed top mat (a higher concentration of chairs and/or a means of spreading concentrated loads in designated load-out areas).
- Carpet mat (rolls of reinforcement) is also a concentrated load and consideration has to be given to access and the forces exerted during rolling.
- Bracing bars to a prefabricated beam element need to be provided to prevent racking.

32.6. Structural behaviour of cages

Reinforcement cage behaviour is dominated by the shear mode deformation. It is not valid to calculate bending resistance according to the parallel axis theorem without taking into account the shear stiffness. As with open, parallel chord truss forms, the shear stiffness depends on the diagonals and vertical post members forming the web elements. In a typical reinforcement cage the only web members are those introduced by the fixers on site. These are normally in the form of light U-shaped bars, which are used to space the mats apart. These form web post members but, due to their low stiffness, lack of end fixity (to pins connections, which are also prone to sliding) and large spacing within the cage, the spacers and chairs provide negligible resistance to shear deformation. Therefore, the cage stiffness will tend to be no greater than the sum of the stiffness of the chord bars. For traditional tied cages

$$\text{cage stiffness} = \sum I_{\text{chord bars}} \qquad (32.1)$$

In this condition, many vertical reinforcement cage forms are slender, flexible structures. Free-standing cantilever cages in particular are prone to vertical buckling under their own weight. Vertical buckling of a cantilever results in a small side sway at the tip of the cantilever. The deformation induced by vertical buckling is then acted on by gravity. This leads to a progressive, creeping side sway and failure of the cage in bending. Tied lap joints in the vertical bars will become overloaded and contribute to the overall failure mode.

32.7. Design solutions – walls

Vertical buckling resistance can be increased by introducing lateral restraint near the free end of the cantilever and at intermediate locations on the tall walls. The magnitude of lateral restraint required to resist vertical buckling is very small. Traditionally, fixers are used to tie the cage into the access scaffold as part of the fixing process. This would create resistance to buckling 'by accident' rather than by design. Two possible consequences of tying into an access scaffold without input from an approved designer are

- High wind loads on the cage will be transferred onto the scaffold structure – the scaffold must be designed to resist this action.
- The sequence of formwork installation, including scaffold adjustment to accommodate the formwork, must be very carefully sequenced to ensure that sufficient lateral cage restraint is always in place. Designers can improve the overall cage rigidity by introducing diagonal bracing bars into the cage.

A practical consideration is the timing and work access for the installation of these bars to maintain a robust structure throughout the assembly process. The temporary works bars may cause lapping congestion and clashes with other items. A fully coordinated temporary and permanent works design, including an assumed assembly sequence, is required.

In the case of wall cages, good practice is to cast the first set of diagonal bars into the concrete base (as part of the starter bar assembly). In a cantilever truss structure subject to wind force the largest load in the diagonals will be at the base of the cantilever. The diagonal bars are typically installed at $c.$ 1 m centres along the cage wall. Loads of 1.5 t per diagonal bar are not uncommon in large free-standing wall cages. The bar must be appropriately sized to resist both tension and compression as wind load and impact loads can occur on either wall face, leading to load reversal. The connection between the diagonal web bars and the chord bars is critical. The splice legs of the diagonals should lap directly onto the vertical chord bars and must be long enough to transmit load. Mechanical clamps such as bulldog grips can be used. Sacrificial welded truss frames are an obvious solution to overcome the difficult assessment of tying wire connections. For 'tied-only' diagonal bar connections, a large number of doubled wire splice ties are required to ensure that diagonal loads are transmitted into the chord bars. A concentration of cruciform ties to the horizontal and vertical bars in the vicinity of each diagonal node point helps to transmit diagonal bar loads into the wall face bars.

For walls the following checks should be made

- *Check 1*: Vertical buckling due to self-weight

$$L = \sqrt[3]{0.795\pi^2 EI/p} \qquad (32.2)$$

where L is the critical length, E is Young's Modulus (205 000 N/mm²), $I = \pi D4/6$ (D is the diameter of the bar in mm) and P is the vertical UDL along the bar (N/mm).
- *Check 2*: Bending capacity due to
  - design wind load
  - working wind load and accidental impact.
- *Check 3*: Combined axial and bending capacity check of the form

$$\frac{f_{ac}}{P_{ac}} + \frac{f_{bc}}{P_{bc}} \leq 1 \tag{32.3}$$

where f_{ac} is the calculated axial compressive stress, P_{ac} is the permissible compressive stress, f_{bc} is the calculated maximum compressive stress due to bending about both principal axes and P_{bc} is the permissible compressive stress in bending.
- *Check 4*: Deflection – include second-order P–Δ effects (especially gravity acting on deformed shape).

Due to the complex nature of reinforcement cages, including workmanship issues, unknown connection stiffness between bars and the eccentricity between lapped bars, it is usual to use simplified analysis models and a permissible stress approach, aligned to BS 5975:2019 (BSI, 2019).

32.8. Wind loading
Three conditions should be considered

1. wind blowing normal to the plane of the wall
2. wind blowing at 10–15° off parallel to the cage plane
3. wind blowing at 45° to the wall plane.

In condition (1), the wind pressure acts on every vertical bar and generates significant force magnitude parallel to the cage plane. In conditions (1) and (3) there is an element of wind shielding from the windward near face onto the leeward far-face mat. In condition (2) and at wall end zones for condition (3) there is unlikely to be any such wind shielding benefit.

Apply wind load to the near-face and far-face reinforcement. The windward reinforcement (near face) will provide some wind shielding to the leeward reinforcement (far face). The shielding factor is dependent on the spacing ratio and the aerodynamic solidity ratio, as defined in BS 5975:2019 (BSI, 2019; annex M). The effective area subject to wind forces, A_{ref}, is the shadow area of the reinforcement.

Shadow area. Reinforcement bars are specified in nominal diameters. These should be increased to the average of the nominal and the rib-to-rib diameter. The shadow area of the vertical and horizontal bar lapping zones should be taken into account. Any solid objects placed in the cage, for example plywood or polystyrene protection to slab couplers, will increase the wind load acting at that location.

The force coefficient c_f is 1.2 for circular members, and the wind pressure acting on the near- and far-face wall mats must be considered.

Dynamic effects. Where a cage is rigid and propped, a dynamic factor of unity is normally appropriate. The application of a larger dynamic factor is likely to be required for flexible, free-standing cantilever reinforcement cages. Analyse the force in both directions.

Critical structures (e.g. next to railway line) require robust measures. Specifically designed bolted or welded subframes, formwork shutters and props should be used in preference to tied-only reinforcement where there is risk (workmanship, ties becoming overloaded). In general, disregard the short-term probability reduction factor in BS 5975:2019 (BSI, 2019).

32.9. On-site inspections

The site team should ensure the designer's instructions are followed and that a comprehensive planning and inspection regime is in place. Emphasis should be placed on the competence and diligence of the inspectors. A thorough understanding of the TWD, PWD and workmanship requirements is needed. In the case of complex reinforcement structures the checks might be a collaborative endeavour. For example, a competent steel fixer, quality inspector and lift supervisor might support the temporary works inspector leading the stability check. Benefits of this approach are: shared knowledge, peer learning, more questioning and reduced complacency.

- Is a safe system of work in place which adequately considers site constraints?
- Has the cage been constructed correctly to the most up-to-date drawings (correct bars, spacing, ties, laps, cover)?
- Have safety issues associated with safe loading and offloading of bundles or rebar and prefabricated cages been considered (e.g. safe access to sling the load, specified slinging arrangement)?
- Are measures in place to prevent accidental impact?
- Do temporary stiffeners and temporary bracing have to be removed or are they sacrificial? If they do need to be removed, has safe access been provided and how is stability maintained during and after removal?
- When large vertical cages are constructed in situ on site a check should be carried out on verticality.
- Has the cage been constructed correctly to acceptable tolerances?
- Has manual handling of individual pieces of reinforcement been assessed?
- Are splices in the correct place and are they staggered and tied correctly?
- Have the correct ties and tie patterns been used?
- Has the quality of any factory welding been inspected?
- Have box-outs been positioned correctly and fixed securely?
- Have sufficient chairs been installed for slabs – are the chairs in the right place?
- If formwork or towers are to be used for stability of wall reinforcement, have they been designed and installed as per the TWD design?
- Has adequate access for work at height been provided?

- Check the tightness of ties and any cast in items.
- What prevents racking?
- If formwork or towers are to be used for stability of wall reinforcement, have they been designed (e.g. for accidental impact when placing the cage) and have they been installed as per the design?
- Have hold points been identified and has a permit been issued in accordance with BS 5975:2019 (BSI, 2019)? Any on-site changes to the design should be approved by the designer.
- Is a lift plan provided, have lifting points been designed and are competent persons available to carry out the lift? If possible, an exclusion zone should be created beneath the lifting operation.
- Have temporary stiffeners been provided to ensure the prefabricated cage does not fold during lifting or collapse when placed in position?
- Has the self-weight been calculated accurately and has the centre of gravity of the cage been identified? Are the lifting arrangements suitable?
- Have loose items been removed from the cage prior to lifting?
- Have wind speeds been checked prior to lifting and guide ropes provided for large cages?
- Has access been considered to release lifting points from vertical cages or can ground release shackles be used?
- Is the lifted prefabricated cage secure and stable before lifting chains/strops are removed?
- After lifting, the cage should be inspected again to ensure no damage has occurred.
- Reinforcement cover and spacing should be checked after lifting, as items may have moved.

32.10. Ties

Ties are non-structural wired bar-to-bar connections used to assemble bars and keep them in place until bounded by setting concrete. Stainless steel wire is normally specified in highways structures and where slab soffits will be visible. Compared to soft black annealed wire, stainless steel wire is particularly sharp and is more likely to unwind under heavy load.

BS 7973-2:2001 provides guidance on cover block design and distribution, spacer (chair) arrangements to keep mats of reinforcement apart and basic tying provisions to ensure bars do not displace during concrete operations. BS 7973-2:2001 ensures the quality of the permanent works but does *not* guarantee the structural integrity of reinforcement cages in their temporary state. A TWD design is required to ensure cage stability throughout the temporary works life cycle. The UK Temporary Works Forum provides guidance on *semi-structural* fixing arrangements for the temporary stability of cages. The Temporary Works Forum is sponsoring ongoing research on this subject and its website should be consulted for the most up-to-date guidance on this matter.

All *semi-structural* tying of reinforcement should be carried out using one of the following wire types

- Soft black annealed tying wire with a minimum diameter of 1.6 mm and a typical strength in the range 280–320 MPa.
- Soft stainless steel tying wire, grade 1.4301 (304-S31), with a minimum diameter of 1.2 mm, a typical strength in the range 600–800 MPa, a minimum tensile strength of 500 MPa and a minimum elongation at fracture of 40%.

REFERENCES

BSI (British Standards Institution) (1999–2017) BS 7121. Code of practice for safe use of cranes. General. BSI, London, UK. (12 parts.)

BSI (2001) BS 7973-2:2001. Spacers and chairs for steel reinforcement and their specification. Fixing and application of spacers and chairs and tying of reinforcement. BSI, London, UK.

BSI (2005) BS 8666:2005. Scheduling, dimensioning, bending and cutting of steel reinforcement for concrete. Specification. BSI, London, UK.

BSI (2019) BS 5975:2019. Code of practice for temporary works procedures and the permissible stress design of falsework. BSI, London, UK.

TWf (Temporary Works Forum) (2013) *Stability of Reinforcement Cages Prior to Concreting*. TWf, London, UK, TWf2013:01. (Addendum, 2014.)

UK Government (1998a) Lifting Operations and Lifting Equipment Regulations 1998. The Stationery Office, London, UK.

UK Government (1998b) Provision and Use of Work Equipment Regulations 1998. The Stationery Office, London, UK.

UK Government (2015) Construction (Design and Management) Regulations 2015. Statutory Instrument 2015/15. The Stationery Office, London, UK.

FURTHER READING

BSI (British Standards Institution) (2005) BS EN 10080:2005. Steel for the reinforcement of concrete. Weldable reinforcing steel. General. BSI, London, UK.

BSI (2005) BS 4449:2005 + A3:2016. Steel for the reinforcement of concrete. Weldable reinforcing steel. Bar, coil and decoiled product. Specification. BSI, London, UK.

BSI (2005) BS 4482:2005. Steel wire for the reinforcement of concrete products. Specification. BSI, London, UK.

BSI (2006) BS EN ISO 17660-1:2006. Welding of reinforcing steel. Load-bearing welded joints. BSI, London, UK.

BSI (2006) BS EN ISO 17660-2:2006. Welding of reinforcing steel. Non-load bearing welded joints. BSI, London, UK.

Tubman J (1995) SPU SP 118 *Steel Reinforcement: A Handbook for Young Construction Professionals*. CIRIA, London, UK.

Useful web addresses

Temporary Works Forum: http://www.twforum.org.uk (accessed 01/08/2018).

Pallett, Peter F and Filip, Ray
ISBN 978-0-7277-6338-9
https://doi.org/10.1680/twse.63389.479
ICE Publishing: All rights reserved

Chapter 33
Needling and forming openings in walls

Ray Filip
Temporary Works Consultant and Training Provider, RKF Consult Ltd

33.1. Introduction

One of the most common activities during a refurbishment project is to form a new opening or enlarge an existing opening in a load-bearing brick or masonry wall. Obviously, making a hole through a wall for a 150 mm wastewater pipe requires some judgement, but this chapter is concerned with more significant openings. The hole could be for a door, window or services, or to make a space 'open plan'. Relatively simple operations have led to problems due to lack of competent people to assess the issues or supervise the work, lack of knowledge/information about the existing structure to be supported, perceived lack of time and money, and unrealistic expectations of how to carry out the work. The Construction Industry Research and Information Association (CIRIA) publication C740 (Gilbertson, 2017) gives guidance for the various parties involved.

Historic buildings and structures may often have been built using traditional 'rules of thumb', which may not conform to modern building regulations, and 'as-built' drawings often do not exist. Buildings may also be in a poor state of repair, which can make the work particularly challenging and risky.

The management of any temporary works that may be required should be in accordance with BS 5975:2019 (BSI, 2019) (see Chapter 2) and demolition should be carried out in accordance with BS 6187:2011 (BSI, 2011) (see Chapter 34).

Further guidance for clients, designers, contractors and so on is also available in CIRIA C740 (Gilbertson, 2017). The series of Building Research Establishment (BRE) Good Building Guides (see Further Reading at the end of this chapter) provide information and guidance for many common needling projects.

33.2. Assessment of the building

Before attempting to form a significant opening or enlarge an existing opening it is imperative that there is adequate planning and that a thorough structural assessment is carried out of the wall, the overall stability of the structure and any water tanks, roof areas, beams or floors supported by the wall. This should be carried out by an experienced and competent engineer. The following generalised sequence should be adopted for the assessment

1. Research any existing documentation (desk study) such as original drawings, original calculations, surveys, soils investigations and so on. Also check for any planning constraints due to listing or conservation. Listed building consent or building control approval may be necessary. Adjacent property owners may need to be informed or consulted as neighbouring properties may rely on the building or structure for support or stability and may require some form of temporary structural support or protection from the weather.
2. From existing information try to understand how the structure 'works', taking into consideration spans, load-bearing supports, foundations and how overall stability is achieved. (However, due to lack of information it is not always possible to get a full understanding.)
3. Ensure the existing structure is safe to access. If not, safe access should be established. There may be some areas that are deemed dangerous and in imminent danger of collapse, and these should be identified and addressed as a matter of urgency to ensure safety before any works are commenced. If the building or structure is to remain in public use then public protection needs to be considered and the additional live loading needs to be considered in any design work.
4. Carry out an investigation into the structure. This involves looking behind finishes, such as ceilings and plaster, to establish how the structure actually 'works'. It is possible this may be different to assumptions made in the desk study. Establish spans, materials used, foundations, hazards, state of repair, what provides overall stability, and so on. Particular attention should be paid to areas of obvious damage or distress (bowing, leaning, water ingress, cracking, etc.) and areas where previous repairs or alterations have been carried out (the quality of workmanship and materials used could be suspect).
5. Even though thorough surveys may have been carried out not all details may have been established or some items may not be as expected. Often further issues may be found only once work on site starts, and hence the design and site management process should allow for dealing with unexpected items. In order to prevent collapse during construction, careful thought must be given to correctly sequencing the demolition and rebuilding, ensuring that permanent supports are not removed without temporary support being provided.

33.3. Is support required?

If the new opening is small then the wall may be able to arch over the new opening, and it is likely that a lintel and temporary works will not be required. However, the ability of the wall to arch over an opening should be assessed by an experienced engineer. Consideration should be given to location, materials used in the wall construction, condition, size of opening, point loads from floors or beams above the opening, position of openings above, and so on. A different approach will be necessary if the opening is within the main body of a load-bearing wall, where arching can be considered (the magnitude of loading to be supported is discussed below). However, if the opening is very large (removal of most of a wall), if the opening is adjacent to a corner, adjacent to an expansion joint or there are beams and openings directly above the new opening, then it will be difficult to justify arching, and the lintel and temporary works are likely to be designed for the full vertical applied load. Beams that span onto the wall above the new opening may be propped separately.

If a new opening is to be formed in a solid wall of substantial thickness and in good condition (no beams or openings above, not near a corner, etc.), it is likely that two permanent steel or pre-cast lintels may be required. These may be installed one at a time by removing half the thickness of the wall and installing the first lintel. The remainder of the wall can then be removed and the second lintel installed. The first lintel to be installed should be designed to support the total applied loading in the temporary condition.

Further consideration should also be given to significant openings in shear walls, where horizontal loading should be considered in permanent and temporary condition to ensure stability of the structure. It is likely that the permanent works may be designed as a 'picture frame' with moment connections and any temporary works are likely to require significant bracing. If in doubt take advice from someone who knows.

33.4. Assessing the loads to be supported

An assessment of the loading to be supported by the new lintel and any associated temporary works should be made by a competent engineer. The new lintel and any temporary works should then be designed for these loads. If the new opening is large and unable to arch over the opening then it will have to be permanently bridged. Typically, permanent support can be provided by using one or more of the following: steel beams, pre-cast concrete lintels, in situ reinforced concrete beams, timber beams or forming an arch. The loading should be calculated as per BS 5977-1:1981 (BSI, 1981), as shown in Figure 33.1, by considering the full weight of brickwork within the 45° load triangle plus an allowance for any additional loading from floors or beams within the 60° interaction zone (unless these are supported separately). The new lintel should span over the new opening and have sufficient bearing on the wall either side to prevent overstressing the brickwork or masonry (minimum 150 mm bearing and padstones may also be required to spread the load locally). The lintel will be installed into a slot formed in the wall, which will be wider than the lintel. The slot will be formed by removing bricks, and so the base of the load and interaction triangles should be taken as the length of the new lintel plus 215 mm either side (i.e. one brick length either side).

Existing openings or columns, above a new opening, need to be identified (they may be partially obscured by an existing floor) as these will affect the ability of the masonry to arch and the main loading will be between the openings.

Once the loads have been calculated any temporary supports can be designed. In order to prevent cracking (or more severe damage) to the wall, an acceptable deflection limit needs to be agreed. The acceptable deflection will be based on the sensitivity of the structure being supported and its condition. The temporary supports should be stiff enough to satisfy the deflection criteria.

Figure 33.1 Lintel design theory to BS 5977-1:1981 for assess loading. (Courtesy of RKF Consult Ltd)

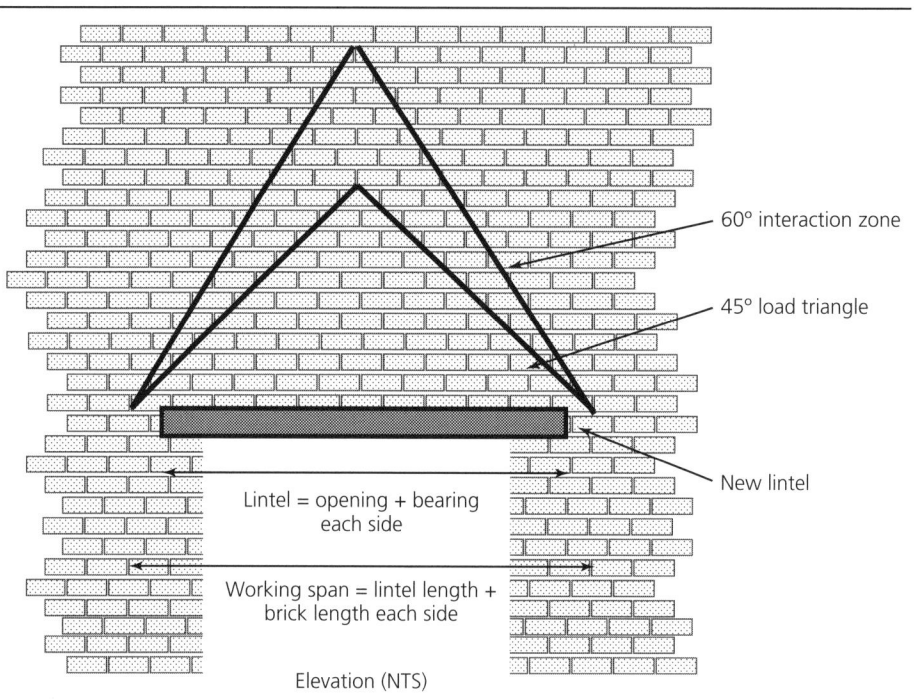

Elevation (NTS)

The Permanent Works Designer should also consider the effect of the new opening on the remaining brickwork either side of the new opening and on the existing foundations. Locally the loading on the foundation will increase and underpinning (or a similar solution) may be necessary.

33.5. Responsibility for temporary works

Clients, principal designers, designers, principal contractors, contractors and equipment suppliers all have duties and responsibilities relating to temporary works (see the Construction (Design and Management) Regulations 2015 (CDM 2015) and BS 5975:2019 (BSI, 2019)). On site the principal contractor is generally responsible for ensuring that any temporary works are classified according to risk, designed (and checked) by competent designers (unless standard solutions are used) and the work is carried out and supervised by competent persons (see BS 5975:2019 for further details). However, CDM 2015 require that Permanent Works Designers should identify and communicate risks and take into account any temporary works that may be required during the construction phase. Selecting an appropriate and competent Temporary Works Designer (and design checker) is critical. The design (and design checking) of temporary works may be carried out by a number of different organisations

- principal contractor's in-house designers
- Permanent Works Designer (structural engineer)

- equipment suppliers' in-house designers
- specialist subcontractors' in-house designers
- specialist Temporary Works Designers.

Often the responsibility for various aspects of the design may be split between different organisations (e.g. a structural engineer may be responsible for calculating the magnitude of the load to be supported and an acceptable deflection limit). An equipment supplier or the principal contractor's in-house designers may then be responsible for ensuring a suitable scheme is provided to support the loading. Whichever organisation is selected to carry out any design it is essential that the principal contractor's Temporary Works Coordinator ensures that any 'grey areas' of responsibility are appropriately managed.

33.6. Simple temporary works solutions

If an opening is to be formed within the main body of a single or double skin of brickwork (which is in good condition), temporary support is required. There are two simple temporary works solutions that are often used

- Proprietary horizontal steel frames (e.g. 'no more props') can be cut into mortar joints above the new opening and then bolted in position. The frame is tensioned and acts as a temporary lintel to support the brickwork until the new permanent lintel has been installed, after which the temporary frame is removed. This type of system does not require props and is normally suitable for openings up to around 2.5 m wide.
- If some arching can be assumed, the wall is relatively thin (relatively light loading from single or double skin of brickwork) and the opening is relatively small, the Strongboy proprietary support system can be used. Strongboy props are steel plates ($c.$ 150 × 350 mm) that fit onto the head of a standard adjustable steel prop (Acrow). The plates can be cut into the mortar joint above the new lintel. They will be positioned at relatively close centres (less than 900 mm apart) as they rely on the brickwork arching between them. They can typically support loads up to 350 kg (as the applied load is eccentric to the supporting prop) and are generally used for openings up to around 3 m wide with heights less than around 3 m. They should be treated as a very low or low risk solution as they are not appropriate for high risk openings.

33.7. Needling schemes

For larger openings where arching can be considered, temporary supports known as 'needles' will be required. Needles are relatively short beams which are positioned through the wall above the new lintel position and supported either side of the wall. The ability of the brickwork to span should be assessed but generally the needles should be at a maximum of 1 m centres as it may be difficult to justify arching in most brickwork or masonry for greater spans. If the wall is in poor condition or the loads to be supported are large the spacing should be reduced. The load on the needles should be calculated using the principles shown in Figure 33.2. The needle beams can be timber, pre-cast concrete beams, proprietary soldier beams or steel beams (depending on the magnitude

Figure 33.2 Typical needling scheme with backpropping. (Courtesy of RKF Consult Ltd)

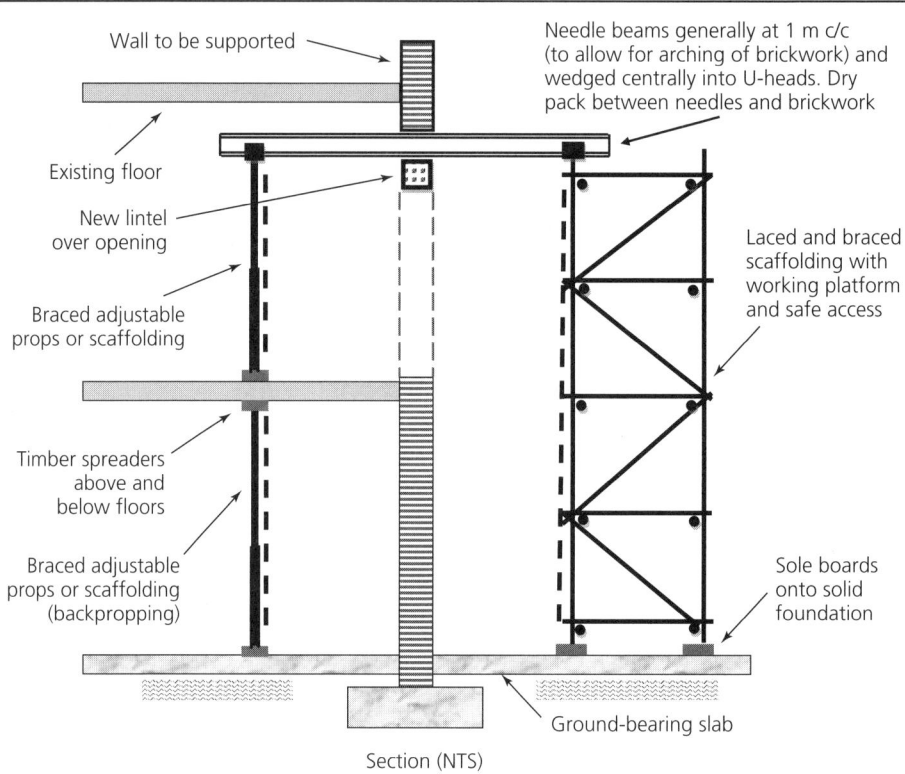

Section (NTS)

of loading) supported on timber props, proprietary adjustable props, scaffolding, proprietary soldiers or steel column sections. If a new opening requires several needles for support then, for ease and to prevent mistakes being made on site, all needles are generally kept the same. Dry pack (dry sand–cement mix) is placed between the needle beam and the wall above to ensure even bearing.

Designs should *not* ask 'Will it deflect or settle?' but rather 'How much deflection or settlement is acceptable?' Needle beams will deflect due to bending, props will elastically shorten under loading and foundations will settle. These movements are cumulative and may result in cracking of the supported brickwork. Reducing spans (or support centres) or providing stiffer supports will reduce deflection and settlement. For larger loads pre-loading or pre-deflecting can be done using the adjustable jacks on props or by installing 'hydraulic flat jacks'. The amount of the pre-load should be carefully assessed and controlled to avoid 'lifting the structure', which can also cause damage. As a general rule, many designers would limit the pre-load to around two-thirds of the applied vertical dead loading of the structure (excluding the live loading).

In general, deflections of around 3 mm on many types of brickwork can be acceptable. However, very 'sound' brickwork may be able to accommodate slightly larger deflections. An assessment should be made and deflection limits agreed prior to commencement of design, and on-site monitoring and remedial actions should be agreed if these deflection limits are exceeded.

Brickwork bearing stresses should be checked (in the permanent and temporary condition). Load spread through pad stones (in the permanent condition) and dry packing between the brickwork and needles (in the temporary condition) can be considered. If the stresses are deemed to be excessive then the spacing of the needles may need to be reduced to reduce the loading, or greater spread may be required using wider needle beams or more packing.

If there are openings above the new openings (windows or doors on the floors above) then the load will be in the brickwork between the openings. Needles should be positioned beneath the loads from above. Positioning needles directly beneath windows or doors is not sensible, as minimal support may be required and this may be achieved by using Strongboy props with the needles beneath the brickwork 'piers'. Heavily loaded beams, and occasionally floors or slabs, which span into the wall (just above the position of the new opening) may need to be supported independently of the wall. Scaffolding or propping can be used to support these vertical loads.

Design rules from BS 449:1969 (BSI, 1969; standard withdrawn) are still often used to design steel needles, with the effective length being determined by the restraint conditions (these rules are also still used in BS 5975:2019 (BSI, 2019)). Often the effective length of the needles is taken as $1.2l + 2D$ (BS 449:1969; clause 26 – also still used in BS 5975:2019), where l is the span of the needle beam from support to support and D is the depth of the needle beam.

Other restraint conditions may be appropriate. A check should also be carried out at points of significant load concentration, as the steel beam may require web stiffeners.

Self-weight (structure, finishes and services) and imposed loading to be supported should be calculated. The construction operations recommended imposed loading is taken as 1.5 kN/m^2 (service class 2) where there are small tools in use, equipment and small quantities of materials. Access-only imposed loading is taken as 0.75 kN/m^2 (service class 1). If there is significant storage of materials or heavy machinery the imposed loading should be assessed accurately. Environmental loading from wind on external walls and rain or snow may also be considered if a roof is to be supported. The new lintel should be positioned at the base of the wall before the props are installed. The designer should then consider how the lintel is to be lifted. If it is too heavy to lift by hand then block and tackle or an electric winch can be secured to the needle to aid lifting. The design of the needles (and their supports) should include an allowance for the self-weight of the lifting equipment and the lintel.

For most needling schemes the support of the vertical loading is the main consideration. However, horizontal loading (e.g. wind) in external walls or shear walls should also be considered and sufficient stability provided. Care should be exercised as occasionally an internal wall may be subjected to wind loading (e.g. during partial demolition of a structure, where an internal wall may become an external one until the structure has been rebuilt). Particular care should be taken when a very significant proportion of a structure is to be removed as there is a possibility of the entire wall being able to move. In this circumstance the needles and supports may need to be designed for this degree of movement (possibly in any direction) or some form of additional temporary support (flying or raking shores) provided. Some degree of continuity of loading can also be considered by the designer.

Often a particular challenge is the installation and removal of any temporary works. Access (how operatives will get to the workplace and work at height) and working space need to be considered. Typically 1000 mm of working space either side of the wall should be sufficient. The amount of working space determines the span of the needles. Inclining the supporting props reduces the span of the needles, but consideration of the horizontal reaction is required. The temporary works are installed prior to main demolition and will remain in place until the new lintel has been installed. The temporary works need to be positioned in such a way as to provide the required amount of support, while allowing the work to be carried out with as little obstruction as possible.

Three examples of needling techniques for forming larger openings or for walls of substantial thickness are described below.

When needling upper floors, backpropping may be necessary to carry the vertical loads down to the ground. If the new lintel is within the depth of the existing floor then it is likely that the needles will have to be placed above the floor and the floor joists will have to be supported independently. A useful practical 'trick' is to place the new lintel inside the line of the props before the props are installed otherwise access can be very difficult. When forming new openings in shear walls any temporary works must allow for the transfer of horizontal loading until the permanent works are capable of doing so.

When backpropping is not possible or practicable 'balanced needles' can be used. The vertical load is carried back into the load-bearing wall by the lower needle beams and hence backpropping is not required (Figure 33.3). Similarly, 'scissor needles' can be used when the work is to be carried out from one side of the wall only (Figure 33.4). With this scheme bricks should be prevented from falling outwards by using netting or anchors and rope inside the structure. A similar concept is to use welded steel brackets formed into a C shape. These brackets can be placed into the wall with the 'legs' of the brackets above and below the new lintel to support the wall. A slot is then formed in the wall to accommodate the new lintel and the new lintel positioned. The brackets can then be removed and the opening below the lintel formed.

To avoid needles and vertical supports the Abbey Pynford method can be used (Figure 33.5). Small holes are formed through the wall (as above) and steelwork 'stools' are installed into the holes and dry packed above and below. The holes are joined together to form a slot and reinforcement is placed around the stools. Simple formwork is then positioned and concrete is poured to form a reinforced concrete beam (concrete

Figure 33.3 Balanced needling arrangement without backpropping (e.g. when floors below are still occupied). (Courtesy of RKF Consult Ltd)

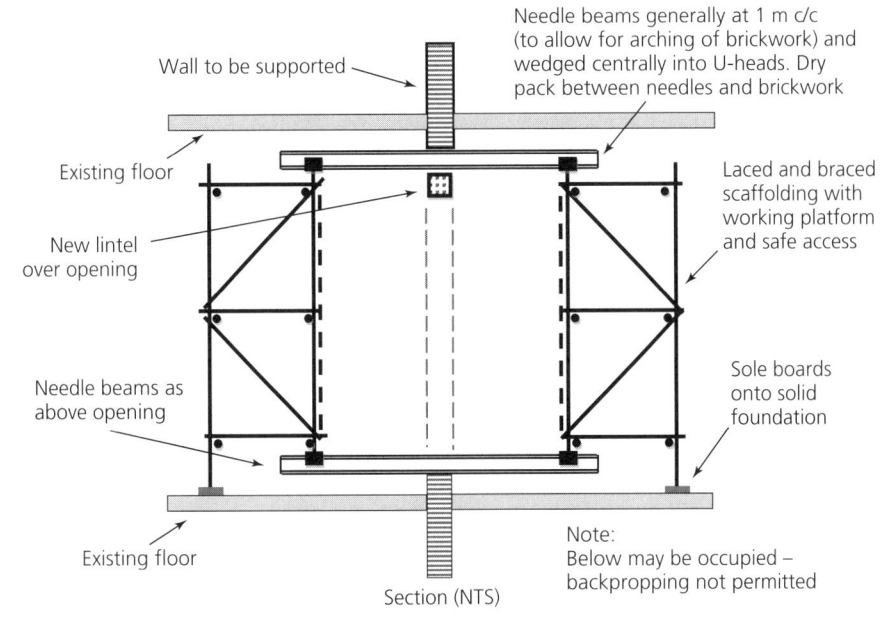

Figure 33.4 Scissor needling arrangement without backpropping (e.g. when working from only one side of the wall and floors below are still occupied). (Courtesy of RKF Consult Ltd)

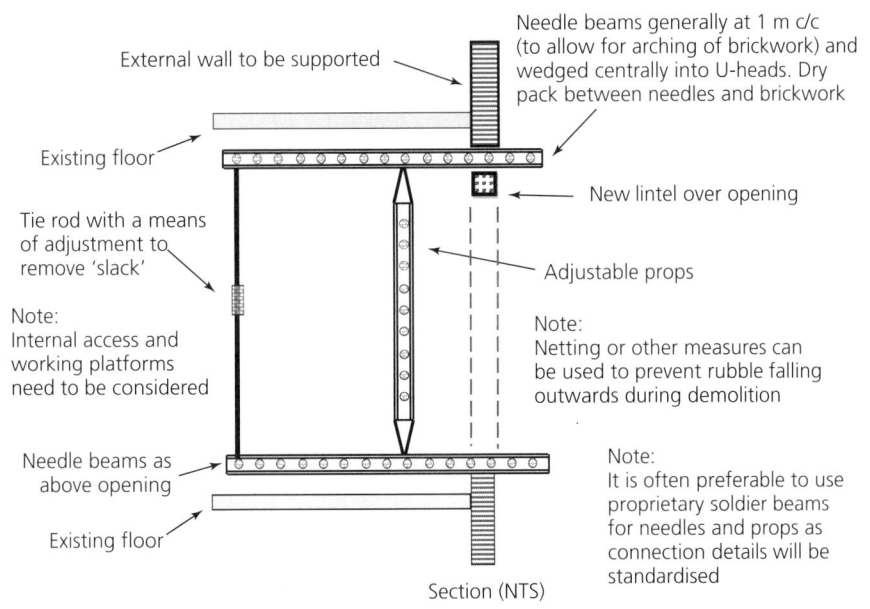

Figure 33.5 Abbey Pynford method for forming an opening in a wall without needles. (Courtesy of RKF Consult Ltd)

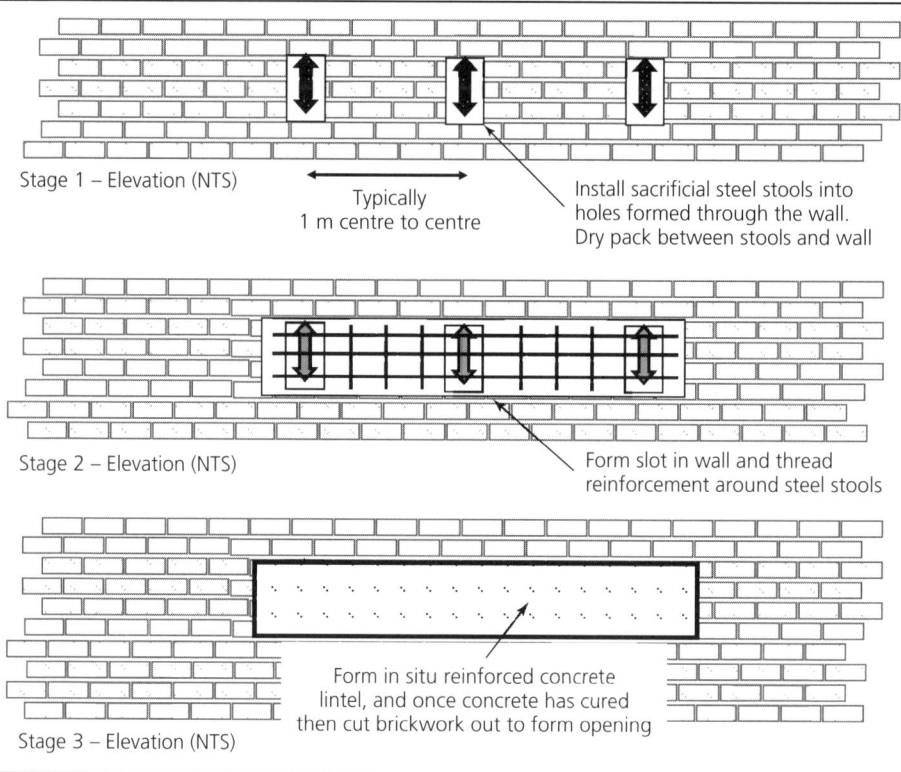

poured to within 50–75 mm of the wall above). The gap is then dry packed, and once the concrete has achieved its required strength the new opening is cut out below this in situ reinforced concrete lintel.

33.8. Propping to needles

Propping beneath needles will generally be vertical but could also be inclined to reduce the span of needles or to avoid obstructions.

Construction tolerances should be considered in the design of the propping

- Eccentricity of the applied load onto props (generally 25 mm), unless concentric loading can be guaranteed.
- Props being installed 'out of plumb' (generally 1.5°).

The props should have some form of adjustment (screw or hydraulic jacks) to allow for safe installation and removal. If the vertical loads are very large it may become difficult to release loads from props and hence remove them. Techniques such as sand boxes (steel boxes filled with dry sand) can be used, whereby the sand is allowed to run out of the box, thereby releasing the load (in a controlled manner) and facilitating removal of the props. Hydraulic jacks can also be used for this purpose. The system should have some mechanism to prevent the pressure being accidentally released.

For temporary works the effects of thermal expansion and contraction are rarely considered (perhaps with the exception of large temporary props to a major excavation). Fatigue is also rarely considered, as the props should be inspected for signs of distress prior to use. Accidental impact loading on temporary works should generally be avoided and on-site measures should be provided to prevent accidental impact. However, it is prudent for the designer to consider the risk of accidental impact loading. Even for extremely large schemes where large machinery is to be used, this load is unlikely to exceed 10 kN. For most schemes this magnitude of impact load would be catastrophic, and a more realistic figure should be determined by risk assessment (the magnitude can be small and a degree of redundancy should be considered to prevent progressive collapse) and control measures provided to avoid impact loading.

Once the magnitude of the load and the effective length of the prop have been determined a suitable structural member can be designed. The prop could be bespoke (timber, steel, aluminium, etc.) or proprietary propping equipment or scaffolding. In addition to simply supporting the vertical load, the designer and site team should also consider the following

- The sequencing of the works, including a means of adjustment so the prop can be installed and removed. The removal sequence should be identified (when and how the temporary works will be removed, see Section 33.9).
- For punching shear, packing and spreaders may be considered, but if timber is used then crushing should be considered. Also any existing connections should be checked.
- If props are founded on a suspended floor (timber joists and floorboards) or a reinforced concrete slab an assessment should be made of the ability of the floor or slab to support the point loads. Often timber spreaders (scaffold boards or sleepers) or steel spreader beams can be used. For floors the timber spreader should allow the point load to span between joists. Backpropping with spreaders below the props may be required to carry the loads down through the structure to an adequate foundation. For slabs a check should also be carried out on punching, and spreaders or backpropping provided as necessary.
- There is likely to be a localised increase in foundation loading (either side of the new opening). If necessary, the existing foundations may need to be increased by localised underpinning or similar techniques. A design is required for the foundation beneath the props to prevent excessive settlement; the existing foundations may be utilised or new temporary foundations may be required. Mass concrete foundations, steel spreader beams or timber spreaders can be used. For extremely large loads, bearing piles may be necessary. New foundations should not undermine existing foundations. When the prop is positioned above a suspended slab or floor then backpropping may be required.
- It should be determined whether crash decks (or netting) will be required to prevent items falling during demolition and also how will rubble be removed.

Excessive demolition rubble should not be allowed to accumulate on existing floors or slabs as they may become overloaded unless backpropping is provided.
- Screens around the working area may be required to control noise and dust during demolition.
- Elastic shortening of the prop and foundation settlement under loading will lead to movement in the item being supported, and this movement may be significant enough to cause damage. The degree of acceptable movement should be agreed by the parties involved. Deflection can be minimised by using stiffer supports and pre-loading by using the mechanical advantage of threads or jacking systems. As a general rule the author uses a maximum pre-load of around two-thirds of the applied vertical dead load.
- Sufficient working space and safe access for any work at height should be provided for handling the prop (if the use of lifting equipment is not possible then props should be light enough to be 'manhandled').
- The overall stability of a structure may become an issue if shear walls are wholly or partially removed. Propping systems should be braced to provide the required stability.
- If it is not possible to place props directly beneath needles (perhaps to avoid services at ground level) then a header beam can be used to bridge between the props.
- The designer should ensure any temporary works that are provided are robust and be designed so that progressive collapse is prevented.

33.9. Sequence of removal of needles

The sequence and order of removing the needles and transferring the load into the new lintel is important, and if this is not carried out correctly then damage can be caused to the permanent structure. Generally the removal of temporary works should follow the load path of the permanent works, which means for a simply supported lintel the removal of the supports should start mid-span and work should proceed progressively towards the lintel supports. In this way the new lintel takes up its designed deflected shape. If the removal process is started from the ends and progresses towards the centre, the central supports could become overloaded and could create tension cracks in the structure above as the brickwork hogs over the remaining central supports. A safe means of releasing the load from the temporary props may need to be determined, especially when the applied vertical loads are large. Complex systems may also require a load-transfer system to safely transfer loading from the temporary works onto the permanent works. The sequence should be agreed between the Permanent Works Designer and the site Temporary Works Coordinator, and needs to be communicated to the site team to avoid mistakes.

33.10. On-site checklist

- Has a thorough site survey and assessment been made of the actual conditions and loads to be supported? Repairs may be required prior to the main work commencing.
- If site conditions change compared to those assumed by the designer – has the designer been informed?

- Set up exclusion zones for safe working and protection measures to prevent significant accidental loading.
- Check for existing services, and avoid, protect or reposition them.
- Have a risk assessment and method statement for the installation and removal of the temporary works been provided?
- Is the correct equipment being used and has the quality of the equipment been inspected? (Temporary works equipment will be used multiple times and can be subject to abuse by the users.)
- Have the needles been set out and positioned correctly to suit the loading?
- Are the correct props being used and have they been installed to acceptable construction tolerances?
- Has the new lintel been delivered and positioned ready for lifting?
- Have working space, access, preventing falls from height and so on been considered?
- How will the holes be formed in the wall and how will the needles be positioned?
- How will the opening be formed and how will the demolition rubble be removed?
- Are measures in place to prevent accidental impact and damage to the temporary works?
- How is overall stability of structure and temporary works achieved?
- How will deflection and settlement be monitored, and is a contingency plan in place in case of problems?
- If an in situ reinforced concrete beam is provided has it achieved the required strength before the temporary works are removed?
- If steel, pre-cast or timber lintels are provided has the gap between the lintel and the wall above been packed before the temporary works are removed?

REFERENCES

BSI (British Standards Institution) (1969) BS 449:1989-2:1969. Specification for the use of structural steel in building. Metric units. BSI, London, UK (withdrawn).

BSI (1981) BS 5977-1:1981. Lintels. Method for assessment of load. BSI, London, UK.

BSI (2011) BS 6187:2011. Code of practice for full and partial demolition. BSI, London, UK.

BSI (2019) BS 5975:2019. Code of practice for temporary works procedures and the permissible stress design of falsework. BSI, London, UK.

Gilbertson A (2017) *Structural Stability of Buildings during Refurbishment*. CIRIA, London, UK, C740.

HSE (2015) *Construction (Design and Management) Regulations*. HSE, London, UK.

FURTHER READING

BRE (Building Research Establishment) (1991) *Repairing Brick and Block Masonry*. BRE, London, UK, Digest 359.

BRE (1991) *Why do Buildings Crack?* BRE, London, UK, Digest 361.

BRE (1992) *Repair and Replacing Lintels*. BRE, London, UK, Good Building Guide GG1.

BRE (1992) *Assessing Loads above Openings in External Walls*. BRE, London, UK, Good Building Guide GG10.

BRE (1992) *Providing Temporary Support during Openings in External Walls*. BRE, London, UK, Good Building Guide GG15.

BRE (1995) *Assessment of Damage to Low-rise Buildings*. BRE, London, UK, Digest 251.

BRE (1999) *Removing Internal Load Bearing Walls in Older Buildings*. BRE, London, UK, Good Building Guide GG20.

BRE (1999) *Supporting Temporary Openings*. BRE, London, UK, Good Building Guide GG25.

BSI (British Standards Institution) (1999) BS EN 1065:1999. Adjustable telescopic steel props. Product specifications, design and assessment by calculation and tests. BSI, London, UK.

BSI (2015) BS 12999:2015. Damage management. Code of practice for the organization and management of the stabilization, mitigation and restoration of properties, contents, facilities and assets following incident damage. BSI, London, UK.

CIRIA (Construction Industry Research and Information Association) (1994) R111 *Structural Renovation of Traditional Buildings*. CIRIA, London, UK.

HSE (Health and Safety Executive) (1990) *Evaluation and Inspection of Buildings and Structures*. HSE, London, UK.

HSE (2004) *Health and Safety in Refurbishment involving Demolition and Structural Instability*. HSE, London, UK, RR204.

HSE (2006) *Avoiding Structural Collapse during Refurbishment*. HSE, London, UK, RR463.

NASC (National Access & Scaffolding Confederation) (2013) *Good Practice Guidance for Tube & Fitting Scaffolding*. NASC, London, UK, tG20:13.

Useful web addresses

Building Research Establishment (BRE): https://bregroup.com (accessed 01/08/2018).

Construction Industry Research and Information Association (CIRIA): https://www.ciria.org (accessed 01/08/2018).

Temporary Works Forum: http://www.twforum.org.uk (accessed 01/08/2018).

Temporary Works, Second edition

Pallett, Peter F and Filip, Ray
ISBN 978-0-7277-6338-9
https://doi.org/10.1680/twse.63389.493
ICE Publishing: All rights reserved

Chapter 34
Temporary works in demolition

Angus Holdsworth
Managing Director, Andun Ltd

The demolition industry covers a wide spectrum of structures. The level of engineering input required can vary drastically, for what would be a simple structure to demolish in a greenfield site can often require significant engineering input in a city centre environment.

The temporary works associated with demolition works are interesting, varied and often unique. The design of these temporary works must follow established engineering principles. It should be noted, however, that unlike to temporary works for construction there are circumstances where serviceability limits can be exceeded; generally demolition is only concerned with ultimate strength and maintaining stability, while ensuring an adequate factor of safety is maintained.

34.1. Introduction
To facilitate the more complex projects, recent advances have seen demolition contractors developing highly specialised plant and techniques to assist in these challenging projects, alongside often complex temporary works designs. An example is the remotely operated excavator known as the Megamuncher (Figure 34.1), developed by the demolition contractor with the plant manufacturer JCB.

Staged demolition often involves extensive temporary works to ensure stability of other parts of the structure. Centre Point in London is an example of complex demolition programming and the use of temporary works in redevelopment (Figure 34.2).

34.2. Understand the structure
Regardless of the demolition works being undertaken it is absolutely vital that the engineer understands the structure to be demolished. A site visit should be undertaken following the soft strip to allow the engineer to review the structure. If the original drawings are available they should be reviewed and used as a starting point for analysis. It is prudent to undertake opening up works to ensure that the drawings match reality. It must be borne in mind that structures are often modified during their lifetime and records of such may not exist. It is also not unusual to come across latent defects in what may generally be a well-built structure. Past experience of this includes

Figure 34.1 Remotely operated excavator. (Courtesy of Coleman & Co.)

Figure 34.2 Complex demolition works as part of Centre Point redevelopment. (Courtesy of John F Hunt)

- Slabs that were shown on drawings to have starter bars from the walls cast in turned out to have nothing at all on two edges and very minimal shear reinforcement to the other two edges.
- Large box-outs formed in the compression face of large concrete beams, meaning that beams provided only a quarter of the capacity identified by referencing the drawings.
- Prefabricated structures where loops and dowels were missing or misplaced.

An example of a reinforcement loop that had missed the pin during the construction of a prefabricated structure is shown in Figure 34.3(a). The fault had been in this condition since constructed 50 years previously, and would have caused a significant issue during demolition if it had not been identified, see Figure 34.3(b), by the demolition contractor.

Figure 34.3 Construction error of a loop missing the connecting pin. (Courtesy of Andun Ltd)

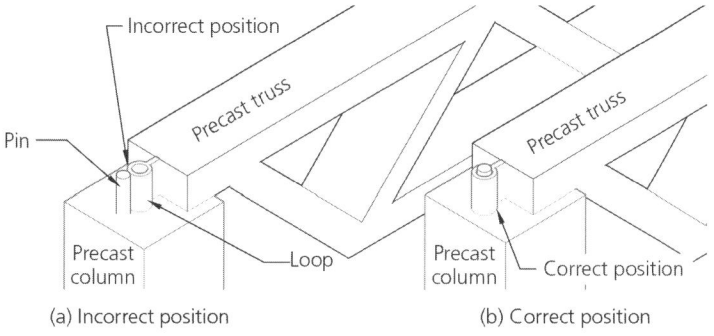

(a) Incorrect position   (b) Correct position

(c) Side view of missing connection as exposed

34.3. Demolition
34.3.1 General
To understand the temporary works associated with demolition it is important to have some understanding of demolition. Demolition covers a wide spectrum, from simple domestic structures to the total demolition of large and complex structures such as a power station. It is often viewed as the poor cousin to construction, but an experienced demolition contractor will often know more about the ultimate strength of materials and building collapse mechanisms than many structural engineers, for during demolition the true strength of a structure and its components are revealed. Demolition should never be started until the demolition contractor knows how the building/structure works in the first place (see Section 34.2).

Wherever temporary stability of the structure and/or temporary works is required in demolition, the safety issues and procedural controls for managing the temporary works will apply (see Chapters 1 and 2). Both BS 6187:2011 (BSI, 2011) and BS 5975:2019 (BSI, 2019) have guidance on controlling temporary works in demolition.

All demolition work can be considered as falling under two broad categories: progressive or deliberate collapse.

34.3.2 Progressive demolition
Progressive demolition involves the controlled removal of portions of the structure while ensuring the remaining structure remains stable. In general terms, the demolition is carried out working from the top to the bottom of the structure, usually working towards the structural core to ensure that stability is maintained. Partial demolition of a structure will follow similar constraints. However, the works will not necessarily be from top to bottom. Progressive demolition can be further subdivided into three broad methods based on the plant used for the works. It is not uncommon for a combination of all three methods to be used on a single project

- Top down: as it sounds, start at the top and work progressively down the building.
- High reach: a specialist excavator that can reach great heights is used to progressively demolish the structure.
- Deconstruction and dismantling: the reverse of construction. Hot cutting or cold cutting is used to release sections.

34.3.3 Deliberate collapse mechanism
Typically, this is referred to as a 'blowdown' or a 'pull down'. The unplanned collapse of Didcot power station in February 2016 during pre-weakening, and prior to a blowdown, clearly highlights the hazards associated with these works. At the time of writing (2018) the final reports are not available and it is hoped that the lessons learned will be shared with the industry.

Planning the deliberate collapse of a structure requires an engineer to have a very good understanding of the structure and experience of how structures behave in extreme

Figure 34.4 Example of pre-weakening design. (Courtesy of Andun Ltd)

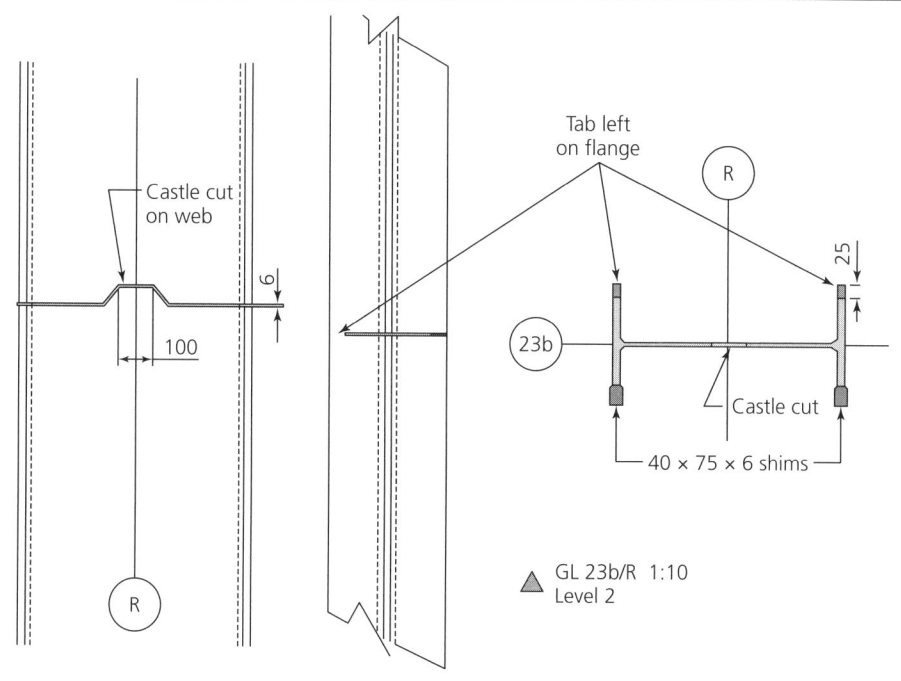

conditions. Although arguably not temporary works, these works are highly specialist in nature and could form a chapter in themselves. The arrangement for steel and concrete pre-weakening are different.

34.3.4 Deliberate collapse of steel structures

To pre-weaken a steel structure for demolition, sit cuts and hinge cuts are formed in the column and beam sections. This can include cutting out bracing and secondary beams. This will affect the structure's stability and must be sequenced and pre-planned. The final cuts should be made only when everything is ready. An example of pre-weakening design for cuts to an I-beam column is shown at Figure 34.4. Explosive or wire rope pulling is then used to trigger the structure to collapse, generally by knocking out columns. Gravity then does the rest. These works need careful planning to minimise the risks of working in a weakened structure. However, the use of cutting charges (shaped charges) is becoming more commonplace, reducing the risk of undertaking the final cuts.

34.3.5 Deliberate collapse of concrete structures

To prepare concrete structures key structural elements are weakened. Explosives are typically placed in holes pre-drilled into the key elements that are to be fragmented. Once detonated the solid explosive converts instantly to a gas of much higher volume, which effectively blows the concrete off the reinforcement, and gravity does the rest.

The temporary works engineer must demonstrate by calculation that the pre-weakened concrete structure will remain stable until the moment the charges are detonated to initiate the planned collapse.

34.3.6 Key points for pre-weakening

When planning the pre-weakening of a structure the following should be taken into consideration

- If shaped charges can be used, then use them. They significantly reduce the hazards associated with the works, although the risk of damaging adjacent buildings with the air overpressure needs to be given careful consideration.
- The simplest collapse mechanism possible should be used to ensure the lowest risk. As is often the case with any design, *keep it simple*.
- Design should utilise the minimum of pre-weakening necessary. Positioning of the cuts should be considered carefully.
- The works should be planned by a suitably experienced engineer (i.e. an engineer who has a track record of undertaking similar works).
- Calculations should be sufficient to justify the design.
- The age of the structure must be recognised and appropriate design factors and material strengths incorporated.
- Clear drawings should be provided detailing all cuts and locations, and workmanship tolerances.
- Stability must be maintained. Release cuts should be made only after all preparations have been completed.
- Accurate setting out of the cuts is critical and must be given particular attention.
- Workmanship is critical when pre-weakening a structure, and only suitably qualified and competent burners should be permitted to undertake this work. There should be a clear specification and method statement.
- Exclusion zone(s) should be planned and enforced based on the worst-case collapse radius. See BS 6187:2011 (BSI, 2011) for detailed guidance on exclusion zones. See also the National Federation of Demolition Contractors (NFDC) guidance on the subject (NFDC, 2014a).

34.4. Temporary works for demolition

Some of the temporary works used in demolition are covered in more depth elsewhere in this book: façade retention (Chapter 26), needling and forming openings (Chapter 33), access and proprietary scaffolds (Chapter 21), site compounds and hoardings (Chapter 3), and management (Chapter 2). Digital aids for communication and inspections are often necessary in demolition and their use is increasing (for more information see Chapter 31). It is not proposed to add anything further here other than a brief discussion on key points for scaffolding and hoarding used during demolition. The NFDC also provides specific guidance on the subject (NFDC, 2014b).

34.4.1 Site perimeter

Given the potential for high hazard works being undertaken on a demolition site the first temporary works to be considered is a secure perimeter. Demolition is often the first operation on a site prior to commencing construction, so the hoarding/fencing to the site perimeter may have to remain in place for an extended period of time. Therefore, consideration of robustness and detailing for longevity should be considered (see Chapter 3). The Temporary Works Forum guidance (TWf, 2012) is an invaluable resource.

34.4.2 Scaffold for demolition

It is common practice to erect scaffold full height to the external façade around structures when they are being deconstructed, particularly in city-centre locations. This scaffold provides a number of functions

- It helps with dust control.
- It provides safe access to external façades when required.
- It helps to prevent material falling to ground level during the demolition works.

Generally the scaffold will need to be stripped in tandem with the demolition works (i.e. following the demolition of a storey of the structure the scaffold will be left projecting one lift above the structure). This projecting scaffold should then be stripped prior to demolishing the next floor. The scaffold designer should be made aware of the intended use in the design brief so that the calculations should reflect the potential for a lift to project without any ties, and effectively cantilever vertically from a tied level; ledger bracing at all standards should be considered. The 'standard solutions' for scaffolds with ties at every lift or at alternate lifts will rarely be acceptable for demolition scaffolds. The risk of unplanned local additional rubble or debris onto the scaffold platforms during demolition work should be taken into account – the traditional service class loads (see Section 21.5.3.1) may not be sufficient.

When forming ties around columns or similar, ensure that tubes do not project further than required. There have been cases where rubble has impacted on projecting ties and has led to partial collapses. An example is shown diagrammatically in Figure 34.5.

The position of scaffold ties and the loads imparted into the partially demolished structure should be considered carefully. When considering the removal of smaller elements, confirm that the scaffold design and the demolition sequence have been coordinated; if the scaffold relies on a structural element for stability then ensure that the element can provide sufficient support. The NFDC provides specific guidance on the subject (NFDC, 2014b).

34.4.3 Working platforms for high-reach machines

High-reach machines start life as excavators. They are then modified with the addition of a counterweight, an increased track base and a new boom that allows the machines to work at high level. They typically start at around 50 te in weight with the ability to reach around 23 m. The largest high-reach machine currently in the UK (Figure 34.6) weighs 225 te and has the ability to reach 70 m into the air; it uses a 2.5 te attachment on the end of the dipper arm to demolish a structure. It is therefore vital that proper consideration is given to the working platform for these machines.

Figure 34.5 Dangers of projecting ties. (Courtesy of Andun Ltd)

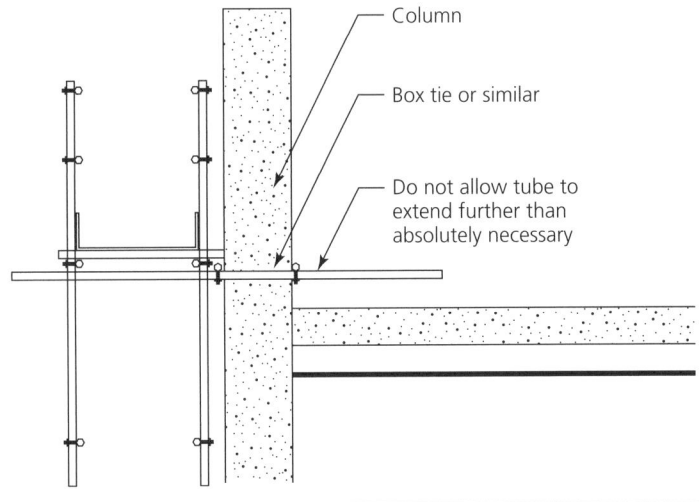

Figure 34.6 High-reach (70 m) machine. (Courtesy of DSM)

The geometry of these machines means that they require a significant plan area in which to operate. The demolition sequence should be planned to ensure the machine can stand off the building sufficiently to prevent falling rubble impacting the machine.

It is of note that the designer is unlikely to receive a set of track pressures, as is common when designing piling platforms. Typically the engineer will have to calculate the track pressures themselves based on the weight and the centre of gravity of the machine components. The basic principles laid out in the Building Research Establishment (BRE) publication BR470 (BRE, 2004) can be followed when designing the platforms. Furthermore, unlike 'piling platforms' there is no standard certification of the platform, but the temporary works procedures adopted by the responsible demolition contractor should have a satisfactory checking system to control the suitability of the platform.

The working platform is typically constructed from demolition arisings (the material already available from the demolition operation); it is very unlikely that a demolition contractor will import stone to construct a platform. The platform may be constructed at ground level as is typical for a piling platform; however, it is more likely that the platform will be used to raise the machines up to reach the upper storeys of the structure being demolished. This can be up to several storeys in height. Loose and compacted pulverised brick and concrete have a very high angle of internal friction and there is often no problem maintaining 45° slopes using this material. It is important that the designer considers the risks of any voids under the platform and/or any surcharging on hidden basement walls.

For further information on the use of these machines see the NFDC guidance on the subject (NFDC, 2012).

If the high-reach machines are to operate on any suspended slabs within the structure then the slabs must be checked and/or propped as required (see following section).

34.4.4 Propping for demolition plant

If access to the structure is limited by adjoining properties, roads or railways then top-down demolition work is likely to be undertaken. The plant will be lifted onto the roof and the structure will then be demolished floor by floor. Generally the core is the last section on each floor to be demolished, thereby ensuring stability. Excavators can be up to 20 te in weight, although they are typically less than 5 te in weight. Sometimes remotely operated excavators, such as the Brokk excavator shown in Figure 34.7, will be used. The resulting rubble is generally cleared away by small skid steer loaders, either wheeled or tracked.

There are two distinct approaches to undertake these works

- The excavator sits on the floor and breaks it out while maintaining a suitable stand-off. The rubble falls to the level below and is then cleared.
- The excavator sits one floor below and breaks the floor above. The rubble falls to the floor the excavator sits on. The rubble must then be cleared regularly.

Figure 34.7 Brokk excavator in operation. (Courtesy of John F. Hunt)

An initial engineering assessment will be undertaken which will identify if the structure needs propping. One, two, three or even all floors may need to be propped. Typically lines of laced and braced steel adjustable props at regular centres (500–1000 mm) are used at the one-third or one-quarter points on the span. A typical arrangement is shown in Figure 34.8. Scaffold boards or similar are used at the head and base of the props. It is good practice to ensure the props are adequately braced or spiked to the slabs to prevent them falling over unexpectedly during movement or vibrations from the ongoing demolition works. Steel props are used as they are inexpensive, less susceptible to damage than aluminium props, easy to erect and offer a good capacity for typical top-down demolition works (see also Chapter 28 on backpropping). Larger machines will typically be used where the structure is assessed as being capable of supporting them without propping.

It is of vital importance that the correct load cases are considered when assessing the structure. Small excavators working hard in a demolition environment can typically end up on tip toes (i.e. the full weight of the machine concentrated over a short length of the tracks). It is also important to consider the skid steer in operation. These machines can impart high wheel loads and are often the driving load case for the assessment, particularly when scooping up rubble. The weight of the rubble should also be taken into account when assessing the structure.

34.4.5 Temporary vertical propping

When carrying out the partial demolition of a structure or working up to a cut line there is often the requirement for temporary vertical propping. The propping should ideally

Figure 34.8 Typical backpropping arrangement. (Courtesy of Andun Ltd)

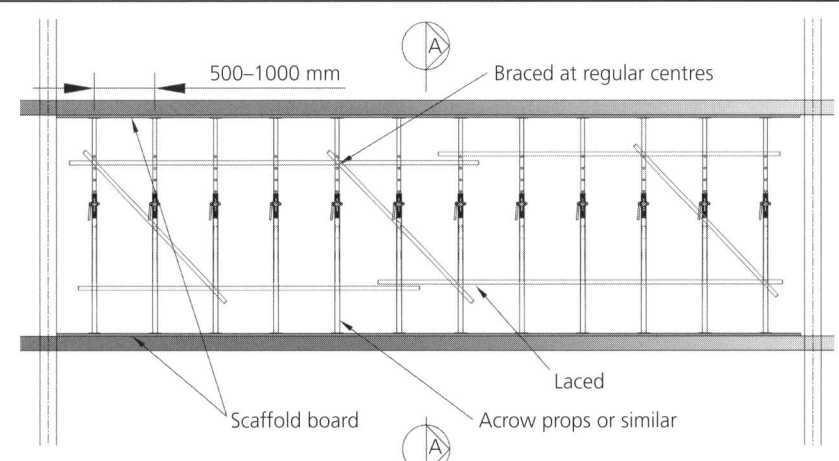

(a) Side elevation of propping to floor

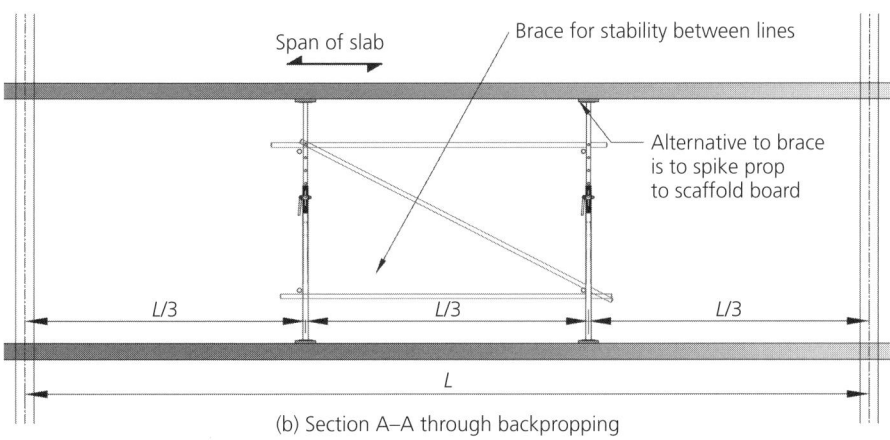

(b) Section A–A through backpropping

replicate as far as possible the original load path. Ensure that the propping is suitable for dead and any envisaged live loads.

It is always preferable to demolish towards a cut line. If a separation cut is necessary the temporary prop arrangement must be designed for the horizontal sway forces and machine breaking loads that will be generated.

34.5. Load testing of slabs

In recent years, load testing of slabs within structures has become established, often motivated by the contractor wishing to use larger plant on the structure. While this form of testing has its place, it can equally lead the contractor into a false sense of security. Where previously a smaller machine may have been used, a heavier machine may be used now; and should the structure contain a latent defect (or have been adapted since first constructed) it is more likely to lead to serious consequences.

If load testing is to be used it should ideally be undertaken only once the capacity of the structure has been calculated by traditional means. Failure to do so can lead to overloading the structure. These load tests should be planned by an engineer, based on the identified likely failure mechanism of the structure; typically, shear at the support in short span structures and failure in bending in long spans. Shear failure is the more dangerous failure mode because it can be sudden with no warning signs. The results should be interpreted by an engineer to ensure that a sufficient factor of safety is in place for the intended plant loadings. Choosing the location of a load test is often difficult, as many structures have been adapted over the years and the changes may make determining representative locations challenging.

34.6. Moving plant between floors

Where cranes or hoists are available the simplest method to move plant between floors is to simply open a hole in the slab and lift the plant down to the next level. However, the alternatives to be considered include the following

- Where the slabs have sufficient capacity, rubble may be placed to form a ramp between the levels.
- In certain circumstances a hinge may be formed at the slab–column interface within one bay of the structure, thereby hinging a complete floor panel down. The plant can then be tracked down to the floor below.
- Ramps formed from steelwork or aluminium are placed between floors. The plant then tracks down.

Whichever option is used it is important that a proper assessment of the structure is undertaken, particularly when it is intended to form ramps from rubble, as the loadings can be significant.

34.7. Structural stability during demolition
34.7.1 General

When demolishing buildings, structural stability throughout the sequence of demolition is of paramount importance. Ideally the sequence should be planned to ensure that any stiff part of the structure, generally the core, is the final element of demolition, and often 'demolition sequence' drawings are produced to supplement the method statement. This may not always be possible; therefore, a design to ensure temporary stability may be required. This applies equally to portions of the structure remaining in the case of partial demolition.

Temporary stability can take the form of push–pull props, fabricated steelwork to form tension–compression bracing, or the installation of DYWIDAG ties (or similar proprietary tension ties) or cross-braces. In certain circumstances it may be that the structure needs only to be restrained in one direction, such as preventing movement towards the site boundary. If this is the case the temporary restraint may safely be designed as a tension-only member. Suitably sized ratchet straps can be used to good effect for temporary local stability.

The lateral restraint system should not only be designed for the known lateral loads and/or notional loads but should also be sufficiently robust to survive impact from rubble; alternatively, there should be sufficient redundancy in the system to allow for the local replacement of bracing if it is damaged. It is important to consider how to carry out the demolition around any temporary bracing without disturbing its effectiveness.

If a cut line between structures is required it may be tempting to separate the structures first with a saw cut. However, this can actually make the situation worse, as it becomes difficult to ensure the stability of the section as it is demolished. It is often better to pulverise and munch up to an offset cut line, then tidy up with a final cut, if required.

34.7.2 Stability of prefabricated and large panel structures

Large panel structure (LPS) residential tower block systems were developed in the 1960s as a quick build solution for housing. They are essentially designed as a gravity structure, where the stability of the walls is provided by the connection to the slabs, which in turn are connected to a stable staircase or lift core. Typically they were erected storey by storey, with pre-cast concrete slabs sitting on top of storey-high pre-cast wall panels. There are many types, some of which are more prevalent to defects, which can pose an increasing risk during demolition. An example of a hidden unknown latent defect was discussed above (see Section 34.2 and in Figure 34.3). Good procedural control and management by a skilled, knowledgeable and experienced demolition contractor will reduce the risk posed by these potential hidden problems.

The shortcomings of the design were realised following the partial collapse of Ronan Point in London in 1968. A minor gas explosion on the 18th storey led to the collapse of 20 storeys of one corner of the structure. Following this collapse many LPS structures were strengthened, however not all. The workmanship when constructing and strengthening these structures is known to have been poor in many instances; typically the floor–wall connections are investigated by breaking out a 600 mm opening. For further information relating to LPS structures see *The Structural Adequacy and Durability of Large Panel System Dwellings, Parts 1 and 2* (Currie *et al.*, 1987a,b).

It is clear from the above that LPS structures pose a significant challenge when being demolished. If an LPS structure is to be dismantled then temporary propping will be required to maintain stability. If LPS structures (or any high-rise structure) are demolished using a high-reach excavator there is a potential to create dominant openings, thereby significantly changing the wind loading to the structure and increasing the chances of panels of the structure being dislodged. Temporary restraints to prevent this should be considered.

34.8. Basement stability and shoring during demolition

It is often the case that the structure being demolished is to be replaced with a structure of a similar footprint but with a deeper basement, often for car parking or underground

Figure 34.9 Use of demolition material for basement construction. (Courtesy of Bowmer & Kirkland)

(a) Stages one, two and three – berm to support basement walls after demolition

(b) Stage four – temporary propping to cast base slab and support side walls

fitness centres or pools. This can require complex temporary works and careful phasing of the works (see Chapter 30 on basement construction).

Typically the original basement walls were designed as propped by the action of the floors, and therefore as the floors are removed there is the potential for instability of the walls. To make the installation of temporary propping simple it is worth considering temporary berms to the retaining walls. The temporary berms can be created from compacted demolition rubble, installed in a controlled sequence to prevent instability of the walls (i.e. break out a bay, compact the berm, then repeat). Good material allows steep batter angles to be used. The addition of a geogrid can increase the safe angle of the berm to stabilise the walls further.

An example of the use of demolition material to construct a basement in stages is shown in Figure 34.9. The stages of construction are

(i) Demolish the superstructure, but leave a temporary berm to stabilise the walls.
(ii) Cast the central section of basement slab (see Figure 34.9(a)).
(iii) Erect inclined propping above the berm to support the basement walls.
(iv) Remove the berm under the inclined props and cast the remainder of the base slab – see Figure 34.9(b).
(v) The temporary works for the ground-floor slab can be installed and the ground-floor slab cast. The slab is cast up to the walls and is the permanent new support.
(vi) Once the ground-floor slab has sufficient strength, and approval has been received, the temporary propping can be carefully removed in sequence.

The props can be founded on temporary foundations or, ideally, directly onto the new permanent works as stage 2. When designing the propping it is important to consider how the original structure worked (i.e. did the basement wall span vertically from floor to floor, or horizontally from pier to pier). The propping should ideally maintain this load path. The anchoring of props into the original basement walls requires careful consideration if raking props are used; the vertical reaction from the props can require significant fixings to restrain adequately.

It may be feasible to use flying shores to brace between the opposing walls in conjunction with braces between the corners. Alternatively, plunge columns with struts, or even king post wall solutions, can be used to ensure stability. The correct solution depends on the existing structure and the proposed permanent works, particularly the position of the core.

34.9. Column removal/structural openings

Openings in masonry walls will typically be formed by needling and propping (see Chapter 33). It may be possible to remove columns without any additional temporary works. However a very good understanding of the structure is required. Typically, goal-post type arrangements of vertical props with header beams can be used.

Figure 34.10 Typical hanging column. (Courtesy of Andun Ltd)

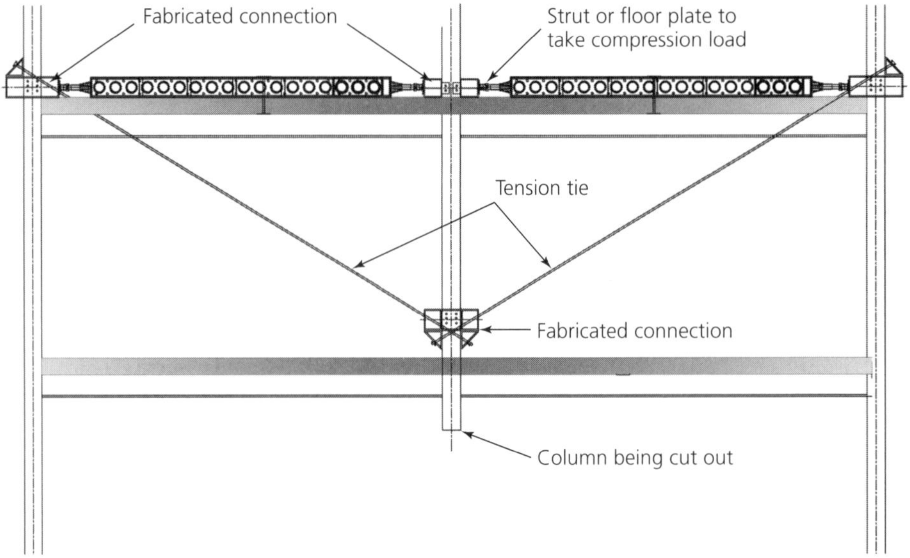

A solution often adopted in demolition, and in the right structure, is where a truss can be formed in the storeys above to allow the column below to be removed (Figure 34.10). The floor plate (or temporary struts at slab level) becomes the compression chord while tension members raking from the top of the column being removed to the floor plate above are used to transfer the column loads. Suitable exclusion zones, or permissible live loads to the floors, should be identified.

34.10. Conclusion

Temporary works for demolition pose an interesting and technical challenge for engineers. The ability to understand an existing structure, often with limited information, and develop a scheme for demolishing the building in a safe and controlled manner provides a challenge unlike any other.

This chapter has emphasised the need for engineering judgement, good communications, careful records and knowledge of similar structures, so that the use of temporary works in demolition can lead to safe methods of work in the unique environment of deconstruction.

Demolition is rarely taught as a subject and it relies on specialists, so in conclusion the golden rule is simple: 'Don't assume, always seek advice'.

REFERENCES

BRE (Building Research Establishment) (2004) *Working Platforms for Tracked Plant*. BRE, London, UK, BR470.

BSI (British Standards Institution) (2011) BS 6187:2011. Code of practice for full and partial demolition. BSI, London, UK.

BSI (2019) BS 5975:2019. Code of practice for temporary works procedures and the permissible stress design of falsework. BSI, London, UK.

Currie RJ, Armer CST and Moore JFA (1987a) *The Structural Adequacy and Durability of Large Panel System Dwellings: Part 1. Investigations of Construction*. BRE, London, UK.

Currie RJ, Armer CST and Moore JFA (1987b) *The Structural Adequacy and Durability of Large Panel System Dwellings: Part 2. Guidance on Appraisal*. BRE, London, UK.

NFDC (National Federation of Demolition Contractors) (2012) *High Reach Demolition Rig Guidance Notes*. NFDC, Hemel Hempstead, UK.

NFDC (2014a) *Demolition Exclusion Zones*. NFDC, Hemel Hempstead, UK, DRG 110:2014.

NFDC (2014b) *Demolition Scaffolding*. NFDC, Hemel Hempstead, UK, DRG 102:2014.

TWf (Temporary Works Forum) (2012) *Hoardings a Guide to Good Practice*. TWf, London, UK, 2012:01.

FURTHER READING

Clarke R (2010) Role of the structural engineer in demolition. *The Structural Engineer* **88(11)**.

Useful web addresses

Concrete Society: http://www.concrete.org.uk (accessed 01/08/2018).

CONSTRUCT (Concrete Structures Group): http://www.construct.org.uk (accessed 01/08/2018).

Institute of Demolition Engineers (IDE): http://www.ide.org.uk (accessed 01/08/2018).

National Federation of Demolition Contractors (NFDC): http://www.demolition-nfdc.com (accessed 01/08/2018).

Temporary Works Forum: http://www.twforum.org.uk (accessed 01/08/2018).

Index

Page numbers in italics refer to illustrations. Page numbers followed by a t refer to tables.

A23 Coulsdon Town Improvement Scheme, 397
Abbey Pynford method, 486, *488*
Access and Egress from Scaffolds via Ladders and Stair Towers etc. (2014a), 297
Access and Egress from Scaffolds via Ladders and Stair Towers etc. (NASC, 2014a), 297
access mats, 83
accommodations, on site, 51, *52–53*, 54
alternatives
  to shafts, 208
  to sheet piling, 141
  to temporary façade retention, 370
  to trenches, 159
American Society of Mechanical Engineers (ASME), 277
anchors, 220, 441
ArcelorMittal
  Piling Handbook (2016), 142, *143*, 148, 152, 223, 439
artificial ground freezing (AGF)
  in general, 117–118
  advantages/disadvantages, 120
  systems for
    brine, 118–119, *119*
    design of, 120–123
    liquid nitrogen, 118–119, *119*
    monitoring of, 125
    secondary effects of, 123–125
    for shaft construction, *120*
    for tunnels, 122–123, *123*, *124*
The Assessment of Highway Bridges and Structures (HA, 2001), 236, 257
augmented reality, 453, *455*
Avoiding Danger from Overhead Power Lines (HSE, 2013), 277

backpropping

in general, 403–404
calculations for
  in general, 416–417
  slabs, 412–413, *414*, 415
  slabs, with one level of backpropping, 415–416
  slabs, with two levels of backpropping, 416
definition of, 403
in demolition, 502, *503*
heavy constructions
  silo construction, 417, *418*, 419
loads/loadings, 406–407
multi-storey construction
  in general, 409
  calculations for. *see* calculations (above)
  left-in-place prop-and-panel systems, 409–410
  nomenclature of, *406*
  one-for-one or 50% fewer, 411–412, *411*
  and slab stiffness, 412
  and slab strength, 412
  struck-and-moved props, 410–411
pre-loading of, 408–409
research on, 407–409
theory of, 404–406, *404*
Bailey bridges, 252
barges, 240–243, *241*
basements
  construction of
    in general, 433–434
    beneath existing building, 437, *438*
    next to adjacent buildings, 437, *438*
    in open cut, 435–436
    planning of, 434–435, *435*
    sequences in, 442
    in supported excavation, 436–437
    temporary retaining walls, 437–440

511

basements (*continued*)
    temporary retaining walls, support
        schemes for, 440–442, *442*
    top-down, 443
    use of demolition material in, *506*
  design considerations for, 443–444
  stability of, during demolition, 505, 507
bearing piles
  in general, 219
  applications for, 217–218
  classification of, 217
  design principles for
    ground parameters, 221–223, *222*
    loadings, 224, *224*
    resistance of, 218, 222–223, *222*
  drive chart, *222*
  installation of
    in general, 220–221
    for platforms and jetties, 234
    site constraints, 225
    tolerances, 225
  types of, 219–220
bentonite support fluids, 178–179, *180*
Best Practice Guide: Early Age Strength Assessment of Concrete on Site (BCA, 2000), 352
Best Practice Guide to Tremie Concrete for Deep Foundations (EFFC/DFI, 2016), 182, 201
BIM (building information modelling) . *see* building information modelling
bored continuous flight auger piles, 219
bored rotary piles, 219–220
bracing, to scaffolding, 293–294
Bragg Report, 3–4, 7, 15, 16–17, 25, 301–302, 312
bridge erection techniques
  in general, 385–386
  for decks as single units. *see* deck erection techniques
  for partial deck erection schemes. *see* deck erection techniques
  preparation of, 386
  segmental, 399–400
  selection of installation techniques for, 386–387
  by self-propelled modular transporter. *see* deck erection techniques
  by tunnelling and mining
    in general, 396–397
    base requirements for, 397
    design considerations, 397–398
    risks and opportunities, 397
bridges. *see* bridge erection techniques; military bridges; panel bridges; temporary bridges
Bridges: Design for Improved Buildability (CIRIA, R155), 339
brine freezing systems, 118–119, *119*
British Cement Association, 322
Brokk excavators, 501, *502*
BS 449-1:1970 (BSI, 1970)
  on steel design, 303
BS 449:1969 (BSI, 1969)
  on needle design, 485
  on sheet piling, 152
BS 5950:1990 (BSI, 1990)
  on sheet piling, 152
BS 5975:1982
  on falsework, 3, 302
BS 5975:2008 + A1:2011 (BSI, 2011)
  on falsework, 16
  on organisations, 23
  on personnel, 25
  on scaffolding, 281–283, 291
  on temporary works, 1
BS 5975:2011 (BSI, 2011)
  on soffit formwork, 341, 343–345
  on temporary façade retention, 377, 380
BS 5975:2019 (BSI, 2019)
  on adjustable telescopic props, 306
  application of, 302
  on basement construction, 434
  on bearing piles, 224
  on demolition, 485, 496
  on falsework, 302–303, 306, 307–310, 312–316
  on formwork, 339
  on needling, 485
  on needling schemes, 485
  on permissible stress, 303
  on rebar stability, 465, 474–476
  and reinforcement cages, 475, 476
  on temporary bridges, 255–256, 262
  on temporary façade retention, 370
  on temporary works, 3, 5, 15–17, 19t, 20–21, 23–26, 28, 29t, 30, 31t, 32, 33–34, 450, 479, 482
  on tower crane bases, 58
  on trenching, 160
  on wind loading, 236

BS 5977-1:1981 (BSI, 1981)
   on lintel design, 481, *482*
BS 6187:2011 (BSI, 2011)
   on demolition, 382, 479, 496, 498
BS 6349-1-1:2013 (BSI, 2013)
   on jetties, 236
BS 6349-1-3:2012 (BSI, 2012)
   on jetties, 236
BS 6349-1-4:2013 (BSI, 2013)
   on jetties, 236
BS 7121 (BSI, 1999–2017)
   on cranes, 470
BS 7973-2:2001 (BSI, 2001)
   on tying of scaffolding, 476
BS 8002: (BSI, 2015)
   on trenching, 160
BS 8002:2015 (BSI, 2015)
   on diaphragm walls, 187
BS 8004:2015 (BSI, 2015)
   on bearing piles, 203
BS 8102:2009 (BSI, 2009)
   on ground water control, 203
   on groundwater control, 435
BS 8666:2005 (BSI, 2005)
   on traditional chairs, 471
BS EN 39:2001 (BSI, 2001)
   on scaffold tube struts, 286, 287t
BS EN 805:2000 (BSI, 2000)
   on pressure testing, 423t, 424, 425–426, 427
BS EN 1065:1999 (BSI, 1999)
   on telescopic steel props, 306
BS EN 1536:2010 + A1:2015 (BSI, 2010)
   on piling, 203
BS EN 1538:2010 (BSI, 2010)
   on diaphragm walls, 182, 189
BS EN 1991-1-4:2005 + A1:2010 (BSI, 2005)
   on falsework, 307, 308
   on loading, 366
   on scaffolding, 291
   on wind loading, 377
BS EN 1991-2:2003 (BSI, 2003)
   on loading, 257
BS EN 1991-04 (BSI, 2005)
   on wind loading, 236
BS EN 1992-1-1:2004 + A1:2014 (BSI, 2004)
   on diaphragm walls, 189
BS EN 1993:2003 (BSI, 2003)
   on sheet piling, 148, 150, 152
BS EN 1997-1: 2004 + A1:2013 (BSI, 2004)
   on diaphragm walls, 187, 189

BS EN 1997-1:2004
   on trenching, 160
BS EN 12063:1999 (BSI, 1999)
   on sheet piling, 149
BS EN 12716:2001 (BSI, 2001)
   on jet grouting, 111
BS EN 12810-1:2003 (BSI, 2003)
   on scaffolding, 295
BS EN 12810-2:2003 (BSI, 2003)
   on scaffolding, 295
BS EN 12811-1:2003 (BSI, 2003)
   on scaffolding, 5, 282, 285, 290, 293
   on working platforms, 8
BS EN 12812 (BSI, 2008)
   on density of concrete, 307
   on falsework, 5, 16, 302–304, 307, 309–310, 312, 314, 315
   on formwork, 339
   on limit state, 303–304
BS EN 13670 (BSI, 2009)
   on formwork, 321, 334, 339
building information modelling (BIM)
   basics of, 447–448
   and communication, 448–449, *449*, 462
   mention of, 400
   . *see also* visual planning
Building Research Establishment (BRE), 79, 348, 479

caissons
   cutting edge of, 211, *212*, 213
   out of alignment, 213
   sinking of, 210–211, 213–215, *214*
   . *see also* jet grouting
cantilever method, 260, *260*
cantilevered soffit formwork, 344–345, *345*
CDM 2015. *see* Construction (Design and Management) Regulations 2015
cement, ground stabilisation with, 99–100, 101–102, *102*
Channel Tunnel Rail Link Contract (CTRL) 342, 395, *396*
checking/inspections
   of climbing formwork, 366
   of falsework, 316–317
   of falsework design, 310, 312–314, *314*
   of formwork, 335
   of reinforcement cages, 474–475
   of scaffolding, 297
   of sheet piling, 154

checking/inspections (*continued*)
   of soffit formwork, 352
   of temporary bridge design, 258
   of temporary façade retention, 383
   of tower crane foundations, 65–66
   by TWC, 28–29, 32
   of wall openings, 490–491
Checklist for the Assembly, Use and Striking of Formwork (CS, 2003), 335
CIRIA. *see* Construction Industry Research and Information Association
CISRS (Construction Industry Scaffolders Record Scheme), 283
CITB (Construction Industry Training Board), 6
Civil Engineering Specification for the Water Industry (UKWIR, 2011), 9, 321, 334, 339, 422
clients, 20–21, 37
climbing formwork
   in general, 355
   checking/inspections of, 366
   design of, 358, 360–361, *361*
   economy of, 356–357
   inspections of, 366
   protection screens
     in general, 364, *364*
     design considerations, 365–366
     types of, 364–365, *365*
   selection of, 359
   sequences in, 359–360, *360*
   viability assessment, 357–358
   . *see also* formwork; jump-form formwork; slip forms; soffit formwork
climbing protection screens, 364–366, *364*
coarse-grained soils, 130–131, *130*, 133–134, 135t
cofferdams, 145–148, *147*
cohesive soils, 160
columns
   design of, 333–334
   hanging, *508*
   removal of, 507–508
communications
   BIM and, 448–449, *449*, 462
   on site compound, 47–48
A Comprehensive Guide to Good Practice for Tube and Fitting Scaffolding (NASC, 2013a), 282, 293, 294, 297
concrete

   density of, 307
   use of
     in diaphragm walls, 182
     in piled walls, 200–201
     . *see also* slabs; *under specific forms of formwork*
Concrete Pressure on Formwork (CIRIA, R108), 326–327
Concrete Society
   Checklist for the Assembly, Use and Striking of Formwork (2003), 335
   Falsework checklist (1999), 316
   Falsework: Report of the Joint Committee (1971), 301
   Formwork: A Guide to Good Practice (2011), 322–323, 331–332, 334, 337, 339, 341, 343, 346, 348
   Formwork: A Guide to Good Practice (2012), 307, 355, 364, 366, 410, 413, 417
   Good Concrete Guide 6: Slipforming of Vertical Structures (2008), 356, 362
   Plain Formed Concrete Finishes (1999), 321
   mention of, 408
Concrete Structures Group (CONSTRUCT)
   Guide to Flat Slab Formwork and Falsework (2003), 307, 337, 341, 345, 348, 352, 407, 413, 415–416
   A Guide to the Safe Use of Formwork and Falsework (2008), 335, 352
   National Structural Concrete Specification for Building Construction (2010), 9, 321, 334, 339
CONIAC (Construction Industry Advisory Committee), 6
consent
   for ground water control, 92
   for platforms and jetties, 234
Construction (Design and Management) Regulations 2015 (CDM 2015)
   on backpropping, 403
   on basement construction, 433, 436
   on bridges, 261
   on formwork, 331
   on pressure testing, 429
   on rebar stability, 465
   on scaffolding, 282, 285
   on sheet piling, 154
   on site compounds, 37
   on temporary works, 5–7, 16–17, 20–21, 27–28, 33, 482

on tower crane bases, 58
on trenching, 162
Construction Industry Advisory Committee (CONIAC), 6
Construction Industry Research and Information Association (CIRIA)
  Bridges: Design for Improved Buildability (R155), 339
  Concrete Pressure on Formwork (R108), 326–327
  Design for Movement in Buildings (TN107), 379
  Embedded Retaining Walls – Guidance for Economic Design (C580), 236
  Guidance on Embedded Retaining Wall Design (C760), 204
  Guide to the Design of Thrust Blocks for Buried Pressure Pipelines (R128), 424–425, 427
  The Observational Method in Ground Engineering: Principles and Applications (R185), 190
  Retention of Masonry Façades: Best Practice Guide (C579), 372, 377–379, 380–382
  Retention of Masonry Façades: Best Practice Site Handbook (C589), 369, 370, 381
  Structural Renovation of Traditional Buildings (R111), 370, 372
  Structural Stability of Buildings during Refurbishment (C740), 433, 434, 479
  Tower Crane Foundation and Tie Design (C761), 65
  Tower Crane Stability (C654), 65
  Trenching Practice (97), 168–171, *171*
Construction Industry Scaffolders Record Scheme (CISRS), 283
Construction Industry Training Board (CITB), 6
construction materials
  in general, 18
  for falsework
    adjustable telescopic props, 306
    proprietary systems, 303, *304*, 305
    scaffolding, 306
  for formwork
    bearers, 323
    face contact material, 322
    formwork ties, 323–324, *324*
    proprietary panels, 324, *325*
    release agents, 324–325

  soldiers, 323
  for scaffolding
    aluminium scaffold tubes, 288
    fittings, 288
    proprietary scaffolds, 289
    steel scaffold tubes, 286–288, 287t
  on site compound, 54–55
  for site roads, 82–84
  for working platforms, 82–84
construction phase, in general, 37–38
contractors, 7, 37
contractual obligations, 9–10
Control of Asbestos Regulations 2012, 37
Control of Pollution Act 1974, 144
costs. *see* economy
crane barges, 245
cranes
  used in heavy moves
    crawler-mounted lattice boom cranes, 265–266, *265*
    and electricity regulations, 277, 277t
    lifting design of, 274–277, 276t
    selection of, 266–267
    truck-mounted telescopic boom cranes, 265
  . *see also* tower cranes
cruciform base foundations, 59–61, *60*, *61*, *62*, *63*, 218
cylinder jacks, 268

Darcy's law, 94
deck erection techniques
  for decks as single unit
    in general, 391, *392*
    base requirements for, 392
    design considerations, 393–394
    risks and opportunities, 392–393, *393*
    self-propelled modular transporters (SPMTs), 394–396
  for pre-cast concrete arches
    in general, 389
    base requirements for, 389
    design considerations, 389–391, *390*
    risks and opportunities, 389
  by self-propelled modular transporter
    in general, 394–395, *396*
    base requirements for, 395
    design considerations, 395–396
    risks and opportunities, 395
  for single/multispan bridges
    in general, 387

deck erection techniques (*continued*)
    base requirements for, 387
    design considerations, 387–388, *388*
    risks and opportunities, 387
  by tunnelling and mining
    in general, 396–397
    base requirements for, 397
    design considerations, 397–398
    risks and opportunities, 397
deep wells, 89–91, *90*
demolition
  in general, 493, 496
  basement stability during, 505, 507
  column removal/structural openings, 507–508, *508*
  loads testing in, 503–504
  plant moving between floors, 504
  pre-weakening in, 498
  propping in
    in general, 501–502
    vertical, 502–503
  scaffolding for, 499
  site perimeter, 498–499
  structural stability during
    in general, 504–505
    of prefabricated and large panel structures, 505
  types of
    deliberate collapse mechanism, 496–498, *497*
    progressive, 496
    understanding of structure, 493, *494*, 495, *495*
Demolition Exclusion Zones (NFDC, 2014a), 498
Demolition Scaffolding (NFDC, 2014b), 498, 499
design
  briefs, 26–28, 30
  certificates, 22–23
  custom and practice, xxvii
  . *see also* limit state design; permissible stress design; *under specific structures*
Design for Movement in Buildings (CIRIA, TN107), 379
Design Manual for Roads and Bridges (HA, 1991), 104
design transport weight (DTW), 278
designated individual (DI), 23
designers, 6–7
diaphragm walls
  in general, 175
  applications for, 175–176
  in basement construction, 440
  construction of
    concreting, 182
    guide walls, 178
    hydraulic grabs, 182, 183–185, *184*
    hydromills, 184–185, *185*
    planning of, 176
    reinforcement cages, 179, 181–182, *181*
    rope grabs, 182–183, *183*
    sequences in, *177*
    site preparation, 176, 178
    stop ends, 186, *186*
    support fluids, 178–179, *180*
    working platforms, 178
  design of
    in general, 186–189
    and geotechnical model, 187
    observational approach to, 190
    reinforcement requirements, 189–190
    vertical capacity in, 189
    and watertightness, 190
  . *see also* guide walls; piled walls; post-and-lagging walls; walls
Digest 251 (BRE, 1995), 379, 381
digital project delivery
  in general, 447
  . *see also* building information modelling; visual planning
DIPRA (Ductile Iron Pipe Research Association), 426
drag boxes, 164
drainage, 84
driven pre-cast concrete piles, 220
driven steel piles, 220
Ductile Iron Pipe Research Association (DIPRA), 426
dynamic amplification factor (DAF), 275–276, 276t, 278

economy
  climbing formwork, 356–357
  of formwork, 320–321
  of slip forms, 356–357
edge protection barriers, for bridges, 257
electricity, 48–50
Electricity at Work Regulations 1989, 277
Embedded Retaining Walls – Guidance for Economic Design (CIRIA, C580), 236

energy supply, 48–50
environmental impact survey, 44–45
erection tolerances, in falsework, 315
Eurocode 1 (BSI, 2002-2005), 307
Eurocode 2 (BSI, 2004), 189, 201
Eurocode 3 (BSI, 2003), 150, 152
Eurocode 5 (BSI, 2004), 153
Eurocode 7 (BSI, 2004), 65, 129, 131–133,
    148–149, 153, 154, 187, 189, 201, 202,
    225, 443
Eurocodes. *see under corresponding British
    Standards (BS)*
European Concrete Building Project (ECBP),
    407–409
excavators. *see* Brokk excavators; mini
    excavators
expendable base foundations, 62–65, *63*, *64*

façades. *see* temporary façade retention
failures
  of falsework, 15–16, 301
  of reinforcement cages, 465, *466*
  of temporary slopes, 129
  of temporary works, 2–3
falsework
  in general, 301–302
  Bragg Report on, 3–4
  codes for, 3, 5
  construction materials for
    adjustable telescopic props, 306
    proprietary systems, 303, *304*, 305
    scaffolding, 306
  design of
    in general, 310, *311*
    checking of, 310, 312–314, *314*
    and erection tolerance, 315
    fully or partially braced, 315
    method of analysis for, 310
  failures of, 3, 15–16, 301
  Health and Safety Executive research into,
    4–5
  inspections of, 316–317
  loads/loadings on
    in general, 307–308, 310, *311*, 313
    environmental, 308–309
    and horizontal disturbing force, 309
    and horizontal imposed load, 309
    indirect, 309
  standards choice in, 303–304
  workmanship in, 316

Falsework checklist (CS, 1999), 316
Falsework: Report of the Joint Committee
    (CS/ISE, 1971), 3, 321
fencing, for site compound, 55–56
fill rock (rip rap), 229
fine-grained soils, 130–131, *130*, 135–136
FlexiArch system, 391
floating plants
  in general, 239–240
  applications for, 239
  barges
    in general, 240–243, *241*
    crane, 245
    hopper barges, 245
    jack-up, 243–244, *244*, 247–248
  design principles for
    jack-up barges, 247–248
    for stability, 245–247
  pontoons
    in general, 240–243, *242*
    lightweight modular, 243, *244*
flowchart, for striking flat slabs, *347*
formwork
  in general, 319
  checking of, 335
  climbing formwork. *see* climbing formwork
  concrete pressure calculation in, 325–327,
    326, *327*, 328t, 329t
  construction materials for
    bearers, 323
    face contact material, 322
    formwork ties, 323–324, *324*
    proprietary panels, 324, *325*
    release agents, 324–325
    soldiers, 323
  design of
    column, 333–334
    double-faced, 330–331, *330*, *331*
    single-faced, 332–333, *332*, *333*
    soffit. *see* soffit formwork
    stability of, 331–332
  deviations in, 322
  economy of, 320–321
  jump-form. *see* jump-form formwork
  pressure calculation in formwork, 330
  slip forms. *see* slip forms
  soffit. *see* soffit formwork
  striking of vertical, 334
  surface finishes, 321, 321t
  tolerances of, 322

517

formwork (*continued*)
  vertical, 319–320, *320*
  workmanship in, 335
Formwork: A Guide to Good Practice (CS, 2011), 322–323, 331–332, 334, 337, 339, 341, 343, 346, 348
Formwork: A Guide to Good Practice (CS, 2012), 307, 355, 364, 366, 410, 413, 417
foundations
  of tower cranes
    construction of, 65–66
    cruciform base, 59–61, *60, 61, 62, 63*, 218
    design principles for, 65
    expendable base, 62–65, *63, 64*
    inspection of, 65–66
    loading on, 59
    on rails, 61–62, *63*
frames. *see* horizontal frames; hydraulic waling frames; inclined frames
free-standing scaffolds, 285

gantry lifting systems, 269, *270*
geosynthetics, 75–78
geotechnical surveys, 44–45
Good Concrete Guide 6: Slipforming of Vertical Structures (CS, 2008), 356, 362
granular materials, 82–84
granular soils, 160
  . *see also* coarse-grained soils; fine-grained soils
gravity sewers, 422
ground heave, 114
ground stabilisation
  in general, 83
  for caisson sinking, 215
  with cement, 99–100, 101–102, *102*
  effectiveness of, 103–104
  with lime, 99–101, *101, 102*
  with lime/cement blend, 102
  mixing of additives for, 105–106, *105*
  test soil for suitability for, 104
  . *see also* artificial ground freezing; ground water control; jet grouting
ground support. *see* diaphragm walls; sheet piling; temporary slopes; trenches
ground water control
  in general, 87–88
  in basement construction, 436, 437
  consents for, 92
  deep wells, 89–91, *90*

dewatering systems
  investigations for, 92–93
  and permeability, 93–95
  pumped well, 95–96, *96*
  in diaphragm walls design, 190
  filters, 91
  maintenance of, 92
  monitoring of, 92
  in piled walls design, 203–204
  recharging, 92
  risk considerations, 91–92
  selection of techniques, 88
  and sheet piling, 146
  sump pumping, 88–89, *88*
  in temporary slopes, 131
  and trenches, 172
  well point systems, 89–91, *89, 90*
grouting. *see* jet grouting
Guidance on Embedded Retaining Wall Design (CIRIA, C760), 204
Guide to Flat Slab Formwork and Falsework (CONSTRUCT, 2003), 307, 337, 341, 345, 348, 352, 407, 413, 415–416
Guide to formwork (CS, 2012), 307
A Guide to Good Practice (CS, 2011), 320, 322–323, 331, 333, 334
Guide to Pressure Testing of Pressure Pipes and Fittings for use by Public Water Suppliers (2015), 424, 426
Guide to the Design of Thrust Blocks for Buried Pressure Pipelines (CIRIA, R128), 424–425, 427
A Guide to the Safe Use of Formwork and Falsework (CONSTRUCT, 2008), 335, 352
guide walls, 178, 198
  . *see also* walls

hand–arm vibration syndrome (HAVS) regulations, 209
health and safety
  in pressure testing, 425, 429–430, *429*
  quick lime, 106
  in temporary works, 5–6, 7–8, 16–17, 21
Health and Safety at Work etc. Act 1974 (HSWA), 8, 37, 295
Health and Safety Executive (HSE)
  in general, 248, 282, 433
  Avoiding Danger from Overhead Power Lines (2013), 277

Investigation into Aspects of Falsework (2001), 4, 17
Preventing Catastrophic Events in Construction (2011), 449
Safety Requirements for Pressure Testing (2012), 424
The Health and Safety (Offences) Act 2008 (HSOA 2008), 8
heavy loads, movement of. *see* heavy moves
heavy moves
  in general, 263–264
  programme savings, 263–264
  techniques used for
    cranes, 264–267, *265*
    design of, crane lifting, 274–277, 276t, 277t
    design of, SPMTs, 277–279
    jacking systems, 268–272, *269*, *270*, *271*
    skidding systems, 272–273, *273*, 274t
    trailers, 267–268, *267*
High Reach Demolition Rig Guidance Notes (NFDC, 2012), 501
high-reach machines, 499, *500*, 501
high-security sites, 45
Highways Agency (HA)
  The Assessment of Highway Bridges and Structures (2001), 236, 257
  Design Manual for Roads and Bridges (1991), 104
  Manual of Contract Documents for Highways Works (2004), 321, 321t
  Specification for Highway Works. Manual of Contract Documents for Highway Works (2006), 9, 339, 346
Highways England contracts, 30
Hiley formulae, 222, *222*
hoardings, for site compound, 55–56
hopper barges, 245
horizontal frames, for retaining walls, 440–441
HSE. *see* Health and Safety Executive
hydraulic grabs, 183–185, *184*
hydraulic pushers, 144, *145*
hydraulic skid shoes, 273, *273*
hydraulic waling frames, 166–168, *167*
hydrofracture, 114
hydrofraises. *see* hydromills
hydromills, 184–185, *185*

ICE Specification for Piling and Embedded Retaining Walls (2017), 149, 179, 182, 187, 203, 225

impact hammers, 144, *145*
inclined frames, for retaining walls, 441
independent tied scaffolds, *284*, 285, 292t, 293
infrastructure, of site compound, 46–47
inspections. *see* checking/inspections
Institution of Structural Engineers, 3, 301
internet access, 46–47
Investigation into Aspects of Falsework (HSE, 2001), 4, 17

jack and pack systems, 268, *269*
jacking systems, 268–272, *269*, *270*, *271*
jack-up barges, 243–244, *244*, 247–248
jet grouting
  in general, 109–110
  specialist, 220
  systems, *110*, *111*
    construction of, 111–112
    design of, 112–114
    and ground heave, 114
    monitoring of, 114
    secondary effects of, 114–115
    spoil disposal, 115
    validation of, 114
jetties
  applications for, *230*
  connections on, 233–234, *234*
  definition of, 229
  floating, 232–324, *233*
  loadings, 234–236
  mooring points on, 233, *233*
  with open structure, 231, *232*, *233*
  and site team, 234
  with solid structure
    mass fill gravity, 229
    of sheet piling, 231, *231*
  wind loadings, 236
jump-form formwork
  in general, 355, 356
  selection of, 359
  sequences in, 359–360, *360*
  viability assessment, 356–357
  . *see also* climbing formwork; formwork; slip forms; soffit formwork

Kylesku Bridge (Scotland), 263, *264*

large panel structure (LPS), 505
lead designer (temporary works), 22

left-in-place prop-and-panel systems, 409
Legal Guidance (HSE, L153), 403
Lifting Operations and Lifting Equipment Regulations 1998, 470
lightweight modular pontoons, 243, *244*
lime, 99–101, *101*, *102*, 106
Limit equilibrium methodology, 150
limit state design
   in general, xxvi–xxvii
   and BS EN 12812, 303–304
   codes for, 5
   versus permissible stress design, 302–303
lintel design theory, 481, *482*
liquid nitrogen freezing systems, 118–119, *119*
loading towers, 290–291
loads/loadings
   in backpropping, 406–407
   on bearing piles, 224–225, *224*
   on bridges, 256
   in demolition, 503–504
   on façade retention, 376–377, 378–379, 378t
   on falsework, 307–310, *311*, 313
   on foundations of tower cranes, 59
   on jetties, 234–236
   on plant platforms, 234–236
   on scaffolds, 287–289, 288t, 290–291
   in sheet piling, 150, 151–152
   on site roads, 70–72
   on soffit formwork, 342–343, 346, *347*, 348–350, *350*
   on temporary bridges, 257
   on temporary roads, 81–82
   and wall opening, 481–482, *482*
   . see also wind loadings
LOK test, 352
low-friction skid interfaces, 273–274, 274t
lugs, 234, *234*

maintenance
   of ground water control, 92
   of sheet piling, 154
   of visual planning, 461–462
management
   of risks. *see* risk considerations
   of temporary works
      in general, 5, 15–17
      key issues in, 449–450
      parties involved in, 20–25
         . see also under specific parties

Management of Health and Safety at Work Regulations 1999, 37
Manual of Contract Documents for Highways Works (HA, 2004), 321, 321t
Marsh Mills Viaduct, 391, *392*
Meyerhof distribution, 235, *235*
military bridges, 252, *253*
mini excavators, *435*, 437
mini piles, 220
Minimum Structural Properties and Test Procedure for TG20 Compliant Prefabricated Structural Transom Units (NASC, 2014b), 294, 296
mixed reality, 455, *455*
mobile elevating working platforms (MEWPs), 465, 469
monitoring
   of artificial ground freezing systems, 125
   of ground water control, 92
   of jet grouting systems, 114
   of temporary façade retention, 382–383
   of temporary slope constructions, 137–138
movement, of heavy loads. *see* heavy moves

National Access & Scaffolding Confederation (NASC)
   in general, 8, 282, 297
   Access and Egress from Scaffolds via Ladders and Stair Towers etc. (2014a), 297
   A Comprehensive Guide to Good Practice for Tube and Fitting Scaffolding (2013a), 282, 293, 294, 297
   Minimum Structural Properties and Test Procedure for TG20 Compliant Prefabricated Structural Transom Units (2014b), 294, 296
   Preventing Falls in Scaffolding Operations (2016a), 282, 297
   TG20:13 Design Guide. A Comprehensive Guide to Good Practice for Tube and Fitting Scaffolding (2013b), 282, 286, 287–288, 290, 291, 295, 297
   TG20:13 eGuide. A Comprehensive Guide to Good Practice for Tube and Fitting Scaffolding (2013c), 282, 286, 291, 293, 294, 297
National Building Specification 'E20: Formwork for in situ concrete, 9, 321, 339

National Federation of Demolition
    Contractors (NFDC)
  Demolition Exclusion Zones (2014a), 498
  Demolition Scaffolding (2014b), 498, 499
  High Reach Demolition Rig Guidance Notes
    (2012), 501
National Structural Concrete Specification for
    Building Construction (NSCS), 9, 321,
    334, 339
needling schemes
  in general, 483–486
  with backpropping, *484*
  with propping to needles, 488–489
  sequence of needle removal, 490
  without backpropping, *487*
Network Rail contracts, 30

The Observational Method in Ground
    Engineering: Principles and Applications
    (CIRIA, R185), 190
open excavations, *128*
  . *see also* temporary slopes

Paddington Bridge Project, 392–393, *393*
panel bridges, 253–254, *254*
parties
  involved in temporary works, 10, 20–25
  . *see also under specific parties*
PAS 8812:2016 (BSI, 2016), 302
pedestrian loadings, 256
permanent works
  definition of, 17
  incomplete, 18
permanent works designer (PWD)
  in general, 20, 22
  and backpropping, 405, 410, 412, 417
  and climbing formwork, 357–358, 366
  and needling, 482
  and reinforcement cages, 467
  theory of, *406*
permeability, measurement of, 93–95
permissible stress design
  in general, xxv
  and BSI, 2018, 303
  versus limit state design, 302–303
  for scaffolding, 289–290
permits, to load/unload, 32–33
piled walls
  in general, 193
  in basement construction, 438, *441*
construction of
  in general, 195, 198
  concreting in, 200–201
  guide walls, 198–199
  pile techniques, 199
  reinforcement cages, 199–200
  site preparation, 198
  working platforms, 198
contiguous
  in general, 193
  applications for, 195
  in basement construction, 440
  construction sequence, *196*
  hard/soft, *194*
design of
  in general, 201, 204
  instrumentation, 201–202
  reinforcement requirements, 202–203
  temporary support, 201, *202*
  vertical capacity in, 204
  and watertightness, 203–204
secant
  applications for, 195
  in basement construction, 440
  construction sequence, *197*
  hard/hard, 194–195, *194*
  hard/soft, 193–194
  . *see also* diaphram walls; guide walls;
    post-and-lagging walls; walls
piles. *see* bearing piles; bored continuous
    flight auger piles; bored rotary piles;
    driven pre-cast concrete piles; driven steel
    piles; mini piles
Piling Handbook (2016; ArcelorMittal), 142,
    143, 148, 152, 223, 439
pipe nominal rated pressure (PN), 426, 427
pipelines
  pressure testing of
    in general, 421–422, *422*
    design of, 424–425
    design of thrust blocks, 427–428, *428*
    internal flanges, 427
    methods for, 422, 423t
    onsite safety, 429–430, *429*
    preventing injury/damage, 425
    with temporary works, 426–427
    without temporary works, 425–426
    with retrained joints, 425–426
Plain Formed Concrete Finishes (CS, 1999),
    321

plant platforms
	applications for, 230
	connections on, 233–234, 234
	definition of, 229
	loadings on, 234–236
	with solid structure
		in general, 229
		mass fill gravity, 229
		of sheet piling, 231, 231
	wind loadings, 236
platforms. *see* plant platforms; working platforms
polyacrylamide polymer support fluids, 178–179
pontoons, 240–243, 242
post and plank vertical H-sections, 168, 169
post-and-lagging walls, 439–440, 439
	. *see also* walls
pre-cast concrete arches, 389–390, 390
pre-cast segmental linings, of shafts, 207–208
Preventing Catastrophic Events in Construction (HSE, 2011), 449
Preventing Falls in Scaffolding Operations (NASC, 2016a), 282, 297
principal contractors (PC), 7, 20, 21, 38
principal designers (PD), 6–7, 20, 21
propping, in demolition, 501–502
proprietary systems
	for falsework, 303, 304, 305
	for panels for formwork, 324, 325
	for scaffolding, 295–296, 295, 296
	for temporary façade retention, 373, 374
Provision and Use of Work Equipment Regulations 1998 (PUWER), 424, 470
public safety, 11
	. *see also* health and safety

Queensferry Crossing, 399–400, 400

rails, tower cranes on, 61–62, 63
rebar stability
	of reinforcement cages
		design of, 471–472
		design of, solutions, 472–474
		in general, 465, 467, 467t
		modes of failure, 470
		potential problems, 467–469
		on-site inspections, 474–475
		solutions, 470–471
		and wind loading, 474–475

	. *see also* stability
Reducing the Risk of Leaking Substructure: A Clients' Guide, 190
refurbishment projects, 451, 452
	. *see also* temporary façade retention
reinforcement cages
	for diaphragm walls, 179
	failures of, 465, 466
	for piled walls, 199–200
	rebar stability of
		design of, 471–472
		design of, solutions, 472–474
		in general, 465, 467, 467t
		modes of failure, 470
		potential problems, 467–469
		on-site inspections, 474–475
		solutions, 470–471
		and wind loading, 474–475
	structural behaviour of, 472
	tying of, 476–477
research
	on backpropping, 407–409
	on falsework, 4–5
	of Health and Safety Executive, 4, 17
	on temporary works, 4–5
	of Temporary Works Forum, 476
Retention of Masonry Façades: Best Practice Guide (CIRIA, C579), 372, 377–379, 380–382
Retention of Masonry Façades: Best Practice Site Handbook (CIRIA, C589), 369, 370, 381
retrained joints, pipelines with, 425–426
rip rap (fill rock), 229
risk classes, in temporary works, 30, 31t
risk considerations
	in bridge erection techniques, 397
	in deck erection techniques, 389, 392–393, 393, 395, 397
	in ground water control, 91–92
	in temporary works, 19t
	in visual planning, 455–456
roads. *see* site roads; temporary roads
robustness, of temporary works, 10–11
rock, 160
rock slopes, 136–137
rollers, 272–273
rope grabs, 182–183, 183

Sale and Supply of Goods Act 1994, 295

scaffold tubes, 286–288, 287t
Scaffolders Record Scheme (CISRS), 283, 297
scaffolding
  in general, 281–282
  codes for, 5
  construction materials for
    aluminium scaffold tubes, 288
    fittings, 288
    proprietary scaffolds, 289
    steel scaffold tubes, 286–288, 287t
  for demolition, 499
  design of
    in general, 289
    bracing, 293–294
    considerations, 296–297
    element strenght and stability in, 292t
    element strength and stability in, 292–293
    limit state, 290
    loadings, 290–291
    permissible stress, 289–290
    proprietary systems, 295–296, 295, 296
    TG20 compliant, 294–295
    tube and fitting scaffolds, 291–292
    tying, 293, 294
  designation of, 285–286
  for falsework, 306
  inspections of, 297
  managing of, 282–283
  selection of, 283–285
  for temporary façade retention, 373, 374
  types of, 284, 285
  workmanship of, 297
security, of site compound, 55–56
selection
  of climbing formwork, 359
  of cranes, 266–267
  of equipment, for soffit formwork, 340–342, 342, 343
  of jump-form formwork, 359
  of scaffolding, 283–285
  of techniques
    for bridge erection, 386–387
    for ground water control, 88
self-propelled modular transporters (SPMTs)
  used for
    in general, 394–395
    bridge erection, 394–396, 396, 399
    heavy moves, 267–268, 267, 277–279
self-weight, 235

sequences
  in basement construction, 442
  in climbing formwork, 359–360, 360
  in construction
    of diaphragm walls, 177
    of piled walls, 196, 197
    of trench boxes, 166
  in sheet piling, 150–151, 152–153
  in soffit formwork, 350
Sewers for Adoption (WRc, 2012), 422, 424
shafts
  in general, 207–208
  alternatives to, 208
  construction of
    caisson sinking, 210–215, 212, 214
    ground stabilisation for, 215
    roof slabs, 215–216
    underpinning, 209–210, 210, 211
  design principles of, 216
  with pre-cast segmental linings, 207–208
  with sprayed concrete lining, 208
sheet piling
  in general, 141
  alternatives to, 141
  cofferdams, 146–148
  design of
    in general, 149–152
    loads/loadings in, 150, 151–152
    partial factors in, 154
    sequences in, 150–151, 152–153
  extraction of, 146, 147
  and ground water control, 146
  inspections of, 154
  installation of
    driving aids used in, 144–146
    sequences in, 150–151
    techniques used in, 142–144, 145
    working platforms for, 79
  maintenance of, 154
  of plastic, 154–155
  platforms and jetties of, 231, 231
  standards/regulations for, 148–149
  types of, 141–142, 142, 143
sheets
  for trenches, 166–168
  . *see also* sheet piling
side rail system, 168, 169
silo construction, 417, 418, 419
single/multispan bridges, 387–388, 388
single-vehicle loading, 256

site compound
  in general, 37–38, 38–44t, 44
  accommodations on, 51, *52–53*, 54
  construction materials on, 54–55
  energy supply for, 48–50
  hoardings and fencing for, 55–56
  infrastructure of, 46–47
  internet access on, 47–48
  layout for
    accommodations, *52–53*
    road project, *46*
    small site, *47*
  location of, 45–46, *46*, *47*
  security of, 55–56
  surveys, 44–45
  visits and inspections, 44
  wastewater disposal for, 51
  water supply for, 50–51
  . *see also* foundations; site roads
Site Management Safety Training Scheme (SMSTS), 283
site roads
  in general, 69–70, *70*
  alignment of, 70
  construction materials for, 82–84
  design of
    stabilised soil, 78
    and subgrades, 73–76, *75*
    unbound granular, 73–75, *77*, 78
    unbound granular with geosynthetics, 75–76, *77*
  drainage of, 84
  ground conditions, 72–73
  traffic loading, 70–72
Site Supervision Safety Training Scheme (SSSTS), 283
SketchUp (Trimble), 453, *454*, *456*, *457*
skidding systems, *270*
slabs
  backpropping calculations for, 412–413, *414*, 415
  stiffness of, 412
  strength of, 412
  striking of, 346, *347*, 348–350, *350*
slip forms
  in general, 355–356, 361–362, *362*
  design of, 358, 363–364, *363*
  economy of, 356–357
  viability assessment, 357–358
slopes. *see* temporary slopes

slurry walls. *see* diaphragm walls
soffit formwork
  in general, 337, *338*
  assessment of concrete strenght in, *351*
  assessment of concrete strength in, 351–352
  cantilevered, 344–345, *345*
  checking/inspections of, 352
  design of
    in general, 337, 339, 343–344, *344*
    equipment selection, 339–342, *341*, *342*
    finishes, 339
  inspection of, 352
  loads/loadings on, 342–343, 346, *347*, *348–350*
  striking of
    in general, 345–346, 345t
    bridge soffits, 346
    sequences in, 350
    slabs, 346, *347*, 348–350, *350*
  . *see also* climbing formwork; formwork; jump-form formwork; slip forms
soils, classification of, 160
  . *see also* cohesive soils; granular soils
Specification for Highway Works. Manual of Contract Documents for Highway Works (HA, 2006), 9, 339, 346
Specification for Tunnelling (BTS, 2004), 207, 208
sprayed concrete lining (SCL), 208
spud stabilisers, 241, *242*, 424
stabilised soil roads, 78
stability
  during demolition, 504–505
  in floating plants, 245–247
  in formwork, 331–332
  in scaffolding, 292–293, 292t
  . *see also* rebar stability
Stability of Reinforcement Cages Prior to Concreting (TWf, 2013), 469
standard axles, vehicles, 71t
standards/regulations. *see under specific names of standards and regulations*
Standing Committee on Structural Safety (SCOSS), 4, 17
Steel Bearing Piles (SCI, 1989), 225
Steel Construction Institute
  Steel Bearing Piles (1989), 225
stop ends, 186, *186*
Storebælt Bridge, 352
strand jacks, 269, *270*, 271–272, *271*

Stratford Olympic Park, 387, *388*
stratigraphical survey, 92
Structural Renovation of Traditional Buildings (CIRIA, R111), 370, 372
Structural Stability of Buildings during Refurbishment (CIRIA, C740), 433, 434, 479
subgrades, of site roads, 73–76, *75*
sump pumping, 88–89, *88*
support fluids, 178–179
surveys, 44–45, 92, 372–373

temporary bridges
   in general, 251–252
   Bailey bridges, 252
   construction of
      methods for, 259–260, *260*
      site planning, 261–262
   design process of
      in general, 259
      checking of, 258
      cross-section and span considerations, 257–258, *258*
      loadings, 256
      programme, 256
      and site constraints, 257–258, *258*
   foundations of, 255–256, *255*
   loadings for, 257
   military bridges, 252, *253*
   panel bridges, 252–253, *254*
   parts of, other applications for, 262
   plate girder bolted, 254
   with prefabricated deck sections, 254, *255*
   of special-purpose design, 255
   transportation of, 259
   truss, 254
temporary façade retention
   in general, 369–370
   alternatives to, 370
   demolition of, 382
   design considerations
      in general, 380
      connections, 381
      deflection criteria, 380–381
      restraints, 381–382
      stability, 380
   general principles in, 370–372
   inspections of, 383
   loads/loadings on
      in general, 376–377, 378–379, 378t

      wind, 377–378, *379*
   monitoring of, 382–383
   and party walls, 372
   survey of existing building, 372–373
   types of schemes
      fabricated steelwork, 373
      horizontal frames, 375–376, *376*
      proprietary systems, 373, *374*
      scaffolding, 373, *374*
      timber shoring, 373
      vertical towers, 373, 375, *375*
temporary roads
   construction materials for, 82–84
   drainage of, 84
   loads/loadings, 81–82
   . see also site roads
temporary slopes
   in general, 127, *128*, 129
   in basement construction, 436
   construction of
      monitoring of, 137–138
      principles of, 129–130
   design of
      and coarse-grained soils, 130–131, *130*, 133–134, 135t
      and fine-grained soils, 130–131, *130*, 135–136
      and geotechnical categories, 131–133, *133*
      and ground water control, 131
      medium sized, 133–135, 135t
      rock slopes, 136–137
      small sized, 133, *134*
   failures of, 129
temporary substation, Thames Tideway project, 456, *456*, *457*, 458, *458*
temporary works
   in general, xxiii–xxiv, 1
   changes in construction industry and, 4
   code of practice on, 3
   construction materials
      in general, 18
      for falsework. *see* falsework
      for formwork. *see* formwork
      for scaffolding. *see* scaffolding
      on site compound, 54–55
      for site roads, 82–84
      for working platforms, 82–84
   contractual obligations, 9–10
   definition of, xxiv–xxv, 1, 17–18

temporary works (*continued*)
  in demolition. *see* demolition
  design of, xxv–xxvii
    . *see also* design
  examples of, 2, 18
  failure of. *see* failures
  fundamental aspects of, 2
  Health and Safety Executive research into, 4–5
  management of
    in general, 5, 15–17
    key issues in, 449–450
  parties involved in, 10, 20–25
    . *see also under specific parties*
  responsibilities in, 10, 482–483
  risk classes in, 30, 31t
  risk considerations in, 19t
  robustness of, 10–11
  . *see also under specific temporary works*
temporary works coordinator (TWC)
  in general, 20, 24–25
  activities of
    design briefs by, 26–28, 30
    design checks by, 28–29
    permits to load/unload, 32–33
    on-site checking by, 32
    on-site supervision by, 30
    temporary works register, 26–27
  and backpropping, 415–416
  and climbing formwork, 357, 366
  and temporary bridge design, 257–258
temporary works designer (TWD)
  and backpropping, 412, 416
  and climbing formwork, 356, 366
  and needling, 483
  and reinforcement cages, 471, 472
Temporary Works Forum
  research of, 476
  Stability of Reinforcement Cages Prior to Concreting (2013), 469
temporary works register, 26–27
temporary works supervisor (TWS), 26
TG20:13 Design Guide. A Comprehensive Guide to Good Practice for Tube and Fitting Scaffolding (NASC, 2013b), 282, 286, 287–288, 290, 291, 295, 297
TG20:13 eGuide. A Comprehensive Guide to Good Practice for Tube and Fitting Scaffolding (NASC, 2013c), 282, 286, 291, 293, 294, 297

Thames Tideway project, temporary substation, 456, *456*, *457*, 458, *458*
3D/4D modelling, 452–453, *454*, 456, *456*, *457*, 458, *458*
thrust blocks, 427–428, *428*
timber shoring, 373
timbering, of trenches, 163, 164
topographical surveys, 44–45
Tower Crane Foundation and Tie Design (CIRIA, C761), 65
Tower Crane Stability (CIRIA, C654), 65
tower cranes
  in general, 57–58
  climbing of, 65
  foundations of. *see* foundations
  on rails, 61–62, *63*
  types of, 58–59
towers. *see* loading towers; tower cranes; vertical towers
trailers, 267–268, *267*
Trapezoidal distribution, 235, *235*
trench boxes, 163–164, *164*, *165*, *166*
trenchcutters, 184–185, *185*
trenches
  alternatives to, 159
  battered, 160–161
  definition of, 159
  design of
    in general, 161–162
    and CIRIA 97, 168–171, *171*
    and ground water control, 172
    standards/regulations, 168–171
  techniques used for
    drag boxes, *164*
    hydraulic waling frames, 166–168, *167*
    post and plank vertical H-sections, 168, *169*
    timbering, 163
    trench boxes, 163–164, *164*, *165*, *166*
    trench sheets, 166–168
    vertical shores, 164, 166
Trenching Practice (CIRIA, 97), 168–171, *171*
tube struts, 287–288, 287t
Tunnel Lining Design Guide (BTS, 2004), 208
tying, to scaffolding, 293, *294*

UK National Foreword for the European Falsework Code (BS EN 12812:2008), 16
UK National Guidance (BSI, 2015), 366

UK Water Industry Research (UKWIR)
 Civil Engineering Specification for the Water Industry (2011), 9, 321, 334, 339, 422
unbound granular roads, 73–76, *77, 78*
Unifloat pontoons, 242

validation, of jet grouting systems, 114
vehicles, standard axles, 71t
vertical formwork, 319–320, *320*
vertical towers, 373, 375, *375*
vibrators, 143–144, 146
virtual reality, 453
visual planning
 and BIM
  basics of, 447–448
  and communication, 448–449, *449*, 462
  mention of, 400
  and initial baseline information, 451
  and key issues in temporary works, 449–450
  managing risks, 455–456
  operations and maintenance, 461–462
 techniques for
  in general, 450–451
  3D/4D modelling, 452–453, *454*
  3D/4D modelling, example of, 456, *456, 457,* 458, *458*
  augmented reality, 453, *455*
  high-quality animation, 458, *459, 460, 461*
  initial baseline information, *452–453*
  mixed reality, 455, *455*
  virtual reality, 453
 training and support, 462

walls
 opening of
  in general, 479
  assessment of building, 479–480
  load assessments, 481–482, *482*
  needling schemes for, 483–486, *484, 487,* 488–490

 on-site checklist, 490–491
 support required, 480–481
 temporary works solution, 483
 . see also diaphragm walls; guide walls; piled walls; post-and-lagging walls
wastewater disposal, 50–51
water jetting, 145–146
water supply, 50–51
Water UK
 Guide to Pressure Testing of Pressure Pipes and Fittings for use by Public Water Suppliers (2015), 424, 426
well point systems, 89–91, *90*
wheel-wash facilities, 83
wind loadings
 on façade retention, 377–378, *379*
 on jetties, 236
 on plant platforms, 236
 and reinforcement cages, 474–475
 on scaffolding, 291
Work at Height (Amendment) Regulations 2007, 37
Work at Height Regulations 2005, 7–8, 282, 291, 297, 331
working platforms
 in general, 69, 78
 construction materials for, 82–84
 for construction of
  diaphragm walls, 178
  of piled walls, 198
  sheet piling, *79*
 drainage of, 84
 for high-reach machines, 499, *500,* 501
 mobile elevating, 465, 469
 for tracked plant, 79–81
 unbound granular platforms, 79–81, *80*
Working Platforms for Tracked Plant (BRE, 2004), 79, *80*, 501